Studies in Turbulence

T.B. Gatski S. Sarkar C.G. Speziale
Editors

Studies in Turbulence

With 182 Figures

Springer-Verlag
New York Berlin Heidelberg London Paris
Tokyo Hong Kong Barcelona Budapest

Thomas B. Gatski
Theoretical Flow Physics Branch
NASA Langley Research Center
Hampton, VA 23665-5225
USA

Sutanu Sarkar
ICASE
NASA Langley Research Center
Hampton, VA 23665-5225
USA

Charles G. Speziale
ICASE
NASA Langley Research Center
Hampton, VA 23665-5225
USA

Library of Congress Cataloging-in-Publication Data
Studies in turbulence
/Thomas B. Gatski, Sutanu Sarkar, Charles G. Speziale, editors.
p. cm.
Includes bibliographical references.
ISBN 0-387-97613-2 (acid-free paper). — ISBN 3-540-97613-2 (acid-free paper)
1. Turbulence. I. Lumley, John L. (John Leask), 1930–
II. Gatski, T.B. III. Sarkar, Sutanu. IV. Speziale, C.G.
(Charles G.), 1948–
QA913.S87 1991
532′.0527 – dc20 91-17250

Printed on acid-free paper.

Production managed by Francine Sikorski.
Camera-ready copy provided by the editors.
Printed and bound by Braun-Brumfield, Ann Arbor, MI.
Printed in the United States of America.

9 8 7 6 5 4 3 2 1

ISBN 0-387-97613-2 Springer-Verlag New York Berlin Heidelberg
ISBN 3-540-97613-2 Springer-Verlag Berlin Heidelberg New York

This volume is dedicated to John L. Lumley by his colleagues and friends on the occasion of his sixtieth birthday in recognition of his pioneering research in turbulence spanning four decades.

John L. Lumley

DEDICATION

It is indeed a pleasure to dedicate this volume in honor of Professor John L. Lumley on the occasion of his 60th birthday. In his nearly four decades of research on turbulence, he has made significant strides toward the understanding of atmospheric and aerodynamic turbulent flows through his unique blend of mathematical analysis and physical insight. This compilation of contributions from his colleagues, friends and students provides the opportunity to acknowledge John's scientific accomplishments as well as to reflect and reminisce about the influence he has had on us through the years.

John Lumley was born on November 4, 1930, in Detroit, Michigan. After graduating from the Detroit University School, he entered Harvard College in 1948 where he received a Bachelor of Arts Degree in 1952 in Engineering Sciences and Applied Physics. With the encouragement of Professor George Wislicenus, he entered The Johns Hopkins University in 1952 and received an M.S.E. degree in Mechanical Engineering in 1954. During this period, his interest turned to the study of turbulence and stochastic processes under the influence and supervision of Professor Stanley Corrsin. In 1957, he received a Ph.D. in Aeronautics. His study of turbulence continued at Johns Hopkins as a Post-Doctoral fellow until 1959. During this period, Professor Wislicenus had moved on to The Pennsylvania State University in the Aeronautical Engineering Department and was anxious to have John Lumley join him there. The association with Penn State would last for nearly twenty years. In 1963, John Lumley became the youngest person to attain the rank of Professor at the University and in 1974 became the youngest person to hold a chaired position at the University as an Evan Pugh Professor of Aerospace Engineering. His research during this period spanned the fields of atmospheric turbulence, non-Newtonian fluids and random wave propagation. In 1977, John Lumley joined the faculty of Cornell University in the Sibley School of Mechanical and Aerospace Engineering as the Willis H. Carrier Professor of Engineering. During his tenure at Cornell, he has focused his research on the development of Reynolds stress closure models and the application of dynamical systems theory to the study of turbulent flow structures.

It is certainly not possible in this short space to detail all of John Lumley's scientific work. Throughout his career he has been in the forefront with his research on the leading topical issues facing turbulence at the time. These have included his development of the Proper Orthogonal Decomposition Theorem for analyzing turbulent flows, his innovative application of invariant tensor theory and realizability constraints in developing more sophisticated turbulence models, and his utilization of dynamical systems theory in the study of turbulent flow structures. His honors and awards are numerous. He is a member of the National Academy of Engineering and a Fellow of the American Physical Society, the American Academy of Arts and Sciences, and the American Academy of Mechanics. In addition, he is a charter member of the Johns Hopkins Society of Scholars, and an honorary member of Sigma Gamma Tau. His awards and prizes include Medallion, Universite de Liege, Belgium (1972); AIAA Fluid and Plasma Dynamics Award (1982); Haute Distinction Honoris Causa - Ecole Centrale de Lyon (1987) and the American Physical Society Fluid Dynamics Prize (1990). He is the author and co-author of over 150 journal articles and three books on the subject of turbulence - *Stochastic Tools in Turbulence* (1970), *Structure of Atmospheric Turbulence* (with H. A. Panofsky, 1964), and *A First Course in Turbulence* (with H. Tennekes, 1972). He has had a close association with scientists throughout the world, particularly in France and the Soviet Union, for the past twenty-five years. His sabbatical leaves have been highlighted by his selection as a Guggenheim Fellow at the Laboratoire de Mecanique des Fluides, Ecole Centrale de Lyon and the Institut de Mecanique Statistique de la Turbulence, Universite de Provence, France, and as Fulbright Senior Lecturer at the Universite de Liege, Belgium. He has served in an editorial capacity for several Russian publications. He was the Technical Editor of the English translations of the texts *Statistical Fluid Mechanics* (by A. S. Monin and A. M. Yaglom; 1971, 1975) and *Variability of the Oceans* (by A. S. Monin, V. M. Kamenkovich and V. G. Kort; 1977). In addition, he is a member of the Editorial Board of Fluid Mechanics: Soviet Research, and Chairman of the Technical Editorial Board of Izvestiya: Atmospheric and Oceanic Physics. Domestically, he has twice been Chairman of the Division of Fluid Dynamics of the American Physical Society, and an Associate Editor (1976 - 1986) and then Co-Editor (since 1987) of the *Annual Review of Fluid Mechanics*. He has been the Principal on two films produced by the National Committee for Fluid Mechanics Films. One in 1963 on the *Deformation of Continuous Media* and the other in 1968 on

Eulerian-Lagrangian Frames. Added to this extensive list of accomplishments is the supervision of ten Masters and twenty-eight Ph.D. students.

As one scans through John's career over the last four decades, one realizes the impact he has had on the field of turbulence as well as on his students and colleagues. We are all impressed with his knowledge and innovation in all aspects of his research. He is as adept at the mathematical aspects and the physical insights of a problem as he is in the design of an experiment. He is a man who leads by example and a man who instills in his students and others self-discipline and self-motivation. In sum, he is a man whom we respect for his professionalism and his ability. John's technical interests have long been balanced by his interests in working on old English cars and his appreciation for fine wines and French cuisine, each of which he continues to enjoy and cultivate with the passing years. Suffice it to say that John's enthusiasm for his work, his hobbies, and life has not diminished, and this sixtieth birthday volume is but a time to honor his career to date and to look forward to another decade and beyond for still further achievements.

Thomas B. Gatski
Sutanu Sarkar
Charles G. Speziale

PREFACE

This book contains contributions by former students, colleagues and friends of Professor John L. Lumley, on the occasion of his 60th birthday, in recognition of his enormous impact on the advancement of turbulence research. A variety of experimental, computational and theoretical topics, including turbulence modeling, direct numerical simulations, compressible turbulence, turbulent shear flows, coherent structures and the Proper Orthogonal Decomposition are contained herein. The diversity and scope of these contributions are further acknowledgment of John Lumley's wide ranging influence in the field of turbulence.

The large number of contributions by the authors, many of whom were participants in *The Lumley Symposium: Recent Developments in Turbulence* (held at ICASE, NASA Langley Research Center on November 12 & 13, 1990), has presented us with the unique opportunity to select a few numerical and theoretical papers for inclusion in the journal *Theoretical and Computational Fluid Dynamics* for which Professor Lumley serves as Editor. Extended Abstracts of these papers are included in this volume and are appropriately marked. The special issue of *TCFD* will appear this year and will serve as an additional tribute to John Lumley.

As is usually the case, the efforts of others have significantly eased our tasks. We would like to express our deep appreciation to Drs. R. G. Voigt and M. Y. Hussaini of ICASE for their encouragement and support for the symposium, as well as this volume, and to Ms. Emily Todd for her tireless efforts in assisting with the symposium arrangements and the volume preparation. Likewise, our special thanks go to the Cornell University members of the Symposium Organizing Committee – Drs. S. Leibovich and Z. Warhaft – for their assistance with this project. But most of all, we would like to thank the authors for their enthusiastic support and cooperation. This volume represents their expression of respect and admiration for the outstanding accomplishments of John L. Lumley.

Thomas B. Gatski
Sutanu Sarkar
Charles G. Speziale

CONTENTS

ACKNOWLEDGMENTS FOR PERMISSIONS

Springer-Verlag wishes to thank the publishers and authors listed below for their copyright permission and endorsement to use their previously published figures in this book. Their invaluable help in this matter has made the publication of this volume possible. The figures listed below are reprinted with permission from the following:

"Lumley's PODT Definition of Large Eddies and a Trio of Numerical Procedures," by Fred R. Payne
Figures 3a and 3b reprinted with permission from University of Texas, Arlington, from *Interaction of the Large Eddy Structure in the Inner-Half of a Turbulent Boundary Layer*, by Myung-Seong Kim (1989).

"Structural Genesis in Wall-Bounded Turbulent Flows," by Sidney Leibovich
Figure 2 reprinted with permission from Cambridge University Press, from *Journal of Fluid Mechanics*, vol. *73*, p. 158 (Mackrodt).

"Characterization and Compression of Turbulent Signals and Images Using Wavelet-Packets," by Lareef Zubair, K. R. Sreenivasan, and M. Victor Wickerhauser
Figure 1 reprinted with permission from MIT Press, from *A First Course in Turbulence*, by Tennekes/Lumley, 1972.

"Self-Preservation of Temperature Fluctuations in Isotropic Turbulence," by William K. George
Figure 1 reprinted with permission from Cambridge University Press, from *Journal of Fluid Mechanics*, vol. *88*, p. 660 (Warhaft/Lumley).

PART 1

TURBULENCE MODELING AND SIMULATIONS

PARAMETERIZING TURBULENT DIFFUSION IN THE ATMOSPHERIC BOUNDARY LAYER

J. C. Wyngaard, C.-H. Moeng, J. C. Weil

National Center for Atmospheric Research[1]
Boulder, Colorado

ABSTRACT

The statistical approach to atmospheric turbulence, elucidated by John Lumley in *The Structure of Atmospheric Turbulence*, his pathbreaking 1964 monograph with Hans Panofsky, is particularly useful with large-eddy simulation. We discuss the progress we have made along these lines in parameterizing the diffusion of conservative scalars in the convective boundary layer of the atmosphere.

1. Introduction by JCW

It was my good fortune to be a graduate student at Penn State in the early 1960s when the paths of Hans Panofsky and John Lumley crossed. John was an up-and-coming member of the Aerospace Engineering faculty; Hans was in Meteorology, which happily was located not more than a few hundred meters away. Hans soon heard of John's prowess with turbulence. In a recent memorial volume for Hans, John told the story of their collaboration (Wyngaard, 1989):

> In the early sixties, Hans invited me to join him in the writing of what was to become *The Structure of Atmospheric Turbulence*. I was a very young assistant professor and

---

[1] The National Center for Atmospheric Research is sponsored by the National Science Foundation.

was quite flattered at the attention of such a distinguished man. At the same time, I did not know what I was getting into because I had never worked closely with anyone outside of my own little specialty. At about the same time, Hans started sitting in on my turbulence course and dragged along with him whatever other people from the meteorology department could not escape....

The book was a lot of fun to write. In addition to totally different backgrounds, our approaches are very different. Hans had in mind a book primarily covering Monin-Obukhov similarity theory, then relatively new and unfamiliar in the West. I had in mind something to alleviate what seemed to me the shocking ignorance in the meteorological community. Each of us wrote a first draft of his own part, and then we exchanged parts. Then the telephone calls started.... He was concerned, and rightly so, that meteorologists would not understand what I, still wet behind the ears, had written.... On the other hand, I found his part filled with things that seemed to me logically unfounded. When each of us had had enough of this, the manuscript was sent to two colleagues and mentors, one of mine and one of HAP's. Predictably, HAP's colleague found nothing to comment on in HAP's part, other than its unusual clarity, but found my part unnecessarily obscure, mathematical and difficult. My colleague found nothing to comment on in my part, other than *its* clarity, but found HAP's part untidy, confused, reaching unjustified conclusions, etc. By this process we crept together, finally arriving just close enough to touch....

John's half of the monograph carefully set out the mechanics and statistical representation of atmospheric turbulence in what my engineering-student colleagues and I found to be clear and compelling terms. I found it directly useful in my thesis research. I began thumbing through *Journal of the Atmospheric Sciences* on visits to the library. When I began to look for jobs I considered both the traditional variety, in laboratory fluid mechanics, but also some in micrometeorology. I chose one of the latter and began a career that was to take me into meteorology. I had been converted.

2. Understanding Diffusion in the Convective Boundary Layer

A. Some History

By the 1950s convection experiments—both in the laboratory and in the lower atmosphere—had made it evident that the upward heat flux in mid and upper levels was maintained in the presence of a vanishing or even positive (i.e., counter) vertical gradient of mean temperature. Deardorff (1966) interpreted this in the context of the balance of temperature variance of the form

$$\frac{\partial \overline{\theta^2}}{\partial t} = -2\overline{w\theta}\frac{\partial \Theta}{\partial z} - \frac{\partial \overline{w\theta^2}}{\partial z} - \epsilon_\theta, \qquad (1)$$

where Θ and θ are mean and fluctuating (potential) temperature, z is the vertical coordinate, w is fluctuating vertical velocity, and ϵ_θ is the molecular destruction rate. Mixing-length arguments imply that the flux and mean gradient have opposite signs, implying in turn that in steady conditions the mean-gradient term in (1) is a source of $\overline{\theta^2}$. Since the divergence term only redistributes $\overline{\theta^2}$-stuff in space, ϵ_θ must be the principal sink. Deardorff had enough detailed data to confirm that the sign reversal of $\partial\Theta/\partial z$ was indeed associated with its becoming a sink of temperature variance, rather than a source. Deardorff used (1) to estimate the magnitude of this counter mean gradient.

During the 1970s Deardorff and colleague Glen Willis did a number of experiments of diffusion from line sources in a laboratory convection tank. This work, summarized in Deardorff (1985) and Lamb (1982), clearly illuminated an important fact: the diffusion characteristics were strongly dependent on the release height. For example, the centerline or locus of maximum mean concentration ascended from a surface source but descended from an elevated release. In parallel with these experiments, Lamb (1978, 1982) calculated diffusion in the convective boundary layer with a Lagrangian statistical approach using Deardorff's large-eddy-simulation (LES) turbulence fields. He reproduced the results from the laboratory experiments and extended them to higher sources. Lamb showed that the variation of the rms particle displacement with source height was asymmetric about the midplane of the boundary layer.

In atmospheric applications one often needs information on species diffusing through the boundary layer from large-area sources. Since diffusion is mathematically a linear problem, in principle one can generate area-source solutions by integrating results from line or point sources. It is not clear whether this was ever done with the convection-tank results, but they certainly did stimulate interest in the odd diffusion properties of convective turbulence.

The one-dimensional (i.e., infinite-area source) vertical diffusion of a conservative scalar constituent in the convective boundary layer occurs in two geometries. In one, which we have labeled "top-down diffusion," the constituent enters the boundary layer by entrainment from the stably stratified capping layer at the top. An example is the entrainment of ozone into the boundary layer from above. In "bottom-up diffusion" the scalar enters through the surface—as in the case of water vapor entering through transpiration, for example. In the early 1980s we asked ourselves: how different are the diffusion properties in these two problems?

B. Some Results from Large-eddy Simulation

We attempted to isolate the properties of bottom-up and top-down diffusion through large-eddy simulation. These processes are schematically defined in Fig. 1. The vertical profile of scalar flux $\overline{cw}$ in each is

$$\text{bottom} - \text{up}: \quad \overline{cw} = \overline{cw}_0(1 - z/h), \tag{2}$$

$$\text{top} - \text{down}: \quad \overline{cw} = \overline{cw}_1(z/h). \tag{3}$$

We considered horizontally homogeneous diffusion where the mean concentration $\overline{C}$ is governed by

$$\frac{\partial \overline{C}}{\partial t} + \frac{\partial \overline{cw}}{\partial z} = 0. \tag{4}$$

In quasi-steady conditions $\partial \overline{C}/\partial z$ becomes independent of time, so that $\overline{cw}$ becomes linear. Any such problem can be represented as the superposition of these two processes. Thus, top-down and bottom-up diffusion are canonical problems.

We soon found that it was difficult to isolate physically the top-down and bottom-up processes. We could simulate top-down diffusion by allowing the convective layer to entrain a scalar from above (or, of course, to diffuse a scalar into the free flow above;

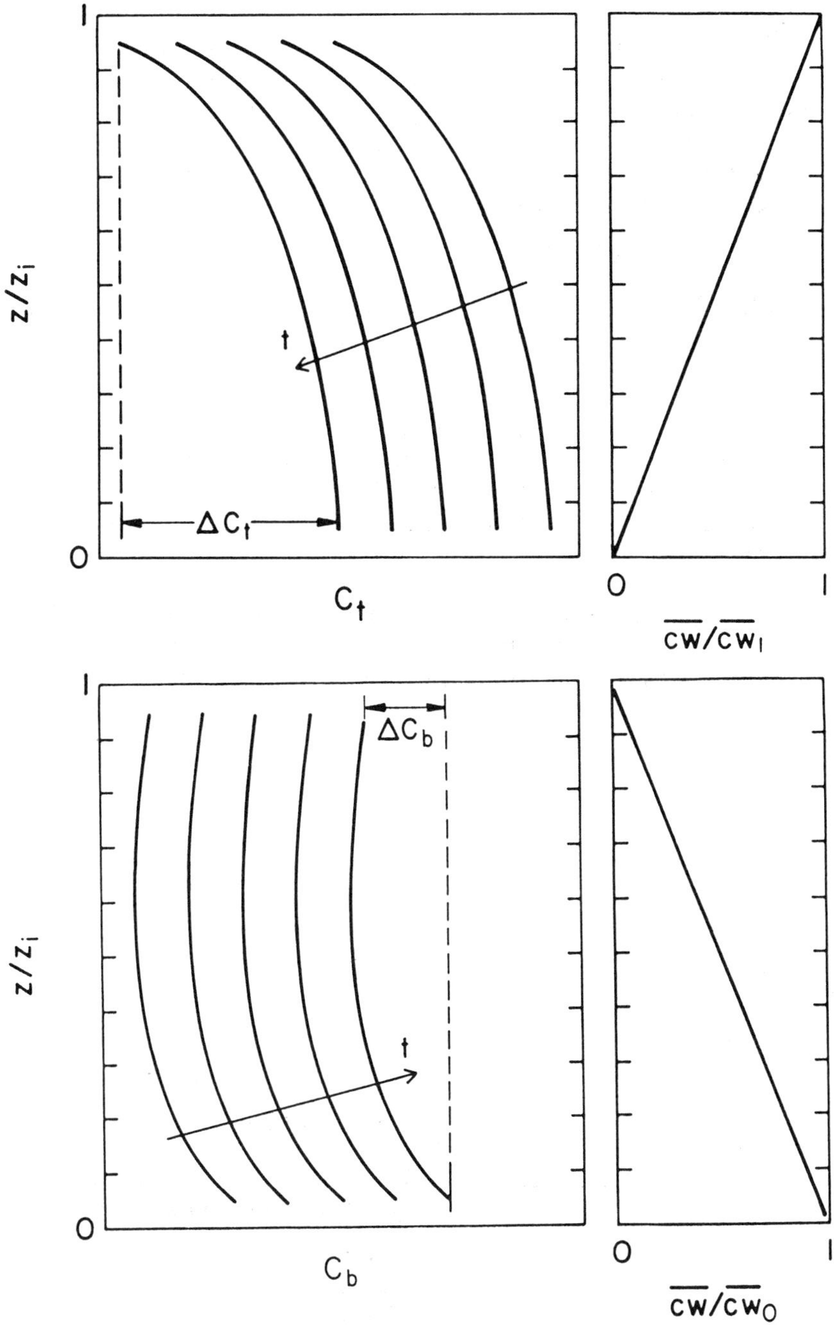

Figure 1. Schematic of top-down (top panel) and bottom-up (bottom panel) diffusion of a scalar c over depth z_i. The scalar flux profile is on the right; the time evolution of the mean concentration is on the left.

the sign of flux is not important) while maintaining a zero-flux condition at the rigid surface below. If we attempted to simulate the bottom-up case by emitting a scalar from the surface, however, it would soon generate a top-down process in response to the mean concentration jump that developed across the capping inversion. As a result we represented a conservative scalar field C as the sum of top-down and bottom-up parts,

$$C(\boldsymbol{x}, t) = C_t(\boldsymbol{x}, t) + C_b(\boldsymbol{x}, t), \tag{5}$$

and isolated the top-down and bottom-up properties statistically from runs made with three diffusing scalars.

In this way we found a large difference between the top-down and bottom-up diffusivities K—where we use the traditional meaning of K, the negative of the ratio of flux and mean gradient. These results are shown in Figure 2. We also found a considerable difference between the variances of the two processes (Moeng and Wyngaard, 1984).

The large LES "data base" we generated allowed us to study in some detail—more than has yet been revealed experimentally—the Reynolds equations for the top-down and bottom-up fields. We

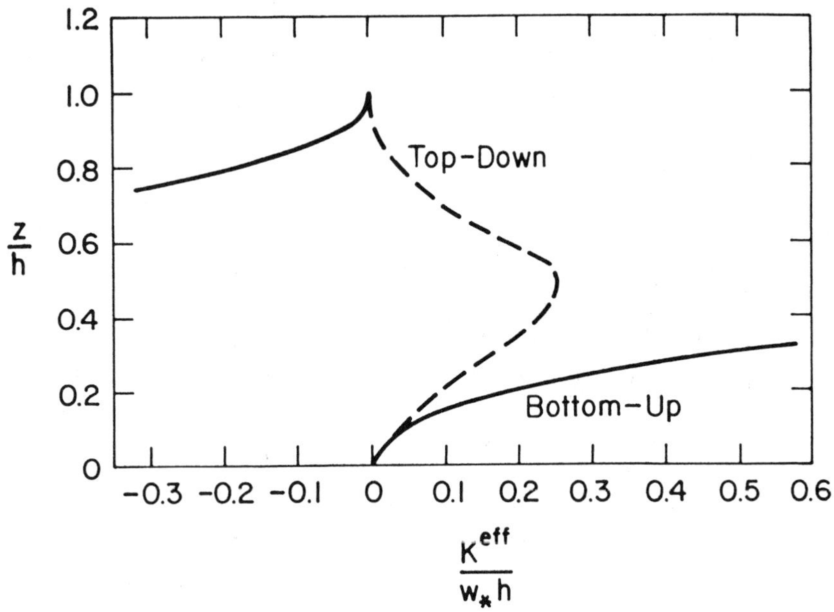

Figure 2. LES results for effective diffusivity, scaled with layer depth h and convective velocity scale w_*.

found that different closures are required for the two processes, and that contemporary versions do not fare uniformly well against our results (Moeng and Wyngaard, 1988).

We also calculated the joint probability densities of the top-down and bottom-up fields with w with the results shown in Figure 3. These were calculated in mid-PBL and in the same velocity field. The difference in the shapes is striking.

These LES results indicate that there are profound differences between the top-down and bottom-up diffusion processes in convective turbulence. These processes occur routinely in the lower atmosphere, upper ocean, and in cumulus clouds. Current large-scale meteorological and oceanographic models do not parameterize them accurately, or even differentiate between them. Let us take a fresh look at this parameterization issue.

3. Parameterizing Vertical Diffusion

A. A Constitutive Equation for Small-time-scale Turbulence

We recently found an interesting, if seemingly unlikely, link between a statistical result for turbulent diffusion and our LES results for the convective boundary layer.

Lumley (1975) derived an expression for the turbulent flux of a passive, conservative scalar quantity in a turbulent flow. In one-dimensional form and for arbitrary point z_0, it is

$$\text{flux} = \sum_{n=0}^{\infty} \frac{(-1)^n}{(n+1)!} \frac{\partial^n}{\partial z_0^n} \left[C(z_0, 0) \overline{\dot{\zeta^{n+1}}} \right]. \tag{6}$$

This is based on an accounting at time zero and again at time t; ζ is the difference between the locations of a particle at these two times, the overbar represents an ensemble average, and the dot represents a time derivative.

Wyngaard and Weil (1990) suggested that a constraint be added when using (6)—that it be invariant to the addition of a constant to C. To see how this affects the result, we expand (6):

$$\text{flux} = C(z_0, 0)\overline{\dot{\zeta}} - \frac{1}{2}\frac{\partial}{\partial z_0}\left[C(z_0, 0)\overline{\dot{\zeta^2}}\right] + \frac{1}{6}\frac{\partial^2}{\partial z_0^2}\left[C(z_0, 0)\overline{\dot{\zeta^3}}\right] + \cdots. \tag{7}$$

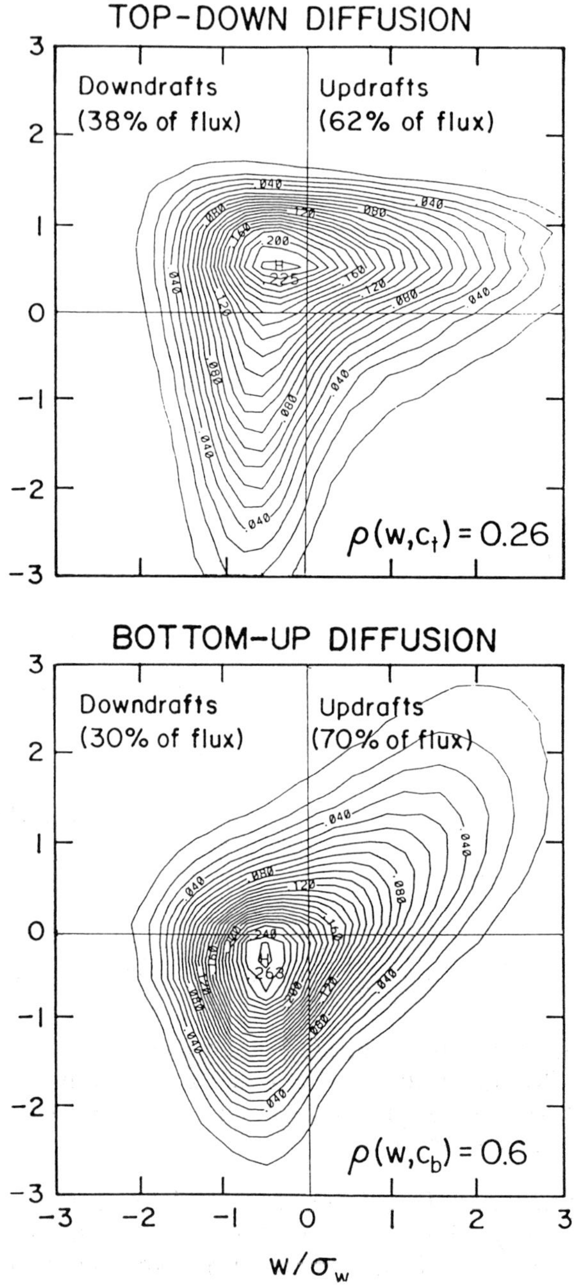

Figure 3. The joint probability densities of vertical velocity w and scalar concentration c for top-down and bottom-up diffusion. Note the much higher correlation coefficient in the bottom-up case, and the dominance of updraft transport in both cases.

Differentiating and grouping terms gives

$$\begin{aligned}\text{flux} = &\left[\dot{\overline{\zeta}} - \frac{1}{2}\frac{\partial}{\partial z_0}\dot{\overline{\zeta^2}} + \frac{1}{6}\frac{\partial^2}{\partial z_0^2}\dot{\overline{\zeta^3}} + \cdots\right] C + \\ &\left[-\frac{1}{2}\dot{\overline{\zeta^2}} + \frac{1}{3}\frac{\partial}{\partial z_0}\dot{\overline{\zeta^3}} - \frac{1}{8}\frac{\partial^2}{\partial z_0^2}\dot{\overline{\zeta^4}} + \cdots\right]\frac{\partial C}{\partial z_0} + \\ &\left[\frac{1}{6}\dot{\overline{\zeta^3}} - \frac{1}{8}\frac{\partial \dot{\overline{\zeta^4}}}{\partial z_0} + \cdots\right]\frac{\partial^2 C}{\partial z_0^2} + \ldots \end{aligned} \tag{8}$$

The first bracketed term must vanish if (8) is to be invariant under the addition of a constant to C. This implies

$$\dot{\overline{\zeta}} = \frac{1}{2}\frac{\partial}{\partial z_0}\dot{\overline{\zeta^2}} + \cdots, \tag{9}$$

an extension of the usual relation for mean displacement velocity (Monin and Yaglom, 1971).

Thus (8) becomes, accepting Lumley's replacement of C by its ensemble mean,

$$\begin{aligned}\text{flux} = &\left[-\frac{1}{2}\dot{\overline{\zeta^2}} + \frac{1}{3}\frac{\partial}{\partial z_0}\dot{\overline{\zeta^3}} - \frac{1}{8}\frac{\partial^2}{\partial z_0^2}\dot{\overline{\zeta^4}} + \cdots\right]\frac{\partial \overline{C}}{\partial z_0} + \\ &\left[\frac{1}{6}\dot{\overline{\zeta^3}} - \frac{1}{8}\frac{\partial \dot{\overline{\zeta^4}}}{\partial z_0} + \cdots\right]\frac{\partial^2 \overline{C}}{\partial z_0^2} + \cdots, \end{aligned} \tag{10}$$

which is a constitutive equation for scalar flux.

We considered the simple case of skewed, homogeneous turbulence. This is a crude model of the w field in the convective mixed layer. There—i.e., away from immediate vicinity of the surface and the capping inversion—the vertical velocity field w is strongly skewed ($S = \overline{w^3}/(\overline{w^2})^{3/2} \sim 1$); Wyngaard, 1988); and yet its intensity varies by only about 10%. Skewness implies that $\dot{\overline{\zeta^n}}$, the *time derivative* of the n-th moment of displacement, does not vanish for odd $n > 1$—as it would in a Gaussian field, for example. Homogeneity implies that the *spatial derivatives* of displacement moments do vanish.

Our constitutive equation (10) now reads, for the flux $\overline{cw}$ in skewed, homogeneous turbulence,

$$\overline{cw} = -\frac{1}{2}\dot{\overline{\zeta^2}}\frac{\partial \overline{C}}{\partial z} + \frac{1}{6}\dot{\overline{\zeta^3}}\frac{\partial^2 \overline{C}}{\partial z^2} - \frac{1}{24}\dot{\overline{\zeta^4}}\frac{\partial^3 \overline{C}}{\partial z^3} + \cdots. \quad (11)$$

There is a simple physical interpretation of (11). To first approximation, the concentration fluctuation c is produced by displacement in the presence of a mean concentration gradient so that $c \sim -\zeta \partial\overline{C}/\partial z$. Thus, the flux $\overline{cw}$ is

$$\overline{cw} \sim -\overline{\zeta w}\frac{\partial \overline{C}}{\partial z} \sim -\overline{\zeta\dot{\zeta}}\frac{\partial \overline{C}}{\partial z} \sim -\frac{1}{2}\dot{\overline{\zeta^2}}\frac{\partial \overline{C}}{\partial z}, \quad (12)$$

the first term in (11). The next term in a higher approximation to c involves $\partial^2\overline{C}/\partial z^2$, with a coefficient $\overline{\zeta^2\dot{\zeta}}/2 = \dot{\overline{\zeta^3}}/6$; this is the second term in (11).

We traditionally write for $\dot{\overline{\zeta^2}}$

$$\dot{\overline{\zeta^2}} = 2\overline{\zeta\frac{d\zeta}{dt}} = 2\int_0^t \overline{w(t)w(t')}dt' = 2\overline{w^2}\int_0^t R(t-t')d(t-t'), \quad (13)$$

where R is the Lagrangian correlation function. If the Lagrangian integral time scale T_L is sufficiently small (i.e., if $t \gg T_L$) we can write (13) as

$$\dot{\overline{\zeta^2}} = 2\overline{w^2}T_L = 2K, \quad (14)$$

with K the "eddy" diffusivity. We can also write

$$\dot{\overline{\zeta^3}} = 3\overline{\zeta^2\frac{d\zeta}{dt}} = 3\int_0^t\int_0^t \overline{w(t)w(t')w(t'')}dt'dt''. \quad (15)$$

Following (13), it is reasonable to assume that

$$\int_0^t\int_0^t \overline{w(t)w(t')w(t'')}dt'dt'' = \int_0^t\int_0^t Q(t-t',t-t'')dt'dt'', \quad (16)$$

where Q is a triple correlation function that depends only on time differences. Since $Q(0,0) = \overline{w^3}$, it is reasonable to assume further that

$$\int_0^t\int_0^t Q(t-t',t-t'')dt'dt'' = A\overline{w^3}T_L^2 = AS(\overline{w^2})^{3/2}T_L^2, \quad (17)$$

where S is the skewness of the transporting turbulence and A is an $O(1)$ constant. For $t \gg T_L$ (14) implies that $\overline{\zeta^2} \sim \overline{w^2} T_L t$, while (17) implies that $\overline{\zeta^3} \sim (\overline{w^2})^{3/2} T_L^2 t$. It follows that the skewness of ζ goes as $(T_L/t)^{1/2}$ and therefore vanishes for $t \gg T_L$, as required by its Gaussianity (Lumley, 1972). In skewed, homogeneous turbulence the expansion (11) then becomes, using (14)–(17),

$$\overline{cw} = -K\frac{\partial \overline{C}}{\partial z} + \frac{ASK\sigma_w T_L}{2}\frac{\partial^2 \overline{C}}{\partial z^2} + \cdots, \tag{18}$$

where $\sigma_w = (\overline{w^2})^{1/2}$.

B. Diffusion Through a Layer of Depth h

Let us now consider the implications of (18) for the mean concentration structure in diffusion through a layer of depth h of skewed, homogeneous turbulence. We write (18) in the form

$$\overline{cw} = -K\frac{\partial \overline{C}}{\partial z} + \frac{SK\sigma_w T_L}{2}\frac{\partial^2 \overline{C}}{\partial z^2} + D\frac{\partial^3 \overline{C}}{\partial z^3} + \cdots, \tag{19}$$

where D represents the coefficient of the third derivative term in the expansion (11) and we have taken $A = 1$. In order to avoid possible inconsistencies (Corrsin, 1974) that could result from truncating (19), we treat (19) as a differential equation for $\partial \overline{C}/\partial z$, with the flux $\overline{cw}$ as the forcing term. Its solution can also be written as a series expansion:

$$\begin{aligned}\frac{\partial \overline{C}}{\partial z} &= C_1 e^{2z/S\sigma_w T_L} - \frac{\overline{cw}}{K} - \left(\frac{S\sigma_w T_L}{2K}\right)\frac{\partial \overline{cw}}{\partial z} \\ &\quad - \left(\frac{(S\sigma_w T_L)^2}{4K} + \frac{D}{K^2}\right)\frac{\partial^2 \overline{cw}}{\partial z^2} + \cdots,\end{aligned} \tag{20}$$

where C_1 is a constant. We require that $\partial \overline{C}/\partial z$ vanish when $\overline{cw}$ and its derivatives do, so $C_1 = 0$.

A linear $\overline{cw}$ profile truncates the expansion (20) for us. This occurs in the horizontally homogeneous vertical diffusion of a scalar field from a large, plane source, where the mean concentration balance is

$$\frac{\partial \overline{C}}{\partial t} + \frac{\partial \overline{cw}}{\partial z} = 0. \tag{21}$$

Although (21) indicates that $\overline{C}$ is unsteady, it has *quasi-steady* solutions with $\partial \overline{C}/\partial z$ (and higher derivatives of $\overline{C}$) independent of time; in these cases (21) shows that $\overline{cw}$ is linear in z. In another problem, a balance between z-independent mean advection and flux divergence allows a steady solution:

$$U\frac{\partial \overline{C}}{\partial x} + \frac{\partial \overline{cw}}{\partial z} = 0. \tag{22}$$

Again, the $\overline{cw}$ profile is linear in z. In these cases, therefore, where the flow has depth h, (20) reduces to

$$\frac{\partial \overline{C}}{\partial z} = -\frac{\overline{cw}}{K}\left[1 + \left(\frac{S\sigma_w T_L}{2h}\right)\frac{h}{\overline{cw}}\frac{\partial \overline{cw}}{\partial z}\right]. \tag{23}$$

It follows from (23) that the "effective" scalar eddy diffusivity in these flows with a linear flux profile is

$$K^{eff} = \frac{-\overline{cw}}{\partial \overline{C}/\partial z} = \frac{K}{1 + (\frac{S\sigma_w T_L}{2h})\frac{h}{\overline{cw}}\frac{\partial \overline{cw}}{\partial z}}. \tag{24}$$

If the skewness S and the flux profile are such that the denominator is less than 1, $K^{eff} > K$; if it exceeds 1, $K^{eff} < K$. We refer to this dependence of the effective eddy diffusivity on the shape of the scalar flux profile as "transport asymmetry."

C. Applicability to the Convective Boundary Layer

There are strong similarities between our prediction (24) for eddy diffusivity K in skewed, homogeneous turbulence of small time scale and our LES results for the convective boundary layer. In the bottom-up case the LES results show the mid-layer singularity in K^{eff} predicted by (24) (see Figure 1).

This similarity is intriguing but puzzling. In the convective boundary layer $S \sim 1$, so that explaining the mid-layer simgularity in K^{eff} with the kinematic model result (24) requires $\sigma_w T_L \sim h$. This is consistent with the structure of the convective boundary layer, since a single displacement can carry a particle across the layer. However, eddy-diffusion theory is usually held to require that $\sigma_w T_L$ be much smaller than the length scale over which $\overline{C}(z)$ changes significantly (Corrsin, 1974). Therefore, (24) would seem to require that $T_L \ll h/\sigma_w$—i.e., that a fluid particle transporting the scalar species undergoes many independent displacements in traveling across the fluid layer.

Wyngaard and Weil (1990) attempted to shed some light on this inconsistency through use of Weil's Lagrangian particle model. Weil (1990) had previously shown that a Lagrangian particle model could reproduce the LES diffusion results. By adjusting the profiles of $\overline{w^2}$, S, and T_L, he could create a mid-layer singularity in K with nonskewed but inhomogeneous turbulence at large T_L. He found that a (positively) skewed w-field suppresses this singularity in the top-down case and amplifies it in the bottom-up one, creating a transport asymmetry much like that in the LES results. He concluded that the w-skewness played a key role in modifying the eddy diffusivity—as Lamb (1978) had earlier speculated.

Wyngaard and Weil (1990) found, first, that the effect of skewness on K was qualitatively the same for homogeneous and inhomogeneous turbulence. Second, we found that the magnitude of the change in K indeed scaled with the dimensionless group $(S\sigma_w T_L/h)$, as predicted by (24). T_L is difficult to measure directly, but observations in the convective atmospheric boundary layer suggest that $T_L \sim h/\sigma_w$. Since $S \sim 1$ there, this group is $O(1)$ as well. Thus, the kinematic model result (24) does seem to account for the unusual behavior of K in the convective boundary layer—even though the usual conditions on the applicability of such models would not seem to be met there.

REFERENCES

[1] Corrsin, S., 1974, *Limitations of gradient transport models in random walks and in turbulence*, Adv. in Geophysics, **18A**, F. N. Frenkiel and R. E. Munn, eds., 25.

[2] Deardorff, J. W., 1966, *The counter-gradient heat flux in the atmosphere and in the laboratory*, J. Atmos. Sci., **23**, 503.

[3] Deardorff, J. W., 1985, *Laboratory experiments on diffusion: The use of convective mixed layer scaling*, J. Climate Appl. Meteor., **24**, 1143.

[4] Lamb, R. G., 1978, *A numerical simulation of dispersion from an elevated point source in the convective planetary boundary layer*, Atmos. Environ., **12**, 1297.

[5] Lamb, R. G., 1982, *Diffusion in the convective boundary layer*, Atmospheric Turbulence and Air Pollution Modelling, F. T. M. Nieuwstadt and H. Van Dop, eds., Reidel, 159.

[6] Lumley, J. L. and H. A. Panofsky, 1964, *The Structure of Atmospheric Turbulence*, Interscience.

[7] Lumley, J. L., 1972, *Application of central limit theorems to turbulence problems*, Lecture Notes in Physics, **12**, M. Rosenblatt and C. Van Atta, eds., 1.

[8] Lumley, J. L., 1975, *Modeling turbulent flux of passive scalar quantities in inhomogeneous turbulent flows*, Phys. Fluids, *18*, 619.

[9] Moeng, C.-H. and Wyngaard, J. C., 1984, *Statistics of conservative scalars in the convective boundary layer*, J. Atmos. Sci., **41**, 3161.

[10] Moeng, C.-H. and Wyngaard, J. C., 1988, *A study of turbulent transport and pressure covariances in second-order modeling*, J. Atmos. Sci., **46**, 2311.

[11] Monin, A. S. and Yaglom, A. M., 1971, *Statistical Fluid Mechanics*, J. L. Lumley, ed., MIT Press.

[12] Weil, J. C., 1990, *A diagnosis of the asymmetry in top-down and bottom-up diffusion using a Lagrangian stochastic model*, J. Atmos. Sci., **47**, 501.

[13] Wyngaard, J. C. and Weil, J. C., 1990: *Transport asymmetry in skewed turbulence*, to appear, Phys. Fluids.

[14] Wyngaard, J. C., 1988, *Structure of the PBL*, Lectures on Air Pollution Modeling, A. Venkatram and J. C. Wyngaard, eds., Amer. Meteorol. Soc., 9.

[15] Wyngaard, J. C., 1989, *Hans Panofsky, 1917–1988*, Boundary-Layer Meteor., **47**, 1.

MODELING THE DISSIPATION-RATE EQUATION WITH THE AID OF DIRECT SIMULATION DATA

W. Rodi[1] and *N. N. Mansour*[2]
[1]University of Karlsruhe, F.R. Germany
[2]NASA Ames Research Center, Moffett Field, California

ABSTRACT

The ϵ-budget was computed from the direct simulation data (DNS) of Kim [10] for developed channel flow at $Re_\tau = 395$. The relative magnitude of the terms in the ϵ-equation is shown with the aid of scaling arguments, and the parameter governing this magnitude is established. The modeling of the terms in the equation is then addressed in the context of eddy-viscosity k-ϵ models. Some existing models for the sum of all source and sink terms in the ϵ-equation are tested against DNS data, and an improved model is proposed on the basis of these data.

1. Introduction

The dissipation rate, ϵ, of the turbulent kinetic energy is an important parameter in many turbulence models, since it appears as sink term in the equation governing the kinetic energy k and in modeled Reynolds-stress equations. In addition, it is often needed to characterize the turbulence through a time scale (k/ϵ) or a length scale ($k^{3/2}/\epsilon$) in modeling turbulent transport. When applying these turbulence models, the distribution of ϵ over the flow field must be determined. This is usually achieved by solving a transport equation for ϵ. An exact equation for the transport of $\epsilon = \nu\overline{u'_{i,j}u'_{i,j}}$ has been derived from the Navier-Stokes equations (see *e.g.* Davydov[1], Harlow and Nakayama[2]), but it cannot be used directly in turbulence models because of the appearance of a number of unknown correlations in it. Model assumptions have to be introduced in formulating a usable ϵ-equation. The pioneering work of Professor Lumley in this area over the last twenty years has contributed much to the progress achieved in this area (see *e.g.* Lumley[3], Tennekes and Lumley [4], Lumley and Khajeh-Nouri [5], Lumley [6], Zeman and Lumley [7]).

The various model forms suggested in the literature for the ϵ-equation have not been derived on the basis of data, but rest on various modeling ideas, and so far they have only been subjected to indirect testing by calculating various flows with the models. Direct numerical simulations (DNS) are now available with which the individual model assumptions can be tested directly. The data can also be used as a basis for the development of improved models. The DNS data available are still for flows at fairly low Reynolds numbers, but this makes them suitable for examining the near-wall behavior of models and for aiding the development of realistic models in this region. For high-Reynolds (high-Re) number situations, where the energy-containing and dissipative motions are very different in scale, the exact ϵ-equation provides little if any guidance. However, in low-Reynolds (low-Re) number regions near walls the situation is different because energy-containing and dissipative motions overlap in scales, and the exact ϵ-equation is useful for identifying and describing the various near-wall influence on ϵ (Launder [8]). As low-Reynolds number versions of turbulence models are now increasingly used instead of bridging the viscous sublayer with the aid of wall functions, proper modeling of the near-wall influences is important. This issue is addressed in this paper for the ϵ-equation used in low-Reynolds-number eddy-viscosity $k-\epsilon$ models with the aid of DNS data. The data evaluated are for developed channel flow at two Reynolds numbers (Kim *et al.* [9], Kim [10]). For the lower Reynolds number ($Re_\tau = 180$ based on friction velocity and channel half width), Mansour *et al.* [11] have already provided the ϵ-budget and tested some model approximations. Here, the ϵ-budget is presented for Kim's [10] new channel flow calculations at $Re_\tau = 395$. Some of the modeled ϵ equations in low-Reynolds-number k-ϵ models considered more successful in Patel *et al.*'s [12] review paper are tested against these data, and new model proposals are made for the source/sink terms in the ϵ-equation.

2. Model ϵ-equation used in low-Reynolds-number k-ϵ models

The modeled ϵ-equation used in the various low-Re k-ϵ models proposed so far can be written for 2D shear layers in the following form

$$\frac{D}{Dt}\tilde{\epsilon} = \left[(\nu + \frac{\nu_t}{\sigma_\epsilon})\tilde{\epsilon}_{,y}\right]_{,y} + C_{\epsilon 1} f_1 \frac{\tilde{\epsilon}}{k}\nu_t(U_{,y})^2 - C_{\epsilon 2} f_2 \frac{\tilde{\epsilon}^2}{k} + E \quad (1)$$

where,

$$\tilde{\epsilon} = \epsilon - D \quad (2)$$

The various models differ through the use of different functions f_1, f_2 and different terms D and E. Some models use as turbulence time scale k/ϵ and solve an equation for ϵ itself (effectively putting D to zero), while other models use as time scale $k/\tilde{\epsilon}$ and solve an equation for $\tilde{\epsilon}$ which, in contrast to ϵ, goes to zero at the wall. The function f_2 in the $\tilde{\epsilon}$-equation is usually effective only very close to the wall and is introduced to simulate the change in the decay rate of homogeneous turbulence as the Reynolds number $Re_t(= k^2/\nu\epsilon)$ becomes small. The extra term E or alternatively the function f_1 were introduced to increase the ϵ-production near the wall. A general discussion on the functions and extra terms in the various models proposed until 1984 can be found in Patel *et al.* [12]. Here, the performance of the three low-Re k-ϵ models rated best in the review of Patel *et al.* [12] is examined with the aid of DNS data for channel flow. These are the models due to Launder and Sharma [13] (hereafter referred to as LS), Lam and Bremhorst [14] (hereafter referred to as LB) and Chien [15] (hereafter referred to as CH). The constants and functions used in these three models are compiled in Table 1.

All of the models considered are of the form such that when f_1, f_2 are set to 1, and terms D and E are set to zero, the standard high-Re version of the k-ϵ model is recovered.

Model	Code	D	$\tilde{\epsilon}_w$ - B.C.	$C_{\epsilon 1}$	$C_{\epsilon 2}$	σ_ϵ	E
Launder-Sharma	LS	$2\nu\left(\sqrt{k}_{,y}\right)^2$	0	1.44	1.92	1.3	$2\nu\nu_t\left(U_{,yy}\right)^2$
Chien	CH	$2\nu k/y^2$	0	1.35	1.8	1.3	$-2\nu\tilde{\epsilon}/y^2 \exp(-y^+/2)$
Lam-Bremhorst	LB	0	$\nu k_{,yy} = 0$ or $\epsilon_{,y} = 0$	1.44	1.92	1.3	0

Table 1a. Summary of Constants, terms and boundary conditions used in existing low-Reynolds number k-ϵ models.

Model	Code	f_1	f_2
Launder-Sharma	LS	1.	$1 - 0.3\exp(-Re_t^2)$
Chien	CH	1.0	$1 - 0.22\exp[-(Re_t/6)^2]$
Lam-Bremhorst	LB	$1 + (0.05/f_\mu)^3$	$1 - \exp(-Re_t^2)$

Table 1b. Summary of damping functions used in existing low-Reynolds number k-ϵ models ($Re_t = k^2/\nu\epsilon$, $y^+ = U_\tau y/\nu$, $y^* = k^{1/2}y/\nu$).

3. ϵ-budget

The exact transport equation for ϵ $(= \nu\overline{u'_{i,j}u'_{i,j}})$ derived from the Navier-Stokes equations can be written in tensor form as

$$\frac{D}{Dt}\epsilon = P^1_\epsilon + P^2_\epsilon + P^3_\epsilon + P^4_\epsilon + T_\epsilon + \Pi_\epsilon + D_\epsilon - \Upsilon \qquad (3)$$

where the individual terms on the right-hand side are defined and identified as

$P^1_\epsilon = -\nu 2\overline{u'_{i,j}u'_{k,j}}S_{ik}$	Mixed production
$P^2_\epsilon = -\nu 2\overline{u'_{i,k}u'_{i,m}}S_{km}$	Production by mean velocity gradient
$P^3_\epsilon = -\nu 2\overline{u'_k u'_{i,m}}U_{i,km}$	Gradient production
$P^4_\epsilon = -\nu 2\overline{u'_{i,k}u'_{i,m}u'_{k,m}}$	Turbulent production
$T_\epsilon = -\nu(\overline{u'_k u'_{i,m}u'_{i,m}})_{,k}$	Turbulent transport
$\Pi_\epsilon = -\nu\frac{2}{\rho}(\overline{p'_{,m}u'_{k,m}})_{,k}$	Pressure transport
$D_\epsilon = \nu\epsilon_{,kk}$	Viscous diffusion
$\Upsilon = \nu^2 2\overline{u'_{i,km}u'_{i,km}}$	Dissipation

and $S_{ij} = (U_{i,j} + U_{j,i})/2$ is the mean strain rate. For developed channel flow, the left-hand side is zero so that the terms on the right-hand side should balance each other. Mansour *et al.* [11] have calculated the ϵ-budget from the DNS data for the low-*Re* channel flow. The ϵ-budget evaluated from the data for the channel with $Re_\tau = 395$ is given in Fig. 1. All terms have been made dimensionless with U_τ^6/ν^2. The imbalance in the budget (crosses) is also given; this is a measure of the errors in the budget terms due to discretization and limited sample

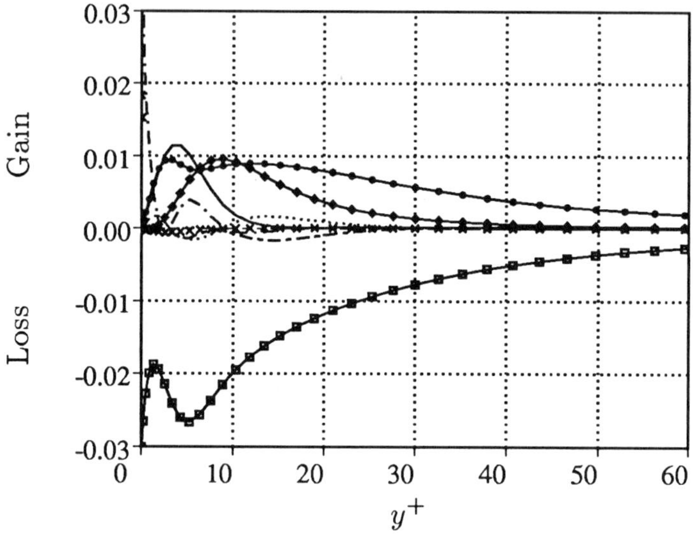

FIGURE 1. ϵ-budget for channel flow $Re_\tau = 395$. ——— P_ϵ^1; ◇◇◇ P_ϵ^2; ········ P_ϵ^3; ●●● P_ϵ^4; – – – T_ϵ; - - - - Π_ϵ; —·— D_ϵ; ⊟⊟⊟⊟ Υ; × × × × sum of all terms.

size. As can be seen, the imbalance is fairly small except very close to the wall ($y^+ < 8$) so that the ϵ-budget determined from the DNS data can be considered accurate for $y^+ > 8$. As was to be expected from the order of magnitude analysis of Tennekes and Lumley [4], the turbulent production rate P_ϵ^4 due to vortex stretching and the viscous destruction Υ dominate the balance equation in the high-Reynolds-number region away from the wall. However, near the wall, the production terms P_ϵ^1 and P_ϵ^2 become equally important, and at the wall itself viscous destruction is balanced by viscous diffusion and pressure transport. The smaller terms P_ϵ^3, T_ϵ and Π_ϵ are shown in Fig. 2 on an expanded scale. The imbalance can be seen to be small even compared with these small terms for $y^+ > 8$, and the pressure diffusion Π_ϵ can be seen to be negligible everywhere. On the other hand, the relatively small production term P_ϵ^3 is of the same order of magnitude as the turbulent diffusion T_ϵ.

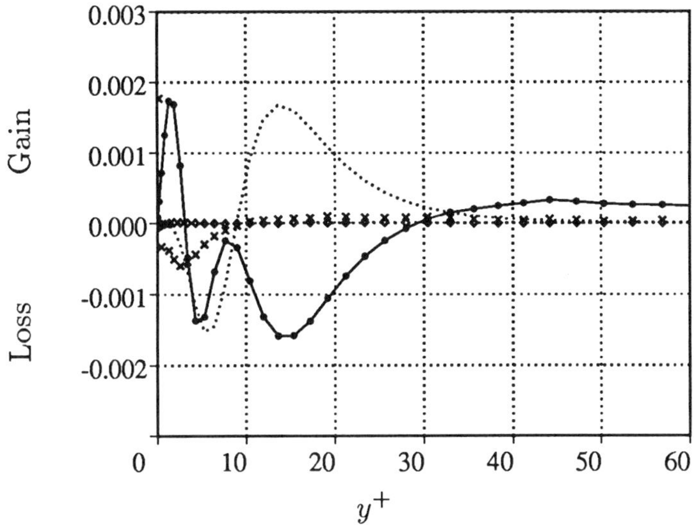

FIGURE 2. Small terms in ϵ-budget for channel flow $Re_\tau = 395$. ········ P_ϵ^3; ●—●—● T_ϵ; ◇—◇—◇ Π_ϵ; × × × × sum of all terms.

The terms P_ϵ^4 and Υ increase with increasing Reynolds number, but their difference remains independent of Reynolds number (once this is sufficiently high); the latter is true also for the rate of change and transport terms, of which in the channel flow situation only the diffusion term is non-zero. According to Tennekes and Lumley [4], the terms P_ϵ^1 and P_ϵ^2 relative to the difference $(P_\epsilon^4 - \Upsilon)$ are of order $1/Re^{1/2}$. Fig. 3 shows the sum $P_\epsilon^1 + P_\epsilon^2$ and the difference $P_\epsilon^4 - \Upsilon$ for channel flow at both Reynolds numbers investigated. With the non-dimensionalization chosen, there is no noticeable Reynolds-number influence on either group of terms away from the wall. Near the wall, both groups go up somewhat with the Reynolds number. Also, it is clear that the destruction term Υ adjusts to the increase of ϵ-production due to P_ϵ^1 and P_ϵ^2 near the wall and in fact tends to overreact somewhat. The sum of all source and sink terms (i.e. $P_\epsilon^1 + P_\epsilon^2 + P_\epsilon^3 + P_\epsilon^4 - \Upsilon$) is also included in Fig. 3. This sum, which is little influenced by the Reynolds number, is very small compared with the actual terms in the ϵ-equation. It is this sum which, in general, balances the rate

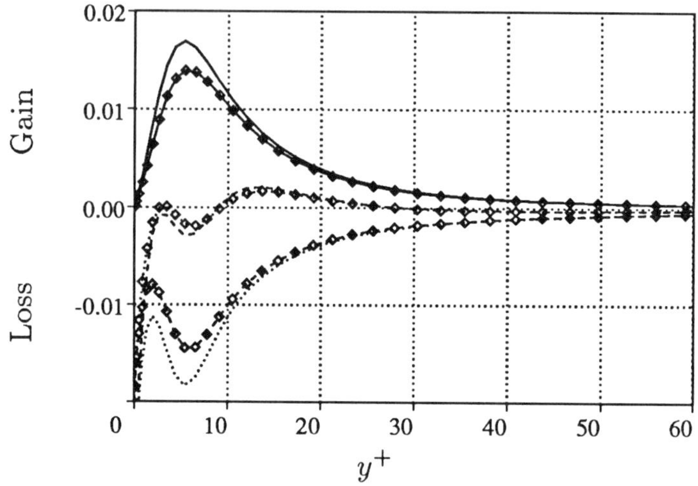

FIGURE 3. Distribution of $P_\epsilon^1 + P_\epsilon^2$, $P_\epsilon^4 - \Upsilon$ and sum of all source and sink terms in channel flow. ——— $P_\epsilon^1 + P_\epsilon^2$, $Re_\tau = 395$; ········ $P_\epsilon^4 - \Upsilon$, $Re_\tau = 395$; ---- $P_\epsilon^1 + P_\epsilon^2 + P_\epsilon^3 + P_\epsilon^4 - \Upsilon$, $Re_\tau = 395$; ◆◆◆ $P_\epsilon^1 + P_\epsilon^2$, $Re_\tau = 180$; ◆◇◆ $P_\epsilon^4 - \Upsilon$, $Re_\tau = 180$; ◇·◇·◇ $P_\epsilon^1 + P_\epsilon^2 + P_\epsilon^3 + P_\epsilon^4 - \Upsilon$, $Re_\tau = 180$.

of change, convective and diffusive transport terms in the ϵ-equation and therefore governs the magnitude of ϵ. Hence it is only this sum that really matters and has to be modeled. Because of the small magnitude of the net source compared with the original terms in the ϵ-equation, the usefulness of the exact ϵ-equation has sometimes been considered doubtful. However, even though this source is small it is still finite and balances the rate of change and transport of ϵ. At high Reynolds numbers, where an inertial subrange exists, the source/sink terms are given by the integral over the low-wave number part of the spectrum of the spectral transfer function multiplied by the wave number squared (see *e.g.* Rodi [16]).This shows that ϵ is governed by the larger-scale turbulent motions which are independent of Reynolds number.

4. Scaling arguments

Tennekes and Lumley [4] made an order of magnitude analysis of the terms in the vorticity-fluctuation equation. The order of magnitude was expressed in terms of the velocity scale u, the macro-length-scale l and the Taylor micro-scale λ. They found that the relative magnitude of the individual terms and hence also the importance of the production terms P_ϵ^1 and P_ϵ^2 depends on the Reynolds number $Re_l = ul/\nu$. In a study of homogeneous shear flow, Bardina [17] argued that the Reynolds number is not the only parameter determining the relative importance of the P_ϵ^1 and P_ϵ^2 production terms but that the mean shear number $S_n = Sk/\epsilon$ plays also a role (S is the mean shear rate). His conclusions are of interest here, but his derivation seems not to be quite correct and also does not allow direct insight into the Re-dependence of the terms in the ϵ-equation. Hence, scaling arguments are elaborated here once more.

Because of the close relation between the dissipation rate ϵ and the fluctuating vorticity $\overline{\omega_i'\omega_i'}$ (with $\epsilon = \nu\overline{\omega_i'\omega_i'}$ in homogeneous flows) the scaling arguments of Tennekes and Lumley can be applied directly to the ϵ-equation. When the strain rate in the terms P_ϵ^1 and P_ϵ^2 is not expressed as u/l but is retained as a strain-rate parameter S (which in channel flow is equal to the shear rate $\partial U/\partial y$) and with $\epsilon \propto \nu u^2/\lambda^2 \propto u^3/l$ and the velocity scale $u = k^{1/2}$, there follows the order of magnitude of the various terms as

$$P_\epsilon^1, \quad P_\epsilon^2 = O\left(S\frac{k^{3/2}}{l}\cdot\frac{\lambda}{l}\right) \tag{4}$$

$$P_\epsilon^4, \quad \Upsilon = O\left(\frac{k^2}{l^2}\cdot\frac{l}{\lambda}\right) \tag{5}$$

$$P_\epsilon^4 - \Upsilon, \quad \frac{D}{Dt}\epsilon, \quad T_\epsilon = O\left(\frac{k^2}{l^2}\right) \tag{6}$$

Since $l/\lambda \propto Re_l^{1/2} \propto Re_t^{1/2}$, the terms P_ϵ^1 and P_ϵ^2 decrease as Re_t increases while the terms P_ϵ^4 and Υ increase. The difference of the latter terms, $P_\epsilon^4 - \Upsilon$, and also the rate of change and transport terms are independent of Reynolds number, as was mentioned already. The magnitude of the production terms P_ϵ^1 and P_ϵ^2 relative to the main terms in the model ϵ- equation (difference $P_\epsilon^4 - \Upsilon$ and transport terms) can now be established as:

$$\frac{P_\epsilon^1, \quad P_\epsilon^2}{P_\epsilon^4 - \Upsilon} = O\left(\frac{Sk}{\epsilon} \cdot \frac{1}{Re_t^{1/2}}\right) \tag{7}$$

The relative order of magnitude of P_ϵ^1 and P_ϵ^2 is therefore determined by the parameter

$$R = \frac{Sk}{\epsilon} \frac{1}{Re_t^{1/2}} = \frac{S}{(\epsilon/\nu)^{1/2}} \tag{8}$$

which indeed involves the parameter $S_n = Sk/\epsilon$. Relation (8) shows that the parameter R is the ratio of the time scale of the dissipative motion to the time scale of the mean strain field. Bardina [17] also arrived at this parameter, but his derivation concerns the ratio of order of magnitudes of terms P_ϵ^1 and P_ϵ^2 to P_ϵ^4 and Υ, while here it represents the magnitude relative to the difference of the latter terms. The scaling arguments show that the production terms P_ϵ^1 and P_ϵ^2 are important when the parameter $R > 1$. Bardina [17] examined two sets of homogeneous shear-flow data obtained by direct numerical simulations. For low shear cases with typically $R < .3$ the terms P_ϵ^1 and P_ϵ^2 were found small compared with P_ϵ^4 while for the case with high shear (and low Reynolds number) characterized by $R = 6 - 15$, the terms P_ϵ^1 and P_ϵ^2 were found to be larger than P_ϵ^4.

The distribution of the parameter R in the channel flow with $Re_\tau = 395$ is shown in Fig. 4. It can be seen that in the bulk of the channel flow the parameter is substantially below 1 ($\approx .25$). In the near-wall region, where the terms P_ϵ^1 and P_ϵ^2 become important, the parameter R increases strongly and reaches a maximum value of 2.4.

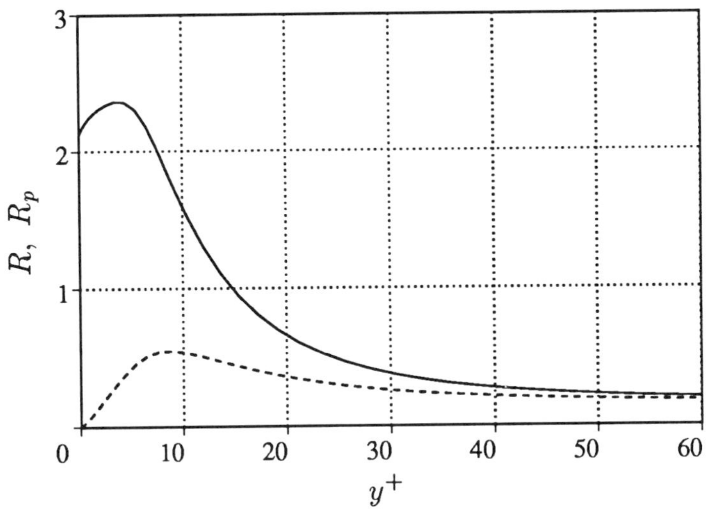

FIGURE 4. Distribution of parameters R and R_p in channel flow at $Re_\tau = 395$. ——— R; ---- R_p.

5. Modeling the terms in the ϵ-equation

In order to turn the exact ϵ-equation into an equation that can be used in a turbulence model, the source and sink terms as well as the turbulent diffusion term need to be modeled. As was mentioned already, the sum of the source and sink terms is much smaller than the actual terms but it is still finite and in channel flow it is balanced by the equally small diffusion term. Models for the source and sink terms are considered first, and the starting point is the widely used basic model for high Reynolds number (or low R) situations where only $P_\epsilon^4 - \Upsilon$ is left. This difference is often modeled as (see e.g. Launder *et al.* [18]):

$$P_\epsilon^4 - \Upsilon = \left(C_{\epsilon 1} \frac{P}{\epsilon} - C_{\epsilon 2} \right) \frac{\epsilon^2}{k} \qquad (9)$$

where P is the production of turbulent energy, that is the energy input into the low-wave-number part of the spectrum. The model relation (9) is compared in Fig. 5 with $P_\epsilon^1 + P_\epsilon^2 + P_\epsilon^4 - \Upsilon$

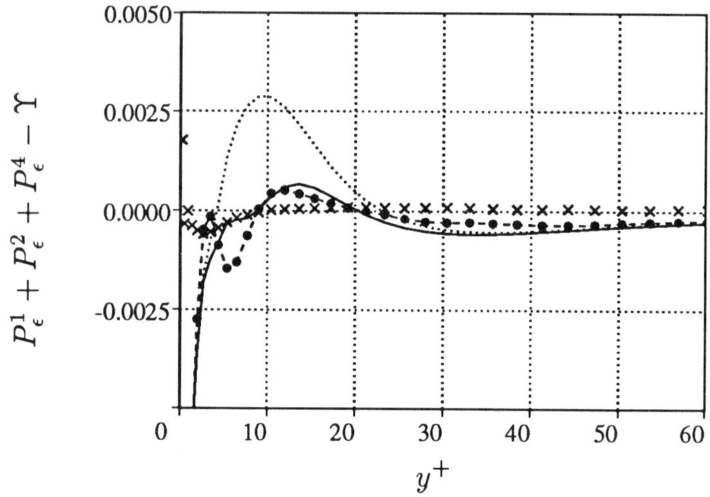

FIGURE 5. New model for source/sink terms $P_\epsilon^1 + P_\epsilon^2 + P_\epsilon^4 - \Upsilon$ compared with DNS data ($Re_\tau = 395$). •-•-• DNS data; —— new model; ········ High-Re model; × × × sum of all terms.

and in Fig. 7 with the sum of all source and sink terms ($C_{\epsilon 2}$ is multiplied by the damping function f_2 according to Table 1, but f_2 is effective only for $y^+ < 5$). Depending somewhat on the constants $C_{\epsilon 1}$ and $C_{\epsilon 2}$ used, the basic model can be seen to simulate fairly well the sum of source and sink terms away from the wall, where P_ϵ^1, P_ϵ^2 and P_ϵ^3 are unimportant. The question is now how to bring in the influence of the production terms P_ϵ^1, P_ϵ^2 and P_ϵ^3 and the consequential increase in the destruction term Υ. The terms P_ϵ^1 and P_ϵ^2 involving the mean strain rate (first derivatives of velocities) are treated separately from the term P_ϵ^3 involving second derivatives.

5.1 Modeling the effect of P_ϵ^1 and P_ϵ^2.

One possibility would be to add modeled terms of P_ϵ^1 and P_ϵ^2. P_ϵ^1 can be expressed as $-\epsilon_{ij} U_{i,j}$, where ϵ_{ij} is the dissipation rate of the Reynolds- stress component $\overline{u_i' u_j'}$; this term is therefore closely related to $\epsilon^2/k \times P/\epsilon$ and hence effectively increases the value of the coefficient $C_{\epsilon 1}$ in (9), depending on

P/ϵ. With such modeling of P_ϵ^1, and similar one for P_ϵ^2, the sink term would have to be cranked up drastically in order to account for the adjustment of Υ due to the extra production by P_ϵ^1 and P_ϵ^2. As an alternative, the suggestion is therefore made here to model the influence of the combined effect of P_ϵ^1 and P_ϵ^2 and Υ, which is to increase somewhat the sink term in the ϵ- equation. First the time-scale ratio R seemed to be a suitable parameter as it governs the relative importance of the terms P_ϵ^1 and P_ϵ^2, but it was found not to correlate too well. The close relation of P_ϵ^1 to the ratio of production to dissipation, P/ϵ, suggests that this ratio is an important parameter and was therefore taken to replace the strain-rate parameter kS/ϵ in (8). Of course the influence of the Reynolds number Re_t must be retained. The resulting parameter is

$$R_p = \frac{P/\epsilon}{.3Re_t^{1/2}} = \frac{P/k}{.3(\epsilon/\nu)^{1/2}} \qquad (10)$$

which is also a time-scale ratio, namely the ratio of the time scale of the dissipating motion to the time scale P/k involving the production of turbulence. The variation of the parameter R_p in channel flow is shown in Fig. 4.

The combined effect of P_ϵ^1, P_ϵ^2 and the increase in Υ is accounted for in the new model by multiplying the coefficient $C_{\epsilon 2}$ in the sink term of (9) by the following function:

$$f_3 = \exp(2R_p^3) \qquad (11)$$

This was obtained by fitting the exponential function to the DNS data of Fig. 5. The effect of this function can be seen in Fig. 5; it is restricted to the near-wall region of $y^+ < 30$. The new model simulates quite well the distribution of the source and sink terms $P_\epsilon^1 + P_\epsilon^2 + P_\epsilon^4 - \Upsilon$ down to $y^+ \approx 8$, below which the DNS results are not very reliable anyway.

5.2 Modeling of P_ϵ^3.

The production term P_ϵ^3 involving second derivatives of the mean velocity is small compared with the other source and sink

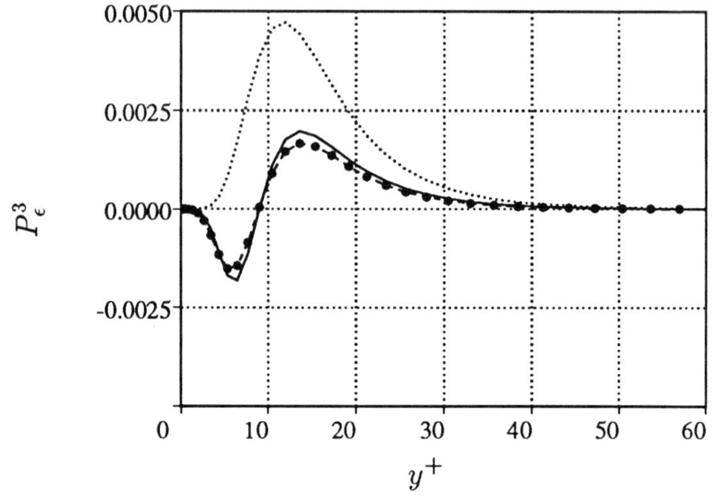

FIGURE 6. Model for P_ϵ^3-production compared with DNS data ($Re_\tau = 395$). ●-●-● DNS data; —— new model; ········ LS model.

terms, but it is comparable with their sum and also with the turbulent diffusion term (see Fig. 2). Hence it is important to represent realistically also this term in a model. Hanjalić and Launder [19] used a generalized gradient approximation for the fluctuating velocity gradients $u'_{i,j}$ appearing in the turbulence correlation in the P_ϵ^3 term and expressed these gradients in terms of second derivatives of the mean velocity. They arrived at a model expression which reads for the special case of channel flow:

$$P_\epsilon^3 = C_{\epsilon\nu} 2\nu \frac{\overline{v'^2} k}{\epsilon} \left(U_{,yy}\right)^2 \tag{12}$$

With $\overline{v'^2}$ replaced by k and $\nu_t \propto k^2/\epsilon$, there results the model used by LS:

$$P_\epsilon^3 = 2\nu\nu_t \left(U_{,yy}\right)^2 \tag{13}$$

The P_ϵ^3-distribution resulting from this model is compared in Fig. 6 with the DNS data. The model can be seen to have two problems: One is that the level of P_ϵ^3 is significantly overpredicted; this could be fixed by using a different constant. The other more fundamental problem is that the LS model predicts P_ϵ^3 to be always positive while the DNS data show that P_ϵ^3 goes negative near the wall. It can be shown from a series expansion of the fluctuating velocities u' and v' as given in Mansour *et al.* [11] that the correlation $\overline{v'u'_{,y}}$, which is the only contribution to P_ϵ^3 in channel flow, should behave as $(\overline{u'v'})_{,y}$ near the wall, i.e. as y^2, and should be negative. This confirms the behavior resulting from the DNS data.

Because of the fundamental problems with the LS model for P_ϵ^3, an improved model was developed. To this end, an exact equation for the correlation $\overline{u'_k u'_{i,j}}$ appearing in the P_ϵ^3-definition was derived by manipulating the Navier-Stokes equations (see Rodi and Mansour [20]). For modeling purposes, it was then assumed that the correlation is related to the source terms in the exact equation involving mean velocity derivatives. For channel flow only the correlation $\overline{v'u'_{,y}}$ is of interest, and the main source terms in the equation for this correlation are

$$-\overline{v'^2}U_{,yy} - \frac{1}{2}\overline{v'^2}_{,y}U_{,y} \tag{14}$$

Multiplying these terms by a time scale $k/\tilde{\epsilon}$ for dimensional reasons and assuming $\overline{v'^2} \propto k$ and $k^2/\tilde{\epsilon} \propto \nu_t$ in the context of a k-ϵ eddy-viscosity model, and allowing for different multiplying constants for each of the terms in (14), the following model results for shear-layer flows:

$$P_\epsilon^3 = -2\nu\overline{v'u'_{,y}}U_{,yy} = C_1^3 2\nu\nu_t\left(U_{,yy}\right)^2 + C_2^3\nu\frac{k}{\epsilon}k_{,y}U_{,y}U_{,yy} \tag{15}$$

The first term can be seen to be the model of LS which is always positive. The second term is negative near the wall as desired and turns positive further away from the wall. This term also has the correct near-wall behavior, namely it approaches the wall as y^2. Adjusting the constants C_1^3 and C_2^3 to best fit the DNS data ($C_1^3 = 0.5$, $C_2^3 = 0.006$) there follows the curve given in Fig. 6. The fit can be seen to be very good.

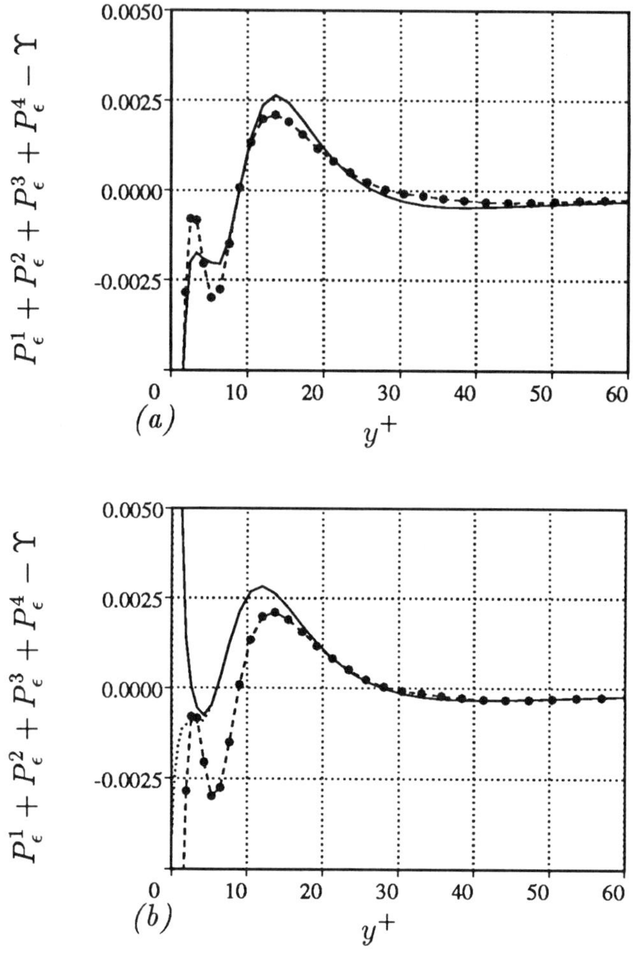

FIGURE 7. Models for all source/sink terms in ϵ-equation compared with DNS data for channel flow ($Re_\tau = 395$) ●-●-● DNS data; —— model. *(a)* New model. *(b)* CH model; ········ CH model with $E = 0$. *(c)* LS model; ········ LS model with $E = 0$. *(d)* LB model; ········ LB model with $f_1 = 1$.

5.3 Model performance for sum of all source/sink terms.

The model for P_ϵ^3 is now put together with the previously discussed model for the other source and sink terms, including the R_p-function (11). The performance of the resulting model

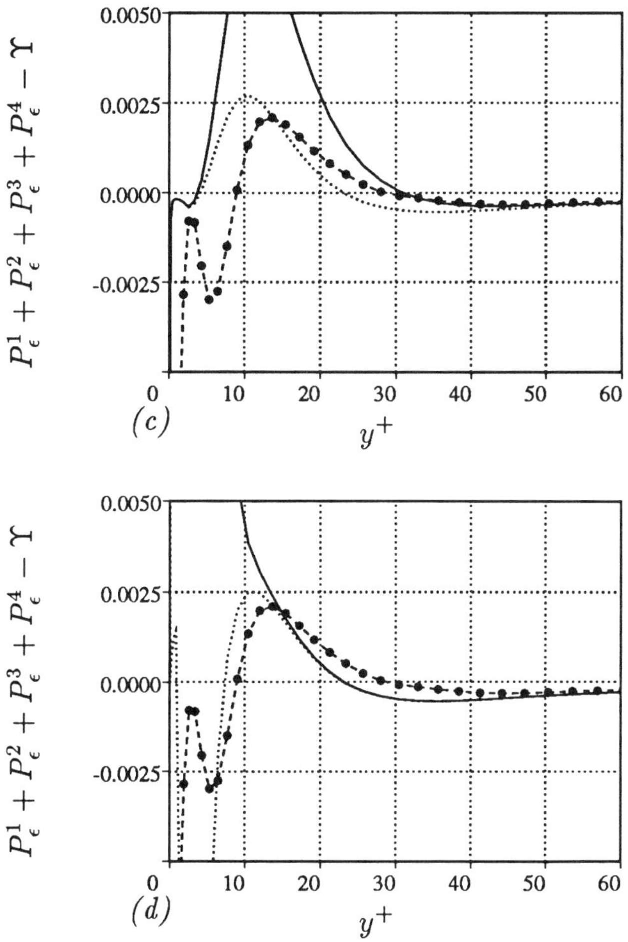

FIGURE 7. Continued.

for the sum of all source and sink terms is shown in Fig. 7a. The agreement between the model prediction and the DNS data is good down to $y^+ \approx 8$, below which the DNS data are not so reliable anyway.

It is of interest to see how this performance compares with that of the three selected existing models CH, LS and LB. Hence, in Figs. 7b - d the models for the sum of all source and sink terms in the ϵ-equations due to CH, LS and LB are compared with the DNS data. For $y+ \geq 40$, all models behave basically the same, because here the extra terms E and functions f_1 and f_2 are not effective. In the range $20 \leq y^+ \leq 40$, the CH model is superior because it uses different constants $C_{\epsilon 1}$ and $C_{\epsilon 2}$ than the other models (see Table 1), which seem to be more suitable in channel flow. However, these constants may not be so suitable for other flows, *e.g.* free shear layers, for which the constants $C_{\epsilon 1}$ and $C_{\epsilon 2}$ used in the other models were optimized. Even below $y^+ = 20$, the CH model is quite reasonable, but it is not as accurate as the new model, for which results are shown in Fig. 7a. The E-term in the CH model is effective only below $y^+ \approx 5$ which shows that the basic model of equation (9) without an extra E-term or f_1-function is quite reasonable, especially when suitable $C_{\epsilon 1}$ and $C_{\epsilon 2}$ constants are chosen. Fig. 7c shows that the LS model predicts far too high values of the source/sink terms near the wall which is due to the E-term in their model ϵ-equation representing the P_ϵ^3-production (see also Fig. 6). Overall, the model behavior is better without this term. Similarly, the f_1-function in the LB model which increases the production of ϵ has the wrong effect because f_1 assumes very large values near the wall. Again, the model behavior is better without this function.

5.4 Diffusion model.

Finally, the diffusion model generally used in the ϵ-equation is tested against the DNS data in Fig. 8. In the channel flow considered, the diffusion model reads

$$T_\epsilon = \left(\frac{\nu_t}{\sigma_\epsilon} \epsilon_{,y} \right)_{,y} \tag{16}$$

and the adjustable constant σ_ϵ is normally taken as 1.3. Fig. 8 shows that this model simulates the distribution of the diffusion term fairly well, even though the accuracy is marginal near the wall.

6. Conclusions

The ϵ-budget was determined from Kim's [10] data for channel flow at $Re_\tau = 395$. This was found reliable down to a

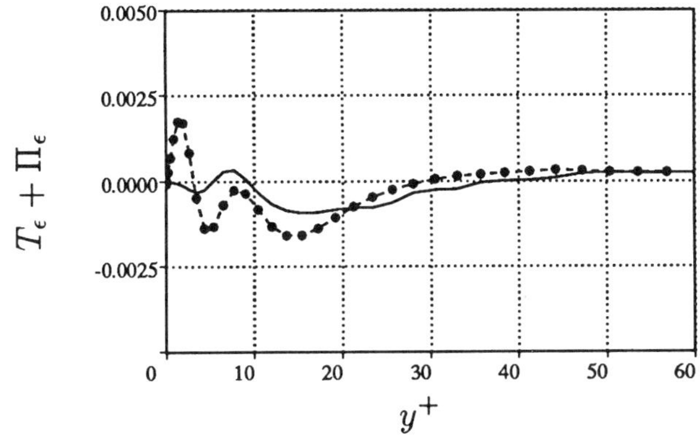

FIGURE 8. Model for diffusion term $T_\epsilon + \Pi_\epsilon$ in ϵ-equation compared with DNS data ($Re_\tau = 395$). ●-●-● DNS data; ——— model.

wall distance of $y^+ \approx 8$, as the calculated imbalance term is very small for $y^+ > 8$. As expected, the main terms in the ϵ-budget are the vortex-stretching- production term P_ϵ^4 and the viscous destruction term Υ, but near the wall the production terms P_ϵ^1 and P_ϵ^2 involving the mean strain rate are of similar magnitude. These production terms were found to cause the viscous destruction term to increase near the wall so that the sum of all source and sink terms is small compared with the main individual source and sink terms over the whole channel depth. The turbulent diffusion and the P_ϵ^3 term involving second derivatives of the mean velocity are of the same small magnitude. Through scaling considerations it was shown that the difference of the main source/sink terms, $P_\epsilon^4 - \Upsilon$, and the transport terms (here only turbulent diffusion) are independent of Reynolds number and also that the ratio of P_ϵ^1 and P_ϵ^2 to these terms is given by a parameter R involving the strain rate and the Reynolds number Re_t. This parameter represents the ratio of the time scale of the dissipating motion to the time scale of the mean strain field.

A new model was proposed and tested against the DNS channel data which simulates the net effect of the production terms

P_ϵ^1 and P_ϵ^2 and the consequential increase in the destruction term Υ. In this new model, the sink term in the ϵ-equation is increased slightly near the wall through a parameter R_p involving the ratio of production to dissipation, P/ϵ, and the turbulent Reynolds number Re_t. A new model for the source term P_ϵ^3 was also derived, based on the production terms in the exact equation for the turbulence correlation appearing in P_ϵ^3. Altogether, the new models simulate the sum of all source and sink terms in the channel flow very well down to $y^+ \approx 8$. Judging from the comparison with the DNS data, the new model is better than the existing models investigated. The CH model is not as accurate near the wall, but it is still quite reasonable, while the LS and LB models produce too large source terms near the wall. The new model proposals should now be tested in actual flow calculations.

Acknowledgments

The authors are grateful to Dr. J. Kim for providing the unpublished direct simulation data for the channel flow at $Re_\tau = 395$. The first author (WR) likes to acknowledge the generous support of the Center for Turbulence Research, Stanford University/NASA Ames research center.

References

[1] Davydov, B. I., 1961, *On the statistical dynamics of an incompressible turbulent fluid.* Doc. Acad. Nauk SSSR. **136,** 47-50 (Soviet Physics – Doklady, **6** (1), 10-12, 1961).

[2] Harlow, F. H. & Nakayama, P. I., 1967, *Turbulent transport equations.* Phys. Fluids. **10,** 2323.

[3] Lumley, J. L.,, 1970, *Toward a turbulent constitutive relation.* J. Fluid Mech. **41,** 413-434.

[4] Tennekes, H. & Lumley, J.L., 1972, A First Course in Turbulence. The MIT Press, Cambridge, Mass.

[5] Lumley, J. L. & Khajeh-Nouri, B., 1974, *Computation of turbulent transport.* Adv. Geophys. **A18,** 169-192

[6] Lumley, J. L., 1978, *Computational modeling of turbulent flows.* in Advances in Applied Mech. **18,** editor C. S. Sih, Academic Press, New York, 123-176.

[7] Zeman, O. & Lumley, J. L., 1979, *Buoyancy effects in entraining turbulent boundary layers: a second-order closure study.* Turbulent Shear Flows. **I,** Editors F. Durst *et al.*, Springer, Heidelberg.

[8] Launder, B. E., 1986, *Low-Reynolds-number turbulence near walls.* Report TFD/86/4,UMIST, Manchester, U.K.

[9] Kim, J., Moin, P. & Moser, R., 1987, *Turbulence statististics in fully developed channel flow at low Reynolds number.* J. Fluid Mech. **177,** 133-166.

[10] Kim, J., 1990, private communication.

[11] Mansour, N. N., Kim, J. & Moin, P., 1988, *Reynolds-stress and dissipation rate budgets in a turbulent channel flow.* J. Fluid Mech. **194,** 15-44.

[12] Patel, V.C., Rodi, W. & Scheuerer, G., 1985, *Turbulence models for near-wall and low- Reynolds-number flows: A review.* AIAA J. **23**(9), 1308-1319.

[13] Launder, B.E. & Sharma, B.I., 1974, *Application of the energy-dissipation model of turbulence to the calculation of flow near a spinning disk.* Letters in Heat and Mass Transfer. **1,** 131-138.

[14] Lam, C.K.G. & Bremhorst, K.A., 1981, *Modified form of the k-ϵ model for predicting wall turbulence.* J. Fluids Eng. **103,** 456-460.

[15] Chien, K.-Y., 1982, *Predictions of channel and boundary-layer flows with a low-Reynolds- number turbulence model.* AIAA J. **20,** 33-38.

[16] Rodi, W., 1971, *On the equation governing the rate of turbulent energy dissipation.* Report TM/TN/A/14, Imperial College of Science and Technology, Dept. of Mech. Eng., London, U.K.

[17] Bardina, J., 1988, *Turbulence modeling based on direct simulation of the Navier-Stokes equations.* First National

Fluid Dynamics Congress, Cincinnati, Ohio, AIAA 88-3747-CP.

[18] Launder, B.E., Reece, G.J. & Rodi, W., 1975, *Progress in the development of a Reynolds-stress turbulence closure.* J. Fluid Mech. **68,** 537-566.

[19] Hanjalić, K. & Launder, B.E., 1976, *Contribution towards a Reynolds-stress closure for low-Reynolds-number turbulence.* J. Fluid Mech. **74,** 593-610.

[20] Rodi, W. & Mansour, N. N., 1990, *Low-Reynolds number k-ϵ modeling with the aid of direct simulation data.* Proceedings of the Center for Turbulence Research, Summer 1990, Ames research center, Moffett Field, CA 94035.

SECOND-ORDER MODELING OF TURBULENT DIFFUSION IN AN ATMOSPHERIC SURFACE LAYER

V. M. Tsarenko and A. M. Yaglom[1]

Institute of Atmospheric Physics, USSR Academy of Sciences
Moscow, USSR

ABSTRACT

The semiempirical theories of passive admixture diffusion in the thermally stratified atmospheric surface layer are developed on the basis of second-order closures. Four different models of equations for the second-order moments including concentration fluctuation are investigated. All these models lead to gradient transport hypotheses corresponding to specific forms of the eddy-diffusivity tensor $K_{ij} = K_{ij}(z)$. The dependence of dimensionless eddy diffusivities $k_{ij} = K_{ij}(z)/K_M(z)$ (where $K_M(z) = u_*^2/[dU/dz]$ is the vertical eddy viscosity) on the dimensionless height $\zeta = z/L$ (where L is the Obukhov length) is determined for all the considered second-order closures.

1. Introduction

Let $c = C + c'$ (where $C =< c >$ and angular brackets designate the probabilistic averaging) be the concentration of some passive admixture in a turbulent flow. We shall assume that the Reynolds number is so high that turbulence can be considered to be locally isotropic and the molecular transfers are negligibly small as compared with the turbulent transfers. Then the equation for the mean concentration $C(\mathrm{x}) = C(x_1, x_2, x_3, t)$ has the form

$$D_t C + \frac{\partial S_i}{\partial x_i} = 0, \quad S_i =< u_i' c' >, \tag{1}$$

where Einstein's summation convention is used, $D_t = \partial/\partial t + U_i \partial/\partial x_i$, U_i, $i = 1, 2, 3$, are the components of the mean velocity $\mathbf{U} =< \mathbf{u} >$

[1]Dedicated to Prof. John L. Lumley on the occasion of his 60th birthday.

and $u'_i = u_i - U_i$ are the components of the velocity fluctuation. Equation (1) includes four unknown functions (the mean concentration C, and three components $S_i =< u'_i c' >$ of the admixture flux vector $\mathbf{S}$). Therefore this equation can be used for the determination of the mean concentration C only if it is supplemented by some approximate closure hypotheses which permit one to express admixture fluxes in terms of the mean concentration. The theories of turbulent diffusion that use some closure hypotheses (including numerical coefficients determined from experimental data) are called semiempirical theories.

The simplest widely used closure hypothesis supplementing (1) is the *gradient transport hypothesis* for the linear dependence of the admixture flux vector $\mathbf{S} = (S_1, S_2, S_3)$ on the mean concentration gradient $\nabla C = (\partial C/\partial x_1, \partial C/\partial x_2, \partial C/\partial x_3)$. The most general form of this hypothesis is given by the equations

$$S_i =< u'_i c' >= -K_{ij}\frac{\partial C}{\partial x_j}, \quad i = 1,2,3, \tag{2}$$

where K_{ij} is the eddy diffusivity tensor. The turbulent diffusion computations with the aid of (1) and (2) require additional assumptions about the functional form of the eddy diffusivities $K_{ij} = K_{ij}(\mathrm{x})$. The applicability conditions for the gradient transport hypothesis and the estimation of its accuracy are discussed by Corrsin (1974), Lumley (1978b), Sreenivasan, Tavoularis and Corrsin (1982) and other authors. It is shown in these papers that this hypothesis can be valid only under severe limitations; however, meteorological practice shows that it is often quite useful. In fact, the parabolic diffusion equations implied by (2) are widely used in air pollution modeling; moreover, the form of the eddy diffusivity tensor is often restricted here by the condition that it is diagonal (i.e., it is assumed that $K_{ij} = 0$ for $i \neq j$). However, the selection of the non-zero eddy diffusivities is usually based in practical studies on rather artificial arguments applied only to some very special turbulent flows (see, e.g., Jaffe 1967, Pasquill 1974, and Berlyand 1975).

More general (and apparently more accurate) semiempirical theories of turbulent diffusion are based on the use of so-called second-order closures. In the second-order closures of the theory of turbulence the exact equations for velocity, temperature, and/or admixture concentration are used and are supplemented by the dynamic equations for the second-order moments of turbulent fluctuations entering the equations for the mean fields. Of course, the equations for the second-order moments include also some additional unknowns which make the system of equations for the mean field and second-order moments unclosed. In the second-order closure theories the approximate closure hypotheses are applied only to the supplementary unknowns in the equations for the second-order moments. Different hypotheses for these unknowns provide different semiempirical theories. The first relatively crude second-order closures were proposed in the forties and fifties by Kolmogorov, Chou, Rotta, and Davidov (see the references in the works by Monin and Yaglom 1971 and Lumley 1978a); later such theories were substantially improved and modernized by many authors (in particular, by J. L. Lumley who played a very important role in the development of modern second-order closures). At present, an enormous literature is devoted to such closures; see e.g., Lumley and Khajeh-Nouri (1974), Mellor and Yamada (1974, 1982), Reynolds (1976), Lewellen (1977), Launder (1978), Lumley (1978a, 1983), Zeman (1981). The applications of the second-order closures to the problems of turbulent diffusion (different from the problem studied in this paper) are discussed by Lewellen and Teske (1978), Lumley (1978b), Newman, Launder, and Lumley (1981), Lumley and Mansfield (1984), Lumley and Van Cruyningen (1985), Sykes, Lewellen and Parker (1986) and some others. Note that the second-order closures imply also the general approach to the derivation of the gradient transport model since it is easy to indicate which closure hypotheses related to the equations for turbulent fluxes $S_i =< u_i'c' >$, lead to relations (2). Such a derivation of the gradient transport model permits one to determine explicitly all the eddy diffusivities K_{ij} and, if desirable, to refine the obtained model by replacing the closure hypotheses leading to equations (2) by more general assumptions. Then the diffusion computations based on equations (2) can be compared with the result following from the more general closure hypotheses to obtain a rough idea of the accuracy of the gradient transport model (see, e.g., Tsarenko 1984).

The turbulent diffusion in the atmospheric surface layer (ASL) is studied in the present paper. The ASL is the lowest air layer of the thickness of the order of 50 m where the turbulent fluxes of momentum $\tau = \rho < -u'w' >= \rho u_*^2$ and temperature $Q =< w'\theta' >$ (where ρ is the air density and $u' = u_1', w' = u_3'$ and θ' are the turbulent fluctuations of the longitudinal and vertical wind components and temperature) are practically independent of time and spatial coordinates. If the underlying surface is flat and homogeneous enough and diffusion is considered within relatively short time intervals (of the order of one hour or less), then the air flow within the ASL can be considered to be plane-parallel and steady. In this case all the statistical parameters of surface-layer turbulence depend only on z and satisfy the similarity theory developed by Monin and Obukhov (and described in detail by Monin and Yaglom 1971, Chap. 4). This paper generalizes the results of a recent paper (Tsarenko and Yaglom 1991, hereafter referred to as I) devoted to the second-order theory of diffusion within the ASL in neutral stratification. As in I, only some relatively simple second-order models leading to the gradient transport equations (2) are considered. The values of all the non-zero eddy diffusivities $K_{ij} = K_{ij}(z)$ obtained from four different second-order models (closely related to those considered in I) are determined and compared with each other. The comparison shows that the different models lead to relatively close values of the eddy diffusivities K_{ij} whose differences can hardly affect the result of the computations of the mean concentration distribution. Note also that in the case of a neutrally stratified ASL, the diffusion computations based on more complicated second-order closures that contradict the gradient transport model, lead to results which are rather close to those implied by the gradient transport assumption (see Tsarenko 1984 and the summary of his results in I). Therefore one can hope that the eddy diffusivities $K_{ij}(z)$ implied by any of the four considered second-order models will lead to sufficiently accurate estimates of the air pollution distributions.

2. Local Equilibrium Second Order Closures

According to Launder (1978) the equation for the admixture flux $S_i =< u_i'c' >$ in a high Reynolds number turbulent flow with the mean velocity components $U_i(\mathrm{x}) = U_i(x_1, x_2, x_3)$, $i = 1, 2, 3$, has the form

$$D_t S_i + \frac{\partial S_{ij}}{\partial x_j} = - < u_i' u_j' > \frac{\partial C}{\partial x_j} - < u_j' c' > \frac{\partial U}{\partial x_j} + \beta_i < \theta' c' > - \Pi_i. \tag{3}$$

Here $\partial S_{ij}/\partial x_j$ describes the spatial transport of the moment S_i due to velocity and pressure fluctuations, $S_{ij} =< u_i' u_j' c' > + \rho^{-1} < p'c' > \delta_{ij}$, (where p' is the pressure fluctuation and δ_{ij} is the Kronecker symbol), $\Pi_i = -\rho^{-1} < p'\partial c'/\partial x_i >$ is the pressure-concentration-gradient correlation and $\beta_i = -\beta\delta_{i3}$ is the ith component of the buoyancy vector $\beta = (0, 0, -\beta)$, where $\beta = -\alpha g$ is the buoyancy parameter (g is the acceleration due to gravity and α is the coefficient of thermal expansion which is equal to $1/T_0$ for an ideal gas, T_0 being the mean temperature for the ASL); the x_3 axis is directed upward. Supplementary unknowns in (3) are combinations of the third-order moments S_{ij} and Π_i (the pressure fluctuation p' is known to be of the second order with respect to velocity fluctuations). Moreover, (3) includes also a new second-order moment $S_0 =< \theta' c >$, satisfying the equation

$$D_t S_0 + \frac{\partial S_{0j}}{\partial x_j} = - < u_j'\theta' > \frac{\partial C}{\partial x_j} - < u_j' c' > \frac{\partial T}{\partial x_j} - \epsilon_0 \qquad (4)$$

where $S_{0j} =< u_j' c' \theta' >$, T is the mean temperature and $\epsilon_0 = (k_t + k_c) < (\partial\theta'/\partial x_j)(\partial c'/\partial x_j) >$ is the dissipation rate for the moment S_0 which depends upon the molecular diffusivities k_t and k_c for the temperature and concentration. Equation (4) includes supplementary unknowns S_{0j} and ϵ_0 which must be replaced by some semiempirical models.

It is easy to show that if the left-hand parts of equations (3) and (4) and the supplementary unknowns Π_i and ϵ_0 on their right-hand sides are approximated by some linear functions of the second-order moments $S_i =< u_i' c' >$ and $S_0 =< \theta' c' >$ (depending linearly on the derivatives $\partial C/\partial x_i$), then the model equations obtained for S_i and S_0 will have solutions of the form

$$S_i = -K_{ij}\frac{\partial C}{\partial x_j}, \quad S_0 = -K_{0j}\frac{\partial C}{\partial x_j}. \qquad (5)$$

Hence such an approximation automatically leads to the gradient transport assumption (2) supplemented by the similar relations for the concentration-temperature correlation S_0. Second-order models leading to equations (5) are often called the *algebraic flux models* since the turbulent fluxes S_i, $i = 1, 2, 3$, are determined here by simple algebraic (not differential) equations. The most widely used algebraic flux models are the so-called *local equilibrium models*, where the left-hand parts of (3) and (4) (describing the time evolution, advection and turbulent transfer of the second-order moments) are neglected. The application of the local equilibrium second-order models described in the available literature (referred to in I) leaves the impression that the accuracy of such models is usually sufficient for most practical aims. (As for the algebraic models, which are not in local equilibrium, they have not been used yet because of lack of data permitting one to choose reasonable non-zero approximations, which are linear with respect to S_i, for the left-hand parts of (3) and (4).) In this paper only the local equilibrium models are considered, i.e., the left-hand sides of (3) and (4) are always replaced by zeroes.

In the case of local equilibrium second-order models, the approximations are needed only for the terms Π_i and ϵ_0 in the left-hand sides of (3) and (4). The simplest widely used approximation for ϵ_0 has the form

$$\epsilon_0 = \frac{1}{\tau_0}\frac{\epsilon}{q^2} < \theta' c' > \tag{6}$$

where $\epsilon = (\nu/2)\sum_{i,j=1}^{3} < (\partial u_i'/\partial x_j + \partial u_j'/\partial x_i)^2 >$ is the rate of energy dissipation (ν being the molecular viscosity) and $q^2 =< u_i' u_i' >= < u'^2 > + < v'^2 > + < w'^2 >$ is the doubled turbulent energy. The available (not too numerous) measurements of ϵ_0 in different turbulent flows agree satisfactorily with the approximate constancy of the coefficient τ_0 whose value is usually close to 0.4. Therefore equation (6) with $\tau_0 = 0.4$ is used in the computations below.

Assuming that the local equilibrium approximation can also be applied to the equation for the moment $< c'^2 >$, we obtain

$$- < u_i' c' > \frac{\partial C}{\partial x_i} - \epsilon_c = 0, \tag{7}$$

where $\epsilon_c = k_c \sum_{i=1}^{3} < (\partial c'/\partial x_i)^2 >$ is the dissipation rate for $< c'^2 > /2$. Using now the approximation

$$\epsilon_c = \frac{1}{\tau_c}\frac{\epsilon}{q^2} < c'^2 > \tag{8}$$

(where τ_c is the empirical constant), which is similar to (6), we obtain from (7), (8), and (2) the equation

$$< c'^2 >= -\tau_c \frac{q^2}{\epsilon} < u_i' c' > \frac{\partial C}{\partial x_i} = \tau_c \frac{q^2}{\epsilon} K_{ij} \frac{\partial C}{\partial x_i}\frac{\partial C}{\partial x_j}. \tag{9}$$

This equation shows that the variance of the concentration fluctuations $< c'^2 >$ can approximately be estimated from the values of the mean concentration $C(\mathrm{x}, t)$ and the eddy diffusivities K_{ij}, if the constant τ_c is also known. The variance $< c'^2 >$ is crucial in the estimation of the probabilities of instantaneous concentration values greatly exceeding the mean concentration $C =< c >$; these probabilities are often of great practical interest.

Let us now consider the approximations for pressure-concentration-gradient correlations Π_i which will be used in this paper. For a special case of the neutrally stratified ASL a number of models for Π_i are considered in I. In particular, four different models for Π_i are especially singled out in I; these models generalize the suggestions formulated by Launder, Reece, and Rodi (1975), Gibson and Launder (1978), Zeman and Lumley (1979), Rodi (1980), Zeman (1981), and some other authors. When the second-order models are applied to thermally stratified turbulent flows, the influence of stratification on Π_i is usually taken into account by adding a term proportional to $\beta_i < \theta' c' >$ to the approximation for Π_i; see, e.g., Launder (1978), Zeman and Lumley (1979), or Zeman (1981). Approximations for Π_i are given in I for the general three-dimensional turbulent flow. However, in the particular case of a steady plane-parallel flow in the direction of the x-axis with mean velocity $U(z)$, these approximations are significantly simplified and take, in the presence of thermal stratification, the form

$$\begin{aligned} \Pi_1 &= b_1 \frac{\epsilon}{q^2} < u'c' > - b_3 \frac{dU}{dz} < w'c' >, \\ \Pi_2 &= b_1 \frac{\epsilon}{q^2} < v'c' >, \\ \Pi_3 &= b_2 \frac{\epsilon}{q^2} < w'c' > - b_4 \frac{dU}{dz} < u'c' > - b_5 \beta < \theta' c' >, \end{aligned} \tag{10}$$

where u', v', and w' are the components of the velocity fluctuation and b_1, b_2, b_3, b_4, and b_5 are the dimensionless coefficients which can depend on the ratio ℓ/z. Here ℓ is the length scale of turbulence at the given point, which can be defined in different ways, and z is the distance from this point to the wall at $z = 0$. The coefficients $b_i = b_i(\ell/z)$, $i = 1, \ldots, 5$, must be chosen, in principle, from the comparisons with the experimental data. However, at present there are no data which could be used for this aim. Therefore, four special models are introduced in I which generalize the previous suggestions of various authors, but at the same time restrict, significantly, possible values of the coefficients b_1, b_2, b_3, and b_4 (coefficient b_5 is not considered in I). These four models are described by the following conditions:

Model 1. $b_2 = b_1,\ b_4 = 0.$

Model 2. $b_2 = b_1,\ b_4 = -b_3/4.$

Model 3. b_1 = constant (does not depend on ℓ/z), $b_2(0) = b_1(0), b_4 = 0.$

Model 4. $b_1 = \text{constant}, b_2(0) = b_1(0),\ b_4 = -b_3(0)/4 = \text{constant}.$

The same four models of $b_1, \ldots, b_4$ are considered in the present paper. As for the coefficient b_5, at present there is no data which permits one to estimate its dependence on ℓ/z even roughly. Therefore, here b_5 is assumed to be constant (the same assumption is used by all the authors who have considered this coefficient before).

In the case of a steady plane-parallel flow with the mean velocity and temperature gradients dU/dz and dT/dz, equations (3), (4), (6), and (10) lead, in the local equilibrium approximation, to equations

$$\begin{aligned}
&< u'^2 > \frac{\partial C}{\partial x} - u_*^2 \frac{\partial C}{\partial z} + (1-b_3)\frac{dU}{dz} < w'c' > \\
&\qquad + b_1 \frac{\epsilon}{q^2} < u'c' >= 0, \\
&< v'^2 > \frac{\partial C}{\partial y} + b_1 \frac{\epsilon}{q^2} < v'c' >= 0, \\
&-u_*^2 \frac{\partial C}{\partial x} + < w'^2 > \frac{\partial C}{\partial z} - (1+b_5)\beta < \theta'c' > + b_2 \frac{\epsilon}{q^2} < w'c' > \\
&\qquad - b_4 \frac{dU}{dz} < u'c' >= 0, \\
&< u'\theta' > \frac{\partial C}{\partial x} + Q\frac{dC}{dz} + \frac{dT}{dz} < w'c' > + \frac{1}{\tau_0}\frac{\epsilon}{q^2} < \theta'c' >= 0.
\end{aligned} \tag{11}$$

These equations can be used for finding the expressions which determine the second-order moments $S_0 =< \theta'c' >$, $S_1 =< u'c' >$, $S_2 = < v'c' >$, and $S_3 =< w'c' >$ in terms of the derivatives $\partial C/\partial x$, $\partial C/\partial y$, and $\partial C/\partial z$. The expressions obtained have the form (5) and give the values of the eddy diffusivities K_{ij}, $i = 1,2,3$, and the coefficient K_{0j}. However, the direct computation of these values is possible only if the exact values of the coefficients b_i, $i = 1,\ldots,5$, on the right-hand side of (10) and of the gradients $dU/dz, dT/z$ and turbulence characteristics $< u'^2 >, < v'^2 >, < w'^2 >, < u'\theta >$ are known.

3. Determination of the Eddy Diffusivities K_{ij}

Selection of the Values for b_i

Assuming that the model equations (10) are valid for any steady plane-parallel turbulent flow we can apply them to the free plane-parallel flows which are not affected by solid walls. In this case, we must assume that $z = \infty$ and $\ell/z = 0$ (for any definition of ℓ). Note that Model 1 coincides with Model 3 and Model 2 with Model 4, if $\ell/z = 0$. In this case all the models include only three undetermined coefficients $b_2(0)$, $b_3(0)$, and b_5, the latter being unimportant in the absence of thermal stratification. An example of a free plane-parallel turbulent flow in neutral stratification is given by sheared homogeneous turbulence with a constant velocity gradient dU/dz. Such homogeneous turbulence was modeled in the laboratory experiments by Tavoularis and Corrsin (1981, 1985) and some others. The values of the eddy diffusivities K_{33} and K_{13} measured in the experiments by Tavoularis and Corrsin are used in I for the estimate of the values $b_2(0)$ and $b_3(0)$ for four models described above. The results obtained are presented in the Table. It is also indicated in I that these models imply the values of all the other non-zero diffusivities K_{ij} which

Table I. Values of coefficients b's in equations (10) for homogeneous sheared turbulence ($\ell/z = 0$) and neutral ASL ($\ell/z = \kappa$).

	$\ell/z = 0$		$\ell/z = \kappa$			
Model	1 and 3	2 and 4	1	2	3	4
b_1	10.26	8.41	7.65	5.47	10.26	8.41
b_2	10.26	8.41	7.65	5.47	7.65	3.87
b_3	-1.04	-0.48	-1.12	-0.28	-2.14	-1.42
b_4	0	0.12	0	0.07	0	0.12

agree satisfactorily with the measurements of K_{22} by Tavoularis and Corrsin (1985) and with the values of the diffusivities K_{ij} determined by Rogers, Mansour, and Reynolds (1989) from the direct numerical simulation of sheared homogeneous turbulence.

When the values of $b_i(0)$, $i = 1, 2, 3, 4$, are determined for all the four models, only the values of $b_2(\ell/z)$, $b_3(\ell/z)$ and b_5 remain undetermined for any fixed value of ℓ/z. In the absence of thermal stratification, the coefficient b_5 plays no role and hence only the values of b_2 and b_3 need determination. These two values can easily be calculated for the neutrally stratified ASL from the values of eddy diffusivities $K_{33}(z)$ and $K_{13}(z)$ which have been measured often in such ASL. The corresponding calculation is described in I and its results are also given in the Table. The value of ℓ/z for the neutral ASL clearly depends on the definition of ℓ. The dissipation scale $\ell_0 = q^3/\epsilon$ is often used in the literature dealing with the second-order closures. However, for our aims the modified dissipation scale $\ell = u_*^3/\epsilon$ (used, e.g., by Gibson and Launder 1978) is more convenient than ℓ_0. Note that the Monin-Obukhov similarity theory implies that in the ASL both the ratios ℓ_0/z and ℓ/z are single-valued functions of $\zeta = z/L$ (where $L = -u_*^3/\kappa Q\beta$ is the Obukhov length and $\kappa \approx 0.4$ is the von Karman constant); hence ℓ_0/z and ℓ/z are uniquely related to each other. In neutral stratification $\ell/z =$ constant for any sensible definition of ℓ; if $\ell = u_*^3/\epsilon$, then $\ell/z = \kappa$ (since $\epsilon = u_*^3/\kappa z$ in the neutral stratification).

Unfortunately, at present there are no reliable data permitting the values of $b_i(\ell/z)$ for $\ell/z \neq 0$ and $\ell/z \neq \kappa$ to be determined. Therefore it seems reasonable to use the simpler linear extrapolation formula of the form

$$b_i\left(\frac{\ell}{z}\right) = b_i(0) + [b_i(\kappa) - b_i(0)]\frac{\ell}{\kappa z}, \quad \ell = \frac{u_*^3}{\epsilon} \tag{12}$$

to determine the values of $b_i(\ell/z)$, $i = 1, \ldots, 4$, as a first approximation. It has already been indicated that lack of data makes it reasonable to assume that $b_5 =$ constant. The numerical value of b_5 affects the result of the calculation. Pilot computations on the algebraic equations for the temperature fluxes, used in computations of the universal functions of the Monin-Obukhov similarity theory (see Tsarenko 1987a) with the value $b_5 = -0.5$ used by Launder (1975), leads to a slightly better agreement with the experimental data than the value $b_5 = -1/3$ used by Launder (1978), Gibson and Launder

(1978), Zeman and Lumley (1979) and Zeman (1981). Therefore, the value $b_5 = -0.5$ is used in our computations.

Turbulence Characteristics of the ASL

The Monin-Obukhov similarity theory implies that all the dimensionless (normalized by the appropriate combinations of parameters u_*, Q, and β) one-point turbulence characteristics are represented in the ASL by the universal functions of the dimensionless height $\zeta = z/L$, where $L = -u_*^3/\kappa Q\beta$. Therefore,

$$\frac{dU}{dz} = \frac{u_*}{\kappa L}\phi(\zeta), \frac{dT}{dz} = -\frac{Q}{\kappa u_* L}\phi_T(\zeta),$$
$$< u'^2 >= u_*^2\psi_1(\zeta), < v'^2 >= u_*^2\psi_2(\zeta), < w'^2 >= u_*^2\psi_3(\zeta), \quad (13)$$
$$q^2 = u_*^2\psi(\zeta), < u'\theta' >= Q\psi_4(\zeta), \epsilon = \frac{u_*^3}{\kappa L}\psi_5(\zeta)$$

where $\phi(\zeta), \phi_T(\zeta), \psi_i(\zeta), i = 1, \ldots, 5$, and $\psi(\zeta) = \psi_1(\zeta) + \psi_2(\zeta) + \psi_3(\zeta)$ are some universal functions (κ is included in some of the equations (13) in accordance with the micrometeorological tradition).

All the indicated universal functions have been determined more than once from micrometerological data. Unfortunately, such data remain till now to be rather scattered (see, e.g., the discussion of this problem by Yaglom 1977). Therefore all the available experimental approximations for the universal function (13) are not very reliable. Nevertheless, the majority of the data related to the function $\phi(\zeta)$ (which has been measured especially often) agree satisfactorily with the equations

$$(\zeta\phi)^4 - 14\zeta(\zeta\phi)^3 - 1 = 0 \quad \text{for} \quad \zeta \leq 0,$$
$$\zeta\phi = 1 + 5\zeta \quad \text{for} \quad \zeta \geq 0 \quad (14)$$

(see, e.g., Monin and Yaglom 1971, Yaglom 1977, Gibson and Launder 1978). The values of $\phi(\zeta)$ determined from equations (14) are used in the present paper.

Numerous data related to the function $\phi_T, \psi, \psi_1, \ldots, \psi_5$ can be found, e.g., in the books by Monin and Yaglom (1971) and Panofsky and Dutton (1984) and the papers by Wyngaard and Coté (1971),

Wyngaard, Coté and Isumi (1971), Yaglom (1974, 1977), Businger (1984), Kader and Yaglom (1990) and some others. However, these data are considerably more scattered than those related to $\phi(\zeta)$. Since the local equilibrium approximation is used in this paper in the equations for turbulent fluxes S_i, it is natural to assume that the same approximation can also be applied to the energy budget equation (i.e., the vertical transfer of the turbulent kinetic energy can be neglected). Then the energy budget equation implies the relation

$$\psi_5(\zeta) = \phi(\zeta) - 1 \tag{15}$$

which agrees satisfactorily with the available data (Wyngaard and Coté 1971). As for the functions $\phi_T(\zeta)$ and $\psi_i(\zeta)$, $i = 1, \ldots, 4$, the experimental data related to them are so scattered that at present it is difficult to use the data for choosing the best approximation for the functions mentioned. Therefore the computed values of these functions are used in this paper. These values are obtained from the local equilibrium approximation applied to the equations for the moments $< u'^2 >, < v'^2 >, < w'^2 >$, and $< u'\theta' >$, supplemented by the model approximations for the correlations involving pressure fluctuations similar to (10). The corresponding computations are due to Tsarenko (1987a, 1989); the universal functions used here are those that correspond to Tsarenko's Model 2 which leads to the best agreement of the computational results with the available data.

Solution of Equations (11)

It has been indicated above that the system of linear equations (11) for the moments $S_i, i = 0, 1, 2, 3$, has the solution of the form (5). It is easy to see that $K_{12} = K_{21} = K_{32} = K_{23} = K_{02} = 0$ for this solution and that the non-zero coefficients K_{ij} by virtue of (13) can be represented in the form

$$K_{ij}(z) = \frac{u_*^2}{dU/dZ} k_{ij}(\zeta) \quad \text{for } i = 1, 2, 3; \quad K_{0j}(z) = \frac{Q}{dU/dz} k_{0j}(\zeta) \tag{16}$$

where $k_{ij}(\zeta)$ and $k_{0j}(\zeta)$ are universal functions of $\zeta = z/L$. (Equations (16) can also be rewritten as $K_{ij}(z) = \kappa u_* L D_{ij}(\zeta), i = 1, 2, 3$; $K_{0j}(z) = \kappa(QL/u_*)D_{0j}(\zeta)$, where $D_{ij}(\zeta) = k_{ij}(\zeta)/\phi(\zeta)$ for all i and j, but this representation is less convenient than (16). Note that $u_*^2[dU/dz]^{-1} = K_M$ is the vertical eddy viscosity in the ASL; hence the tensor $k_{ij}(\zeta)$ is in fact the Prandtl number tensor.) Substituting (5), (13), (15), and (16) into (11) and solving the linear equations for the unknowns $k_{ij}(\zeta)$, we obtain the following equations

$$\begin{aligned}
k_{11} &= \Delta^{-1}\left[b_2\frac{\psi_1(\phi-1)}{\psi\phi}+b_0+b\left(b_0\psi_4+\frac{\phi_T\psi_1\psi}{\phi(\phi-1)}\right)\right],\\
k_{22} &= \frac{\psi\psi_2\phi}{b_1(\phi-1)},\\
k_{33} &= \Delta^{-1}\left[b_1\frac{\psi_3(\phi-1)}{\psi\phi}-b_4-\frac{bb_1}{\phi}\right],\\
k_{13} &= -\Delta^{-1}\left[\frac{b_2(\phi-1)}{\psi\phi}+b_0\psi_3-b\left(b_0-\frac{\phi_T}{\phi}\right)\frac{\psi}{\phi-1}\right],\\
k_{31} &= -\Delta^{-1}\left[\frac{b_1(\phi-1)}{\psi\phi}-b_4\psi_1+\frac{bb_1\psi_4}{\phi}\right],
\end{aligned} \tag{17}$$

where

$$\Delta=\frac{b_1b_2(\phi-1)^2}{\psi^2\phi^2}+b_0b_4+\frac{bb_1\phi_T}{\phi^2},\; b=\tau_0(1+b_5),\; b_0=1-b_3. \tag{18}$$

Equations (17) and (18) generalize the results given in I for the case of the ASL in neutral stratification. Similar equations can also be obtained for the function $k_{01}(\zeta)$ and $k_{03}(\zeta)$; cf. Tsarenko (1987b).

Note that according to (12) and (15)

$$b_i=b_i(\zeta)=b_i(0)+\frac{b_i(\kappa)-b_i(0)}{\zeta(\phi-1)} \quad \text{for } i=1,2,3,4, \tag{19}$$

where $b_i(0)$ and $b_i(\kappa)$ are given in the Table shown previously (and $b_5=-0.5=$ constant). The computed values of the non-zero functions $k_{ij}(\zeta)$, $i=1,2,3$, of the argument $\zeta=z/L$, which correspond to the four models for Π_i considered in this paper, are given in Figures 1-5. Since $b_4=0$ for models 1 and 3 and b_2 takes the same values for these two models at any ℓ/z, the values of k_{33} and k_{31} corresponding to these two models coincide with each other. The figures show that in fact all the considered models lead to values of the eddy diffusivities $K_{ij}(z)=u_*^2[dU/dz]^{-1}k_{ij}(\zeta)$ which do not differ too much; in some cases the curves corresponding to some of the models are practically indistinguishable, even when they do not coincide exactly with each other. (Too much importance should not be attached to the non-monotonicity of the function $k_{33}(\zeta)$ for $\zeta < 0$, since the functions can be made monotone with the aid of small variations in the shapes of the universal functions ϕ, ϕ_T, ψ, and ψ_3 used.) Note that all the measurements of concentration for the air pollutants are rather inaccurate until now; therefore the variations in the computed concentration distributions due to the use of different models for k_{ij} shown in Figures 1-5 will apparently be of no importance from a practical point of view.

References

[1] Berlyand, M. E., 1975, *Modern Problems of Atmospheric Diffusion and Air Pollution*, Gidrometeoizdat, Leningrad. (Engl. transl. by USA Environment Protection Agency, Raleigh, NC, 1976.)

[2] Businger, J. A., 1984, "Equations and concepts," Atmospheric Turbulence and Air Pollution Modelling (E. T. M. Nieuwstadt and H. van Dop, eds.), 1-36, Reidel, Dordrecht.

[3] Corrsin, S., 1974, "Limitations of gradient transport models in random walks and in turbulence," Adv. Geophys. **18A**, 25-60.

[4] Gibson, M. M. and Launder, B. E., 1978, "Ground effects on pressure fluctuations in the atmospheric boundary layer," J. Fluid Mech. **86**, 491-511.

[5] Jaffe, S., 1967, "A three-layer diffusion model as applied to unstable atmospheric conditions," J. Appl. Meteor. **6**, 297-302.

[6] Kader, B. A. and Yaglom, A. M., 1990, "Mean fields and fluctuation moments in unstably stratified turbulent boundary layers," J. Fluid Mech. **212**, 637-662.

[7] Launder, B. E., 1975, "On the effect of gravitational field on the turbulent transport of heat and momentum," J. Fluid Mech. **67**, 569-581.

[8] Launder, B. E., 1978, "Heat and mass transfer," Turbulence (Topics in Applied Physics, Vol. 12; P. Bradshaw, ed.), 231-287, Springer, Berlin.

[9] Launder, B. E., Reece, G. L., and Rodi, W., 1975, "Progress in the development of a Reynolds-stress turbulence closure," J. Fluid Mech. **68**, 537-566.

[10] Lewellen, W. C., 1977, "Use of invariant modeling," Handbook of Turbulence. Volume 1: Fundamentals and Applications (W. Frost and T. H. Moulden, eds.), 237-280, Plenum, NY.

[11] Lewellen, W. C. and Teske, M., 1978, "Second-order closure modeling of diffusion in the atmospheric boundary layer," Boundary-Layer Meteor. **10**, 69-90.

[12] Lumley, J. L., 1978a, "Computational modeling of turbulent flows," Adv. Appl. Mech. **18**, 123-176.

[13] Lumley, J. L., 1978b, "Simulating turbulent transport in urban air-pollution models," Advances in Environmental Sciences and Engineering **1** (J. R. Pfafflin and E. N. Zeigler, eds.), 103-127, Gordon and Breach, NY.

[14] Lumley, J. L., 1983, "Turbulence modeling," J. Appl. Mech. **50**, 1097-1103.

[15] Lumley, J. L. and Khajeh-Nouri, 1974, "Computational modeling of turbulent transport," Adv. Geophys. **18A**, 169-192.

[16] Lumley, J. L. and Mansfield, P., 1984, "Second order modeling of turbulent transport in the surface mixed layer," Boundary-Layer Meteor. **30**, 109-142.

[17] Lumley, J. L. and Van Cruyningen, I., 1985, "Limitations of second order modeling of passive scalar diffusion," Frontiers in Fluid Mechanics (S. H. Davis and J. L. Lumley, eds.), 199-218, Springer, Berlin.

[18] Mellor, G. and Yamada, T., 1974, "A hierarchy of turbulence closure models for planetary boundary layers," J. Atmos. Sci. **31**, 1791-1806.

[19] Mellor, G. and Yamada, T., 1982, "Development of a turbulent closure model for geophysical fluid problems," Rev. Geophys. Space Phys. **20**, 851-875.

[20] Monin, A. S. and Yaglom, A. M., 1971, *Statistical Fluid Mechanics*, Volume 1, MIT Press, Cambridge, MA.

[21] Newman, G. R., Launder, B. E., and Lumley, J. L., 1981, "Modelling the behaviour of homogeneous scalar turbulence," J. Fluid Mech. **111**, 217-232.

[22] Panofsky, H. A. and Dutton, J. A., 1984, *Atmospheric Turbulence. Models and Methods for Engineering Applications*, Wiley, NY.

[23] Pasquill, F., 1974, *Atmospheric Diffusion*, 2nd Edition, E. Horwood, Chichester.

[24] Reynolds, W. C., 1976, "Computation of turbulent flows," Ann. Rev. Fluid Mech. **8**, 183-208.

[25] Rodi, W., 1980, "Environmental turbulent flows," Prediction Methods for Turbulent Flows (W. Kollmann, ed.), 259-349, Hemisphere, NY.

[26] Rogers, M. M., Mansour, N. N., and Reynolds, W. C., 1989, "An algebraic model for the turbulent flux of a passive scalar," J. Fluid Mech. **203**, 77-101.

[27] Sreenivasan, K. R., Tavoularis, S., and Corrsin, S., 1982, "A test of gradient transport and its generalizations," Turbulent Shear Flows **3** (L. J. S. Bradbury et al., eds.), 96-112, Springer, Berlin.

[28] Sykes, R. I., Lewellen, W. C., and Parker, S. E., 1986, "A Gaussian plume model of atmospheric dispersion based on second-order closure," J. Climate Appl. Meteor. **25**, 322-331.

[29] Tavoularis, S. and Corrsin, S., 1981, "Experiments in nearly homogeneous turbulent shear flow with a uniform temperature gradient. Part 1," J. Fluid Mech. **104**, 311-347.

[30] Tavoularis, S. and Corrsin, S., 1985, "Effects of shear on the turbulent diffusivity tensor," Intern. J. Heat Mass Transfer **28**, 265-276.

[31] Tsarenko, V. M., 1984, "Dispersion of a contaminant cloud in the neutral atmospheric surface layer," Izv. Akad. Nauk SSSR, Ser. Fiz. Atmos. i Okeana **20**, 485-494 (463-494 in English transl. of the journal).

[32] Tsarenko, V. M., 1987a, "Semiempirical description of turbulence in the atmospheric surface layer," Izv. Akad. Nauk SSSR, Ser. Fiz. Atmos. i Okeana **23**, 470-479 (353-359 in English transl. of the journal).

[33] Tsarenko, V. M., 1987b, "Model calculation of eddy diffusivity tensor in stratified flows," Izv. Akad. Nauk SSSR, Ser. Fiz. Atmos. i Okeana **23**, 1041-1048 (774-779 in English transl. of the journal).

[34] Tsarenko, V. M., 1989, "Model calculations of the Monin-Obukhov universal functions," Izv. Akad. Nauk SSSR, Ser. Fiz. Atmos. i Okeana **25**, 115-124 (87-93 in English transl. of the journal).

[35] Tsarenko, V. M. and Yaglom, A. M., 1991, "Semiempirical theories of turbulence in a neutrally stratified atmospheric surface layer," Phys. Fluids A (in the press).

[36] Wyngaard, J. C. and Coté, O. R., 1971, "The budgets of turbulent kinetic energy and temperature variance in the atmospheric surface layer," J. Atmos. Sci. **28**, 190-201.

[37] Wyngaard, J. C., Coté, O. R., and Izumi, Y., 1971, "Local free convection, similarity, and the budgets of shear stress and heat flux," J. Atmos. Sci. **28**, 1171-1182.

[38] Yaglom, A. M., 1974, "Data on turbulence characteristics on the atmospheric surface layer," Izv. Akad. Nauk SSSR, Ser. Fiz. Atmos. i Okeana **10**, 566-586 (341-352 in English transl. of the journal).

[39] Yaglom, A. M., 1977, "Comments on wind and temperature flux-profile relationships," Boundary-Layer Meteor. **11**, 89-102.

[40] Zeman, O., 1981, "Progress in the modeling of planetary boundary layers," Ann. Rev. Fluid Mech. **13**, 253-272.

[41] Zeman, O. and Lumley, J. L., 1979, "Buoyancy effects in turbulent boundary layers: A second order closure study," Turbulent Shear Flows 1 (F. Durst et al., eds.), 274-295, Springer, Berlin.

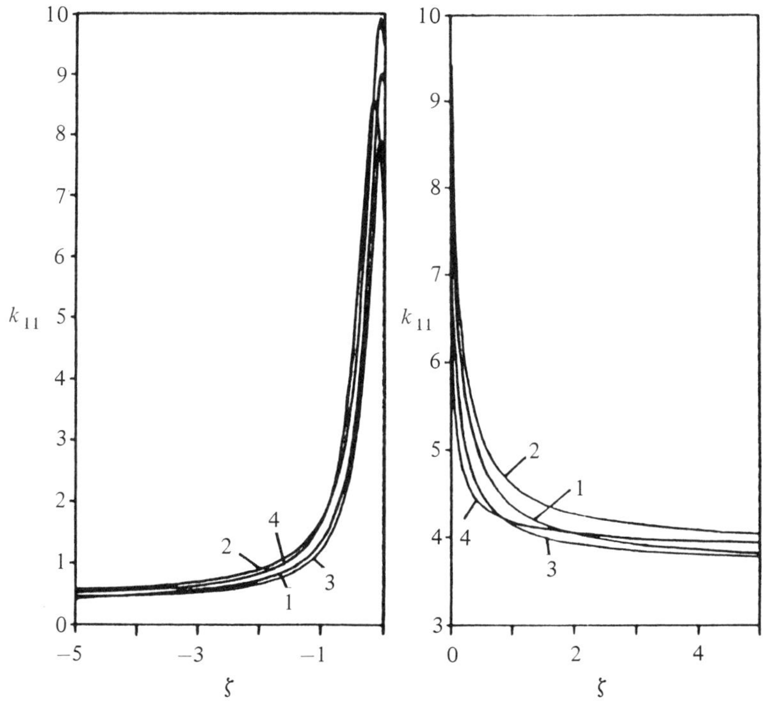

Figure 1. Function $k_{11}(\zeta)$ according to four models (curve numbers are the numbers of the corresponding models).

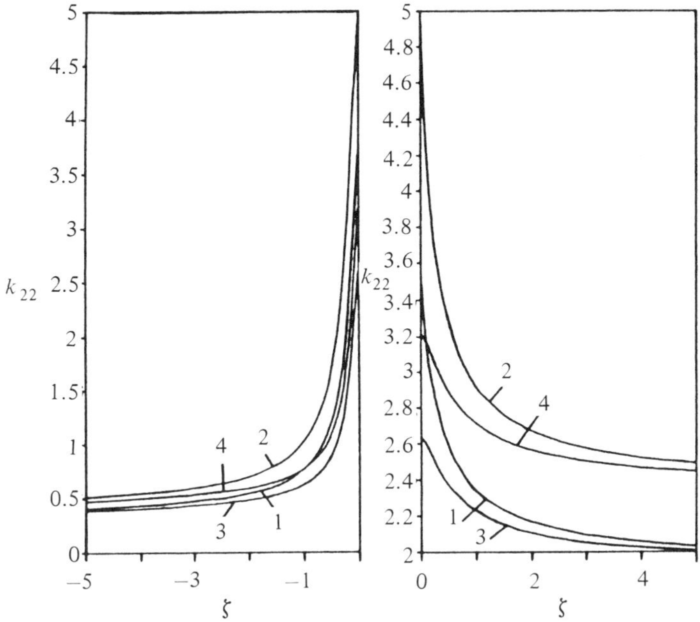

Figure 2. Function $k_{22}(\zeta)$.

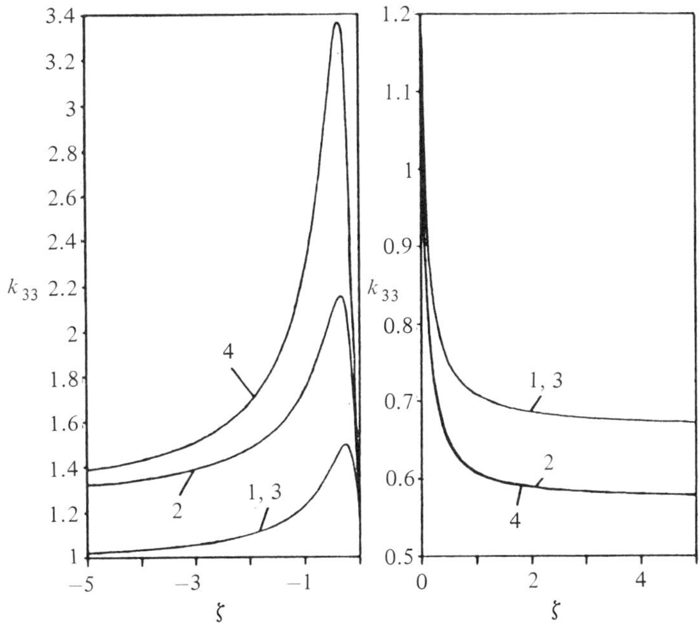

Figure 3. Function $k_{33}(\zeta)$.

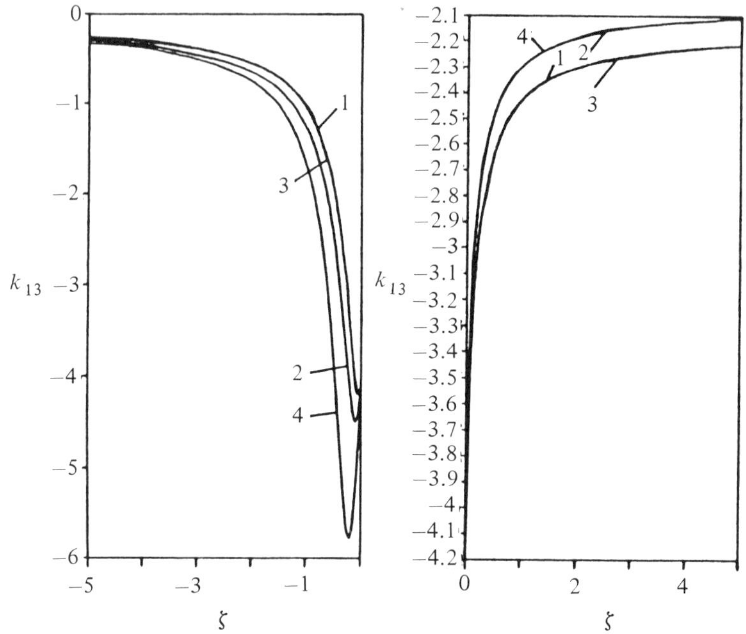

Figure 4. Function $k_{13}(\zeta)$.

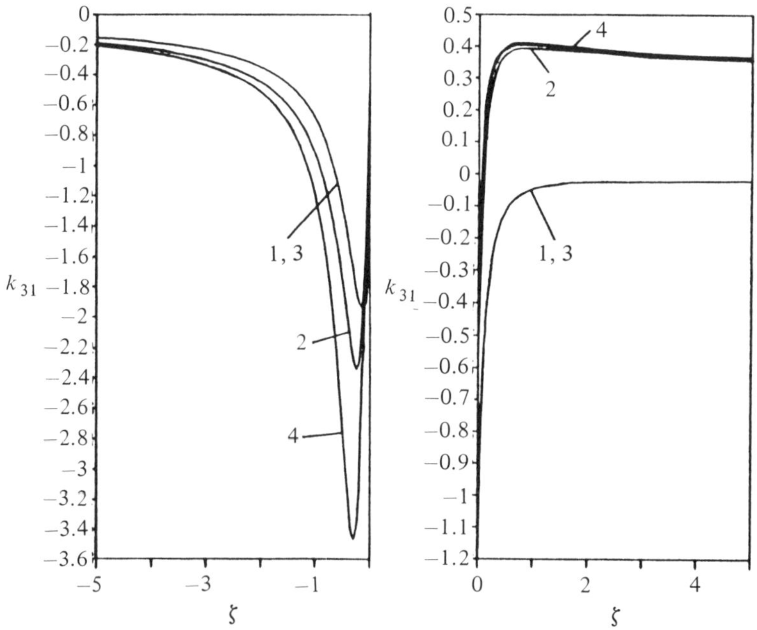

Figure 5. Function $k_{31}(\zeta)$.

THEORETICAL AND COMPUTATIONAL STUDY OF ROTATING ISOTROPIC TURBULENCE

N. N. Mansour,[a] *C. Cambon*[b] and *C. G. Speziale*[c]
Center for Turbulence Research
[a]NASA Ames Research Center, Moffett Field, CA 94035

ABSTRACT

Direct numerical simulations of isotropic turbulence in a rotating frame are conducted and comparisons are made with the predictions of generalized Eddy Damped Quasi Normal Markovian (EDQNM) approximations. It is found that for increasing rotation rates the non-linear triad interactions are reduced causing a reduction in the energy cascade and the turbulence decay rate. For turbulence Rossby numbers $Ro < 0.01$, the transfer is essentially shut-off causing a linear viscous decay of the turbulence with a substantially reduced dissipation rate. While the rotation does not produce any significant anisotropy in the Reynolds stress tensor or the dissipation rate tensor, it does alter the isotropy of the integral length scales at intermediate Rossby numbers. The rapid rotation rate case where $Ro << 1$ is in excellent agreement with Rapid Distortion Theory (RDT) – for large elapsed times – and, hence, does not approach a two-dimensional turbulence as might be expected by the Taylor-Proudman theorem. The RDT solution violates a necessary condition for the Taylor-Proudman theorem to be valid.

1. Introduction

The problem of isotropic turbulence in a rotating frame has been the subject of experimental, theoretical and computational studies during the past few decades. This popularity can be attributed to the fact that it represents the most basic

[b]Ecole Centrale de Lyon, 69130 Ecully, France
[c]ICASE, NASA Langley Research Center, Hampton, Virginia 23665

turbulent flow whose structure is altered by a system rotation, unencumbered by the complications introduced by the presence of mean strains or flow inhomogeneities. Of course, a better understanding of the alteration of the turbulence structure by a system rotation can have a wide range of important applications in the analysis of turbomachinery and geophysical flows. Consequently, fundamental studies of rotating turbulent flows continue to be of considerable interest.

The first successful experiment on rotating isotropic turbulence was conducted by Traugott [1] who concluded that the primary effect of rotation was to *decrease* the decay rate of the turbulent kinetic energy. This result was later questioned by the experimental data of Ibbetson & Tritton [2] who found an increase in the decay rate of the turbulent kinetic energy for rotating isotropic turbulence. However, the subsequent experiments of Wigeland & Nagib [3] strongly confirmed the results of Traugott [1] indicating a reduction in the turbulence decay rate for increasing rotation rates. More recently, Jacquin *et al.* [4] took detailed measurements in rotating isotropic turbulence that document the effects of rotation on the development of various length and time scales.

With the dramatic increases in computer capacity achieved during the late 1970's, direct numerical simulations of the Navier-Stokes equations became possible for basic turbulent flows. Rogallo [5] conducted the first direct numerical simulation of rotating isotropic turbulence. This work was extended by Bardina, Ferziger & Rogallo [6] a short time later by using large-eddy simulation techniques. These studies clearly showed that a system rotation has two major effects on isotropic turbulence: *(a)* there is a *decrease* in the decay rate of the turbulence and *(b)* there is an increase in the integral length scales, particularly those along the axis of rotation. (While there is a mild increase in the integral length scales, the Reynolds stress and dissipation rate tensors remain isotropic.) The increase in the integral length scales along the axis of rotation were thought to be a prelude to a Taylor-Proudman reorganization to two-dimensional turbulence which is often encountered in rapidly rotating flows [6]. Speziale, Mansour & Rogallo [7] conducted a direct numerical simulation of rotating isotropic turbulence for a large rotation rate. They found that the turbulence remained *isotropic and three-dimensional*; however, the rapid rotation essentially shut-off the energy transfer yielding a linear viscous decay. A reduction in the transfer was observed by Teissèdre & Dang [8] who carried out direct simulations of the same flow. They also showed that under rotation the spectral energy gets concentrated near the equatorial plane perpendicular to the rotation axis.

The purpose of the present paper is to provide a more detailed picture of the various flow regimes that occur in rotating isotropic turbulence and to gain a better understanding of their associated turbulence structure. Direct numerical simulations are conducted for moderate and rapid rotation rates, and comparisons are made with the predictions of generalized EDQNM (Eddy Damped Quasi-Normal Markovian) approximations (see Cambon & Jacquin [9]). It will be shown that an analysis of direct simulations of rotating isotropic turbulence, guided by EDQNM, can provide many new insights into the understanding of this basic, yet subtle, turbulent flow.

2. The Governing Equations and Other Formalism

The effect of rotation on turbulence is known to be important in many flows of practical interest. It is known that at small Rossby numbers uniform (solid-body) rotation causes plane waves of the form $\exp(ik_j x_j)$ to propagate with phase speed $2\Omega_j k_j/k$ which gives the decomposition of the flow field into Fourier modes some physical significance (Rogallo [5]).

2.1 The equations of motion in a uniformly rotating frame.

Without loss of generality, we take our reference frame such that the x_3 coordinate is aligned with the axis of rotation ($\Omega_i =$

$[0, 0, \Omega]$). The equation of motions in this rotating frame are given by

continuity

$$u_{i,i} = 0 \tag{1}$$

momentum

$$u_{i,t} - \nu u_{i,jj} - 2\varepsilon_{ij3}\Omega u_j = -p_{,i} - H_i \tag{2}$$

where u_i is the velocity vector in the rotating frame, ε_{ijk} is the alternating tensor, ν is the kinematic viscosity, p is the pressure, and $H_i = (u_i u_j)_{,j}$ represents the non-linear terms. We consider a statistically homogeneous velocity field and define its Fourier transform as follows:

$$\hat{u}_i = \frac{1}{(2\pi)^3} \int u_i \exp(-\mathrm{i} k_j x_j) d^3\mathbf{x}; \quad \mathrm{i} = \sqrt{-1}$$

We Fourier transform Eqs. (1) and (2), and introduce a rapid pressure,

$$p^{(r)} = -\mathrm{i}\ 2\varepsilon_{ij3}\Omega \frac{k_i}{k^2} \hat{u}_j \tag{3}$$

where $k^2 = k_j k_j$, as well as a slow pressure

$$p^{(s)} = \mathrm{i}\ \frac{k_i}{k^2} \hat{H}_i \tag{4}$$

where $\hat{H}_i$ is the Fourier transform of the nonlinear terms. We then integrate the viscous term with an integrating factor $F = \exp(\nu k^2 t)$, and eliminate the pressure between the transformed continuity and momentum equations to yield,

$$\hat{v}_{i,t} - 2P_{ij}\varepsilon_{jl3}\Omega \hat{v}_l = P_{ij} F \hat{H}_j \tag{5}$$

where $P_{ij} = \delta_{ij} - k_i k_j / k^2$ is the classical projection operator, and $\hat{v}_i = F\hat{u}_i$. The left hand side of Eq. (5) is diagonalized by introducing new dependent variables (see Cambon & Jacquin [9])

$$\xi_\varepsilon = \frac{1}{k_{12}} \varepsilon(\mathrm{i} k_2 \hat{v}_1 - \mathrm{i} k_1 \hat{v}_2 - \varepsilon k \hat{v}_3); \qquad \varepsilon = \pm 1 \tag{6}$$

where $k_{12} = \sqrt{k_1^2 + k_2^2}$. The governing equations of ξ_ε are easily derived from Eq. (5):

$$\left[\exp\left(-\mathrm{i}2\varepsilon\Omega\frac{k_3}{k}t\right)\xi_\varepsilon\right]_{,t} = R_\varepsilon \tag{7}$$

where the nonlinear terms are lumped into the term R_ε.

2.2 Rapid distortion theory.

Under rapid distortion theory (RDT) assumptions the nonlinear terms, R_ε, in Eq. (7) are neglected. This leads to the solution,

$$\exp\left(-\mathrm{i}2\varepsilon\Omega\frac{k_3}{k}t\right)\xi_\varepsilon = \text{constant} \tag{8}$$

This result shows that the RDT solution does not undergo a Taylor-Proudman reorganization, since the Fourier transform of the velocity is of the form,

$$\hat{u}_i \propto A(k)\exp(\pm \mathrm{i}k_3\Omega t/k) \tag{9}$$

and, hence, in the limit as $\Omega \to \infty$,

$$\frac{1}{\Omega}\omega_{i,t} = O(1) \tag{10}$$

where ω_i is the vorticity vector. This term needs to vanish in the limit as $\Omega \to \infty$ for a Taylor-Proudman reorganization to occur [7]; Eq. (10) shows that under RDT, the time term remains $O(1)$. Hence, as long as the RDT solution persists, a Taylor-Proudman reorganization to a two-dimensional state cannot occur in rotating isotropic turbulence.

2.3 *Direct numerical simulation.*

In direct numerical simulations (DNS) Fourier transforms are approximated by a truncated Fourier series. The nonlinear terms on the right hand side of Eq.(7) are computed in real space and the time derivative is approximated by a finite difference scheme. In the DNS code that we use to produce the results reported in this work, Rogallo [10] uses a different diagonalization of the equations, but the final outcome is the same:
(i) the code reproduces the viscous-RDT solution exactly, and
(ii) the numerical-stiffness associated with high rotation rates is avoided.

2.4 *Eddy Damped Quasi-Normal Markovian approximation.*

In the EDQNM approximation, an equation for the second-order spectral tensor $\hat{U}_{ij} = \langle \hat{u}_i^* \hat{u}_j \rangle$ is derived from Eq. (5). Taking into account homogeneity and incompressibility, Cambon & Jacquin [9] split the second-order spectral tensor using the two quadratic modes,

$$e = \frac{1}{2}\hat{U}_{ii} \tag{11}$$

and

$$\begin{aligned} z &= \frac{1}{2}\xi_1\xi_1 \\ &= \frac{1}{2}\hat{U}_{ij}N_i^*N_j^* \end{aligned} \tag{12}$$

as follows:

$$\hat{U}_{ij} = eP_{ij} + \mathrm{Real}[zN_iN_j] \tag{13}$$

where N_i and N_i^* are the eigenvectors of the rotation matrix in the plane orthogonal to the wave vector. They show that in the rotating frame, the set (e, z) is governed by the following system of equations:

$$\begin{aligned} (Fe)_{,t} &= T_e \\ \left[\exp(-4i\Omega\frac{k_3}{k}t)Fz\right]_{,t} &= T_z \end{aligned} \tag{14}$$

Re_λ	Ro	$2(t_f - t_0)\epsilon_0/q_0^2$
24.4	∞	7.3
24.4	0.4	5.57
24.4	0.04	4.43
24.4	0.008	8.35

Table 1. Direct numerical simulations of rotating isotropic turbulence. All cases started with the same initial isotropic turbulent flow field; $\nu = 0.005$, $q_0^2 = 2.44$, $\epsilon_0 = 9.96$.

where T_e (energy transfer) and T_z take into account non-linear interactions through triple correlations. The modeling in the EDQNM approximation is applied on these terms. For more details on EDQNM modeling the interested reader is referred to the paper of Cambon & Jacquin [9].

3. Results

In order to gain insight into the effects of rotation, direct numerical simulations of initially isotropic turbulence at various Rossby numbers ($Ro = \epsilon/\Omega q^2$, where $q^2 = \langle u_i u_i \rangle$ is twice the turbulent kinetic energy and ϵ is its dissipation rate) were carried out. Four cases were simulated as shown in Table 1. The case of $Ro = \infty$ is the isotropic decay simulation without rotation; the lowest Rossby number simulated was at $Ro = 0.008$. In addition, a non-isotropic EDQNM simulation was carried out where initially $Re_\lambda = 70$ based on the Taylor microscale, and $Ro \approx 0.5$ (corresponding to $\Omega = 12.57$ sec^{-1}).

3.1 Low Rossby number effects.

In this section we consider the results from a DNS simulation with an initial Reynolds number of $Re_\lambda = 24.4$ and Rossby number of $Ro = 0.008$. This Rossby number is more than an order of magnitude smaller than those which have been considered previously. The direct simulation yielded some surprising results. As has been shown in previous numerical simulations

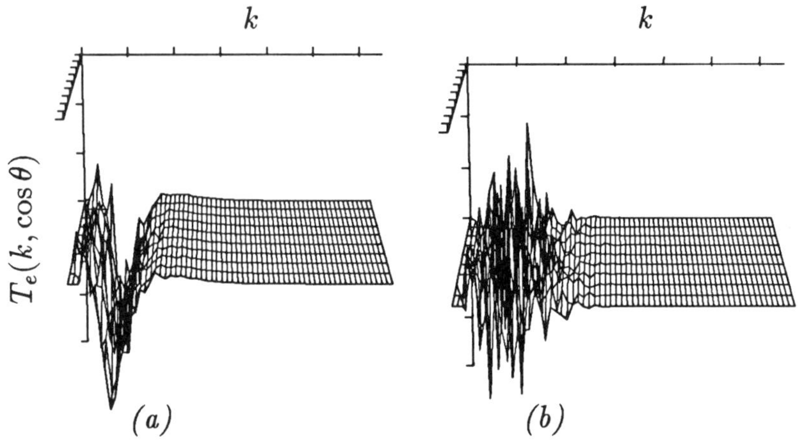

FIGURE 1. Two-dimensional transfer spectra ($T_e(k, \cos\theta)$): *(a)* at $2(t-t_0)\epsilon_0/q_0^2 = 7.3$, $Ro = \infty$. *(b)* at $2(t-t_0)\epsilon_0/q_0^2 = 6.0$, $Ro = 0.008$. The scale on the vertical (T_e) axis is between $-.3 \times 10^{-5}$ and $.3 \times 10^{-5}$; The scale on the horizontal (k) axis is in increments of 10; The third axis ($\perp$ to the other two) is for $\cos\theta$ (between 0.05 and .95).

and experiments (see Bardina, Ferziger and Rogallo [6], and Wigeland and Nagib [3]), the turbulence remained isotropic after the rotation was imposed. The rotation killed the energy transfer and the turbulence underwent a pure viscous decay as would be predicted by RDT. Under rotation, an initially isotropic turbulence will remain axisymmetric with respect to the axis of rotation. In analyzing the flow field statistics we will average over the azimuthal direction in a spherical coordinate system. Figure 1 shows the energy transfer spectra ($T_e(k, \cos\theta) = 1/2\pi \int_0^{2\pi} T_e d\phi$, where θ is the polar angle [with respect to k_3] in k-space) for the isotropic decay case ($Ro = \infty$) and the low Rossby number case ($Ro = 0.008$). We find that while in the isotropic case we have a clear transfer of energy from the low wave numbers to the high wave numbers, the transfer spectrum in the low Rossby number case has been scrambled by the rotation. On the average (for example with respect to $\cos\theta$) the transfer has been essentially shut-off. This is reflected in the fact that the energy spectrum ($E(k,t) = \int_0^{\pi} \int_0^{2\pi} e(\mathbf{k}, t) k^2 \sin\theta d\theta d\phi$) decayed in time in good

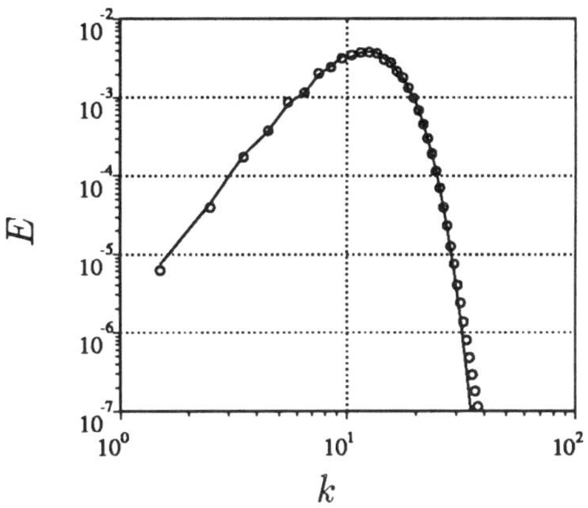

FIGURE 2. Energy spectrum: o o o DNS data at $2(t - t_0)\epsilon_0/q_0^2 = 8.35$ ($Ro = 0.008$). —— Eq. (15).

agreement (see Fig. 2) with the formula

$$E(k,t) = E(k,t_0)\exp\left[-2\nu k^2(t-t_0)\right] \tag{15}$$

which is a rigorous consequence of the Navier-Stokes equations when there is no energy transfer. Equation (15) is obtained from RDT (setting $T_e = 0$) for this problem. The surprising finding was that Eq. (15) remained a good approximation even after the turbulent kinetic energy had decayed to only 3% of its initial value (RDT is usually only formally valid for short elapsed times). As a result of the energy transfer being suppressed, the turbulence decayed at a slower rate in the rotating frame (see Fig. 3). There was no discernible Taylor-Proudman reorganization to a two-dimensional flow. The tensor,

$$A_{ij} = \frac{\nu}{\epsilon}\langle u_{k,i}u_{k,j}\rangle \tag{16}$$

which is normalized by the dissipation rate ϵ, remained isotropic (under a complete Taylor-Proudman reorganization, the velocity gradient along the axis of rotation $u_{,3} \to 0$ as the rotation rate $\Omega \to \infty$ and, hence we would expect $A_{33} << A_{11}, A_{22}$). The integral length scales remained isotropic (see Fig. 4a). All

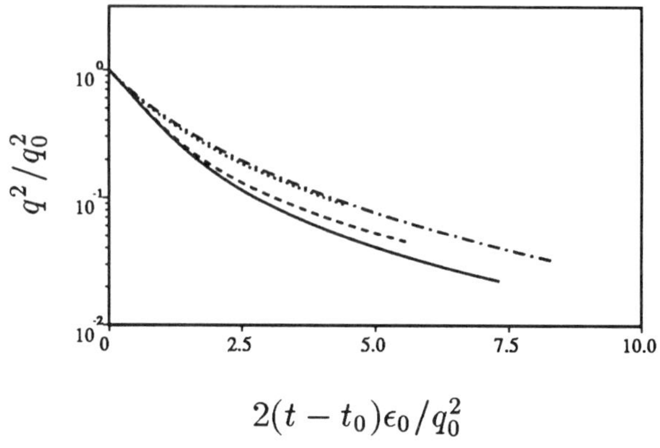

FIGURE 3. Effect of rotation on the decay of the turbulent kinetic energy. —— Isotropic decay, $Ro = \infty$; ---- $Ro = 0.4$; ······· $Ro = 0.04$; —·— $Ro = 0.008$.

of the indicators that we have examined do not show a tendency toward a Taylor-Proudman reorganization. In particular, histograms of the angle of the vorticity vector did not show any sign of alignment of the vorticity with respect to the axis of rotation (see Fig. 5a). It appears that no Taylor-Proudman reorganization will occur in a rapidly rotating isotropic turbulence. Previous investigators (*c.f.*, Bardina, Ferziger and Rogallo [6]) had speculated that such a reorganization to a two-dimensional flow could occur at low Rossby numbers. Their speculation was based on results from intermediate Rossby numbers. We will show in the next section that there is a weak tendency toward a Taylor-Proudman reorganization at intermediate *Ro*.

In isotropic turbulence $e(\mathbf{k}, t)$ is equally distributed on spheres of radius k, and thus $e(\mathbf{k}, t) = E(k, t)/4\pi k^2$. In addition, the term z is rigorously null for isotropic turbulence. Starting from isotropic initial data, the flow remains isotropic if the non-linear terms conserve isotropy. Weak signs of a Taylor-Proudman reorganization are detected at intermediate Rossby numbers. Two point correlations in the axial direction show a

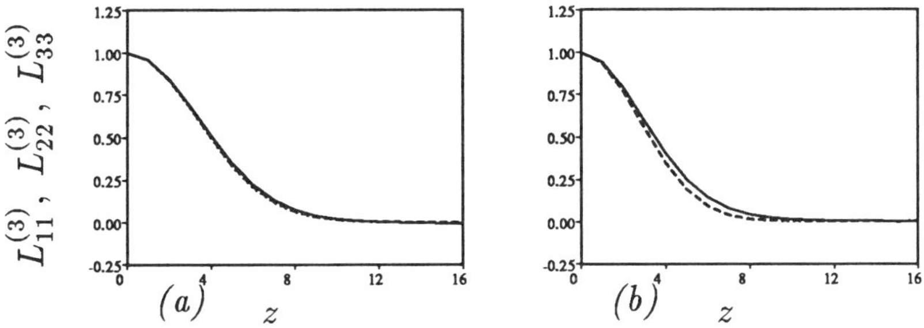

FIGURE 4. Axial integral length scales. —— $L_{33}^{(3)}$; ---- $L_{11}^{(3)}$; ········ $L_{11}^{(3)}$: *(a)* at $2(t-t_0)\epsilon_0/q_0^2 = 8.35$, $Ro = 0.008$. *(b)* at $2(t-t_0)\epsilon_0/q_0^2 = 4.43$, $Ro = 0.04$.

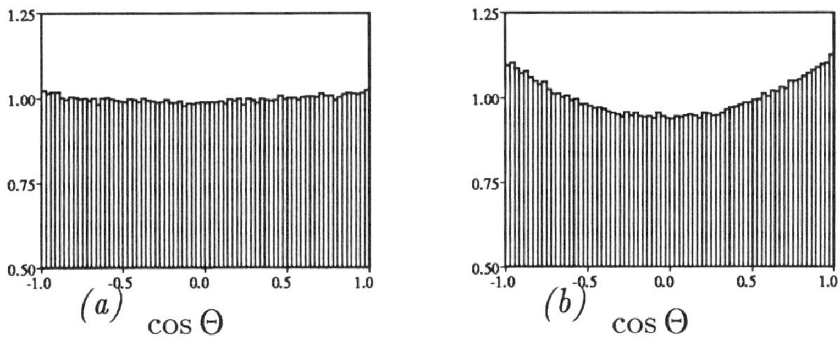

FIGURE 5. Distribution of the orientation of the vorticity vector with respect to the axis of rotation. $\cos\Theta = \omega_3/|\omega|$: *(a)* $Ro = 0.008$. *(b)* $Ro = 0.04$.

mild increase in the integral scale (see Fig. 4b). The vorticity vectors become mildly aligned in the axial direction (see Fig. 5b). Thus for high Rossby numbers ($Ro >> 1$), the isotropy is preserved. In the opposite case for low Rossby numbers ($Ro << 1$), $T_e = 0$ and the spectrum changes through pure viscous decay. For intermediate Rossby numbers, T_e and T_z can be strongly anisotropic, but their order of magnitude is very weak, and the initial isotropy is essentially preserved.

3.2 Intermediate Rossby number effects.

To quantify the departure from isotropy by one-point correlations in the intermediate Rossby number range, Eq. (12) was split into three parts as follows:

$$\begin{aligned}\hat{U}_{ij} &= \frac{E}{4\pi k^2}P_{ij} + (e - \frac{E}{4\pi k^2})P_{ij} + \text{Real}[zN_iN_j] \\ &= \hat{U}^{iso}_{ij} \quad + \hat{U}^{e}_{ij} \quad + \hat{U}^{z}_{ij}\end{aligned} \tag{17}$$

The reason for the splitting becomes apparent when one converts to one-point correlations by integrating Eq. (17) over all wave numbers, yielding

$$\langle u_i u_j \rangle = q^2 \left(\frac{1}{3}\delta_{ij} + b^e_{ij} + b^z_{ij} \right) \tag{18}$$

In terms of the anisotropy tensor $b_{ij} = \langle u_i u_j \rangle / q^2 - \delta_{ij}/3$, we have

$$b_{ij} = b^e_{ij} + b^z_{ij} \tag{19}$$

where:

$q^2\delta_{ij}/3 = \int\int\int E/4\pi k^2 \; P_{ij} d^3\mathbf{k}$ yields the isotropic contribution to the Reynolds stress tensor. (For an isotropic flow $b^e_{ij} = b^z_{ij} = 0$.)

$q^2 b^e_{ij} = \int\int\int U^e_{ij} d^3\mathbf{k}$ ($= -1/2 \; q^2 y_{ij}$ in Reynolds [11] notation) yields the contribution to the anisotropy tensor coming from the "dimensionality" of the flow. Reynolds [11] distinguishes between 3-dimensional, 2-component turbulence and 2-dimensional, 3-component turbulence since realizability conditions can be different for these two flows.

$q^2 b^z_{ij} = \int\int\int U^z_{ij} d^3\mathbf{k}$ yields the contribution to the anisotropy tensor coming from the components of the flow.

The rapid distortion effects (weak nonlinear terms, $T_e = T_z = 0$ in Eq. (14)) on the one-point correlation terms are such that q^2 and b^e_{ij} are unaffected by the rotation. The phase

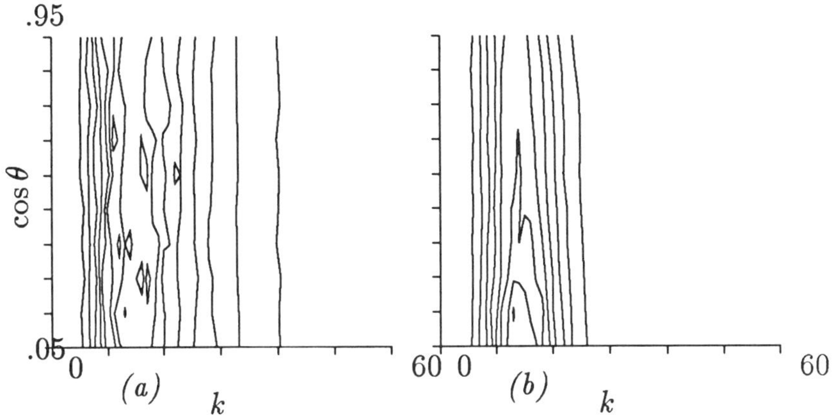

FIGURE 6. Iso-levels of the two-dimensional energy spectra ($E(k, \cos\theta)$): *(a)* at $2(t-t_0)\epsilon_0/q_0^2 = 2.08$, $Ro = \infty$. *(b)* at $2(t-t_0)\epsilon_0/q_0^2 = 4.43$, $Ro = 0.04$. Levels start at 0.00025 with 0.00025 increments.

scrambling of z by the rotation tends to "kill" b_{ij}^z. Computations using EDQNM and DNS show a weak axisymmetric trend at intermediate Rossby numbers,

$$\frac{1}{q^2}(\langle u_3 u_3\rangle - \frac{\langle u_1 u_1\rangle + \langle u_2 u_2\rangle}{2}) = \frac{3}{2} b_{33} > 0 \qquad (20)$$

At intermediate Rossby numbers, the phase scambling effects still hold and the b_{ij}^z term remains quasi-null. This is an indication that the trend observed ($b_{33} > 0$) can only be due to "dimensionality" since $b_{33} \propto b_{33}^e$ when b_{33}^z is small. The creation of $b_{33} > 0$ is due to nonlinear transfer effects. This trend reflects a relative concentration of energy $e(k, \cos\theta)$ in the transverse wave-plane, $\cos\theta = 0$. (according to the definition $b_{33}^e = -\int\int\int e(k, \cos\theta)\cos^2\theta d^3\mathbf{k}$). Figure 6 shows the development of the distribution of $e(k, \cos\theta)$ from the DNS for isotropic decay and for a rotation at $Ro = 0.04$. Both simulations start with the same initial spectrum. The distribution along spheres ($k = \text{const}$) is uniform for the case of isotropic decay. There is a definite concentration of energy around $\cos\theta = 0$ for the case with rotation. Nevertheless, the effects of anisotropy are weakly reflected by the one-point correlations which correspond to 2D integrals (over k and $\cos\theta$) of the basic 2D spectra.

Quantities connected with the integral length scales reflect in a more spectacular way the non-isotropic, non-linear effects of the Coriolis forces. The axial length scales are of particular interest:

$$L_{ij}^{(3)} = \frac{\int_0^\infty \langle u_i(\mathbf{x})u_j(\mathbf{x}+\mathbf{n})\rangle r dr}{\langle u_i u_j\rangle}; \quad n_i = \delta_{i3} \tag{21}$$

In axisymmetric turbulence, two integral length scales are relevant

$$\begin{aligned} \langle u_3 u_3\rangle L_{33}^{(3)} &= \pi \int\int \hat{U}_{33}|_{k_3=0} d^2\mathbf{k} \\ \langle u_\perp u_\perp\rangle L_\perp &= \frac{\pi}{2}\int\int (\hat{U}_{11} + \hat{U}_{22})|_{k_3=0} d^2\mathbf{k} \end{aligned} \tag{22}$$

According to their derivation from the spectral tensor, they display the spectral shape at $\cos\theta = 0$ (transverse wave-plane). The experimental and theoretical works of Jacquin & Cambon [9] show that $\langle u_3 u_3\rangle L_{33}^{(3)}$ is only moderately affected by rotation, but that $\langle u_\perp u_\perp\rangle L_\perp$ show a dramatic change. It is also possible to connect $\langle u_\perp u_\perp\rangle L_\perp$ to the spectral mode which corresponds to a pure two-dimensional turbulence.

In order to quantify the generation of anisotropy, an anisotropy parameter A given by,

$$\begin{aligned} A &= \frac{\langle u_3 u_3\rangle L_{33}^{(3)} - 2\langle u_\perp u_\perp\rangle L_\perp}{\langle u_3 u_3\rangle L_{33}^{(3)}(0)} \\ &= 4\pi^2 \int_0^\infty z(k, \cos\theta = 0) k^2 dk \end{aligned} \tag{23}$$

was computed from both the EDQNM and the DNS results. It is clear that such a quantity is only affected by the non-linear Coriolis effects through T_z because $k_3/k = 0$ (see Eq. (14)).

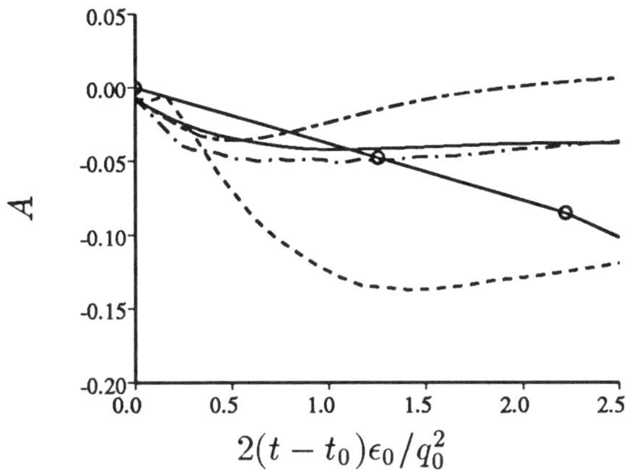

FIGURE 7. Evolution of the anisotropy parameter (Eq. 23). —-— Isotropic decay, $Ro = \infty$ DNS; ——— $Ro = 0.4$ DNS; - - - - $Ro = 0.04$ DNS; —·— $Ro = 0.008$ DNS; -o-o-o $Ro = 0.5$, EDQNM simulation.

Both the EDQNM and DNS results show an increase in the anisotropy in the Rossby number range $1 < Ro < 0.04$ (see Fig. 7). At low Rossby numbers $Ro < 0.04$ the flow remains quasi-isotropic, consistent with the fact that the nonlinear effects are small. While we have qualitative agreement between the DNS and the EDQNM results, there are large quantitative differences between the two simulations due to differences in Reynolds numbers and initial spectra.

4. Summary and Conclusions

We have shown that at high rotation rates the nonlinear transfer terms remain small and that the development of the spectrum is through pure viscous decay. This shows that the effect of rapid rotation on the turbulent kinetic energy is through the shutting off of the production term (vortex stretching by the small scales) in the dissipation rate equation.

At moderate and relatively rapid rotation rates ($1 < Ro < 0.04$), the rotation causes a discernible anisotropy to develop in the integral length scales with a mild trend toward a two-dimensionalization of the flow. But as the flow decays, the Rossby number decreases leading again to the shutting off of the nonlinear transfer terms.

For extremely rapid rotation rates ($Ro < 0.01$) the anisotropies in the integral length scales are small and the DNS results are in good agreement with the RDT solution since the non-linear transfer has been shut-off. The RDT solution was shown to violate a necessary condition for a Taylor-Proudman reorganization to occur. The DNS results do not show a trend toward two-dimensionalization for the Reynolds numbers considered in this study. It is still an open question as to whether such a two-dimensionalization can occur at much higher Reynolds numbers.

Future research will be directed toward the application of these results to the development of improved turbulence models – particularly for the turbulence dissipation rate.

Acknowledgments

We are grateful to Dr. R. S. Rogallo for generously providing us with the DNS code, and for many valuable discussions on rotating flows. We are also grateful to Dr. M. Rogers for his code that was modified to produce the vorticity histograms (figure 5). CC and CGS were supported by the Center for Turbulence Research, NASA Ames research center, during the course of this work.

References

[1] Traugott, S. C., 1958, *Influence of Solid-Body Rotation on Screen-Produced Turbulence. NACA Technical Note 4135.*

[2] Ibbetson, A. & Tritton, D. J., 1975, *Experiments on Turbulence in a Rotating Fluid.* J. Fluid Mech. **68,** 639-672.

[3] Wigeland, R. A. & Nagib, H. M., 1978, *Grid-Generated Turbulence With and Without Rotation about the Streamwise Direction.* IIT Fluids and Heat Transfer, Rep. R78-1,

Illinois Inst. of Tech., Chicago, Illinois.

[4] Jacquin, L., Leuchter, O., Cambon, C. & Mathieu, J., 1990, *Homogeneous Turbulence in the Presence of Rotation.* J. Fluid Mech. **220,** 1-52.

[5] Rogallo, R. S., 1981, *Numerical Experiments in Homogeneous Turbulence.* NASA TM 81315, Ames research center, Moffett Field, Calif.

[6] Bardina, J., Ferziger, J. H. & Rogallo, R. S., 1985, *Effect of Rotation on Isotropic Turbulence: Computation and Modeling.* J. Fluid Mech. **154,** 321-336.

[7] Speziale, C. G., Mansour, N. N. & Rogallo, R. S., 1987, *The Decay of Isotropic Turbulence in a Rapidly Rotating Frame.* Proceedings of the 1987 Summer Program, Report CTR-S87, Center for Turbulence Research, NASA Ames research center, Moffett Field, CA.

[8] Teissèdre, C. & Dang, K., 1987, *Anisotropic Behavior of Rotating Homogeneous Turbulence by Numerical Simulation.* AIAA 87-1250, AIAA 19th Fluid Dynamics, Plasma Dynamics and Lasers Conference, Hawaii.

[9] Cambon, C. & Jacquin, L., 1989, *Spectral Approach to Non-Isotropic Turbulence Subjected to Rotation.* J. Fluid Mech. **202,** 295-317.

[10] Rogallo, R. S., 1987, Private communication.

[11] Reynolds, W. C., 1989, *Effects of Rotation on Homogeneous Turbulence.* in the Proceedings of Tenth Australasian Fluid Mechanics Conf. University of Melbourne.

TOWARDS A STRUCTURE-BASED TURBULENCE MODEL

W. C. Reynolds

Stanford University and NASA Ames Research Center
Stanford, CA 94305-3030

ABSTRACT

The author is developing a new type of turbulence model in which a new one-point quantity, the *eddy structure tensor*, carries information about the two-point structure of the turbulence. The model was motivated by the observation that conventional one-point turbulence models based only on the turbulent stresses do not predict the rapid changes in state that are found when anisotropic homogeneous turbulence is subjected to mean rotation, and hence are fundamentally incorrect for rotation. The model appears to give topologically correct predictions for the changes in stress state and structure state under all types of rapid distortions of homogeneous turbulence, which is something no other type of one-point model can do at this time. The definitions, model elements, and provisional model constants for homogeneous turbulence are given in these *author's working notes* so that others may explore the model. The author is indebted to John Lumley for many inspirations and insights that underlie this work.

1. Eddy structure tensor

$$D_{ij} = \overline{\Psi'_{n,i}\Psi'_{n,j}}. \tag{1}$$

Ψ'_n is the *vector stream function* defined by

$$u'_i = \epsilon_{ijk}\Psi'_{k,j} \tag{2a}$$

and the auxiliary condition

$$\Psi'_{i,i} = 0. \tag{2b}$$

Ψ'_i satisfies

$$\Psi'_{i,jj} = -\epsilon_{ijk}u'_{k,j} = -\omega'_i. \tag{3}$$

For homogeneous turbulence,

$$D_{ij} = \int \frac{k_i k_j}{k^2} E_{nn}(\mathbf{k})d^3\mathbf{k}. \tag{4}$$

Note that $D_{ii} = q^2$, and that $D_{\alpha\beta}$ is zero if the turbulence is independent of x_α. R_{ij} carries information on the *componentality* of the turbulence, D_{ij} carries information about its *dimensionality* and eddy structure. For example, if $D_{11} = 0.1$ and $D_{22} = D_{33} = 0.8$, the eddies are long in the x_1 direction and have the same shorter scale in the x_2 and x_3 directions.

2. Thirteen variables tracked by differential equations

Reynolds stresses: R_{ij} , where $R_{nn} = q^2$.
Eddy orientation tensor: a_{ij}, where $a_{nn} = 1$.
Turbulence time scale: τ (eddy turn-over time).

3. Algebraic definitions

Strain rate $S_{ij} = (U_{i,j} + U_{j,i})/2$.
Rotation rate $\Omega_{ij} = (U_{i,j} - U_{j,i})/2$.
Energy distribution tensor $r_{ij} = R_{ij}/q^2$. Note $r_{nn} = 1$.
Structure distribution tensor $d_{ij} = D_{ij}/q^2$. Note $d_{nn} = 1$.
Energy anisotropy tensor $b_{ij} = r_{ij} - \delta_{ij}/3$. Note $b_{nn} = 0$.
Structure anisotropy tensor $y_{ij} = d_{ij} - \delta_{ij}/3$. Note $y_{nn} = 0$.

4. Evolution of a_{ij} in rapid distortions

a_{ij} represents the average eddy direction cosine tensor. For isotropic turbulence, $a_{ij} = \delta_{ij}/3$. Consider a line element in a fluid. Distortion by imposed mean velocity gradients $G_{ij} = U_{i,j}$ will cause the line to stretch and tilt. Under RDT, the direction cosine tensor changes according to

$$\frac{da_{ij}}{dt} = G_{ik}a_{kj} + G_{jk}a_{ki} - 2G_{nm}a_{mn}a_{ij}. \tag{5}$$

5. Structure model

$$d_{ij} = \frac{1}{2}(\delta_{ij} - a_{ij}). \tag{6}$$

It is most convenient to evolve a_{ij} and then calculate d_{ij}. Turbulence is assumed to restore the appropriate structural isotropy; The model equation is

$$\begin{aligned}\frac{da_{ij}}{dt} &= G_{ik}a_{kj} + G_{jk}a_{ki} - 2G_{nm}a_{mn}a_{ij} \\ &-C_a(d_{ik}a_{kj} + d_{jk}a_{ki} - 2d_{nm}a_{mn}d_{ij})/\tau. \end{aligned}\tag{7}$$

6. Rapid pressure-strain model

The the rapid pressure strain term, modeled generally in terms of d_{ij} and r_{ij}, would contain a large number of different tensor forms. However, we have found it possible to capture the essence of a very wide range of two and three-dimensional turbulent flows with a relatively simple model:

$$T_{ij}^{(r)} = B_1[d_{ik}S_{kn}R_{nj} + d_{jk}S_{kn}R_{ni} - 2d_{pk}S_{kn}R_{np}d_{ij}]$$

$$C_1[d_{ik}\Omega_{kn}R_{nj} + d_{jk}\Omega_{kn}R_{ni} - 2d_{pk}\Omega_{kn}R_{np}d_{ij}]$$
$$+C_2[d_{ik}\Omega_{kn}d_{nm}R_{mj} + d_{jk}\Omega_{kn}d_{nm}R_{mi}]$$
$$+C_3[d_{ik}^2\Omega_{kn}d_{nm}R_{mj} + d_{jk}^2\Omega_{kn}d_{nm}R_{mi} - 2d_{pk}^2\Omega_{kn}d_{nm}R_{mp}d_{ij}]$$

$$-\beta X[d_{ik}c_{kn}R_{nj} + d_{jk}c_{kn}R_{ni} - d_{pk}c_{kn}R_{np}f_{ij}] \tag{8}$$

where

$$f_{ij} = (d_{ik}R_{kj} + d_{jk}R_{ki})/d_{nm}R_{mn}$$

and

$$c_{ij} = r_{ij} - (\delta_{ij} - d_{ij})/2$$

characterizes the departure from a "random phase" spectrum, and

$$X = \sqrt{\Omega_{kn}\Omega_{km}a_{nm}}$$

is a reciprocal time scale associated with the randomization of the phases of the Fourier modes caused by mean rotation. Note that $X = 0$ for 2-d turbulence with the axis of independence aligned with the mean rotation rate. The coefficient of d_{ij} in the sum of the C_n terms is zero for 2-d turbulence, which ensures realizability of 2-c 2-d turbulence. The use of f_{ij} in the other terms achieves trace-free forms and preserves the full realizability of the turbulence.

The effect of mean strain-rate is represented by the single B_1 term. For irrotational deformation of initially isotropic 3-d turbulence, where $\det(\mathbf{d}) = 1/27$, $C_1 = 9/5 \approx 2$ matches the exact result from the Navier-Stokes equations, while for 2-d turbulence $C_1 = 2$ provides an exact match. For RDT in the axisymmetric expansion limit, $C_1 = 12/5 \approx 2$. Therefore, C_1 is treated as a weak function of the structure. The *structure parameter*

$$Z = \sqrt{6}\frac{y_{ij}y_{jk}y_{ki}}{(y_{nm}y_{mn})^{3/2}} \tag{9}$$

is used to parameterize B_1. $Z = -1$ for axisymmetric turbulence with long structures as produced by axisymmetric contraction, and and $Z = 1$ for flat structures as produced by axisymmetric expansion. RDT results for irrotational axisymmetric expansion and contraction, and for plane strain, are fit remarkably well using

$$B_1 = \frac{9}{5} + (1 - 27\det(\mathbf{a}))\frac{2+Z}{5} \tag{10}$$

The C_n rotation-rate terms are constructed to have the following properties:

1. When the Fourier coefficients of the velocity are random, then $r_{ij} = (\delta_{ij} - d_{ij})/2$, and $T_{ij}^{(r)} = 2[\Omega_{ik}R_{kj} + \Omega_{jk}R_{ki}]$ (*random phase model*).
2. When the turbulence is 2-c with the axis of independence aligned with the axis of rotation, then $T_{ij}^{(r)} = 2[\Omega_{ik}R_{kj}+\Omega_{jk}R_{ki}]$ and the turbulence is *materially indifferent to rotation.*
3. For nearly isotropic turbulence with $b_{ij} = r_{ij} = O(\epsilon)$, Ω_{ij} does not enter $T_{ij}^{(r)}$ until $O(\epsilon^2)$.

A set of C_n coefficients consistent with these requirements is

$$C_1 = 2 \qquad C_2 = -C_3 = 4\frac{1 - 27\det(\mathbf{d})}{1 - \mathrm{tr}(\mathbf{d}^2)}$$

The β term allows mean rotation to produce a state of random phase and produces the changes under mean rotation not duplicated by models based only on R_{ij}. With the other constants as given, RDT predictions for homogeneous shear flow are fit quite well using $\beta = 4$, which is the currently recommended value. Note that $C_2 = C_3 = 0$ for isotropic structure ($\det(\mathbf{d}) = 1/27$), and hence the model simplifies considerably for weakly anisotropic turbulence.

7. R_{ij} transport equation

The slow pressure-strain term is assumed to restore the appropriate isotropy,

$$T_{ij}^{(s)} = -C_r(d_{ik}R_{kj} + d_{jk}R_{ki} - 2d_{nm}R_{kn}d_{ij})/\tau. \tag{11}$$

The dissipation is assumed to mirror the anisotropy in the R_{ij}. The resulting R_{ij} model is then

$$\dot{R}_{ij} = -G_{ik}R_{kj} - G_{jk}R_{ki} + T_{ij}^{(r)} + T_{ij}^{(s)} - 2R_{ij}\varepsilon/q^2 \tag{12}$$

8. Dissipation model

$$\varepsilon = q^2/\tau \tag{13}$$

9. τ equation model

In the absence of mean deformation, the time scale should increase. Strain will alter the time scale, except for 1-c 2-d turbulence. The model is

$$\frac{d\tau}{dt} = 5d_{nm}r_{mn} + 2d_{nm}S_{mn}\tau + C_B X\tau \tag{14}$$

The coefficients 5 and 2 are set by theoretical arguments. The 5 is set by the energy decay observed when one has a k^2 spectrum in isotropic turbulence; the 2 is set by matching the integral length scale evolution for RDT for isotropic compression/expansion of 3-d isotropic turbulence where $\rho = \rho(t)$, or axisymmetric expansion/contraction of 2-d incompressible isotropic turbulence. The C_B term accounts for the increase in time scale (reduced dissipation arising from reduced cascade rate) caused by phase randomization by mean rotation; $C_B = 0.18$ is suggested by the large eddy simulations of Bardina.

10. Slow model constants

A provisional determination of the constants C_a and C_r by comparisons against simulations and experiments for homogeneous shear flow gave

$$C_a = 12 \qquad C_r = 0.3$$

NOTE: The slow models are just now being addressed and are likely to change before this work is submitted for journal publication. The current values are given here so that interested researchers can begin to explore this new type of modeling.

11. Performance features

The model displays a rapid return to axisymmetry in R_{ij} and a slower return to isotropy, observed in both simulations and experiments. It also displays the right sort of sensitivity to strain rate, and of course is very good for very rapid deformations. For strongly rotated anisotropic turbulence, the model displays the damped oscillations to the random phase state required by RDT, a feature that can not be captured with conventional models in which the pressure-strain term depends only on b_{ij}. It tracks all of the known RDT solutions for homogeneous turbulence and even predicts some strange behavior found in numerical simulations.

NOTE added in press: We have made important advances and changes in this model since the Lumley Conference as a result of our first extensions to inhomogeneous turbulence (channel flow). The first journal version of this model will differ from what is outlined above in its treatment of homogeneous turbulence, and will be extended to included inhomogeneous turbulence.

EFFECT OF MEAN VELOCITY SHEAR ON THE DISSIPATION RATE OF TURBULENT KINETIC ENERGY

Akira Yoshizawa[1]

Institute of Industrial Science, University of Tokyo
Tokyo, Japan

Meng-Sing Liou

NASA Lewis Research Center
Cleveland, Ohio, U.S.A.

ABSTRACT

The dissipation rate of turbulent kinetic energy in incompressible turbulence is investigated using a two-scale DIA. The dissipation rate is shown to consist of two parts; one corresponds to the dissipation rate used in the current turbulence models of eddy-viscosity type, and another comes from the viscous effect that is closely connected with mean velocity shear. This result can elucidate the physical meaning of the dissipation rate (ϵ) used in the current turbulence models and explain part of the discrepancy in the near-wall dissipation rates between the current turbulence models and direct numerical simulation of the Navier-Stokes equation.

1. Introduction

In the one-point turbulence models for incompressible flows, the dissipation rate of turbulent kinetic energy ϵ plays a key role, regardless of the eddy-viscosity or second-order (Reynolds-stress) models. Nevertheless, the theoretical base of the modeled ϵ equation is not so firm as the counterpart for the turbulent kinetic energy (K) itself. This situation comes from the fact that the detailed property of the energy dissipation range has not been elucidated so far, except some simple flows.

---

[1]This work was conducted while the first author was in residence at the Institute for Computational Mechanics (ICOMP), NASA / Lewis Research Center.

Within the framework of the two-equation models represented by the $K - \epsilon$ model, the eddy-viscosity representation plays a central role. It can be derived theoretically using a two-scale DIA (TSDIA) (Yoshizawa 1984) and is shown to be the first-order approximation around the lowest-order field that survives in the absence of mean velocity shear. This fact means that ϵ, which as well as K is used to express the eddy viscosity ν_e, should be the dissipation rate corresponding to the lowest-order field that is assumed to be isotropic. In this work, we shall investigate this point to clarify the physical meaning of ϵ that is used in the current turbulence models.

The above point can be viewed from another aspect. Recent progress in direct numerical simulation (DNS) shed light on the structures of turbulence quantities that are difficult to measure by current experimental methods (Moser and Moin 1987, Kim et al. 1987, Spalart 1988). A representative quantity of them is the dissipation rate of turbulent kinetic energy, which is defined by

$$\epsilon = \nu\left\langle\left(\frac{\partial u_i'}{\partial x_j}\right)^2\right\rangle \tag{1}$$

where u' is the fluctuation of velocity around the ensemble mean denoted by $U(=\langle u\rangle)$, ν is the kinematic viscosity, and the summation convention is applied to repeated indices. The dissipation rate ϵ obtained from the DNS of a turbulent channel flow shows a remarkable difference from the counterpart from various turbulence models in the close vicinity of the wall (Mansour et al. 1988). The latter has the maximum value slightly away from the wall, whereas the former is maximum at the wall. At present, the cause of this critical discrepancy is considered to arise from incompleteness of models, specifically, that of the wall damping functions describing low Reynolds number effects (Monsour et al. 1989). In this work, we attempt to explain part of discrepancy from the viewpoint of a two-scale DIA (TSDIA) (Yoshizawa 1984, 1988).

2. A brief outline of TSDIA

In the TSDIA, we first introduce two space and time scales:

$$\xi(=x), X(=\delta x);\ \tau(=t), T(=\delta t), \tag{2}$$

using a small scale parameter δ. In terms of (2), we write

$$u = U(X;T) + u'(\xi, X; \tau, T), \tag{3}$$

and expand u' as

$$u' = u'_0 + \delta u'_1 + \delta^2 u'_2 + \cdots. \tag{4}$$

The lowest-order field u'_0 obeys the same equation as for homogeneous turbulence, except there is an implicit dependence on the slow variables X and T. Direct effects of the mean velocity U, which is a function of slow variables only, appear in higher-order terms $u'_n (n \geq 1)$. Using (4), ϵ is written as

$$\epsilon = \nu\left[\left\langle\left(\frac{\partial u'_{0i}}{\partial \xi_j}\right)^2\right\rangle + \delta\left(2\left\langle\frac{\partial u'_{0i}}{\partial \xi_j}\frac{\partial u'_{1i}}{\partial \xi_j}\right\rangle + 2\left\langle\frac{\partial u'_{0i}}{\partial \xi_j}\frac{\partial u'_{0i}}{\partial X_j}\right\rangle\right) + \right.$$
$$\left. + \delta^2\left\{\left\langle\left(\frac{\partial u'_{1i}}{\partial \xi_j}\right)^2\right\rangle + R.T.\right\} + \cdots\right]. \tag{5}$$

Here, the first term ϵ_0 given by the $O(\delta^0)$ field u'_0 characterizes the familiar Kolmogorov inertial range, and the residual terms $R.T.$ are related to the first-order X derivatives of u'_0, u'_1, and u'_2. We can use the TSDIA to calculate the Reynolds stress $R_{ij}(= -\langle u'_i u'_j \rangle)$. The calculation up to the first order $[O(\delta)]$ gives (Yoshizawa 1984)

$$R_{ij} = -\frac{2}{3}K\delta_{ij} + \nu_e\left(\frac{\partial U_i}{\partial x_j} + \frac{\partial U_j}{\partial x_i}\right), \tag{6}$$

where K is the turbulent kinetic energy defined by $\langle u'^2 \rangle / 2$, δ_{ij} is the Kronecker delta symbol, and ν_e is the so-called eddy viscosity, which is written in terms of statistical quantities of u'_0. Assuming the inertial-range law for u'_0, we can obtain a familiar expression in the two-equation turbulence models; namely,

$$\nu_e = C_\nu \frac{K^2}{\epsilon_0}, \tag{7}$$

where C_ν is a model constant.

Here we should note the following two points. The first is that what is adopted as dissipation rate is ϵ_0 but not ϵ (the true dissipation rate) in (1) or (5). Within the framework of the TSDIA, ϵ_0 is linked strongly with homogeneous properties of turbulent field, as can be seen from the feature of u'_0 stated above. From this standpoint, K in (7) should be replaced by K_0 [the $O(\delta^0)$ part of K]. This point will be referred to later. Another point is that (7) is not valid near solid walls. The fact is usually called low Reynolds number effects (see Patel et al. 1987 for their comprehensive review, and Nagano and Hishida 1987, Myong and Kasagi 1990 for the more recent study about them). Such effects can be characterized by two nondimensional parameters $y^+(= u_\tau y/\nu;\ u_\tau$ is the friction velocity and y is the distance from the wall) and $R_t(= \nu_e/\nu)$ that is named the turbulent Reynolds number. The effect expressed by y^+ is difficult to be treated using turbulence theories since it is closely connected with boundary conditions. On the other hand, the R_t effect is just the effect of ν irrespective of boundary conditions. Therefore at least part of the effect can be incorporated into ν_e of (7) by considering both the inertial and dissipation ranges of u'_0 [Inage, Shimomura (two independent private communications)]. In what follows, we shall examine the effect of the remaining higher-order terms in (5) and show that it is described by ν and the mean shear rate. This effect is considered to be a kind of low Reynolds number effect similar to the R_t effect.

3. TSDIA analysis of ϵ

We notice that the wall effect is closely associated with generation of mean velocity shear. Therefore let us consider the effect of the mean velocity shear rate on ϵ that is of higher order within the framework of the TSDIA. Such terms of $O(\delta)$ in (5) are linearly dependent on the first-order X derivative of U and do not give contribution to ϵ that is a scalar quantity. The mean-shear effect on ϵ comes from the first term in the $O(\delta^2)$ terms for ϵ. In order to see this point clearly, we consider the Fourier representation of u' :

$$u'(\xi, X; \tau, T) = \int u'(k, X; \tau, T) exp[-ik \cdot (\xi - U\tau)]dk. \tag{8}$$

Hereafter the dependence of u' on (X,T) will not be explicitly written. The first-order field u'_1 is given by

$$u'_{1i}(k;\tau) = \frac{\partial U_m}{\partial X_j} D_{1mn}(k) \int_{-\infty}^{\tau} d\tau_1 F'_{in}(k;\tau,\tau_1) u'_{0j}(k;\tau_1) + R.T., \tag{9}$$

with $D_{1ij}(k) = \delta_{ij} - 2k_i k_j/k^2$ (Yoshizawa 1984). Here F'_{ij} is the response (Green's) function associated with u'_0. The residual terms R.T. are not directly dependent on U. In the TSDIA, we assume the following isotropy with the inhomogeneous effect appearing implicitly through X and T:

$$\frac{\langle u'_{0i}(k;\tau)u'_{0j}(-k;\tau')\rangle}{\delta(0)} = D_{ij}(k)Q(k;\tau,\tau'), \tag{10}$$

$$\langle F'_{ij}(k;\tau,\tau')\rangle = \delta_{ij}F(k;\tau,\tau'), \tag{11}$$

with $D_{ij}(k) = \delta_{ij} - k_i k_j/k^2$ [$\delta(k)$ is the Dirac delta function]. Using (10) and (11), we can calculate the first term of the $O(\delta^2)$ terms in (5), which is combined with ϵ_0 to give

$$\epsilon = \epsilon_0 + \epsilon_2, \tag{12}$$

where

$$\epsilon_0 = 2\nu \int k^2 Q(k;\tau,\tau)dk, \tag{13}$$

$$\epsilon_2 = \frac{2}{3}\left(\frac{\partial U_i}{\partial X_j}\right)^2 \int k^2 dk \int_{-\infty}^{\tau} d\tau_1 \int_{-\infty}^{\tau_1} d\tau_2$$
$$\times F(k;\tau,\tau_1)F(k;\tau,\tau_2)Q(k;\tau_1,\tau_2). \tag{14}$$

In order to estimate the magnitude of ϵ_2, we introduce a simple form for Q and F:

$$Q(k;\tau,\tau') = Q(k)exp[-\omega(k)|\tau-\tau'|], \tag{15}$$

$$F(k;\tau,\tau') = H(\tau-\tau')exp[-\omega_F(k)(\tau-\tau')] \tag{16}$$

[$H(\tau)$ is the unit step function]. Here the spectrum $Q(k)$ and the inverses of the characteristic time scales in the inertial and dissipation ranges $\omega(k)$ and $\omega_F(k)$ are expressed in a general form:

$$Q(k) = C_Q \epsilon_0^{\frac{2}{3}} k^{-\frac{11}{3}} f_Q\left(\frac{k}{k_d}\right),$$

$$\omega(k) = C_\omega \epsilon_0^{\frac{1}{3}} k^{\frac{2}{3}} f_\omega\left(\frac{k}{k_d}\right), \tag{17}$$

$$\omega_f(k) = C_F \omega(k),$$

where C_Q etc. are numerical constants and $k_d[= (\epsilon_0/\nu^3)^{\frac{1}{4}}]$ is the wavenumber characterizing the dissipation range. In (17), the following constraints are imposed on f_Q and f_ω :

$$f_Q(0) = f_\omega(0) = 1,$$

$$\int_{s \gtrsim 1} s^{-\frac{5}{3}} f_Q(s) ds = \frac{1}{2C_Q}. \tag{18}$$

The latter in (18) comes from (13) and the fact that only fluctuations with the wavenumber magnitude around and larger than k_d contribute to the dissipation of energy. Using (17), we have

$$\epsilon_2 = c_{\epsilon 2} \nu \left(\frac{\partial U_i}{\partial x_j} \right)^2, \tag{19}$$

where the numerical coefficient $c_{\epsilon 2}$ is

$$c_{\epsilon 2} = \frac{2}{3} \frac{C_Q}{C_F(1 + C_F)C_Q^2} \int_{|s| \gtrsim 1} s^{-3} \frac{f_Q(s)}{f_\omega^2(s)} ds. \tag{20}$$

From the properties of $Q(k)$ etc., we can see that $c_{\epsilon 2}$ is positive. Therefore, ϵ_2 gives a positive correction to $c_{\epsilon 2}$.

4. Discussions

Let us exemplify the property of ϵ_2 by considering a turbulent channel flow. In the region away from the walls, the mean velocity gradient is small compared with the fluctuation velocity gradient and hence ϵ_2 is negligible. Near the walls, however, ϵ_2 can become important. We consider the logarithmic-velocity layer where $y^+ \gg 1$,

$$\frac{U}{u_\tau} = \frac{1}{\kappa} \log y^+ + A, \tag{21}$$

and

$$\epsilon_0 = \frac{u_\tau^3}{\kappa y} = \frac{u_\tau^4}{\kappa \nu y^+}. \tag{22}$$

Here U is the streamwise mean velocity, κ is the Kármán constant, and A is another constant characterizing wall roughness. We approach to the so-called buffer layer from the upper logarithmic-velocity layer. Then, ϵ_2 can be estimated using (21) to give

$$\epsilon_2 \simeq \nu \frac{c_{\epsilon 2}}{\kappa^2} \frac{u_\tau^2}{y^2} = \frac{u_\tau^4}{\nu} \frac{c_{\epsilon 2}}{(\kappa y^+)^2}. \tag{23}$$

Comparison of (22) with (23) shows that for $y^+ \gg 1$, $\epsilon_0 \gg \epsilon_2$ as expected. These two quantities, however, can be of the same order at $y^+ = O(1)$ or $y = O(\nu/u_\tau)$. Indeed, for $y^+ = O(1)$ we have

$$U \simeq u_\tau y^+, \tag{24}$$

which gives

$$\epsilon_2 \simeq \frac{c_{\epsilon 2} u_\tau^4}{\nu} \tag{25}$$

from (19). Very near the wall with $y^+ \ll 1$, we have (Hinze 1975)

$$K \simeq C_K u_\tau^2 y^{+2}, \tag{26}$$

which is combined with the familiar near-wall expression in the current turbulence models that $\epsilon_0 \simeq \nu(d^2K/dy^2)$ to give

$$\epsilon_0 \simeq 2C_K \frac{u_\tau^4}{\nu}. \tag{27}$$

Here, C_K is a positive constant. Thus, we conclude from (25) and (27) that ϵ_2 is of the same order as ϵ_0 and that the dissipation rate near the wall should be calculated using $\epsilon_0 + \epsilon_2$. In light of this new dissipation effect near the wall, we consider the current two-equation $(K - \epsilon)$ models. The turbulent kinetic energy satisfies

$$\frac{DK}{Dt} = P - \epsilon + D_K, \tag{28}$$

where P and D_K are the production and diffusion (transport) rates, respectively, and ϵ is approximated by (12) in this work. Modeling of the dissipation rate equation is a major stumbling block in the turbulence models. Previously, Yoshizawa (1987, 1988) proposed a systematic method, for recovering major parts of the current dissipation rate equation with the help of the TSDIA. In this method, we calculate K using the TSDIA and simplify it in terms of ϵ_0 and ℓ [the characteristic length scale of the u_0' field (Yoshizawa 1984, 1987)]. Next, we solve this relation with respect to ℓ to represent it in terms of K and ϵ_0. The lowest-order part in ℓ gives an algebraic relationship among ℓ, K, and ϵ_0. The first-order part includes DK/Dt or $D\epsilon_0/Dt$ linearly. We define ℓ by this algebraic relationship. The constraint that the higher-order terms in ℓ vanish leads to a model equation for ϵ_0 :

$$\frac{D\epsilon_0}{Dt} = C_{\epsilon 1}\frac{\epsilon_0}{K}P - C_{\epsilon 2}\frac{\epsilon_0}{K}\epsilon + D_\epsilon, \tag{29}$$

where $c_{\epsilon 1}$ etc. are model constants, and D_ϵ is the diffusion term. In the current models, the low Reynolds number effects usually enter (29) through ν_e in P as well as through a wall damping function attached to the second term (Patel et al. 1987).

Here, we should notice the following two points. One is that K does not contain the direct effects from the first- and higher-order fields in the sense of the TSDIA scale expansion because of the above constraint. As a result, it is not necessary to distinguish K from K_0 (the lowest-order part of K) in (7). In the absence of the algebraic relationship among ℓ, K, and ϵ_0, K in (7) should be replaced by K_0. The relationship is also important for guaranteeing transferability (Yoshizawa 1987, 1988) among the models that are constructed using any two of ℓ, K, ϵ_0, and their combinations. Another point is that the dissipation rate equation in the current two-equation models is the one for ϵ but not for ϵ itself. The dissipation rate ϵ calculated from DNS should be compared with at least the modeled $\epsilon(= \epsilon_0 + \epsilon_2)$. In this context, the usual boundary condition at the wall for ϵ_0 should be replaced by

$$\epsilon_0 = \nu \frac{d^2 K}{dy^2} - c_{\epsilon 2} \nu \left(\frac{dU}{dy} \right)^2 . \tag{30}$$

On incorporating the present finding into a $K - \epsilon$ model (Nisijima and Yoshizawa 1987) where the low Reynolds number effects are taken into account through the y^+ effect, it was confirmed that the dissipation rate at the wall of a channel, which is given by $\epsilon_0 + \epsilon_2$, increases by about fifty percent for $c_{\epsilon 2} = 0.12$, compared with the one obtained without considering the ϵ_2 effect [Nisijima (private communication)]. This fact brings a promising aspect to the improvement of ϵ computed from the current models. Because, introduction of a variety of R_t and y^+ effects into models have improved the prediction of the near-wall behaviors of K, the Reynolds stress, etc., but not ϵ so much, compared with the DNS results (Patel et al. 1987, Nagano and Hishida 1987, Mansour et al. 1989, Myong and Kasagi 1990). Very recently, Mansour et al. (1989) have pointed out that the near-wall ϵ calculated using the $K - \epsilon$ models combined with the eddy-viscosity-damping function is much closer to the DNS one, but still less than the latter. The present implication does not contradict their finding. For, our new effect incorporated into a $K - \epsilon$ model increases ϵ very near the wall, but it hardly affects ϵ in the region ($y^+ > 15$) where ϵ is proportional to y^{-1}. As a result, there is a possibility that on adopting accurate wall damping functions such as are suggested by the DNS database, the present new effect or ϵ_2 makes the computed ϵ closer to the DNS one through its positive contribution.

Finally, we add that the shortfall of the DIA concerning the Galilean invariance does not affect the dependence of various statistical properties, for instance, ϵ on the mean shear [see the appendix B of Yoshizawa (1984)].

References

[1] Hinze, J.O., 1975, Turbulence, p.673, McGraw-Hill, N.Y.

[2] Inage, S. (private communication)

[3] Kim, J., Moin, P., and Moser, R.D., 1987, Turbulence in fully developed channel flow at low Reynolds number, J. Fluid Mech. 177, 133.

[4] Mansour, N.N., Kim, J., and Moin, P., 1988, Reynolds-stress and dissipation-rate budgets in a turbulent channel flow, J. Fluid Mech. 194, 15.

[5] Mansour, N.N., Kim, J., and Moin, P., 1989, Near-wall $k-\epsilon$ turbulence modeling, AIAA J. 27, 1068.

[6] Moser, R.D. and Moin, P., 1987, The effects of curvature in wall-bounded turbulent flows, J. Fluid Mech. 175, 479.

[7] Myong, H.K. and Kasagi, N., 1990, A new approach to improvement of $k-\epsilon$ turbulence model for wall-bounded shear flows, JSME International J. II 33, 63.

[8] Nagano, Y. and Hishida, M., 1987, Improved form of the $k-\epsilon$ model for wall turbulent shear flows, J. Fluids Engineering 109, 156.

[9] Nisijima, S., and Yoshizawa, A., 1987, Turbulent channel and Couette flows using an anisotropic $k-\epsilon$ model, AIAA J. 25, 414.

[10] Nisijima, S. (private communication)

[11] Patel, V.C., Rodi, W., and Scheuerer, 1985, Turbulence models for near-wall low Reynolds number flows, AIAA J. 23, 1308.

[12] Shimomura, Y. (private communication)

[13] Spalart, P.S., 1988, Direct simulation of a turbulent boundary layer up to $R_\theta = 1400$, J. Fluid Mech. 187, 61.

[14] Yoshizawa, A., 1984, Statistical analysis of the deviation of the Reynolds stress from the eddy-viscosity representation, Phys. Fluids 27, 1377.

[15] Yoshizawa, A., 1987, Statistical modeling of a transport equation for the kinetic energy dissipation rate, Phys. Fluids 30, 628.

[16] Yoshizawa, A., 1988, Statistical modelling of passive-scalar diffusion in turbulent shear flows, J. Fluid Mech. 195, 541.

ADVANCES IN MODELING THE PRESSURE CORRELATION TERMS IN THE SECOND MOMENT EQUATIONS

Tsan-Hsing Shih and Aamir Shabbir

Center for Modeling of Turbulence and Transition
NASA Lewis Research Center
Cleveland, OH 44135

ABSTRACT

In developing turbulence models, different authors have proposed various model constraints in an attempt to make the model equations more general (or universal). Most recent of these are the realizability principle (Lumley 1978, Schumann 1977), linearity principle (Pope 1983), rapid distortion theory (Reynolds 1987) and material indifference principle (Speziale 1983). In this paper we will discuss several issues concerning these principles and will pay special attention to the realizability principle raised by Lumley (1978). Realizability (defined as the requirement of non-negative energy and Schwarz' inequality between any fluctuating quantities) is the basic physical and mathematical principle that any modeled equation should obey. Hence, it is the most universal, important and also the minimal requirement for a model equation to prevent it from producing unphysical results. In this paper we will describe in detail the principle of realizability, derive the realizability conditions for various turbulence models, and propose the model forms for the pressure correlation terms in the second moment equations. Detailed comparisons of various turbulence models (Launder et al 1975, Craft et al 1989, Zeman and Lumley 1976, Shih and Lumley 1985 and the one proposed here) with experiments and direct numerical simulations will be presented. As a special case of turbulence, we will also discuss the two-dimensional, two-component turbulence modeling.

1. Introduction

It is commonly accepted that the Navier-Stokes equation is the basic constraint governing turbulence. Hence, without any model approximations, the direct numerical simulation (DNS) should be an ideal approach to study all the aspects of turbulence. For example, using DNS we may study the dynamics of various turbulence structures; the mechanisms of energy transfer; and the interaction between different size of eddies; and we may also look at the various one-point and two-point statistics. Unfortunately, DNS is limited to relatively low Reynolds number and simpler flows due to a wide range of length and time scales that are present in a turbulent flow and due to today's computational limitations (both computer hardware and numerical methods). Therefore, at present, DNS is far from capable of computing practical engineering problems. For solving general engineering problems, turbulence modeling must be used in conjunction with the CFD. The history and the literature of turbulence modeling is described in detail by Lumley[1,2], Reynolds[3], Launder[4] and Rodi.[5] All the model equations were designed to approximate the Navier-Stokes equations. The model coefficients (or constants) have been tuned against some basic turbulent flows. The accuracy of any model equation (if it works) varies from flow to flow. In addition, we must use them with caution because any model equation will miss some of the physics or violate some of the properties of the Navier-Stokes equations because of the approximations made in these model equations. Results in such cases may be misleading in studying the unknown turbulence phenomena. More severely, some of the model equations may end up with unphysical results, such as negative turbulence kinetic energy etc.

For constructing turbulence models, various model constraints have been proposed by different authors in an attempt to make the model equations as general (or universal) as possible. Most recent of these are, the realizability principle (Schumann[6], Lumley[1]), linearity principle (Pope[7]), rapid distortion theory (Reynolds[8]) and material indifference principle (Speziale[9]). However, some of the above mentioned principles are not universal. For example, the material indifference is not valid for general turbulence in which the fluctuating velocities are three dimensional. The material indifference is valid only for the strict 2D-2C (two dimensional and two velocity component) turbulence which rarely occurs and probably does not exist in the physical world which is described by three dimensional space. Speziale et al [10], in their DNS experiments, found that rapid rotation had no influence on the isotropy of the 3D-3C homogeneous isotropic turbulence. So far, there is no evidence which shows a 3D-3C fluctuating velocity field will evolve to a **strict** 2D-2C state under any conditions. Therefore, the material indifference (or frame indifference) need not to be considered in 3D-3C turbulence modeling. In addition, as we will see later, for a strict 2D-2C turbulence the frame indifference principle will be automatically satisfied and does not provide any additional or new constraints for turbulence models. In another study, Reynolds[8] also showed that in 3D-3C turbulence modeling, at the 2D-2C limit, the frame indifference does not provide any new model constraints, and it also seems to be irrelevant to the modeling of the effect of the rotation on turbulence (see the RDT described later).

The principle of linearity is also not universal, as it holds only for passive scalars. As pointed out by Pope[7], since the governing equations for passive scalars are linear, the modeled equations for these quantities should possess the same property. However, some of the models for the pressure correlations are nonlinear. The calculations of slightly heated turbulent jets (in which the temperature is a passive scalar) with such models (Shih et al [11]) show that the violation of this principle does not cause problems since the results agreed well with the experimental data (also see Lumley [2]).

Rapid distortion theory (RDT) is a convenient tool for studying the asymptotic behavior of turbulence under the rapid distortion of the mean flow. For example, the effect of the mean flow (say strain rate) on turbulence. For some simple homogeneous flows RDT can provide analytical solutions (Reynolds[8]), hence enabling us to calculate the model terms of interest in the second moment equations. One may naturally ask if we can use these results to calibrate the model coefficients appearing in the corresponding model terms. Unfortunately, the answer is not very positive. RDT equations are valid only for very high mean strain rate or very low turbulent intensity and in both these cases the turbulence is in a very non-equilibrium state with the mean flow field. On the other hand the second order modeling is based on the assumption that the turbulence is more or less in an equilibrium state and is mainly characterized by one velocity and one length scale (Lumley[2]). Therefore, RDT results are probably not appropriate for tuning the coefficients of second order models. An interesting illustration of this point is that of Shih et al [12] who showed that a model constant in the linear form of a rapid pressure strain model (Launder et al[13]) determined by matching RDT results is much worse than the one determined from measurements. However this does not mean that RDT is useless in turbulence modeling. On the contrary, RDT can provide the asymptotic behavior of turbulence at the RDT limit. In many cases, we may obtain analytical expressions which explicitly show the relations between turbulence and the mean flow field. For example, the RDT results (Reynolds[14]) of pure mean rotation show a clear effect of the mean rotation on the anisotropy of turbulence ($b_{ij} = \overline{u_i u_j}/q^2 - \delta_{ij}/3$) through its second invariant. However, all the existing rapid pressure strain models (including the one which satisfies the material indifference at the 2D-2C limit) fail to exhibit any effect of rotation on the second invariant of b_{ij}. In this case, RDT will certainly guide us in removing this deficiency from the existing models.

Realizability (defined as the requirement of non-negative energy and Schwarz' inequality between any fluctuating quantities) is the basic physical and mathematical principle that any governing equation should obey. Hence, among all the above mentioned model constraints, realizability is the most universal, important and also the minimal requirement for the turbulence model equations to prevent them from producing unphysical results. Realizability can be applied to various turbulence quantities, for example, Reynolds stresses $\overline{u_i u_j}$, scalar fluxes $\overline{u_i \theta}$ and the triple correlations. For one-point second moment equations, realizability for $\overline{u_i u_j}$ and $\overline{u_i \theta}$ is the most important. In this paper (section 3; also see Shih and Lumley[23]) we will describe the principle of realizability, derive the realizability conditions for various model terms, and propose the model forms for pressure correlation terms in the second moment equations. The comparisons of various turbulence models with DNS and measurements will be presented in section 4. However, first we will discuss the 2D-2C turbulence.

2. Two-Dimensional Two-Component Turbulence

Reynolds and Grau[15] argued that if one component of velocity is suppressed and turbulence becomes 2C, the turbulence will not stay in the 2C state unless the turbulence also becomes two dimensional (2D). In the DNS experiment, Speziale et al[10] were expecting that the homogeneous isotropic turbulence would become anisotropic under the action of rapid rotation. However, their results did not show any such trend. Can a wall bounded inhomogeneous turbulence become a **strict** 2D-2C turbulence under a rapid rotation? In a three dimensional space, it seems unlikely that the instantaneous turbulent velocity field will become a **strict or exact** 2D-2C turbulence. However, if we confine a flow to a two dimensional space, like in the "thin soap sheet film", then it may be possible to obtain a 2D-2C turbulence. Numerically it is always possible to simulate such a special "turbulence".

In this section, we will derive a 2D-2C turbulence model for the rapid pressure correlation term in the Reynolds stress equations (also see Reynolds[8]) and will show that the frame indifference principle does not provide any model constraints.

The Reynolds stress equation can be written as

$$\frac{D}{Dt}\overline{u_i u_q} = \ldots + T_{iq} - 2\nu\overline{u_{i,k}u_{q,k}} \tag{2.1}$$

where

$$\frac{T_{iq}}{q^2} \equiv 2U_{p,j}[M_{ijpq} + M_{qjpi}] \tag{2.2}$$

corresponds to the rapid part of pressure correlation term, and $\overline{q^2} = \overline{u_i u_i}$. For homogeneous turbulence, M_{ijpq} can be expressed as

$$M_{ijpq} = \overline{q^2}\int_K \frac{k_p k_q}{k^2}\, E_{ij}(\mathbf{k})\, d^2\mathbf{k} \tag{2.3}$$

where k_i is the wave number vector and E_{ij} is the spectrum tensor.

As usual, we assume that M_{ijpq} is a function of the anisotropic tensor b_{ij}, and then the most general tensor form of M_{ijpq} (which satisfies the symmetry in indices) will be

$$\begin{aligned} M_{ijpq} &= a_1\delta_{ij}\delta_{pq} + a_2(\delta_{ip}\delta_{jq} + \delta_{iq}\delta_{jp}) + a_3\delta_{ij}b_{pq} + a_4\delta_{pq}b_{ij} \\ &+ a_5(\delta_{ip}b_{jq} + \delta_{iq}b_{jp} + \delta_{jp}b_{iq} + \delta_{jq}b_{ip}) \\ &+ a_6 b_{ij}b_{pq} + a_7(b_{ip}b_{jq} + b_{iq}b_{jp}) \end{aligned} \tag{2.4}$$

where,

$$\begin{aligned} b_{ij} &= \frac{\overline{u_i u_j}}{q^2} - \frac{1}{2}\delta_{ij} \\ \delta_{ii} &= 2 \\ II &= -b_{ij}b_{ij}/2 \end{aligned} \tag{2.5}$$

Because of the Cayley-Hamilton theorem in a 2D space, $b_{ij}^2 + II\delta_{ij} = 0$, the higher powers of b_{ij} are not required. Seven coefficients a_i are, in general, function of the invariants of b_{ij}. From the continuity and definition of M_{ijpq}, we have

$$\begin{aligned} M_{pjpq} &= 0 \\ M_{ijpp} &= b_{ij} + \delta_{ij}/2 \end{aligned} \tag{2.6}$$

These give us the following relations between the coefficients:

$$\begin{aligned} a_1 + 3a_2 - IIa_6 - IIa_7 &= 0 \\ a_3 + a_4 + 4a_5 &= 0 \\ 2a_1 + 2a_2 - 2IIa_7 &= 1/2 \\ 2a_4 + 4a_5 &= 1 \end{aligned} \tag{2.7}$$

The coefficients a_1, a_2, a_3 and a_4 can be solved in terms of a_5, a_6, and a_7:

$$\begin{aligned} a_1 &= 3/8 + II(a_7 - a_6/2) \\ a_2 &= -1/8 + IIa_6/2 \\ a_3 &= -1/2 - 2a_5 \\ a_4 &= 1/2 - 2a_5 \end{aligned} \tag{2.8}$$

Therefore, the tensor M_{ijpq} becomes

$$\begin{aligned} M_{ijpq} &= \frac{1}{8}(3\delta_{ij}\delta_{pq} - \delta_{ip}\delta_{jq} - \delta_{iq}\delta_{jp}) + \frac{1}{2}(\delta_{pq}b_{ij} - \delta_{ij}b_{pq}) \\ &+ a_6[b_{ij}b_{pq} - II\delta_{ij}\delta_{pq}/2 + II(\delta_{ip}\delta_{jq} + \delta_{iq}\delta_{jp})/2] \\ &+ a_7(b_{ip}b_{jq} + b_{iq}b_{jp} + II\delta_{ij}\delta_{pq}) \end{aligned} \tag{2.9}$$

where a_5 drops out because of the following zero tensor:

$$\delta_{ip}b_{jq} + \delta_{iq}b_{jp} + \delta_{jp}b_{iq} + \delta_{jq}b_{ip} - 2\delta_{ij}b_{pq} - 2\delta_{pq}b_{ij} \equiv 0 \tag{2.10}$$

Using (2.9), the Eq.(2.2) becomes

$$\frac{T_{iq}}{q^2} = S_{iq} + 4(a_6 + 2a_7)b_{ij}b_{pq}S_{pj} + 2\Omega_{qj}b_{ij} + 2\Omega_{ij}b_{jq} \tag{2.11}$$

where we have used following relation:

$$4IIS_{iq} + [4b_{iq}b_{jp} - 2(b_{ij}b_{pq} + b_{ip}b_{jq})]S_{pj} = 0 \tag{2.12}$$

The equation (2.10) and (2.12) can be easily checked in the principal axes.

For the 2D-2C turbulence, the turbulence equation should be frame-indifferent (Speziale [9]). If we write Eq.(2.1) in a non-inertial coordinate system (designated by *) which is rotating with an angular velocity Ω_k, we obtain

$$\begin{aligned} \frac{D}{Dt^*}\overline{u_i u_q}^* &= 2(U_{p,j})^*[M_{ijpq} + M_{qjpi}]^*\overline{q^2}^* + \nu\overline{u_{i,k}u_{q,k}}^* \ldots \\ &- 2\varepsilon_{jik}\Omega_k\overline{u_j u_q}^* - 2\varepsilon_{jqk}\Omega_k\overline{u_j u_i}^* \\ &- \varepsilon_{pjk}\Omega_k[M_{ijpq} + M_{qjpi}]^*\overline{q^2}^* \end{aligned} \tag{2.13}$$

In order for this equation to be frame-indifferent, the last two lines must cancel each other. This gives us the necessary constraint on the model form of M_{ijpq} and the result is

$$a_4 - a_3 = 1 \tag{2.14}$$

However, this condition is already satisfied by the coefficients (2.8) and hence the frame indifference principle does not provide any additional constraints. Therefore, in general there is still one undetermined coefficient $a_6 + 2a_7$ in (2.11) and it must be determined from experiment, direct numerical simulation or some other sources. However, we may explore its limiting value by using 1C condition. The condition of M_{ijpq} at the limit of $u_1 = 0$ gives $M_{1jpq} = 0$ for any j, p and q (because $E_{1j} = 0$). This condition gives

$$a_6 + 2a_7 = -1 \qquad if \;\; b_{11} = -1/2, \quad (or \;\; u_1 = 0) \tag{2.15}$$

However, we must note that in general 2D-2C flows, the value of $a_6 + 2a_7$ is unknown and not necessarily -1. The limiting value of the above coefficient can also be obtained by using the realizability conditions (3.4 a,b,c) described in the next section. Interestingly enough such an analysis also gives $a_6 + 2a_7 = -1$.

3. Three-Dimensional Three-Component Turbulence

It is commonly thought that the turbulence in a three dimensional space is always three dimensional (Tennekes and Lumley[16]) with three components of velocity. However, there are situations where the turbulence is considerably suppressed in a certain direction. For example when turbulence approaches a solid wall, the fluctuating velocity component normal to the wall will be mostly suppressed and the turbulence will become 3D-2C. However, during this process, all the turbulent energy components should remain non-negative and Schwarz' inequality should always be satisfied. This is the concept of realizability. In this section, we will discuss realizability in detail and discuss the model development of pressure correlation terms in the second moment equations.

3.1 Realizability Concept

In the second order closure, model equations for the turbulent stresses $\overline{u_i u_j}$ and scalar fluxes $\overline{f u_i}$ are proposed. Realizability for both of these quantities has been discussed in Lumley[1]. To see the realizability concept, let us consider any two fluctuating variables a and b, and their correlation matrix:

$$\begin{pmatrix} \overline{a^2} & \overline{ab} \\ \overline{ab} & \overline{b^2} \end{pmatrix}$$

The simple fact that $\overline{a^2} \geq 0$, $\overline{b^2} \geq 0$, and Schwarz's inequality $\overline{a^2}\,\overline{b^2} - \overline{ab}^2 \geq 0$ should be satisfied by any set of equations governing the evolution of $\overline{a^2}$, $\overline{b^2}$, and $\overline{ab}$ is called realizability. A set of equations violating this principle could produce unphysical results. We will show that these three realizability constraints translate to the condition that the eigenvalues of the correlation matrix should remain non-negative. We will now show that if any one of the above three in-equalities becomes an equality, then one of the eigenvalues of the matrix will vanish. Therefore, by ensuring these eigenvalues from becoming negative will automatically ensure $\overline{a^2}$, $\overline{b^2}$, $(\overline{a^2}\,\overline{b^2} - \overline{ab}^2)$ from becoming negative. This implies that while constructing governing equations for a, b and their correlations, we need to check only the eigenvalues of the correlation matrix. The non-negativity of the eigenvalues of the matrix can be easily seen from their expression:

$$\lambda_{1,2} = (1/2)\left\{(\overline{a^2} + \overline{b^2}) \pm [(\overline{a^2} + \overline{b^2})^2 - 4(\overline{a^2}\,\overline{b^2} - \overline{ab}^2)]^{1/2}\right\} \quad (3.1)$$

Eq. (3.1) shows that λ_2 will vanish if one of the following conditions occurs:

$$\overline{a^2} = 0,$$
$$\overline{b^2} = 0,$$
$$\overline{a^2}\,\overline{b^2} - \overline{ab}^2 = 0.$$

Now, let us take the derivative of (3.1) with respect to time t,

$$(\lambda_{1,2})_{,t} = (1/2)\{(\overline{a^2}_{,t} + \overline{b^2}_{,t}) \pm \frac{(\overline{a^2} + \overline{b^2})(\overline{a^2}_{,t} + \overline{b^2}_{,t}) - 2[(\overline{a^2})_{,t}\overline{b^2} + \overline{a^2}\,(\overline{b^2})_{,t} - 2\overline{ab}\,(\overline{ab})_{,t}]}{[(\overline{a^2} + \overline{b^2})^2 - 4(\overline{a^2}\,\overline{b^2} - \overline{ab}^2)]^{1/2}}\} \quad (3.2)$$

To preserve the positive λ_2, we must require

$$(\lambda_2)_{,t} \geq 0, \quad \text{if} \quad \lambda_2 = 0 \quad (3.3a)$$

or

$$\frac{[\overline{a^2}\,\overline{b^2} - \overline{ab}^2]_{,t}}{\overline{a^2} + \overline{b^2}} \geq 0, \quad \text{if} \quad \lambda_2 = 0 \quad (3.3b)$$

From this necessary condition, it is straight forward to show that

$$\overline{a^2}_{,t} \geq 0, \quad \text{if} \quad \overline{a^2} = 0 \quad (3.4a)$$

$$\overline{b^2}_{,t} \geq 0, \quad \text{if} \quad \overline{b^2} = 0 \tag{3.4b}$$

$$(\overline{a^2}\,\overline{b^2} - \overline{ab}^2)_{,t} \geq 0, \quad \text{if} \quad \overline{a^2}\,\overline{b^2} - \overline{ab}^2 = 0 \tag{3.4c}$$

It is clear that (3.4) is the necessary condition for the quantities $\overline{a^2}$, $\overline{b^2}$ and $\overline{a^2}\,\overline{b^2} - \overline{ab}^2$ to remain non-negative, therefore, the condition (3.3a) is the necessary condition for realizability. To find the sufficient condition for realizability, we must examine the higher order derivative of the eigenvalue when it vanishes. For example, we may require

$$(\lambda_2)_{,tt} \geq 0, \quad \text{if} \quad \lambda_2 = 0. \tag{3.5}$$

For more details on this aspect see the review paper by Lumley[2].

Similar arguments can be applied to the turbulent stress tensor $\overline{u_i u_j}$ and to the correlation matrix formed with a scalar f and velocities u_i. Realizability for tensor $\overline{u_i u_j}$ is straight forward, that is, its eigenvalues must be positive, which will ensure positive turbulent energy components and the Schwarz inequity between fluctuating velocities. Realizability for the correlation matrix of scalar and velocity is described by Eq. (3.1) and (3.3) and is not as simple as for the Reynolds stress. Lumley has suggested a more general quantity D_{ij} consisting of turbulent stress and flux: $D_{ij} = \overline{f^2}\,\overline{u_i u_j} - \overline{f u_i}\,\overline{f u_j}$. It can be shown that realizability for the correlation matrix between the velocity and the scalar will be ensured provided that the eigenvalues of D_{ij} are positive.

It is interesting to note that if $u_\alpha \to 0$ then the α direction will become one of the principal axes of $\overline{u_i u_j}$, and $\overline{u_\alpha u_\alpha}$ (no summation on Greek indices) becomes one of the eigenvalue of $\overline{u_i u_j}$. At the same time, $u_\alpha \to 0$ will make $D_{\alpha\alpha} \to 0$, and the α direction also becomes one of the principal axes of D_{ij}. $D_{\alpha\alpha}$ is then a vanishing eigenvalue of D_{ij}. However $D_{\alpha\alpha}$ will also vanish without $u_\alpha \to 0$ if the perfect correlation between f and u_α is reached. Therefore, we conclude that if one of the eigenvalues of $\overline{u_i u_j}$, say $\overline{u_\alpha u_\alpha}$, vanishes, then $D_{\alpha\alpha}$ will become a vanishing eigenvalue of the tensor D_{ij}. However, the reverse is not true i.e. if $D_{\alpha\alpha} \to 0$, it does not mean that $\overline{u_\alpha u_\alpha}$ (or $\overline{f^2}$) must vanish. This is important in the modeling of scalar turbulence, as it indicates that realizability based on $\overline{u_i u_j}$ alone is not enough and we must also consider the joint realizability based on

D_{ij} to ensure that the model equations do not produce unphysical results.

3.2 Realizability Conditions for Second Order Closure Models

The ideas discussed above can now be applied to the second moment equations to obtain the realizability conditions. The general form of the model equations for the second moments can be written as follows

$$\begin{aligned}\frac{D}{Dt}\overline{u_i u_j} = &- [\ \overline{u_i u_j u_k} - C(\overline{q^2 u_i}\delta_{jk} + \overline{q^2 u_j}\delta_{ik})\ ]_{,k} \\ &- (\overline{u_i u_k}\ U_{j,k} + \overline{u_j u_k}\ U_{i,k}) + 2U_{p,q}(X_{pjqi} + X_{piqj}) \\ &- \beta_k(Y_{kij} + Y_{kji}) - (\Phi_{ij} + 2\delta_{ij}/3)\epsilon \\ &+ \beta_i\overline{\theta u_j} + \beta_j\overline{\theta u_i} \end{aligned} \tag{3.6}$$

$$\begin{aligned}\frac{D}{Dt}\overline{\theta u_i} = &- [\overline{\theta u_i u_k} - C'\overline{\theta q^2}\delta_{ik}]_{,k} - (\overline{\theta u_k}U_{i,k} + \overline{u_i u_k}\Theta_{,k}) + \beta_i\overline{\theta^2} \\ &+ 2U_{j,k}X_{ijk} - \beta_k Y_{ik} - \Phi_i\epsilon/\overline{q^2}\end{aligned} \tag{3.7}$$

$$\frac{D}{Dt}\overline{\theta^2} = - (\overline{\theta^2 u_k})_{,k} - 2\overline{\theta u_k}\ \Theta_{,k} - 2\epsilon_\theta \tag{3.8}$$

where we have followed Lumley[1] for the pressure transport model and in combining the anisotropic part of the dissipation terms with the slow part of the pressure correlations. The molecular diffusion terms in the above equations have been neglected. The definitions of the rapid pressure correlation terms X_{pjqi} and X_{ijk}; the buoyancy terms Y_{kij} and Y_{ik}; and the slow terms Φ_{ij} and Φ_i are listed in appendix A. In the above equations β_i is the buoyancy vector, Θ is the mean scalar and θ is the fluctuating scalar.

To impose realizability into above equations we may write the equations of $\overline{u_i u_j}$ and D_{ij} in their principal axes, i.e., the equations for $\overline{u_\alpha u_\alpha}$ and $D_{\alpha\alpha}$. Then in order to ensure that the eigenvalues of $\overline{u_\alpha u_\alpha}$ and $D_{\alpha\alpha}$ remain positive during the evolution of turbulence, following constraints must be satisfied

$$\begin{aligned}\frac{D}{Dt}\overline{u_\alpha u_\alpha} \to 0 \ \ if \ \ \overline{u_\alpha u_\alpha} \to 0 \\ \frac{D}{Dt}D_{\alpha\alpha} \to 0 \ \ if \ \ D_{\alpha\alpha} \to 0\end{aligned} \tag{3.9}$$

$$\frac{D^2}{Dt^2}\overline{u_\alpha u_\alpha} \geq 0 \;\; if \;\; \overline{u_\alpha u_\alpha} \to 0$$
$$\frac{D^2}{Dt^2}D_{\alpha\alpha} \geq 0 \;\; if \;\; D_{\alpha\alpha} \to 0 \qquad (3.10)$$

Substituting (3.6)-(3.8) into (3.9) we obtain the following necessary realizability conditions for X_{pjqi}, X_{ijk}, Y_{ikj} and Y_{ik}:

$$U_{p,q}X_{p\alpha q\alpha} \to 0$$
$$if \;\; \overline{u_\alpha^2} \to 0 \quad (in\ p.a.\ of\ R_{ij})$$

$$U_{p,q}(\overline{\theta^2}X_{p\alpha q\alpha} - \overline{\theta u_\alpha}X_{\alpha pq}) \to 0$$
$$if \;\; D_{\alpha\alpha} \to 0 \;\; (\overline{u_\alpha^2} \neq 0) \quad (in\ p.a.\ of\ D_{ij})$$

$$U_{p,q}\overline{\theta u_\alpha}X_{\alpha pq} \to 0$$
$$if \;\; D_{\alpha\alpha}, \overline{u_\alpha^2} \to 0 \quad (in\ p.a.\ of\ D_{ij}, R_{ij})$$

$$Y_{\alpha\alpha\alpha} - \overline{\theta u_\alpha} \to 0$$
$$if D_{\alpha\alpha}, \overline{u_\alpha^2} \to 0 \quad (in\ p.a.\ of\ D_{ij}, R_{ij})$$

$$Y_{\gamma\alpha\alpha} \;\; (\gamma \neq \alpha) \to 0$$
$$if \;\; D_{\alpha\alpha}, \overline{u_\alpha^2} \to 0 \quad (in\ p.a.\ of\ D_{ij}, R_{ij})$$

$$\overline{\theta^2}Y_{k\alpha\alpha} - \overline{\theta u_\alpha}Y_{k\alpha} \;(any\ k) \to 0$$
$$if \;\; D_{\alpha\alpha} \to 0 \;\; (\overline{u_\alpha^2} \neq 0) \quad (in\ p.a.\ of\ D_{ij})$$

$$\overline{\theta u_\alpha}Y_{\alpha\alpha} - \overline{\theta^2}\;\overline{\theta u_\alpha} \to 0$$
$$if \;\; D_{\alpha\alpha}, \overline{u_\alpha^2} \to 0 \quad (in\ p.a.\ of\ D_{ij}, R_{ij})$$

$$\overline{\theta u_\alpha}Y_{\gamma\alpha} \;(\gamma \neq \alpha) \to 0$$
$$if \;\; D_{\alpha\alpha}, \overline{u_\alpha^2} \to 0 \quad (in\ p.a.\ of\ D_{ij}, R_{ij})$$

$$(\Phi_{\alpha\alpha} + 2/3)\epsilon \to F^a$$

$$if \quad \overline{u_\alpha^2} \;\rightarrow 0 \qquad (in\ p.a.\ of\ R_{ij})$$

$$2\overline{\theta u_\alpha}\Phi_\alpha\epsilon/\overline{q^2} - \overline{\theta^2}(\Phi_{\alpha\alpha} + 2/3)\epsilon - 2\epsilon_\theta\overline{u_\alpha^2} \quad \rightarrow F_d^b$$
$$if \quad D_{\alpha\alpha} \rightarrow 0 \; (\overline{u_\alpha^2} \neq 0) \qquad (in\ p.a.\ of\ D_{ij})$$

$$2\overline{\theta u_\alpha}\Phi_\alpha\epsilon/\overline{q^2} - 2\epsilon_\theta\overline{u_\alpha^2} + \overline{\theta^2}F^a \quad \rightarrow F_d^b$$
$$if \quad D_{\alpha\alpha},\; \overline{u_\alpha^2} \;\rightarrow 0 \qquad (in\ p.a.\ of\ D_{ij}, R_{ij})$$

$$(\overline{u_\alpha^2 u_k} - 2C\overline{q^2 u_\alpha}\delta_{\alpha k})_{,k} \quad \rightarrow \quad 0$$
$$if \quad \overline{u_\alpha^2} \rightarrow 0 \qquad (in\ p.a.\ of\ R_{ij})$$

$$\begin{aligned}&2\overline{\theta u_\alpha}(\overline{\theta u_\alpha u_k} - C'\overline{q^2\theta}\;\delta_{\alpha k})_{,k}\\&-\overline{\theta^2}(\overline{u_\alpha^2 u_k} - 2C\overline{q^2 u_\alpha}\delta_{\alpha k})_{,k} - \overline{u_\alpha^2}(\overline{\theta u_k})_{,k} \quad \rightarrow 0\end{aligned}$$
$$if \quad D_{\alpha\alpha} \rightarrow 0 \; (\overline{u_\alpha^2} \neq 0) \qquad (in\ p.a.\ of\ D_{ij})$$

$$2\overline{\theta u_\alpha}(\overline{\theta u_\alpha u_k} - C'\overline{q^2\theta}\;\delta_{\alpha k})_{,k} - \overline{u_\alpha^2}(\overline{\theta u_k})_{,k} \quad \rightarrow 0$$
$$if \quad D_{\alpha\alpha}, \overline{u_\alpha^2} \rightarrow 0 \qquad (in\ p.a.\ of\ D_{ij}, R_{ij})$$

These constraints will be used in the next section in constructing the models for pressure-correlations.

3.3 Modeling of the Rapid Pressure-Correlation Terms

First let us look at the rapid terms associated with the mean strain. The symmetry property of these terms requires

$$X_{pjqi} = X_{jpqi}, \quad X_{pjqi} = X_{pjiq}, \quad X_{ijk} = X_{jik} \tag{3.11}$$

If we assume X_{pjqi} is a function of $\overline{u_i u_j}$ only and X_{ijk} is a function of both $\overline{u_i u_j}$ and $\overline{\theta u_i}$, but linear in $\overline{\theta u_i}$, then the most general forms of X_{pjqi} and X_{ijk} satisfying the above symmetry constraints are

$$\begin{aligned}X_{pjqi}/\overline{q^2} &= \alpha_1\delta_{qi}\delta_{pj} + \alpha_2(\delta_{pq}\delta_{ij} + \delta_{qj}\delta_{pi}) + \alpha_3\delta_{qi}b_{pj} + \alpha_4\delta_{pj}b_{qi}\\&+ \alpha_5(\delta_{pq}b_{ij} + \delta_{ij}b_{pq} + \delta_{jq}b_{pi} + \delta_{pi}b_{qj}) + \alpha_6\delta_{qi}b_{pj}^2 + \alpha_7\delta_{pj}b_{qi}^2\end{aligned}$$

$$
\begin{aligned}
&+ \alpha_8(\delta_{pq}b^2_{ij} + \delta_{ij}b^2_{pq} + \delta_{jq}b^2_{pi} + \delta_{pi}b^2_{qj}) + \alpha_9 b_{qi}b_{pj} \\
&+ \alpha_{10}(b_{pq}b_{ij} + b_{qj}b_{pi}) + \alpha_{11}b_{qi}b^2_{pj} + \alpha_{12}b_{pj}b^2_{qi} \\
&+ \alpha_{13}(b_{pq}b^2_{ij} + b_{ij}b^2_{pq} + b_{qj}b^2_{pi}) + b_{pi}b^2_{qj}) \\
&+ \alpha_{14}b^2_{qi}b^2_{pj} + \alpha_{15}(b^2_{pq}b^2_{ij} + b^2_{qj}b^2_{pi})
\end{aligned} \tag{3.12}
$$

$$
\begin{aligned}
X_{ikj} = {} & \beta_1\delta_{ik}\overline{\theta u_j} + \beta_2(\delta_{ij}\overline{\theta u_k} + \delta_{jk}\overline{\theta u_i}) + \beta_3 b_{ik}\overline{\theta u_j} \\
&+ \beta_4(b_{ij}\overline{\theta u_k} + b_{jk}\overline{\theta u_i}) + \beta_5(\delta_{ij}b_{kp} + \delta_{kj}b_{ip})\overline{\theta u_p} \\
&+ \beta_6\delta_{ik}b_{jp}\overline{\theta u_p} + \beta_7 b_{ik}b_{jp}\overline{\theta u_p} + \beta_8(b_{ij}b_{kp} + b_{kj}b_{ip})\overline{\theta u_p} \\
&+ \beta_9 b^2_{ik}\overline{\theta u_j} + \beta_{10}(b^2_{ij}\overline{\theta u_k} + b^2_{jk}\overline{\theta u_i}) + \beta_{11}\delta_{ik}b^2_{jp}\overline{\theta u_p} \\
&+ \beta_{12}(\delta_{ij}b^2_{kp} + \delta_{jk}b^2_{ip})\overline{\theta u_p} + \beta_{13}b^2_{ik}b_{jp}\overline{\theta u_p} + \beta_{14}b_{ik}b^2_{jp}\overline{\theta u_p} \\
&+ \beta_{15}(b^2_{ij}b_{kp} + b^2_{kj}b_{ip})\overline{\theta u_p} + \beta_{16}(b_{ij}b^2_{kp} + b_{kj}b^2_{ip})\overline{\theta u_p} \\
&+ \beta_{17}b^2_{ik}b^2_{jp}\overline{\theta u_p} + \beta_{18}(b^2_{ij}b^2_{kp} + b^2_{kj}b^2_{ip})\overline{\theta u_p}
\end{aligned} \tag{3.13}
$$

Where α's are, in general, functions of the second and third invariants (i.e. II and III) of the anisotropic tensor b_{ij} and β's are functions of the invariants formed by b_{ij} and $\overline{\theta u_i}$ in addition to II and III. From the definition of the rapid terms and incompressibility, following properties are obtained:

$$
X_{ppqi} = \overline{u_q u_i}, \qquad X_{pkqk} = 0, \qquad X_{iik} = \overline{\theta u_k}, \qquad X_{ikk} = 0 \tag{3.14}
$$

Using these properties and the realizability conditions given in the last section we can determine the limiting values of all the model coefficients provided we ignore the terms with the fourth powers of b_{ij} (which is a reasonable assumption since the magnitude of b_{ij} is always less than unity). The final expressions of the X_{pjqi} and X_{ijk} reduce to the surprisingly simple forms and are given by

$$
\begin{aligned}
X_{pjqi}/\overline{q^2} = {} & \frac{1}{30}(4\delta_{pj}\delta_{qi} - \delta_{pq}\delta_{ij} - \delta_{qj}\delta_{pi}) - \frac{1}{3}(\delta_{qi}b_{pj} - \delta_{pj}b_{qi}) \\
&+ a_1(\delta_{pq}b_{ij} + \delta_{ij}b_{pq} + \delta_{qj}b_{pi} + \delta_{pi}b_{qj} - \frac{11}{3}\delta_{qi}b_{pj} - \frac{4}{3}\delta_{pj}b_{qi}) \\
&+ a_2(2\delta_{pj}b^2_{qi} - 3b_{pq}b_{ij} - 3b_{qj}b_{pi} + b_{qi}b_{pj})
\end{aligned} \tag{3.15}
$$

$$
\begin{aligned}
X_{ijk} = {} & \frac{2}{5}\delta_{ij}\overline{\theta u_k} - \frac{1}{10}(\delta_{ik}\overline{\theta u_j} + \delta_{jk}\overline{\theta u_i}) \\
&+ C_{D1}b_{ij}\overline{\theta u_k} \\
&+ C_{D2}(b_{ik}\overline{\theta u_j} + b_{jk}\overline{\theta u_i}) + C_{D3}\delta_{ij}b_{kl}\overline{\theta u_l}
\end{aligned} \tag{3.16}
$$

where the limiting values of the coefficients are

$$a_1 = -\frac{1}{10},\ a_2 = \frac{1}{10},\ C_{D1} = \frac{1}{10},\ C_{D2} = -\frac{3}{10},\ C_{D3} = \frac{1}{5} \quad (3.17)$$

Last line in (3.15) and (3.16) are the non-linear contributions, and if neglected, the linear models used by various other workers will be recovered. It is important to note that the above values of the coefficients a_1, a_2, C_{D1}, C_{D2}, and C_{D3} are their limiting values at the realizability limit, i.e. when $\overline{u_\alpha u_\alpha}$, $D_{\alpha\alpha} \to 0$. For general turbulent flows $\overline{u_\alpha u_\alpha}$ and $D_{\alpha\alpha}$ are not zero and hence the values of the coefficients may deviate from their limiting values. As was pointed out before that these are in general functions of the the invariants II and III. Some guidance can be obtained by inspecting the following two useful parameters (see Lumley [1] and Shih and Lumley [17])

$$F = 1 + 27III + 9II \tag{3.18}$$

$$F_d = 9\ d_{ii}^3 - \frac{27}{2} d_{ii}^2 + \frac{9}{2} \tag{3.19}$$

where

$$II = -\frac{1}{2} b_{ii}^2, \quad III = \frac{1}{3} b_{ii}^3$$

$$d_{ij} = \frac{\overline{f^2}\ \overline{u_i u_j} - \overline{f u_i}\ \overline{f u_j}}{\overline{f^2}\ \overline{u_i u_i} - \overline{f u_i}\ \overline{f u_i}}$$

It can be shown that both F and F_d are bounded between 0 and 1, and particularly

$$F \to 0 \ \ when \ \ \overline{u_\alpha u_\alpha} \to 0$$

$$F_d \to 0 \ \ when \ \ D_{\alpha\alpha} \to 0$$

By using this information, it is convenient to write

$$a_1 = -\frac{1}{10}(1 + AF^\alpha) \tag{3.20}$$

$$a_2 = \frac{1}{10}(1 + BF^\alpha) \tag{3.21}$$

$$C_{D1} = \frac{1}{10} + C_1 F_d^\alpha \tag{3.22}$$

$$C_{D2} = \frac{-3}{10} + C_2 F_d^\alpha \tag{3.23}$$

$$C_{D3} = \frac{1}{5} + C_3 F_d^\alpha \tag{3.24}$$

where A, B, C_1, C_2 and C_3 are adjustable constants but α is not as arbitrary as it might seem at first look. The condition (3.10) suggests $\alpha = \frac{1}{2}$ (see Lumley [2] or Shih and Lumley [17]). In the limiting case these coefficients reach their limiting values. Shih et al. [18,19] took $A = 0.8$ and $B = 0.0$ which fit the DNS data quite well. However, it seems more judicious to use homogeneous shear flow experiment to obtain the values of these coefficients. Shih et al. [11] set $C_1 = C_2 = C_3 = 0$ in their computations. The present study indicates that the full form of C_{D1}, C_{D2} and C_{D3} should be used. The experiment of Tavalouris and Corrsin[27] suggests $C_1 = 1.8$, $C_2 = -1.8$, and $C_3 = 4.5$.

Now returning to the modeling of the buoyancy terms Y_{kij} and Y_{ik}, we assume that these are functions of the anisotropic tensor and heat flux. Their minimum forms required to satisfy realizability are as follows

$$\begin{aligned} Y_{ikj} = {} & \beta_1 \delta_{ik} \overline{\theta u_j} + \beta_2 (\delta_{ij} \overline{\theta u_k} + \delta_{jk} \overline{\theta u_i}) + \beta_3 b_{ik} \overline{\theta u_j} \\ & + \beta_4 (b_{ij} \overline{\theta u_k} + b_{jk} \overline{\theta u_i}) + \beta_5 (\delta_{ij} b_{kp} + \delta_{kj} b_{ip}) \overline{\theta u_p} \\ & + \beta_6 \delta_{ik} b_{jp} \overline{\theta u_p} + \beta_7 b_{ik} b_{jp} \overline{\theta u_p} + \beta_8 (b_{ij} b_{kp} + b_{kj} b_{ip}) \overline{\theta u_p} \\ & + \beta_9 b^2_{ik} \overline{\theta u_j} + \beta_{10} (b^2_{ij} \overline{\theta u_k} + b^2_{jk} \overline{\theta u_i}) + \beta_{11} \delta_{ik} b^2_{jp} \overline{\theta u_p} \\ & + \beta_{12} (\delta_{ij} b^2_{kp} + \delta_{jk} b^2_{ip}) \overline{\theta u_p} \end{aligned} \tag{3.25}$$

$$\begin{aligned} Y_{ik} = {} & \gamma_1 \overline{\theta^2} \delta_{ik} + \gamma_2 \overline{\theta^2} b_{ik} + \gamma_3 \overline{\theta u_i}\, \overline{\theta u_k} + \gamma_4 (\overline{\theta u_k} b_{ip} \overline{\theta u_p} + \overline{\theta u_i} b_{kp} \overline{\theta u_p}) \\ & + \gamma_5 \overline{\theta^2} b^2_{ik} + \gamma_6 b_{ip} b_{kq} \overline{\theta u_p}\, \overline{\theta u_q} \\ & + \gamma_7 (b^2_{ip} \overline{\theta u_p}\, \overline{\theta u_K} + b^2_{kp} \overline{\theta u_p}\, \overline{\theta u_i}) \end{aligned} \tag{3.26}$$

The above forms already satisfy the following symmetry conditions

$$Y_{kij} = Y_{ikj}, \quad Y_{ik} = Y_{ki}$$

By definition, and incompressibility, Y_{ik} have the following properties:

$$Y_{kkj} = \overline{\theta u_j}, \quad Y_{kk} = \overline{\theta^2}$$

Using these and the realizability conditions given in the last section, all except one coefficient can be determined exactly. Their final forms are

$$
\begin{aligned}
\beta_1 =& \frac{4}{3}(1-\beta_5)\\
\beta_2 =& \frac{(\beta_5 - 1)}{3}\\
\beta_3 =& \frac{(1 - 11\beta_5)}{3}\\
\beta_4 =& \beta_5 - 1\\
\beta_5 =& \frac{-\,(12II+7)\underline{r^2}}{6II\overline{\theta^2}\,\overline{q^2} - 10\underline{r^2} - 36II\underline{rbr}}\\
\beta_6 =& \frac{(2-4\beta_5)}{3}\\
\beta_7 =& \frac{-7}{6II} + \beta_5\left(\frac{5}{3II} - 1\right)\\
\beta_8 =& 3\beta_5\\
\beta_9 =& \frac{7}{6II} - \beta_5\left(\frac{5}{3II}\right)\\
\beta_{10} =& 0\\
\beta_{11} =& -\,2\beta_5\\
\beta_{12} =& 0\\
\gamma_1 =& \frac{(6\underline{r^2} + 10\overline{\theta^2}\,\overline{q^2})}{9\overline{\theta^2}\,\overline{q^2}}\\
& -\beta_5\,\frac{[18(\underline{r^2} + \overline{\theta^2}\,\overline{q^2})\underline{rbr} + \underline{r^2}(7\overline{\theta^2}\,\overline{q^2} - 15\underline{r^2}) + (36II-10)(\overline{\theta^2}\,\overline{q^2})^2]}{9\overline{\theta^2}\,\overline{q^2}(\underline{r^2} - \overline{\theta^2}\,\overline{q^2})}\\
& +\gamma_6\,\frac{(\overline{q^2}/9)(3\underline{rbr} + 6II\underline{r^2} - 2II\overline{\theta^2}\,\overline{q^2})}{(\underline{r^2} - \overline{\theta^2}\,\overline{q^2})}\\
\gamma_2 =& \frac{\underline{r^2}(6II+28) - \overline{\theta^2}\,\overline{q^2}(6II+7)}{18II(\underline{r^2} - \overline{\theta^2}\,\overline{q^2})} - \frac{7(\underline{r^2})^2}{(6II)\overline{\theta^2}\,\overline{q^2}(\underline{r^2} - \overline{\theta^2}\,\overline{q^2})}
\end{aligned}
$$

$$
\begin{aligned}
& -\beta_5\,\Bigg[\frac{\underline{r^2}(12II+20) + 108II\underline{rbr} + \overline{\theta^2}\,\overline{q^2}(108II^2 - 21II - 5)}{9II(\underline{r^2} - \overline{\theta^2}\,\overline{q^2})}\\
& -\frac{5(3II+1)(\underline{r^2})^2}{3II\overline{\theta^2}\,\overline{q^2}(\underline{r^2} - \overline{\theta^2}\,\overline{q^2})}\Bigg]\\
& +\gamma_6\,\Bigg[\frac{\underline{r^2}\,\overline{q^2}(18II+4) + 9\overline{q^2}\,\underline{rbr} - \overline{\theta^2}\,\overline{q^2}(6II+1)\overline{q^2}}{9(\underline{r^2} - \overline{\theta^2}\,\overline{q^2})}\\
& -\frac{\underline{r^2}}{3\theta^2}\,\frac{\underline{r^2}}{(\underline{r^2} - \overline{\theta^2}\,\overline{q^2})}\Bigg]
\end{aligned}
$$

$$\gamma_3 = \frac{-2}{\overline{q^2}} + \beta_5 \frac{12\underline{rbr} - 5\underline{r^2} + (12II - 1)\overline{\theta^2}\ \overline{q^2}}{\overline{q^2}(\underline{r^2} - \overline{\theta^2}\ \overline{q^2})}$$
$$- \gamma_6 \frac{6II\underline{r^2} + 3\underline{rbr} - 2II\overline{\theta^2}\ \overline{q^2}}{3(\underline{r^2} - \overline{\theta^2}\ \overline{q^2})}$$
$$\gamma_4 = \frac{-3\beta_5}{\overline{q^2}}$$
$$\gamma_5 = \frac{7}{6II} + \beta_5 \frac{(18II - 5)}{3II} + \gamma_6 \left(\frac{\overline{q^2}}{3} - \frac{\underline{r^2}}{\overline{\theta^2}}\right)$$
$$\gamma_6 = undetermined\ coefficient$$
$$\gamma_7 = 0$$

where $\underline{r^2} = \overline{\theta u_i}\ \overline{\theta u_i}$, $\underline{rbr} = \overline{\theta u_i}\ b_{ij}\ \overline{\theta u_j}$. These expressions may seem a bit complicated but they ensure joint realizability. Since γ_6 is undetermined it was taken as zero in the calculations to be presented latter. As an illustration, Appendix B lists the values of all the above coefficients as calculated from the buoyant plume experiment[29].

3.4 Slow Terms

Lumley[1] has suggested the following expression for the slow term of the pressure-strain correlation

$$\Phi_{ij} = \beta b_{ij} + \gamma(b_{ik}b_{kj} + 2II\delta_{ij}/3) \tag{3.27}$$

where β and γ are, in general, functions of the invariants II and III, and the Reynolds number. Most of the workers have used Rotta's hypothesis which is obtained by ignoring the nonlinear term. In LRR model, the β is taken as a constant, whereas Lumley [1] proposed an expression involving II, III and the turbulence Reynolds number. This expression for β was calibrated from experiments in which invariant III was negative. Gence and Mathieu[20] and Choi[21] experimentally studied the return to isotropy of homogeneous flows and found that the return to isotropy is slower when the invariant III is positive. Choi and Lumley [22] suggested keeping the complete form of (3.27) and using their experiments proposed new forms of β and γ. We will use both Lumley [1] and Choi and Lumley[22] models with the presently proposed models for the rapid term to see how the combinations fair.

For the slow part of the pressure temperature-gradient correlation Shih and Lumley [23] proposed

$$\Phi_i = f_1 \overline{\theta u_i} + f_2 b_{ik} \overline{\theta u_k} \tag{3.28}$$

Again most workers ignore the second term and take f_1 as a constant. Shih and Lumley[17] and Shih et al. [11] use a variable form of the f_1 given by

$$f_1 = \frac{\beta}{2} + \frac{\epsilon_\theta}{\overline{\theta^2}} \frac{\overline{q^2}}{\epsilon} - \frac{(\beta - 2) II_d/6}{(II_d/3 + b_{ij} d_{ij}^2 - I_d b_{ij} d_{ij})} + H F_d^{1/2} \tag{3.29}$$

and take f_2 as zero. This expression is based on the joint realizability principle and the parameter H is given in Shih et al.[11] and β is the same as proposed by Lumley[1]. Both of these functions are given in Appendix C. In the present study, we apply the joint realizability condition keeping the full non-linear model for the pressure strain to obtain

$$\begin{aligned} f_1 =& \frac{\beta}{2} + \frac{\epsilon_\theta}{\overline{\theta^2}} \frac{\overline{q^2}}{\epsilon} - \frac{(\beta + \gamma II - 2) II_d/6}{(II_d/3 + b_{ij} d_{ij}^2 - I_d b_{ij} d_{ij})} \\ &+ \frac{\gamma}{2} \frac{(-II\, II_d + b_{ij}^2 d_{ij}^2 - b_{ij}^2 I_d d_{ij})}{(II_d/3 + b_{ij} d_{ij}^2 - I_d b_{ij} d_{ij})} + H F_d^{1/2} \end{aligned} \tag{3.30}$$

where β and γ are the same coefficients as appearing in (3.27). For $\gamma = 0$, Shih and Lumley [17] model is recovered. In the present study we used the β and γ as given by Choi and Lumley[22] and these are reproduced in Appendix C.The above expression seems to give better agreement with the experiments. In the experiments analysed[27,29], the value of f_1 varied from 6.4 - 8.0. The values used by Zeman and Lumley[24], $f_1 = 7.5$, and Newman et. al. [25], $f_1 = 6.6$, fall within this range.

3.5 Other Models

Other researchers approach the models for pressure correlation terms in a similar but slightly different way. Craft et al. [26] have proposed non-linear models for these correlations. They start with the similar form of the models as Shih and Lumely [23], but apply the realizability based only on $\overline{u_\alpha u_\alpha} \to 0$. Their models do not satisfy joint realizability and, therefore, are not fully realizable when a scalar is present. If we impose these conditions on their models, their coefficients must be changed such that their models will reduce to the same form as proposed here. Reynolds [8] obtained a set of model constraints using 2D-2C conditions on X_{pjqi} which also ensure realizability. His model keeps fourth powers of b_{ij} and , therefore, is more general but at the same time is much more complicated. The comparison between his model and the DNS data did not show significant improvement over the other models.

4. Comparison between Models and Experiments

For model evaluation we will use the homogeneous shear flow experiment of Tavoularis and Corrsin [27], DNS of the same flow by Rogers et al. [28] and the buoyant plume experiment of Shabbir and George[29] and Shabbir[31]. Note that for the homogeneous shear flow the buoyancy effects are negligible. For the experiments the pressure terms were obtained as the closing terms in the balances for the Reynolds stress and heat flux equations. The pressure correlations obtained in this fashion will certainly have some error but this is the closest we can come in obtaining these from the experiment. Also a direct comparison between these and the models is still a better approach in establishing their accuracy rather than solving the differential equations and then comparing the results with the experiment.

4.1 Pressure Strain Correlation

First we look at the comparison between the pressure strain models and the experiment of Tavoularis and Corrsin [27]. The velocity field of this experiment is essentially the same as that of Harris et al.[30] and, therefore, the results presented here are applicable to that experiment as well. Figure (1) shows a comparison between the experiment and five different models. The two complete linear models are Launder et al.[13] (LRR) and Zeman and Lumley [24] and these are of the exact same form except the model coefficients. Of the two, Zeman and Lumley model seems to give a better overall agreement with the data. Shih and Lumley[23] model uses linear form for the slow term with β given by Lumley[1] and is off by a large margin. This is because β of Lumley[1] was calibrated against flows with negative III whereas for the experiment under consideration this invariant is positive. When instead Choi and Lumley[22] model is used for the slow term all the pressure strain components are predicted within experimental accuracy. Craft et al.[26] also does a good job in predicting all the components. This is to be expected since the coefficients in their model are determined from the homogeneous shear flow experiment.

Figure (2) compares the slow term models with the DNS[28] results of the homogeneous shear flow. Lumley's[1] linear and the Choi and Lumley [22] nonlinear model give the best agreement with the data. Recall that these two models gave very different results for the experiment of Tavoularis and Corrsin[27]. Unlike that experiment, the Reynolds number in the DNS is much smaller. Both of these models incorporate the similar low Reynolds number behavior and, therefore, give similar overall results. Both the Zeman and Lumley[24] and LRR[13] model are off the data. The non-linear model of Craft et al.[26] performs well only for the Φ_{xx} and Φ_{xy} components. Figure (3) compares the three different models for the rapid pressure term with the DNS of Rogers et. al.[28]. Of the three Shih and Lumley[23] non-linear model gives the best agreement. The LRR[13] model performs reasonably only for the Π_{zz}^{rapid} and Π_{rz}^{rapid} components. We also note that the Craft et. al. model performs better than the LRR model.

Figure (4) compares the total pressure-strain models for the buoyant plume experiment of Shabbir and George[29]. For the streamwise component the Zeman and Lumley[24] model gives the best agreement. Shih and Lumley[23] and LRR[13] models are off the experiment near the centerline of the plume but are within experimental uncertainty for the rest of the flow. Craft et. al.[26] model overpredicts and Shih-Choi-Lumley model underpredicts the experiment. For the radial component Zeman and Lumley[24] model performs the best and, except Shih-Choi-Lumley, all the other models give reasonable agreement with the experiment. For the Π_{rz} component, which is the most important one, Shih-Choi-Lumley, LRR[13] and Craft et. al.[26] models are in excellent agreement with the experiment. Zeman and Lumley[24] model is also within experimental uncertainty where as Shih and Lumley[23] model overestimates the experiment. We note that no single model gives good agreement with all the flows. It is our feeling that source of this is either the coefficients for the slow term or the rapid term (a_2, given by equation (3.20) was assigned its limiting value). Further work is uderway in order to sort this out.

4.2 Pressure Temperature-Gradient Correlation

Now we compare the models for the pressure-temperature gradient correlations with the same three flows. First we look at the experiment of Tavoularis and Corrsin[27] (figure 5). For the streamwise correlation Shih and Lumley[23] and the present model provide the best agreement with the data. The Craft et. al.[26] model overpredicts the data whereas Zeman and Lumley[24] underpredicts it. The linear model of Launder[32] does a reasonable job. For the correlation in the cross-stream direction the present model gives the best result and all the other models underpredict the experiment. Since this experiment was used in adjusting the constants in the present model, this present comparison does not provide a true test for this model.

Figure (6) shows a similar comparison with the DNS of Rogers et. al.[28] of the same flow. For the streamwise correlation, again the Shih and Lumley and present model give the best agreement with the data. The non-linear model of Craft et. al.[26] and Launder[32] also give reasonable results. For the correlation in the cross-stream direction , however, none of the models predict the data, although at the initial stages of the flow the Launder model does a satisfactory job. However,toward the last part of the simulation, where the simulation is more developed, the Launder[32] model underestimates the simulation by 40%. At the same location the present model overestimates the data by the same amount.

The same models are compared with the buoyant plume experiment of Shabbir and George[29] in figure (7). For the streamwise component we note that the present model gives excellent agreement with the experiment. Both the Shih and Lumley[23] and Zeman and Lumley[24] models also do a reasonable job whereas the rest of the models underpredict the experiment. For the radial component, Π_r, the present model somewhat underpredicts the data near the central part of the plume, but produces the correct peak value and gives a very satisfactory overall agreement with the experiment. The rest of the models are off by a large margin. Note that the prediction of the radial component is extremely important for the correct calculation of the mean temperature.

Acknowledgements One of us (A.S.) would like to acknowledge the support of the National Research Council and the NASA Lewis Research Center for carrying out some of the work reported here.

References

[1] Lumley, J.L. Computational modeling of turbulent flows. in Advances in Applied Mechanics, Vol. 18, 1978, Academic Press, New York, pp.123–176.

[2] Lumley, J.L. Turbulence Modeling. ASME Journal of Applied Mechanics, Journal of Applied Mechanics, Vol. 50, 1983, pp 1097-1103.

[3] Reynolds, W.C. Computation of Turbulent Flows. Ann. Rev. Fluid Mech., Vol 8, 1976, 183-208.

[4] Launder, B.E. Phenomenological Modeling: Present and Future ? Whither Turbulence ? Cornell University, March 22-24, 1989.

[5] Rodi, W. Recent Developments in Turbulence Modeling. Proc. 3rd Int. Symp. on Refined Flow Modeling and Turbulence Measurements, Tokyo, July 26-28, 1988.

[6] Schumann,U. Realizability of Reynolds Stress Turbulence Models. Phys. Fluids, 20, 721-725, 1977.

[7] Pope, S.B. Consistent Modeling of Scalars in Turbulent Flows, Phys. Fluids, Vol. 26, 1983, 404-408.

[8] Reynolds, W.C. Fundamentals of Turbulence for Turbulence Modeling and Simulation, Lecture Notes for Von Karman, March 16-17, 1987.

[9] Speziale, C.G. Closure Models for Rotating Two-dimensional Tur-

bulence. Geophys. and Astrophys. Fluid Dynamics, 23,69, 1983.

[10] Speziale, C.G., N.N. Mansour and R.S. Rogallo. The Decay of Isotropic Turbulence in a Rapid Rotating frame. CTR-S87-12, 1987. Stanford University and NASA Ames Research Center.

[11] Shih, T.H., J.L. Lumley and J.Y. Chen. Second Order Modeling of a Passive Scalar in a Turbulent Shear Flow, AIAA Journal, Vol. 28, No. 4, 1990, 610-617.

[12] Shih, T.H., W.C. Reynolds and N.N. Mansour. A Spectrum model for Weakly Anisotropic Turbulence. Phys. Fluids A2(8), 1990.

[13] Launder, B.E., Reece, G.L. and Rodi, W. Progress in the development of a Reynolds–stress turbulence closure. J. Fluid Mech. 68, 1975, pp. 537–566.

[14] Reynolds, W.C. Effect of Rotation on Homogeneous Turbulence. Tenth Australian Fluid Mechanics Conference, December 11-15, 1989.

[15] Reynolds, W.C. and Grau 1988, personal communication.

[16] Tennekes, H. and J.L. Lumley. A First Course in Turbulence, MIT Press, Cambridege, Massachusetts.

[17] Shih, T.H. and J.L. Lumley. Influence of Timescale Ratio on Scalar Flux Relaxation: Modeling Sirivat and Warhaft's Homogeneous Passive Scalar Fluctuations, J. Fluid Mech., 1986, Vol. 162, 211-222.

[18] Shih, T.h., J.L. Lumley and J.Y. Chen. Second Order Modeling of Boundary-Free Turbulent Shear Flows with a New Model Form of Pressure Correlations. FDA-85-07, August 1985, Sibley School of Mech. and Aero. Engg., Cornell University.

[19] Shih, T.H., N.N. Mansour and J.-Y. Chen. Reynolds Stress Models of Homogeneous Turbulence, Proceedings of the Summer Program 1987, Center for Turbulence Research, Stanford University and NASA Ames.

[20] Gence, J.N. and J. Mathieu. The Return to Isotropy of a Homogeneous Turbulence Having Been Submitted to Two Successive Plane Strains. J. Fluid Mech., Vol 101, 556-566.

[21] Choi, K.-S. A Study of the Return to Isotropy of Homogeneous Turbulence. Ph.D. Dissertation, 1983, Cornell University.

[22] Choi, K.-S. and J.L. Lumely. Return to Isotropy of Homogeneous Turbulence Revisited, Turbulence and Chaotic Phenomena in Fluids, ed. T. Tatsumi, Elsevier, 1984, North Holland.

[23] Shih, T.H. and Lumley, J.L. Modeling of pressure correlation terms

in Reynolds–stress and scalar flux equations. Rept. FDA-85-3, Sibley School of Mech. and Aerospace Eng., 1985, Cornell University.

[24] Zeman, O. and J.L. Lumley. Modeling Buoyancy Driven Mixed Layers. J. Atm. Sci., Vol. 33, 1976

[25] Newman, G.R., J.L. Lumley and B.E. Launder. Modeling the Behavior of Homogeneous Scalar Turbulence. J. Fluid Mech., Vol 111,1981, 217-232.

[26] Craft T.J. S. Fu, B.E. Launder and D.P. Tselepidakis. Developments in Modelling the Turbulent Second-Moment Pressure Correlations. Report TFD/89/1, Mech. Eng. Dept.,1989, UMIST.

[27] Tavoularis, S. and S. Corrsin. J. Fluid Mech., Experiments in Nearly Homogeneous Turbulent Shear Flow with a Uniform Mean Temperature Gradient. Part 2. The Fine Structure. Vol 104, 1981, 349-367.

[28] Rogers, M.M., P. Moin and W.C. Reynolds. The Structure and Modeling of the Hydrodynamic and Passive Scalar Fields in Homogeneous Turbulent Shear Flow. Report No. TF-25, 1986, Department of Mech. Engg., Stanford Univ.

[29] Shabbir, A., and W.K. George. Experimental Study of an Axisymmetric Turbulent Buoyant Plume. 1990, Submitted to J. Fluid Mech.

[30] Harris, V.G., J.A.H. Graham and S. Corrsin. Further Experiments in Nearly Homogeneous Shear Flow. J. Fluid Mech., Vol. 81, 1977, 657-687.

[31] Shabbir, A. Experimental Balances of the Second Moment Equations for the Axisymmetric Buoyant Plume and Its Implications on Turbulence Modeling. Submitted to the 8th Symposium on Turbulent Shear Flows to be held in 1991.

[32] Launder, B.E. On the Effects of Gravitational Field on the Turbulent Transport of Heat and Momentum. J. Fluid Mech., Vol 67,569-590, 1975.

[33] Zeman, O. Progress in the Modeling of Planetary Boundary Layers. Ann. Rev. Fluid Mech. 1981, Vol. 13, 253-272.

APPENDIX A

The following are the exact expressions for the rapid and buoyancy contributions to the pressure-strain and pressure scalar-gradient correlations.

$$X_{pjqi} = -(1/4\pi)\int\int\int_v [\overline{u_q(r)u_i(r')}]_{,pj} dv/|r-r'|$$

$$X_{ijk} = -(1/4\pi)\int\int\int_v [\overline{u_k(r)\theta_j(r')}]_{,ji} dv/|r-r'|$$

$$Y_{kij} = -(1/4\pi)\int\int\int_v [\overline{u_j(r)\theta(r')}]_{,ki} dv/|r-r'|$$

$$Y_{ik} = -(1/4\pi)\int\int\int_v [\overline{\theta(r)\theta(r')}]_{,ik} dv/|r-r'|$$

The slow terms are defined by the following expressions

$$-\Phi_{ij}\,\epsilon = (1/\rho)\overline{p_{sl}(u_{i,j}+u_{j,i})} - 2\nu\overline{u_{i,k}u_{j,k}} + 2\epsilon\delta_{ij}/3$$

$$-\Phi_i\,\epsilon/\overline{q^2} = (1/\rho)\overline{p_{sl}\theta_{,i}} - (\nu+\gamma)\overline{\theta_{,k}u_{i,k}}$$

The total (i.e. sum of the slow, rapid and buoyancy parts) pressure strain and pressure scalar-gradient correlation has been denoted by Π_{ij} and Π_i respectively.

APPENDIX B

The following table lists the values of the model coefficients β's and γ's as calculated from the buoyant plume experiment of Shabbir and George (1990).

r/z	β_1	β_2	β_3	β_4	β_5	β_6	β_7	β_8	β_9	β_{11}
0.00	0.47	-0.12	-2.05	-0.35	0.65	-0.20	6.40	1.95	-7.05	-1.30
0.01	0.47	-0.12	-2.04	-0.35	0.65	-0.20	6.20	1.94	-6.85	-1.30
0.02	0.48	-0.12	-2.02	-0.36	0.64	-0.19	5.72	1.93	-6.36	-1.28
0.03	0.49	-0.12	-1.99	-0.37	0.63	-0.18	5.16	1.90	-5.79	-1.27
0.04	0.50	-0.12	-1.96	-0.37	0.63	-0.17	4.66	1.88	-5.29	-1.25
0.05	0.51	-0.13	-1.93	-0.38	0.62	-0.16	4.28	1.85	-4.90	-1.24
0.06	0.52	-0.13	-1.91	-0.39	0.61	-0.15	4.01	1.83	-4.62	-1.22
0.07	0.53	-0.13	-1.88	-0.40	0.60	-0.14	3.83	1.81	-4.44	-1.21
0.08	0.53	-0.13	-1.86	-0.40	0.60	-0.13	3.73	1.80	-4.33	-1.20
0.09	0.54	-0.14	-1.84	-0.41	0.59	-0.13	3.69	1.78	-4.28	-1.19
0.10	0.55	-0.14	-1.83	-0.41	0.59	-0.12	3.68	1.77	-4.27	-1.18
0.11	0.55	-0.14	-1.81	-0.41	0.59	-0.11	3.70	1.76	-4.28	-1.17
0.12	0.56	-0.14	-1.80	-0.42	0.58	-0.11	3.74	1.75	-4.32	-1.16
0.13	0.56	-0.14	-1.79	-0.42	0.58	-0.11	3.78	1.74	-4.36	-1.16
0.14	0.56	-0.14	-1.79	-0.42	0.58	-0.11	3.83	1.74	-4.41	-1.16
0.15	0.56	-0.14	-1.80	-0.42	0.58	-0.11	3.87	1.74	-4.45	-1.16
0.16	0.56	-0.14	-1.81	-0.42	0.58	-0.11	3.91	1.75	-4.49	-1.17
0.17	0.55	-0.14	-1.83	-0.41	0.59	-0.12	3.93	1.77	-4.51	-1.18
0.18	0.54	-0.13	-1.85	-0.40	0.60	-0.13	3.93	1.79	-4.53	-1.19

r/z	γ_1	γ_2	γ_3	γ_4	γ_5
0.00	0.41	-2.82	-0.36	-0.81	-3.15
0.01	0.41	-2.75	-0.35	-0.81	-2.96
0.02	0.42	-2.58	-0.34	-0.80	-2.51
0.03	0.42	-2.39	-0.32	-0.79	-1.99
0.04	0.43	-2.22	-0.30	-0.79	-1.53
0.05	0.43	-2.10	-0.28	-0.79	-1.19
0.06	0.44	-2.01	-0.27	-0.81	-0.95
0.07	0.44	-1.95	-0.27	-0.83	-0.81
0.08	0.44	-1.92	-0.28	-0.87	-0.74
0.09	0.44	-1.91	-0.30	-0.93	-0.72
0.10	0.44	-1.91	-0.33	-1.00	-0.73
0.11	0.44	-1.92	-0.37	-1.09	-0.77
0.12	0.45	-1.94	-0.41	-1.21	-0.83
0.13	0.45	-1.96	-0.47	-1.35	-0.88
0.14	0.44	-1.98	-0.53	-1.53	-0.93
0.15	0.44	-2.00	-0.59	-1.75	-0.97
0.16	0.44	-2.02	-0.64	-2.02	-0.99
0.17	0.43	-2.03	-0.69	-2.35	-0.98
0.18	0.42	-2.05	-0.71	-2.76	-0.95

APPENDIX C

Here we list the various model coefficients for the slow terms used in the present study. The coefficient β used in the Shih and Lumley[23] model is the one suggested by Lumley[1] and is

$$\begin{aligned}\beta =& 2 + exp[-7.77/R_l^{1/2}] \\ & \left\{72/R_l^{1/2} + 80.1ln[1 + 62.4(-II + 2.3III)]\right\}(\frac{1}{9} + 3III + II)\end{aligned}$$

The function H appearing in (3.29) and (3.30) is taken from Shih et. al.[11] and is given by

$$H = 1.1 + 0.55(\beta - 1)tanh[4(\frac{\tau}{\tau_\theta} - 1)]$$

The coefficients β and γ appearing in equation (3.27) and suggested by Choi and Lumley[22] are given by

$$\begin{aligned}\beta =& 2 + \frac{\rho^* F^{1/2}}{(1 + G\chi^2)} \\ \gamma =& \frac{\rho^* F^{1/2}}{(1 + G\chi^2)}\frac{G}{\xi}\end{aligned}$$

where

$$\begin{aligned}\xi =& (III/2)^{1/3}, \quad \eta = (-II/3)^{1/2} \\ \chi =& \frac{\xi}{\eta}, \quad G = -\chi^4 + 0.8\chi^6 \\ \rho^* =& exp[-9.29/R_l^{1/2}]\{(\frac{7.69}{R_l^{1/2}} + \frac{73.7}{R_l}) - [296 - 16.2(\chi + 1)^4]II\}\end{aligned}$$

For the Zeman and Lumely[24] model following set of coefficients were used

$$\begin{aligned}\beta =& \frac{1}{(-15II)}[1 + (1 - 24II)^{1/2}] \\ f_1 =& (\frac{\beta}{2} + \frac{\overline{q^2\epsilon_\theta}}{\epsilon\overline{t^2}})\end{aligned}$$

Note that these revised coefficients are taken from Zeman[33] and are different from the values originally used by Zeman and Lumley[24]. It should also be pointed out that the expression for β is clipped at 6.0.

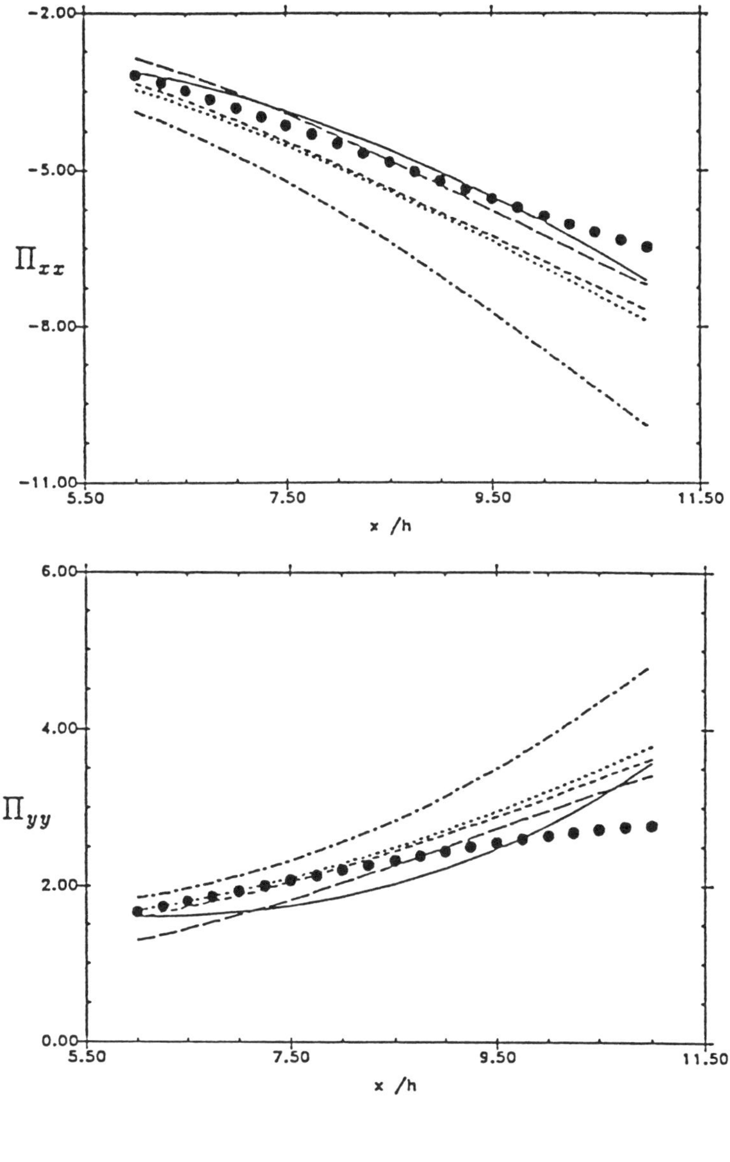

Figure 1. Comparison between models for Π_{ij} and the homogeneous shear flow experiment. ······ LRR model; ---- Zeman and Lumley model (1976); —·—Shih and Lumley (1985) model;——–Craft et. al. (1989) model; —— Shih-Choi-Lumley model; • experiment of Tavoularis and Corrsin (1981).

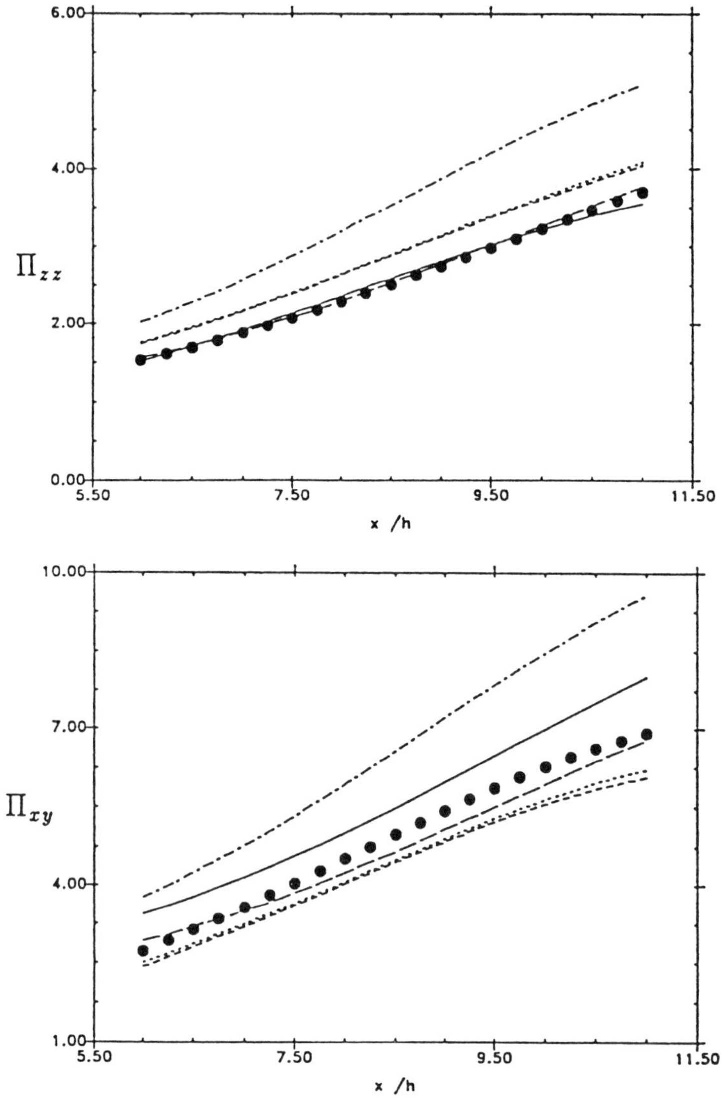

(Figure 1 continued from last page)

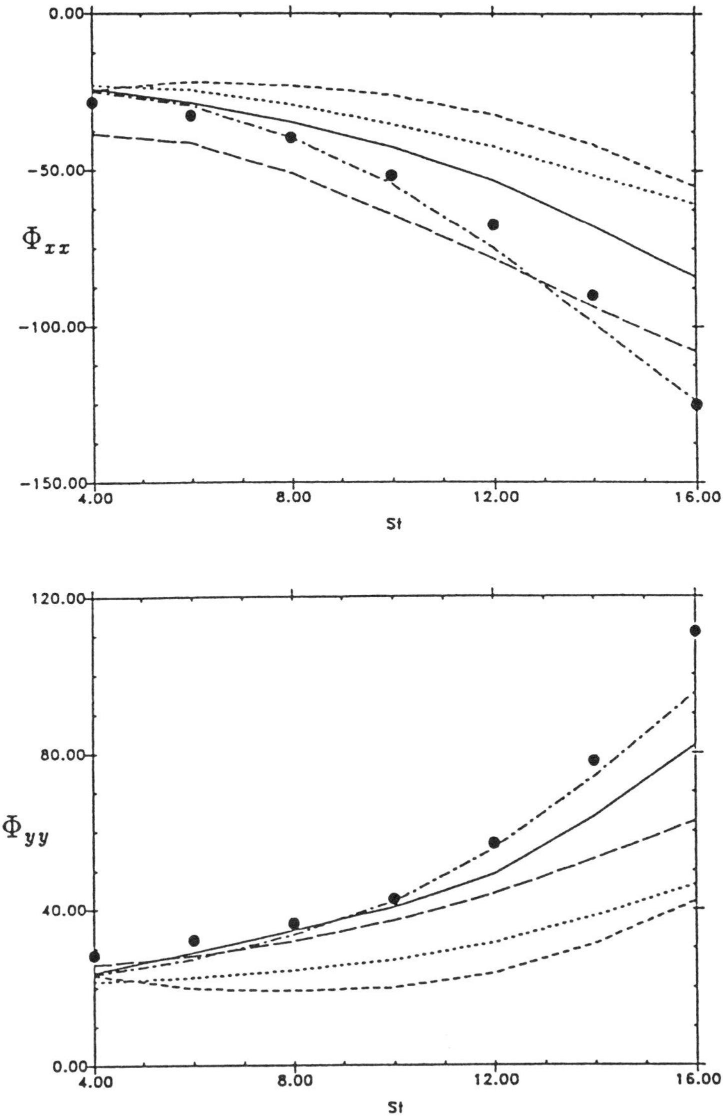

Figure 2. Comparison between the models of the slow part of the pressure-strain and the DNS of homogeneous shear flow. ······· LRR model; ---- Zeman and Lumley model (1976); —·— Shih and Lumley (1985) model; — — Craft et. al. (1989) model; —— Shih-Choi-Lumley model; ● DNS of Rogers et. al. (1986).

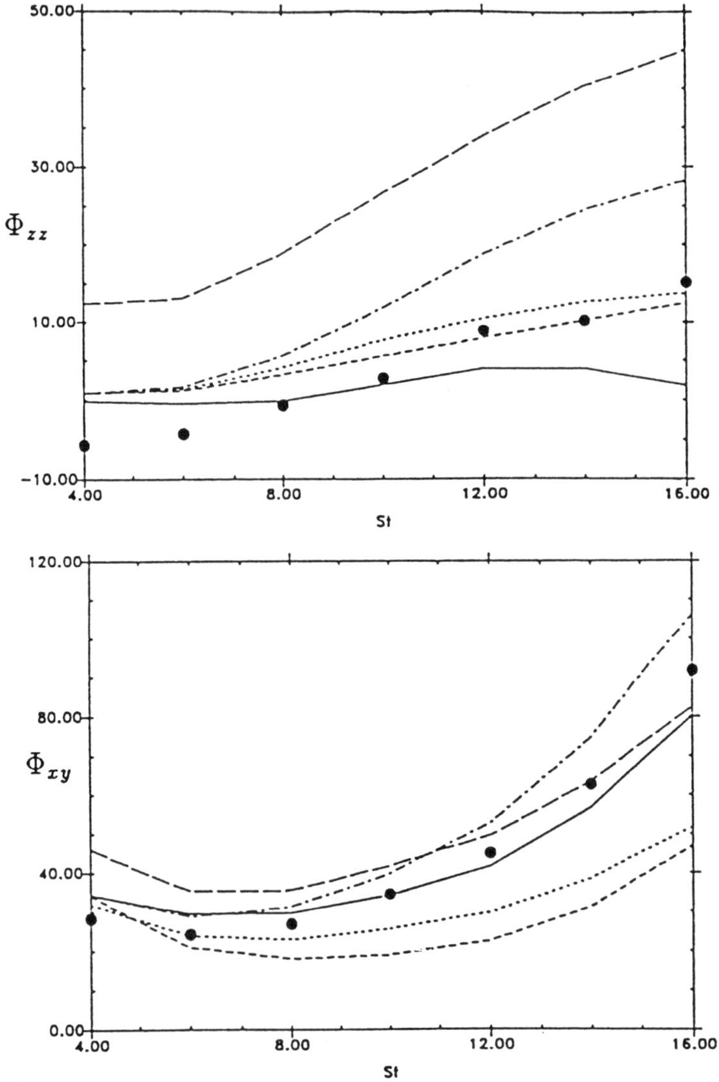

(Figure 2 continued from last page)

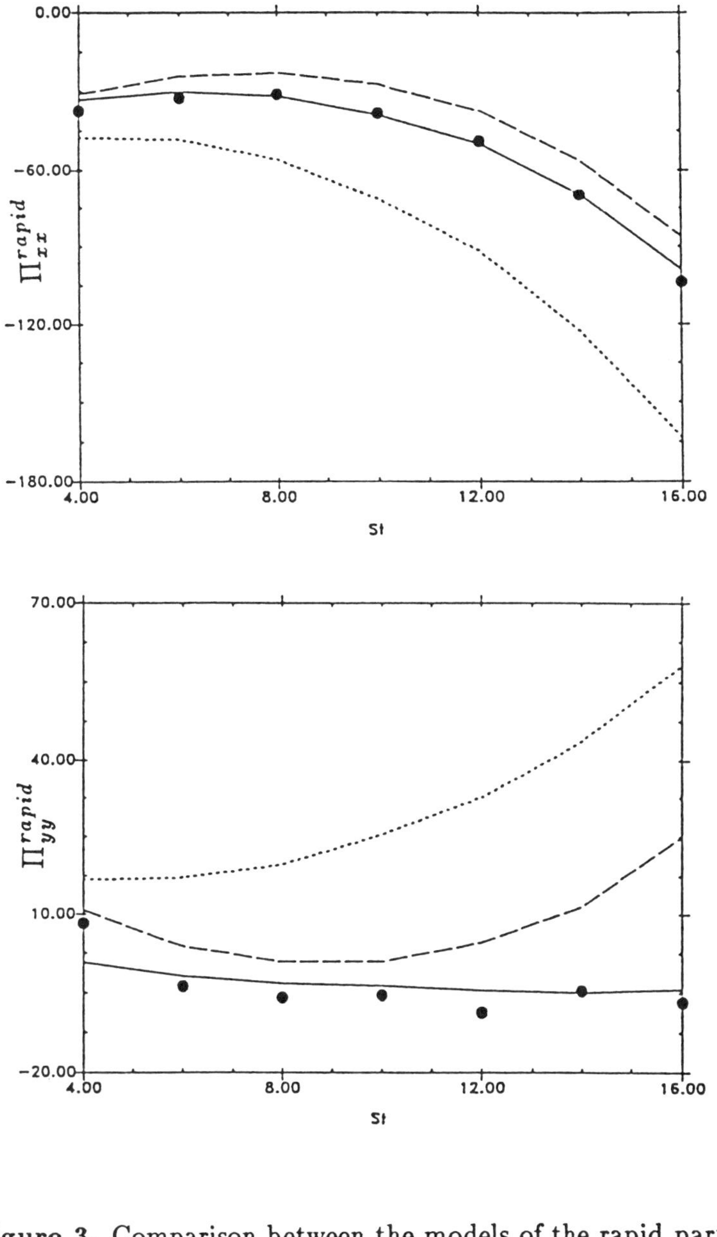

Figure 3. Comparison between the models of the rapid part of the pressure-strain and the DNS of homogeneous shear flow. ----- LRR model; ——— Shih and Lumley (1985) model; ---- Craft et. al. (1989) model; • DNS of Rogers et. al. (1986).

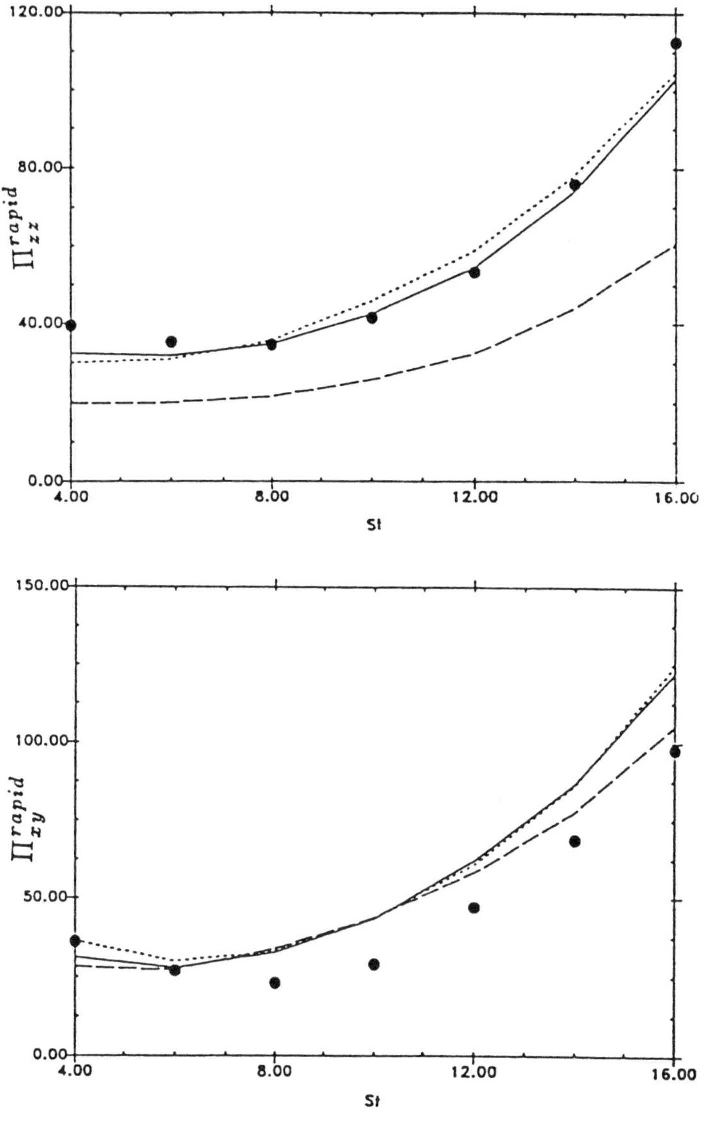

(Figure 3 continued from last page)

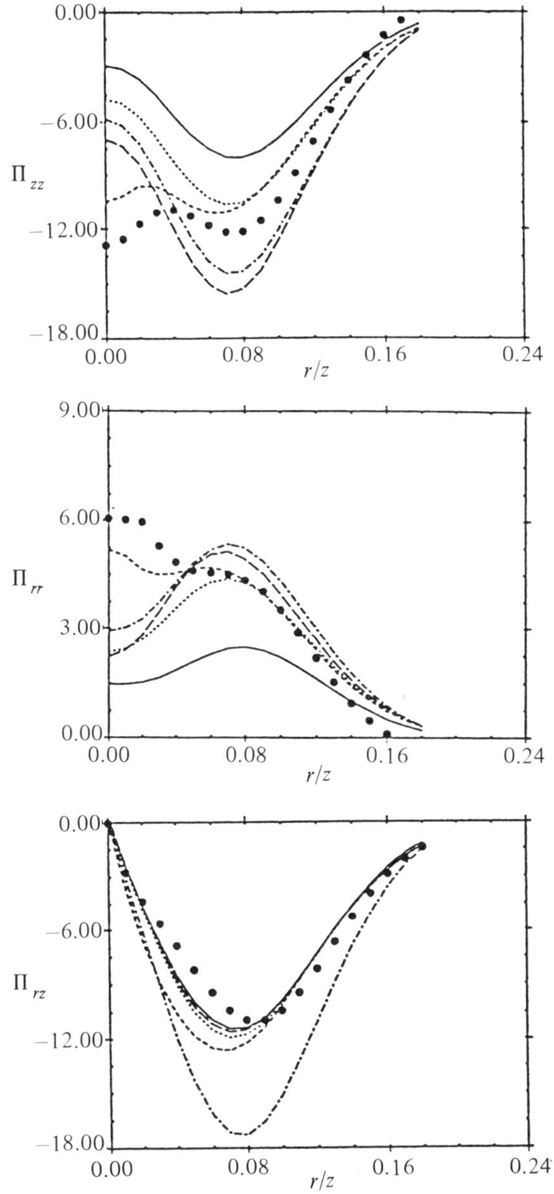

Figure 4. Comparison between models for Π_{ij} and the buoyant plume experiment. .... LRR model; ------ Zeman and Lumley model (1976); -.-.-. Shih and Lumley (1985) model; ——— Craft et. al. (1989) model; ——— Shih-Choi-Lumley model; ● experiment of Shabbir and George (1990).

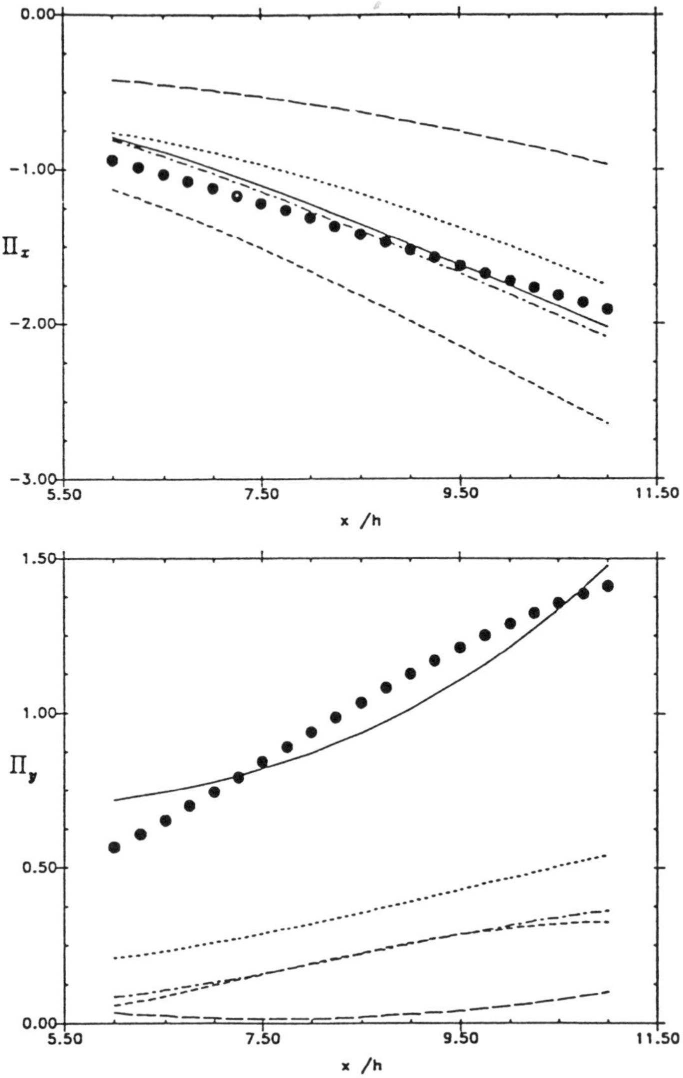

Figure 5. Comparison between models for Π_i and the homogeneous shear flow experiment. ······ Launder (1975) model; ---- Zeman and Lumley model (1976); —·— Shih and Lumley (1985) model; — — — — Craft et. al. (1989) model; ——— present model; • experiment of Tavoularis and Corrsin (1985).

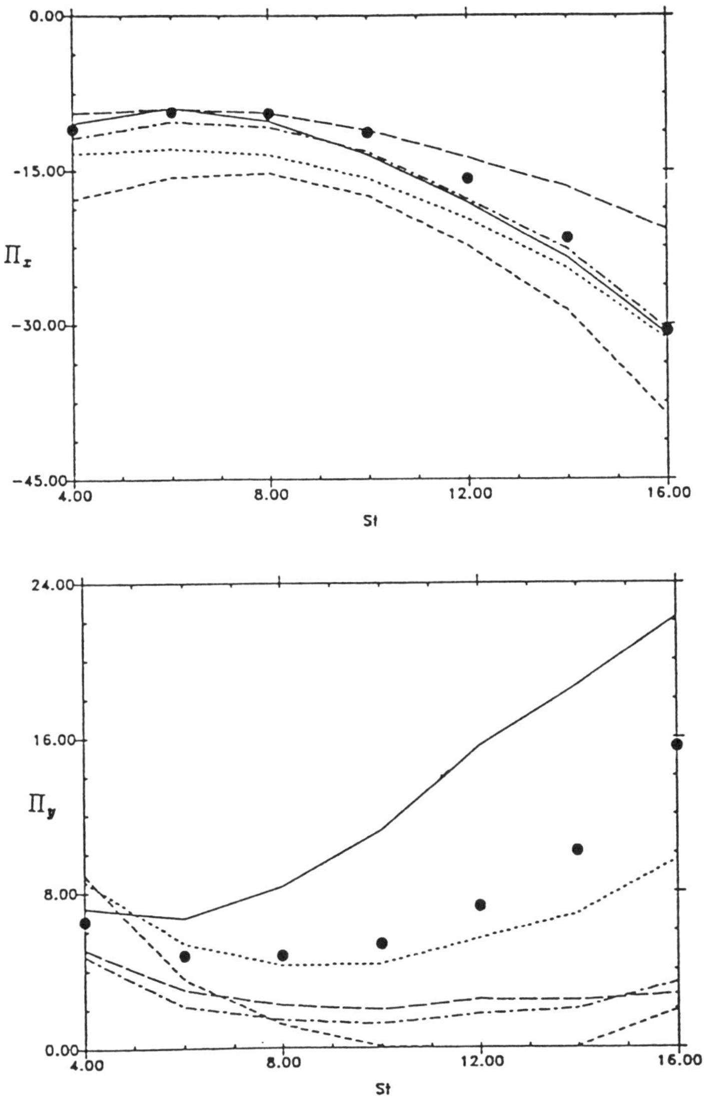

Figure 6. Comparison between models for the Π_i and the DNS of homogeneous shear flow. ·······Launder (1975) model; ---- Zeman and Lumley model (1976); —· —Shih and Lumley (1985) model; — — — — Craft et. al. (1989) model; ——— present model; ● DNS of Rogers et. al. (1986).

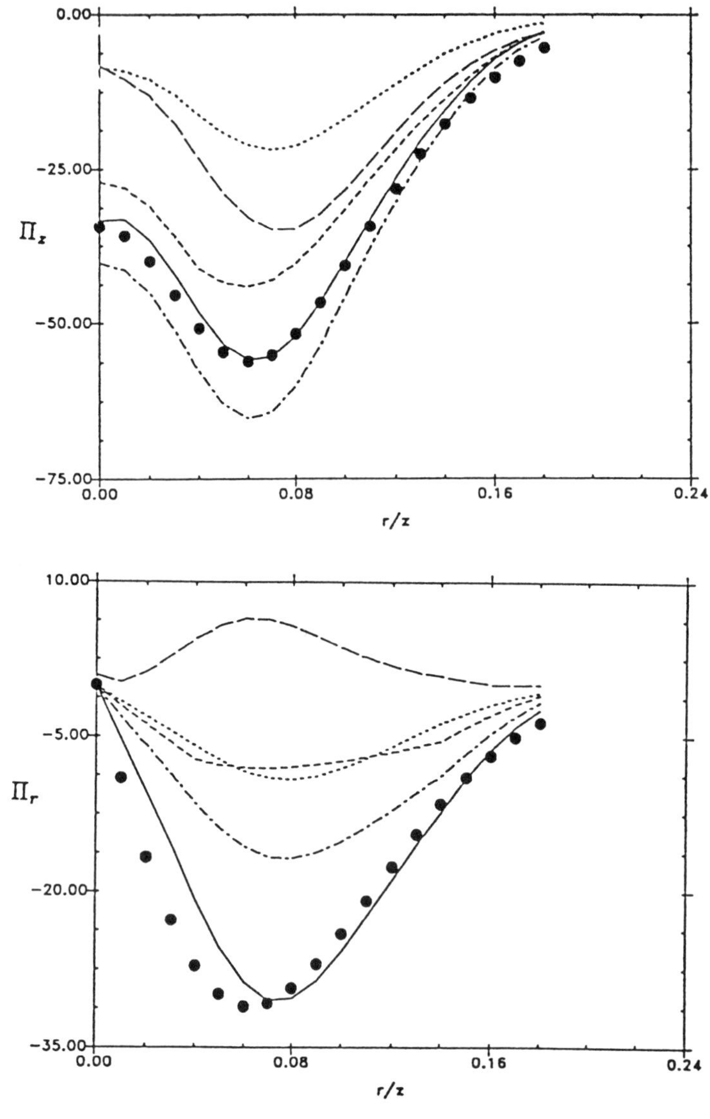

Figure 7. Comparison between models for Π_i and the buoyant plume experiment. ······· Launder (1975) model; - - - - Zeman and Lumley model (1976); —·— Shih and Lumley (1985) model; — — — — Craft et. al. (1989) model; ——— present model; ● experiment of Shabbir and George (1990).

MODELING THE DISSIPATION RATE IN ROTATING TURBULENT FLOWS

Charles G. Speziale[1]

ICASE, NASA Langley Research Center
Hampton, Virginia 23665

Rishi Raj

City College of New York
New York, New York 10031

Thomas B. Gatski

NASA Langley Research Center
Hampton, Virginia 23665

ABSTRACT

A variety of modifications to the modeled dissipation rate transport equation that have been proposed during the past two decades to account for rotational strains are examined. The models are subjected to two crucial test cases: the decay of isotropic turbulence in a rotating frame and homogeneous shear flow in a rotating frame. It is demonstrated that these modifications do not yield substantially improved predictions for these two test cases and in many instances give rise to unphysical behavior. An alternative proposal, based on the use of the tensor dissipation rate, is made for the development of improved models.

1. Introduction

Many turbulent flows of practical importance involve combinations of rotational and irrotational mean strains. The rotational strains can arise from the effects of curvature, a system rotation, or swirl, as well as from shear. Much of the motivation for formulating

[1]This research was supported by the National Aeronautics and Space Administration under NASA Contract No. NAS1-18605 while the author was in residence at the Institute for Computer Applications in Science and Engineering, NASA Langley Research Center, Hampton, VA 23665.

second-order closure models has its origins in the desire to account for the distinct physical effects of rotational and irrotational strains, in a framework where directional information is incorporated. Reynolds stress models of the eddy viscosity type, since they only depend on the mean velocity gradients through their symmetric part (i.e., the mean rate of strain tensor), are oblivious to the presence of rotational strains. Consequently, these models cannot distinguish between the considerably different turbulence physics exhibited in plane shear, plane strain, and rotating plane shear flows – a deficiency that is now well-known [1, 2]. However, the same deficiency still remains with the modeled form of the dissipation rate transport equation that has been commonly used in the turbulence modeling literature starting with the work of Launder and co-workers [3]. This frequently used model has no explicit dependence on rotational strains; rotational strains can only have an indirect effect through the changes that they induce in the Reynolds stress tensor. Such a limited dependence is known to be defective, making it impossible to properly describe rotating isotropic turbulence and swirling jets, among other turbulent flows [4, 5]. Consequently, several independent researchers over the past two decades have tried to develop rotational modifications to the modeled dissipation rate transport equation to resolve this problem. These alterations will be the subject of the present study.

In this paper, a variety of previously proposed modifications to the modeled dissipation rate transport equation will be considered:

(1) the Pope [5] model,

(2) the Hanjalic and Launder [6] model,

(3) the Bardina model [7], and

(4) the Raj [8] model.

These models represent a good cross-section of the rotational modifications to the modeled dissipation rate transport equation that have been proposed during the last fifteen years. Two flows from homogeneous turbulence will be used to test the models: (a) the decay of isotropic turbulence in a rotating frame, and (b) homogeneous shear flow in a rotating frame. These two test cases are chosen since they represent homogeneous flows whose structure is significantly altered by rotation, without the added complication of walls or other sources of mean turbulent diffusion. In this fashion, the question of the interaction of rotational and irrotational strains can be studied in isolation, independent of other complicating features. The performance of each of the models will be documented in detail and specific proposals will be made for the development of improved models. A new model based on the tensor dissipation – formulated in order to include more directional information – will be proposed and tested.

2. The Models and Test Cases to be Considered

We will restrict our attention to the incompressible flow of a viscous fluid with constant properties. The velocity field $\mathbf{v}$ and pressure P are decomposed into mean and fluctuating parts, respectively, as follows:

$$\mathbf{v} = \overline{\mathbf{v}} + \mathbf{u}, \quad P = \overline{P} + p. \tag{1}$$

The exact transport equation for the turbulent dissipation rate in homogeneous flows takes the form [2]

$$\dot{\varepsilon} = \mathcal{P}_\varepsilon - \Phi_\varepsilon \tag{2}$$

where

$$\mathcal{P}_\varepsilon = -2\nu \overline{\frac{\partial u_i}{\partial x_j}\frac{\partial u_k}{\partial x_j}}\frac{\partial \overline{v}_i}{\partial x_k} - 2\nu \overline{\frac{\partial u_j}{\partial x_i}\frac{\partial u_j}{\partial x_k}}\frac{\partial \overline{v}_i}{\partial x_k} - 2\nu \overline{\frac{\partial u_i}{\partial x_k}\frac{\partial u_i}{\partial x_j}\frac{\partial u_k}{\partial x_j}} \tag{3}$$

$$\Phi_\varepsilon = 2\nu^2 \overline{\frac{\partial^2 u_i}{\partial x_j \partial x_k}\frac{\partial^2 u_i}{\partial x_j \partial x_k}} \tag{4}$$

are, respectively, the production and destruction of dissipation terms given that $\varepsilon \equiv 2\nu\overline{\partial u_i/\partial x_j \partial u_i/\partial x_j}$ is the scalar dissipation rate and ν is the kinematic viscosity of the fluid. Here, since the turbulence is homogeneous, the overbar denotes a spatial mean, and the superposed dot denotes an ordinary time derivative.

The modeled version of the transport equation for ε that has been used widely in the turbulence modeling community – which is due to Launder and co-workers [3] – takes the form

$$\dot{\varepsilon} = C_{\varepsilon 1}\frac{\varepsilon}{K}\mathcal{P} - C_{\varepsilon 2}\frac{\varepsilon^2}{K} \tag{5}$$

where $\mathcal{P} = -\overline{u_i u_j}\partial \overline{v}_i/\partial x_j$ is the turbulence production, and $C_{\varepsilon 1}$ and $C_{\varepsilon 2}$ are dimensionless constants that typically assume the values of 1.44 and 1.92, respectively. Equation (5) is obtained from (2) by making the assumption that the production (or destruction) of dissipation is proportional to the production (or destruction) of turbulent kinetic energy, i.e.,

$$\mathcal{P}_\varepsilon \propto \mathcal{P}/\tau_0, \quad \Phi_\varepsilon \propto \varepsilon/\tau_0 \tag{6}$$

where τ_0 is the turbulent time scale that is needed for dimensional consistency. The added assumption that the turbulent time scale τ_0 is given by

$$\tau_0 = \frac{K}{\varepsilon} \tag{7}$$

then yields the modeled equation (5).

Due to the symmetry of the Reynolds stress tensor $\tau_{ij} \equiv \overline{u_i u_j}$, it follows that

$$\mathcal{P} = -\tau_{ij}\overline{S}_{ij} \tag{8}$$

where

$$\overline{S}_{ij} = \frac{1}{2}\left(\frac{\partial \overline{v}_i}{\partial x_j} + \frac{\partial \overline{v}_j}{\partial x_i}\right) \tag{9}$$

is the mean rate of strain tensor. Hence, (5) has *no explicit dependence* on the mean rotation tensor

$$\overline{\omega}_{ij} = \frac{1}{2}\left(\frac{\partial \overline{v}_i}{\partial x_j} - \frac{\partial \overline{v}_j}{\partial x_i}\right). \tag{10}$$

It therefore follows that the commonly used modeled dissipation rate equation can only be affected indirectly by rotational strains through the changes that they induce in the Reynolds stress tensor. As we will show later, this makes it difficult to properly describe certain rotating turbulent flows.

The model suggested by Pope [5] is based on the addition of a mean vortex stretching term and takes the form

$$\dot{\varepsilon} = C_{\varepsilon 1}\frac{\varepsilon}{K}\mathcal{P} - C_{\varepsilon 2}\frac{\varepsilon^2}{K} + C_{\varepsilon 3}\frac{K^2}{\varepsilon}\overline{S}_{ij}\overline{\omega}_{jk}\overline{\omega}_{ki} \tag{11}$$

in homogeneous turbulence where $C_{\varepsilon 1} = 1.44$, $C_{\varepsilon 2} = 1.92$, and $C_{\varepsilon 3} = 0.79$. The standard ε-equation is obtained in the limit as $C_{\varepsilon 3} \to 0$. In the modification proposed by Hanjalic and Launder [6] to "sensitize the dissipation equation to irrotational strains," the ε-equation takes the following form in homogeneous turbulence:

$$\dot{\varepsilon} = C_{\varepsilon 1}\frac{\varepsilon}{K}\mathcal{P} - C_{\varepsilon 2}\frac{\varepsilon^2}{K} - C_{\varepsilon 3}K e_{mij}e_{mkl}\frac{\partial \overline{v}_i}{\partial x_j}\frac{\partial \overline{v}_k}{\partial x_l} \tag{12}$$

where e_{ijk} is the permutation tensor and $C_{\varepsilon 1} = 1.44$, $C_{\varepsilon 2} = 1.92$, and $C_{\varepsilon 3} = 0.27$. Since,

$$\frac{\partial \overline{v}_i}{\partial x_j} = \overline{S}_{ij} + \overline{\omega}_{ij} \tag{13}$$

it is clear that this modification introduces rotational strains; in the limit as $C_{\varepsilon 3} \to 0$ the standard ε-equation is recovered. The Bardina model [7, 9] takes the form

$$\dot{\varepsilon} = C^*_{\varepsilon 1}\frac{\varepsilon}{K}\mathcal{P} - C^*_{\varepsilon 2}\frac{\varepsilon^2}{K} \tag{14}$$

in a homogeneous turbulence, where

$$C^*_{\varepsilon 1} = 1.50 - 0.015\frac{K}{\varepsilon}\left(\frac{1}{2}\overline{\omega}_{ij}\overline{\omega}_{ij}\right)^{\frac{1}{2}} \tag{15}$$

$$C^*_{\varepsilon 2} = 1.83 + 0.15\frac{K}{\varepsilon}\left(\frac{1}{2}\overline{\omega}_{ij}\overline{\omega}_{ij}\right)^{\frac{1}{2}} \tag{16}$$

are functions of the mean rotation tensor. The models (11), (12), and (14)-(16) are written for an inertial frame. In a non-inertial frame, the rotation tensor $\overline{\omega}_{ij}$ must be replaced by the absolute mean rotation tensor

$$\overline{W}_{ij} = \overline{\omega}_{ij} + e_{mji}\Omega_m \tag{17}$$

where Ω_m is the angular velocity of the non-inertial frame relative to an inertial framing (see Speziale [10]).

Raj [8] developed a model based on an analysis of the exact transport equation of the true dissipation. In an inhomogeneous turbulence, this exact transport equation contains terms that depend explicitly on the rotation rate of the reference frame. Based on an empirical argument, Raj [8] developed a simple correction term to account for the effect of rotations which was guided by his turbomachinery research. For homogeneous turbulence, the model proposed by Raj [8] takes the form

$$\dot{\varepsilon} = C_{\varepsilon 1}\frac{\varepsilon}{K}\mathcal{P} - C_{\varepsilon 2}\frac{\varepsilon^2}{K} + C_{\varepsilon 3}\Omega\varepsilon\left(\frac{3\tau_{zz}}{2K} - 1\right) \tag{18}$$

in a rotating turbulent flow, where Ω is the angular velocity of the reference frame – which, for simplicity, is aligned along the z direction – and τ_{zz} is the component of Reynolds stress tensor along the axis of rotation. Here, $C_{\varepsilon 3}$ is a constant of $\mathcal{O}(1)$. In the limited practical applications where this model has been used, $C_{\varepsilon 3}$ has assumed different values ranging from 1 to 5. For simplicity, we will set $C_{\varepsilon 3} = 5$ (the largest of these values); $C_{\varepsilon 1}$ and $C_{\varepsilon 2}$ assume the traditional values of 1.44 and 1.92, respectively. It must be said at the outset that (18) does not constitute a general model and has some internal inconsistencies in the way it was derived [11]. Nonetheless, the model does have some interesting features that are worth testing in the problems to be considered in this study (e.g., the model directly accounts for the effect of the rotationally induced anisotropy in the Reynolds stress tensor, along the axis of rotation, on the dissipation).

In order to test these models in the flows to be considered, a Reynolds stress model is needed. A simple second-order closure model is chosen – namely, the shortened form of the Launder, Reece, and Rodi model which is now referred to as the "Basic Model" by Launder and co-workers [12]. For a homogeneous turbulence in a rotating frame, this Basic Model of Launder and co-workers takes the form [13]:

$$\begin{aligned} \dot{\tau}_{ij} &= (C_2 - 1)\left(\tau_{ik}\frac{\partial \overline{v}_j}{\partial x_k} + \tau_{jk}\frac{\partial \overline{v}_i}{\partial x_k}\right) \\ &\quad +(C_2 - 2)(\tau_{ik}e_{mkj}\Omega_m + \tau_{jk}e_{mki}\Omega_m) \qquad (19) \\ &\quad -C_1\frac{\varepsilon}{K}\left(\tau_{ij} - \frac{2}{3}K\delta_{ij}\right) - \frac{2}{3}C_2\tau_{kl}\frac{\partial \overline{v}_k}{\partial x_l}\delta_{ij} - \frac{2}{3}\varepsilon\delta_{ij} \end{aligned}$$

where C_1 and C_2 are dimensionless constants that assume the values of 1.8 and 0.6, respectively. This model is derived by making two major assumptions – namely that the dissipation rate tensor is isotropic, i.e.,

$$\varepsilon_{ij} = \frac{2}{3}\varepsilon\delta_{ij} \qquad (20)$$

and the pressure strain correlation is of the general form

$$\Pi_{ij} = \varepsilon\mathcal{A}_{ij}(\mathbf{b}) + K\mathcal{M}_{ijkl}(\mathbf{b})\frac{\partial \overline{v}_k}{\partial x_l} \qquad (21)$$

where $\mathcal{A}_{ij}$ and $\mathcal{M}_{ijkl}$ are *linear* functions of the anisotropy tensor **b** (an approximation valid for small anisotropies).

The first problem to be considered as a test case is isotropic turbulence in a rotating frame. Here, an initially decaying isotropic turbulence wherein

$$\tau_{ij} = \frac{2}{3}K_0\delta_{ij}, \quad \varepsilon = \varepsilon_0, \tag{22}$$

at time $t = 0$, is subjected to a solid body rotation with constant angular velocity

$$\Omega_i = (0, 0, \Omega) \tag{23}$$

and vanishing mean velocity gradients

$$\frac{\partial \overline{v}_i}{\partial x_j} = 0. \tag{24}$$

The substitution of (23) - (24) into (19) – along with the use of the initial condition of isotropy (22) – then yields the equation

$$\dot{K} = -\varepsilon \tag{25}$$

where $\tau_{ij} = \frac{2}{3}K\delta_{ij}$. This is the same form as its non-rotating counterpart. Hence, the Launder, Reece, and Rodi model predicts that an initially isotropic turbulence subjected to a rotation decays isotropically – a result that is consistent with the Navier-Stokes equations [14, 15]. For rotating isotropic turbulence, the modeled ε- transport equation takes the standard form

$$\dot{\varepsilon} = -C_{\varepsilon 2}\frac{\varepsilon^2}{K} \tag{26}$$

(where $C_{\varepsilon 2} = 1.92$) for the Pope and the Raj models; consequently, these models do not distinguish the difference between isotropic turbulence in a rotating frame and in an inertial frame. The Bardina model, however, does have a non-zero rotational correction; for rotating isotropic turbulence (14) takes the form

$$\dot{\varepsilon} = -C_{\varepsilon 2}\frac{\varepsilon^2}{K} - C_{\varepsilon 3}\Omega\varepsilon \tag{27}$$

where $C_{\varepsilon 2} = 1.83$ and $C_{\varepsilon 3} = 0.15$. Likewise, so does the Hanjalic and Launder model for which we have the following ε-transport equation in rotating isotropic turbulence:

$$\dot{\varepsilon} = -C_{\varepsilon 2}\frac{\varepsilon^2}{K} - 4C_{\varepsilon 3}K\Omega^2 \tag{28}$$

where $C_{\varepsilon 2} = 1.92$ and $C_{\varepsilon 3} = 0.27$.

The second problem to be considered is homogeneous shear flow in a rotating frame. The mean velocity gradients here take the form

$$\frac{\partial \overline{v}_i}{\partial x_j} = \begin{pmatrix} 0 & S & 0 \\ 0 & 0 & 0 \\ 0 & 0 & 0 \end{pmatrix} \tag{29}$$

which correspond to a plane shear. They are applied in a steadily rotating frame where the axis of rotation is normal to the plane of the shear, i.e.,

$$\Omega_i = (0, 0, \Omega).$$

The substitution of (28) - (29) into (19) yields the Reynolds stress transport equation

$$\dot{\tau}_{ij} = \tau_{ik}A_{jk} + \tau_{jk}A_{ik} - 1.8\frac{\varepsilon}{K}\left(\tau_{ij} - \frac{2}{3}K\delta_{ij}\right) + \frac{2}{3}\left(\frac{3}{5}\mathcal{P} - \varepsilon\right)\delta_{ij} \tag{30}$$

where

$$A_{ij} = \begin{bmatrix} 0 & -\frac{2}{5}S + \frac{7}{5}\Omega & 0 \\ -\frac{7}{5}\Omega & 0 & 0 \\ 0 & 0 & 0 \end{bmatrix} \tag{31}$$

and $\mathcal{P} \equiv -\tau_{12}S$ is the turbulence production. The ε-transport equation for the Pope model is given by

$$\dot{\varepsilon} = -C_{\varepsilon 1}\frac{\varepsilon}{K}\tau_{12}S - C_{\varepsilon 2}\frac{\varepsilon^2}{K} \tag{32}$$

in rotating shear flow. This is the same as the standard model due to the fact that the mean vortex stretching term in (11) vanishes for any plane homogeneous turbulence. The Bardina model for rotating shear flow is obtained by replacing $C_{\varepsilon 1}$ and $C_{\varepsilon 2}$ in (32) with

$$C_{\varepsilon 1}^{*} = 1.50 - 0.015|\tfrac{1}{2}S - \Omega|K/\varepsilon \tag{33}$$

$$C_{\varepsilon 2}^{*} = 1.83 + 0.15|\tfrac{1}{2}S - \Omega|K/\varepsilon. \tag{34}$$

Both the Hanjalic and Launder model and the Raj model take the form

$$\dot{\varepsilon} = -C_{\varepsilon 1}\frac{\varepsilon}{K}\tau_{12}S - C_{\varepsilon 2}\frac{\varepsilon^2}{K} + R \tag{35}$$

in rotating shear flow where $C_{\varepsilon 1} = 1.44$ and $C_{\varepsilon 2} = 1.92$. For the Hanjalic and Launder model

$$R = -C_{\varepsilon 3}K(S - 2\Omega)^2 \tag{36}$$

where $C_{\varepsilon 3} = 0.27$, whereas for the Raj model

$$R = C_{\varepsilon 3} \Omega \varepsilon \left(\frac{3\tau_{zz}}{2K} - 1 \right) \tag{37}$$

where $C_{\varepsilon 3} = 5$.

In rotating shear flow, the transport equations for τ_{ij} and ε are solved subject to the initial conditions of isotropy (22) at time $t = 0$.

3. Discussion of the Results

For the two homogeneous turbulence problems to be considered, the models give rise to a coupled set of first-order nonlinear ordinary differential equations which are solved by a fourth-order accurate Runge-Kutta numerical integration scheme. First, the results for isotropic turbulence in a rotating frame will be presented. For this problem, the solutions depend on Ω, K_0, and ε_0 only through the dimensionless parameter $\Omega K_0/\varepsilon_0$. In Figures 1(a) - (b) the time evolution of the dimensionless turbulent kinetic energy ($K^* \equiv K/K_0$ and $\tau \equiv \varepsilon_0 t/K_0$ is the dimensionless time) obtained for the Bardina model is shown for two moderate rotation rates: $\Omega K_0/\varepsilon_0 = 0.123$ and $\Omega K_0/\varepsilon_0 = 0.469$. These two rotation rates correspond to the test cases considered in the physical experiments of Wigeland and Nagib [15]. It is clear that the agreement between the model predictions and the experiments is good (however, it should be noted that the difference between these results and the predictions of the standard model are small). This is not surprising since the Bardina model was calibrated based on the same experimental data [7]. While the Bardina model performs well for weak rotation rates, it has substantial problems for rapidly rotating isotropic turbulence. In Figure 1(c), the time evolution of the turbulent kinetic energy predicted by the Bardina model for $\Omega K_0/\varepsilon_0 = 69.5$ is compared with the direct simulations of Speziale, Mansour, and Rogallo [14]. It is clear that the model *drastically underpredicts* the decay rate of the turbulent kinetic energy. This arises from a deficiency in the model; it is a simple matter to show that, for $\Omega K_0/\varepsilon_0 > 1.11$, the Bardina model predicts that the turbulence Reynolds number Re_t *increases* during the decay – a result that is unphysical.

The time evolution of the turbulent kinetic energy predicted by the Hanjalic and Launder model is shown in Figure 2(a) along with the predictions of the standard model and the experimental data of

Wigeland and Nagib on rotating isotropic turbulence corresponding to $\Omega K_0/\varepsilon_0 = 0.123$. While the model does not do too bad a job for this case, it yields worse results than the standard model. It then breaks down entirely for appreciable values of $\Omega K_0/\varepsilon_0$. In Figure 2(b), the predictions of the Hanjalic and Launder model are shown for $\Omega K_0/\varepsilon_0 = 0.469$. The model yields extremely bad results for this case and becomes unrealizable when $\tau \doteq 4.2$ (i.e., the dissipation rate becomes negative causing us to terminate the calculations). It is the source term added by Hanjalic and Launder to the right-hand-side of the dissipation rate equation that causes the model to become unrealizable for discernible values of $\Omega K_0/\varepsilon_0$. We will not show the results for the Pope model and the Raj model since they are identical to the predictions of the standard model. Hence, like the standard model, they are unable to predict the reduction in the decay rate of the turbulent kinetic energy that arises from a system rotation of isotropic turbulence [14, 15].

Now, we will discuss the results obtained for rotating shear flow. In Figures 3(a) - (c), the time evolution of the turbulent kinetic energy predicted by the Bardina model is compared with the standard model and the large-eddy simulations of Bardina, Ferziger, and Reynolds [16] for three rotation rates: $\Omega/S = 0, 0.25$, and 0.5. (Again, $K^* = K/K_0$ whereas $t^* = St$). From Figures 3(b) - 3(c) it is clear that the Bardina model has little effect for appreciable rotation rates; like the standard model, the Bardina model drastically underpredicts the growth rate of the turbulent kinetic energy for $\Omega/S = 0.25$ and $\Omega/S = 0.5$. However, what is more serious is that the Bardina modification degrades considerably the results for pure shear flow ($\Omega/S = 0$) as shown in Figure 3(a). The growth rate of the turbulent kinetic energy predicted by the Bardina model is too high. This results from the overprediction of the equilibrium value of the shear anisotropy $(b_{12})_\infty = -0.191$; the standard model yields a value of $(b_{12})_\infty = -0.18$ which is closer to the experimental value of $(b_{12})_\infty = -0.16$.

In Figure 4, the predictions of the Hanjalic and Launder model are compared with the standard model and the large-eddy simulation of Bardina, Ferziger, and Reynolds [16] for homogeneous shear flow ($\Omega/S = 0$). It is clear that the Hanjalic and Launder model predicts much too strong a growth rate for the turbulent kinetic energy in homogeneous shear flow. This results from an underprediction of the dissipation rate. In fact, the calculation is terminated a short time after $St = 2$ due to the model becoming unrealizable through negative dissipation rates. Since the Hanjalic and Launder model yields anomalous results for pure shear flow (the most basic case) we will not bother to show the results for non-zero rotation rates.

In Figures 5(a) - (b), the results predicted by the Raj model for rotating shear flow are compared with those of the standard model as well as with the large-eddy simulations of Bardina, Ferziger, and Reynolds [16]. We do not show the results of the Raj model for pure shear flow ($\Omega/S = 0$) since they are identical to those of the standard model. From Figure 5(a), it is clear that the Raj model yields improved predictions over the standard model for the $\Omega/S = 0.25$ case. However, there are problems with the Raj model for rotation rates that are appreciably larger than $\Omega/S = 0.25$ as can be seen in Figure 5(b). For the case where $\Omega/S = 0.5$, the Raj model gives rise to large amplitude inertial oscillations which appear to be unphysical.

It is clear from these results that simple corrections to the modeled dissipation rate equation obtained by adding a source term that depends on rotational strains can give rise to realizability problems (see Lumley [1]). Since the standard modeled dissipation rate equation is realizable with respect to K and ε (i.e., as shown by Speziale [17], the standard model will always yield positive kinetic energies and dissipation rates in homogeneous turbulent flows), it is clear that these kind of ad hoc corrections are counterproductive. In order to properly describe the turbulent flows discussed in this section, directional and two-point information is needed in the modeling of the dissipation rate. Speziale and Gatski [18] have been recently working on the development of a modeled transport equation for the tensor dissipation rate

$$\varepsilon_{ij} \equiv 2\nu \overline{\frac{\partial u_i}{\partial x_k} \frac{\partial u_j}{\partial x_k}}.$$

This model for the tensor dissipation takes the form

$$
\begin{aligned}
\dot{\varepsilon}_{ij} = & -\varepsilon_{ik}\left(\frac{\partial \overline{v}_j}{\partial x_k} + 2e_{mkj}\Omega_m\right) - \varepsilon_{jk}\left(\frac{\partial \overline{v}_i}{\partial x_k} + 2e_{mki}\Omega_m\right) \\
& + \frac{2}{3}C_{\varepsilon 1}\frac{\varepsilon}{K}\mathcal{P}\delta_{ij} - C_{\varepsilon 2}\frac{\varepsilon}{K}\varepsilon_{ij} + C_{\varepsilon 3}\varepsilon\overline{S}_{ij} \\
& - C_{\varepsilon 4}(\varepsilon_{ik}\overline{W}_{jk} + \varepsilon_{jk}\overline{W}_{ik})
\end{aligned} \tag{38}
$$

in a rotating frame where $C_{\varepsilon 1} = 1.46$, $C_{\varepsilon 2} = 1.83$, $C_{\varepsilon 3} = 1.50$, and $C_{\varepsilon 4} = 1.0$. It was developed for use in conjunction with the SSG model for the pressure-strain correlation derived recently by Speziale, Sarkar, and Gatski [19]. In Figures 6(a) - (c), the results predicted by the tensor dissipation model (used in conjunction with the SSG model) are compared with the predictions of the Launder, Reece, and Rodi model and the large-eddy simulations [16]. It is clear that this new tensor dissipation model does a much better job in capturing the trends of the large-eddy simulations. Most notably, the substantially stronger growth rate of the $\Omega/S = 0.25$ case is, for the most part, captured; furthermore, the model does not restabilize for the $\Omega/S = 0.5$ case – a result that is consistent with linear stability theory as well as with the large-eddy simulations. These improvements are not that surprising in light of the recent findings of Durbin and Speziale [20] which suggest that the dissipation rate tensor can be *anisotropic* in equilibrium homogeneous shear flows at high turbulence Reynolds numbers.

4. Conclusions

Several modifications to the modeled dissipation rate transport equation that have been proposed to account for rotational strains are compared critically for two test flows. The following general conclusions can be drawn:

(1) The Pope [5] model yields the *same* predictions as the standard model for rotating isotropic turbulence and for rotating homogeneous shear flow. These predictions are deficient in that they fail to properly account for the reduction in the turbulence decay rate that results from a system rotation of isotropic turbulence; they also fail to account accurately for the changes in the growth rate of the turbulent kinetic energy that result from a system rotation of ho-

mogeneous shear flow which can either stabilize or destabilize the flow.

(2) The Hanjalic and Launder [6] model has major problems with the violation of realizability. It yields negative dissipation rates in rotating isotropic turbulence for appreciable values of $\Omega K_0/\varepsilon_0$. Furthermore, this modification degrades significantly the predictions for homogeneous shear flow (the predicted growth rate of the turbulent kinetic energy is much too large).

(3) The Bardina model [7] is only able to accurately predict the reduction in the decay rate of the turbulent kinetic energy in rotating isotropic turbulence for weak to moderate rotation rates where this effect is small. For $\Omega K_0/\varepsilon_0 \gg 1$, this model predicts too strong a reduction in the turbulence decay rate. Furthermore, this model does not yield any improvements for rotating homogeneous shear flow and mildly degrades the predictions for pure shear flow.

(4) The Raj [8] model yields the *same* deficient predictions as the standard model in rotating isotropic turbulence. In rotating homogeneous shear flow, this model does yield some improvements for moderate rotation rates for which $\Omega/S < 0.3$. However, for stronger rotation rates the model can give rise to large amplitude inertial oscillations which appear to be unphysical.

In order to properly describe rotating turbulent flows, some directional and two-point information needs to be incorporated into models for the dissipation rate rather than the ad hoc modifications discussed above. The directional information can be incorporated through the use of the tensor dissipation. In this regard, the preliminary calculations presented in Section 3 for rotating shear flow based on a modeled tensor dissipation rate equation appear to be promising. However, the reduction in the decay rate of the turbulent kinetic energy in rotating isotropic turbulence is a two-point phenomena; the inertial waves generated by a system rotation scramble the transfer term in such a way that the phase coherence needed to cascade energy from large to small scales is disturbed. Some limited two-point information needs to be included based on an appropriate integral length scale which responds to rotational strains. This issue is currently under investigation and will be the subject of a future paper.

Acknowledgement

The second author (RR) would like to acknowledge the support provided by an ASEE/NASA Langley Research Fellowship.

References

[1] Lumley, J. L., 1978, *Computational Modeling of Turbulent Flows*, Adv. Appl. Mech. **18**, pp. 123-176.

[2] Speziale, C. G., 1990, *Analytical Methods for the Development of Reynolds Stress Closures in Turbulence*, Ann. Rev. Fluid Mech. **23**, pp. 107-157.

[3] Launder, B. E. and Spalding, D. B., 1974, *The Numerical Computation of Turbulent Flows*, Comput. Methods Appl. Mech. & Engrg. **3**, pp. 269-289.

[4] Launder, B. E., 1990, *Phenomenological Modeling: Present and Future*, Lecture Notes in Physics (J. L. Lumley, ed., Springer-Verlag, New York) **357**, pp. 439-485.

[5] Pope, S. B., 1978, *An Explanation of the Turbulent Round Jet/Plane Jet Anomaly*, AIAA J. **16**, pp. 279-281.

[6] Hanjalic, K. and Launder, B. E., 1980, *Sensitizing the Dissipation Equation to Irrotational Strains*, ASME J. Fluids Eng. **102**, pp. 34-40.

[7] Bardina, J., Ferziger, J. H., and Rogallo, R. S., 1985, *Effect of Rotation on Isotropic Turbulence: Computation and Modeling*, J. Fluid Mech. **154**, pp. 321-336.

[8] Raj, R., 1975, *Form of the Turbulence Dissipation Equation as Applied to Curved and Rotating Turbulent Flows*, Phys. Fluids **18**, pp. 1241-1244.

[9] Bardina, J., 1988, *Turbulence Modeling Based on Direct Simulation of the Navier-Stokes Equations*, Proceedings of the 1st National Fluid Dynamics Congress, Paper No. 88-3747-CP.

[10] Speziale, C. G., 1989, *Turbulence Modeling in Non-Inertial Frames of Reference*, Theoret. & Comput. Fluid Dynamics **1**, pp. 3-19.

[11] Raj, R. and Speziale, C. G., 1990, *A Note on the Dissipation Equation in Rotating Turbulent Flows*, in preparation.

[12] Launder, B. E., Reece, G., and Rodi, W., 1975, *Progress in the Development of a Reynolds Stress Turbulence Closure*, J. Fluid Mech. **68**, pp. 537-566.

[13] Speziale, C. G. and Mac Giolla Mhuiris, N., 1989, *On the Prediction of Equilibrium States in Homogeneous Turbulence*, J. Fluid Mech. **209**, pp. 591-615.

[14] Speziale, C. G., Mansour, N. N., and Rogallo, R. S., 1987, *The Decay of Isotropic Turbulence in a Rapidly Rotating Frame*, Proceedings of the 1987 Summer Program of the Center for Turbulence Research (P. Moin, W. C. Reynolds, and J. Kim, eds., Stanford University Press), pp. 205-211.

[15] Wigeland, R. A. and Nagib, H. M., 1978, *Grid-Generated Turbulence With and Without Rotation about the Streamwise Direction*, IIT Fluids and Heat Transfer Report R78-1, Illinois Institute of Technology.

[16] Bardina, J., Ferziger, J. H., and Reynolds, W. C., 1983, *Improved Turbulence Models Based on Large-Eddy Simulation of Homogeneous, Incompressible Turbulent Flows*, Stanford University Technical Report TF-19.

[17] Speziale, C. G., 1990, *Discussion of Turbulence Modeling: Present and Future*, Lecture Notes in Physics (J. L. Lumley, ed., Springer-Verlag, New York) **357**, pp. 490-512.

[18] Speziale, C. G. and Gatski, T. B., 1990, *A Model for the Tensor Dissipation Rate of Turbulence*, in preparation.

[19] Speziale, C. G., Sarkar, S., and Gatski, T. B., 1989, *Modeling the Pressure-Strain Correlation of Turbulence – An Invariant Dynamical Systems Approach*, J. Fluid Mech., in press.

[20] Durbin, P. A. and Speziale, C. G., 1990, *Local Anisotropy in Strained Turbulence at High Reynolds Numbers*, ASME J. Fluids Eng., submitted for publication.

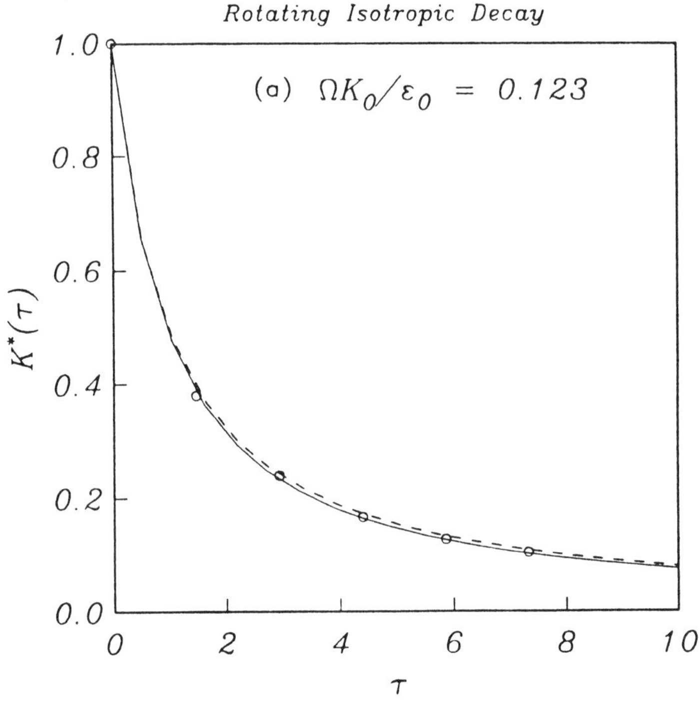

Figure 1. The decay of turbulent kinetic energy in rotating isotropic turbulence: —— Bardina model; - - - standard model; o experimental data [15]; • simulations [14]. (a) $\Omega K_0/\varepsilon_0 = 0.123$, (b) $\Omega K_0/\varepsilon_0 = 0.469$, (c) $\Omega K_0/\varepsilon_0 = 69.5$.

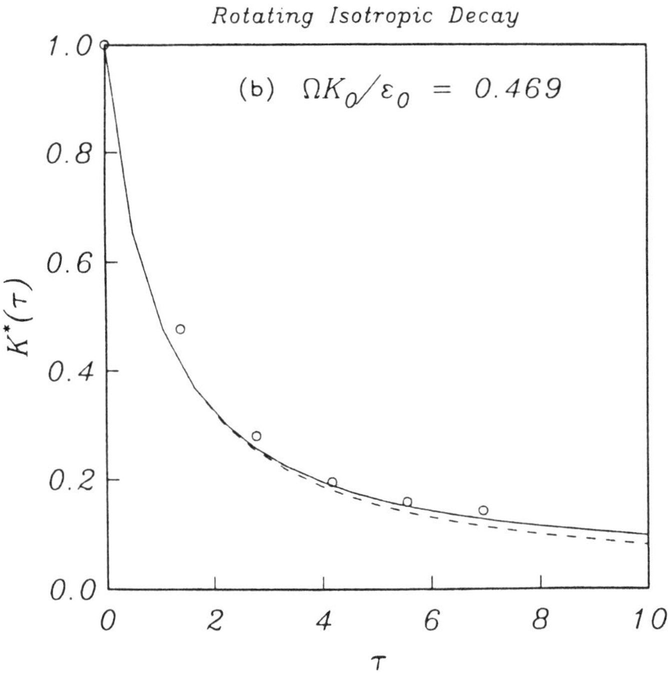

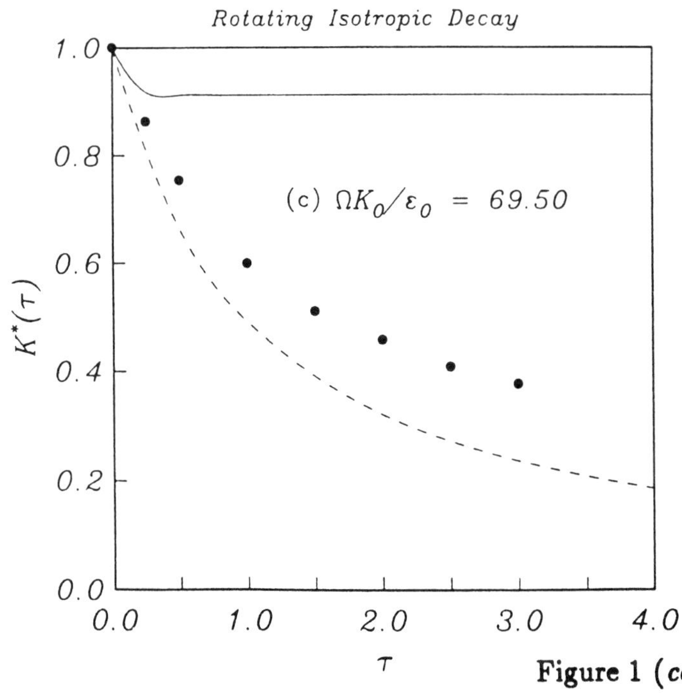

Figure 1 (*cont.*)

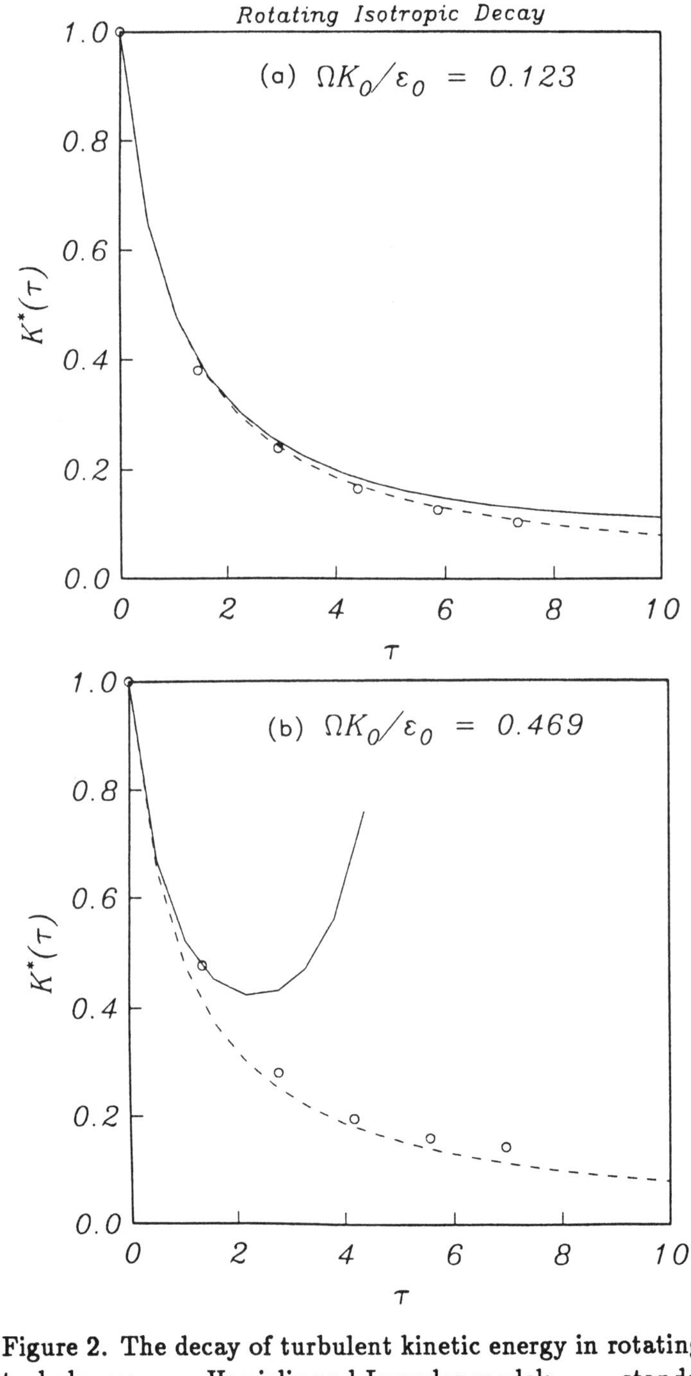

Figure 2. The decay of turbulent kinetic energy in rotating isotropic turbulence: —— Hanjalic and Launder model; - - - standard model; o experimental data [15]. (a) $\Omega K_0/\varepsilon_0 = 0.123$, (b) $\Omega K_0/\varepsilon_0 = 0.469$.

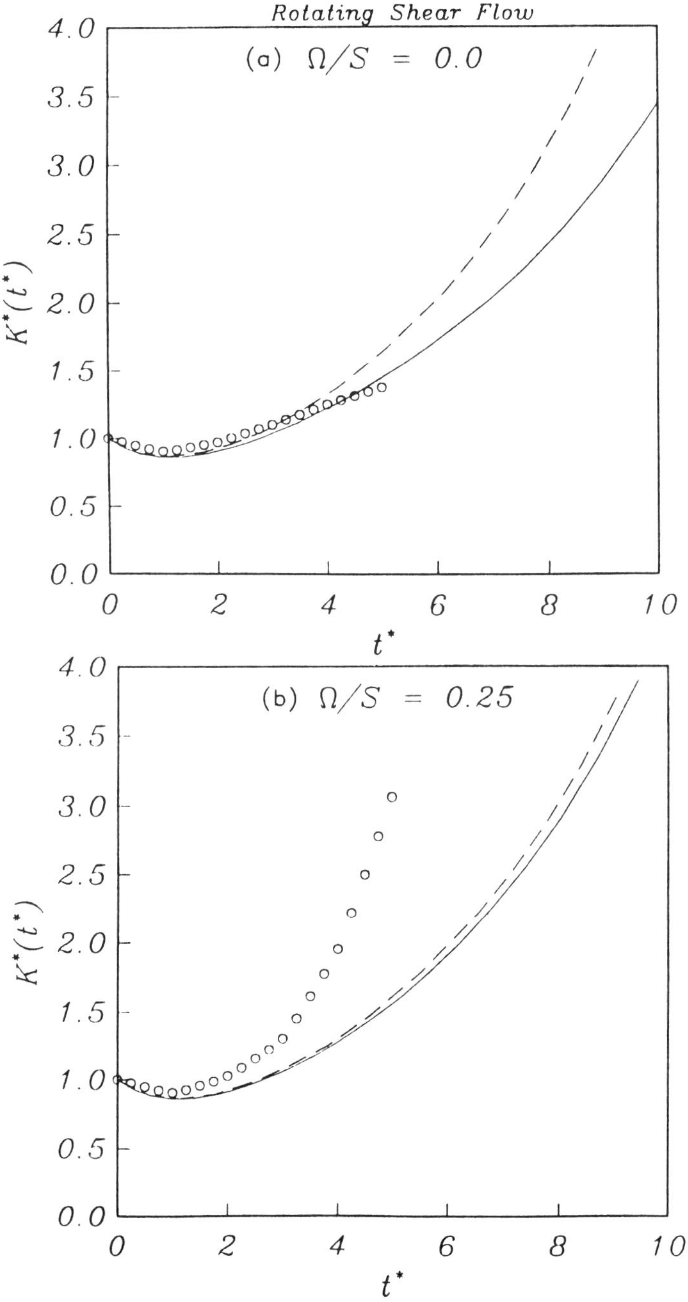

Figure 3. Time evolution of the turbulent kinetic energy in rotating shear flow for $\varepsilon_0/SK_0 = 0.296$: —— standard model; – – – Bardina model; o large eddy simulations [16]. (a) $\Omega/S = 0$, (b) $\Omega/S = 0.25$, (c) $\Omega/S = 0.5$.

Fig. 3 c)

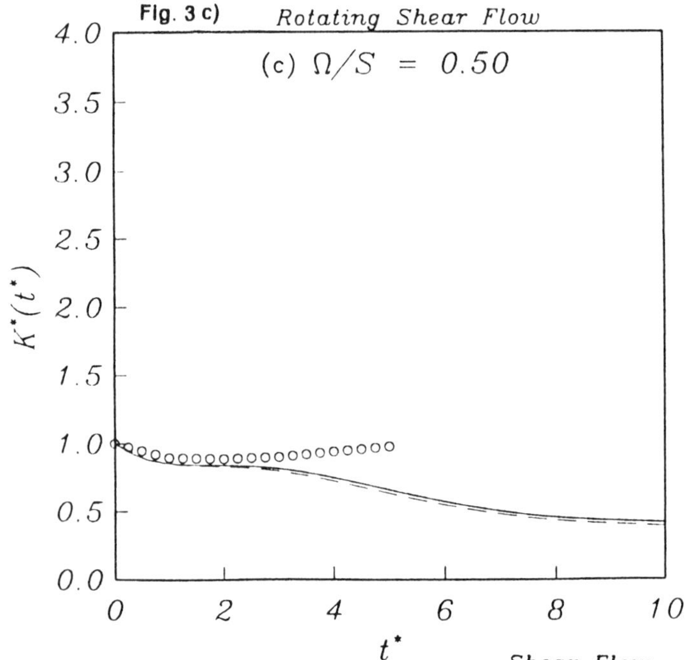

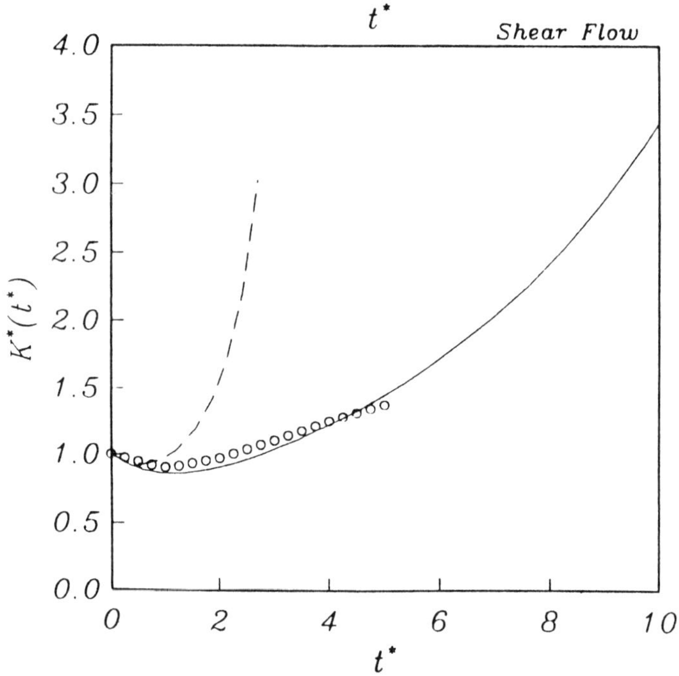

Figure 4. Time evolution of the turbulent kinetic energy in homogeneous shear flow for $\varepsilon_0/SK_0 = 0.296$: —— standard model; - - - Hanjalic and Launder model; o large eddy simulation [16].

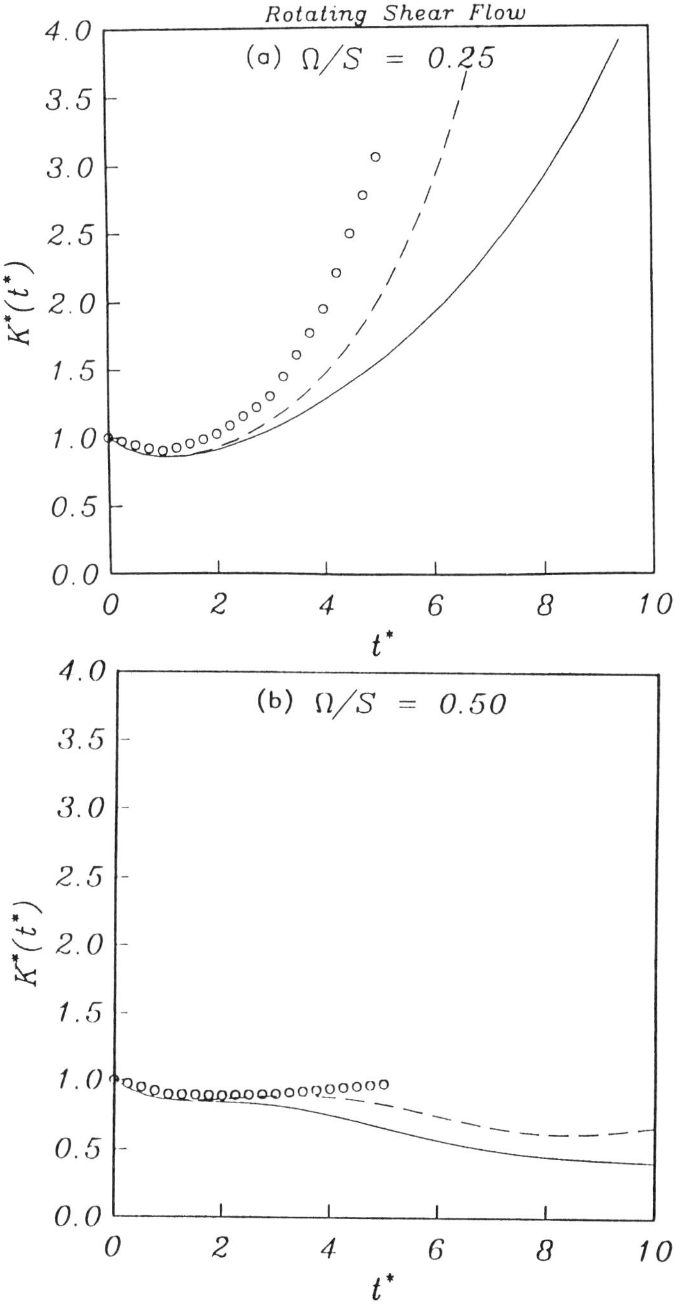

Figure 5. Time evolution of the turbulent kinetic energy in rotating shear flow for $\varepsilon_0/SK_0 = 0.296$: —— standard model; – – – Raj model; o large eddy simulations [16]. (a) $\Omega/S = 0.25$, (b) $\Omega/S = 0.5$.

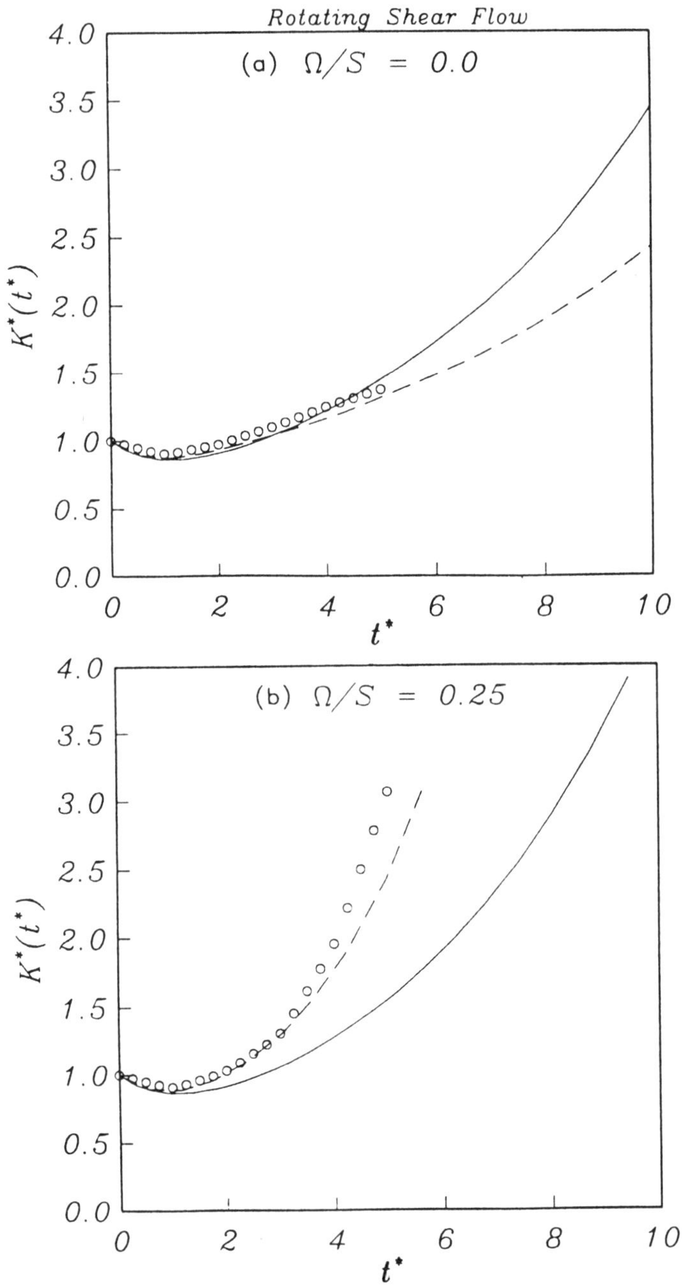

Figure 6. Time evolution of the turbulent kinetic energy in rotating shear flow for $\varepsilon_0/SK_0 = 0.296$: —— standard model; – – – tensor dissipation model [18]; ○ large eddy simulations [16]. (a) $\Omega/S = 0$, (b) $\Omega/S = 0.25$.

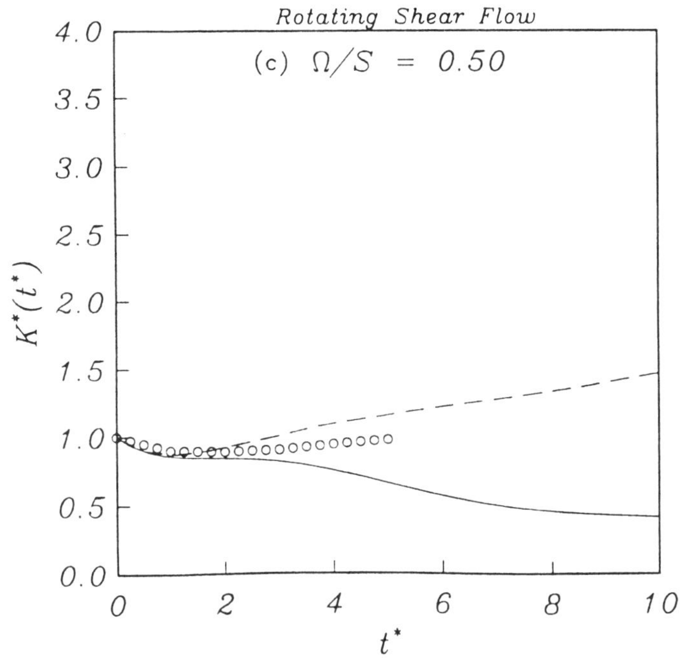

Figure 6. Time evolution of the turbulent kinetic energy in rotating shear flow for $\varepsilon_0/SK_0 = 0.296$: —— standard model; – – – tensor dissipation model [18]; o large eddy simulations [16]. (c) $\Omega/S = 0.5$.

MAPPING CLOSURES FOR TURBULENT MIXING AND REACTION*

S.B. Pope[1]

Cornell University,
Ithaca, New York 14853

ABSTRACT

The mapping closure for the one-point pdf of an inert scalar in homogeneous turbulence is explained and developed. Proposed by Chen, Chen & Kraichnan (1989), the mapping closure formalism is based on fundamentally different physical assumptions than previous closures. It is explained here in terms of three fields. The *turbulent* field is the physical field under study—in this case it is a homogeneous, isotropic, scalar field. At the level of closure considered, the only information known about this field is its one-point pdf. The second field is a specified (and hence completely known) *Gaussian* field. The third field, the *surrogate* field, is constructed by a known amplitude mapping from the Gaussian field. This mapping is constructed so that the one-point pdf's of the turbulent field and the surrogate field are identical. Consequently, at the level of closure, the two fields are indistinguishable. The closure assumption made is that the unknown statistics of the turbulent field are equal to those of the known surrogate field.

For the case of a single inert scalar decaying from a double-delta-function initial pdf, an analytic solution to the mapping closure equation is obtained. This solution is in remarkable agreement with the DNS data of Eswaran & Pope (1988). Again for a single scalar, it is shown that the mapping closure can readily be implemented as a particle method, and hence used in Monte Carlo solutions of modelled pdf equations for inhomogeneous flows. The mapping closure is then extended to many reactive scalars.

[1]This work was supported in part by U.S. Air Force Office of Scientific Research (Grant No. AFOSR-88-0052).

Acknowledgements

It is a pleasure to dedicate this paper to John Lumley on the occasion of his sixtieth birthday. I am grateful to Dr. R.H. Kraichnan for many discussions on mapping closures.

[1] Chen, H., Chen, S. and Kraichnan, R.H. (1989) Phys. Rev. Lett., **62**, 2657

[2] Eswaran, V. and Pope, S.B. (1988) Phys. Fluids, **31**, 506.

---

*The full text of this paper is to appear in *Theoretical & Computational Fluid Dynamics.*

SUBGRID LENGTH-SCALES FOR LARGE-EDDY SIMULATION OF STRATIFIED TURBULENCE*

Ulrich Schumann

DLR, Institute of Atmospheric Physics,
Oberpfaffenhofen, Germany

ABSTRACT

The influence of buoyancy on the length-scales for the dissipation rate of kinetic energy, and for momentum, heat and other scalar transport has to be known for subgrid-scale (SGS) models in a Large-Eddy Simulation (LES). For the inertial subrange, Lilly (1967) has shown that the grid spacing is the relevant length-scale for SGS effects. Deardorff (1980) proposed to reduce all the length-scales for stable stratification. Numerical and experimental data show, however, that the dissipation length-scale may strongly increase in stable layers with little shear. Also Lumley's (1964) theory for the energy spectrum in a stratified fluid suggests such an increase. In this paper we apply analysis of previous algebraic second-order closure SGS-models, parameter studies with different length-scale models in LES, and analysis of direct simulations of sheared and unsheared stably stratified homogeneous turbulence. These analyses suggest that the limiting effect of stratification should only be applied to the length-scales of vertical eddy-diffusivities of heat and scalars but not to those of momentum and dissipation.

References

[1] Deardorff, J. W., 1980, Stratocumulus-capped mixed layers derived from a three-dimensional model. *Boundary-Layer Met.* 18, 495-527.

[2] Lilly, D., K., 1967, The representation of small-scale turbulence in numerical simulation experiments. *Proc. IBM Sci. Comput. Symp. on Environm. Sci.*, Thomas J. Watson Res. Center, Yorktown Heights, N.Y., IBM Form 320-1951, pp. 195-210.

[3] Lumley, J. L., 1964, The spectrum of nearly inertial turbulence in a stably stratified fluid. *J. Atmos. Sci.* 21, 99-102.

---

*The full text of this paper is to appear in *Theoretical & Computational Fluid Dynamics.*

PART 2

COHERENT STRUCTURES AND THE PROPER ORTHOGONAL DECOMPOSITION

UNDERSTANDING TURBULENCE VIA VORTEX DYNAMICS

Fazle Hussain and Mogens V. Melander[+]

University of Houston
Houston, TX 77204

ABSTRACT

Turbulence research in the past half century has seen notable progress, even bursts of hectic activity, on three seemingly independent fronts: *statistical* approaches starting in the forties (e.g. Kolmogorov 1941), *structural* approaches starting in the sixties (e.g. Kline *et al.* 1967; Crow & Champagne 1971; Brown & Roshko 1974) and *dynamical* approaches starting in the eighties (e.g. Berge *et al.* 1984). While researchers typically tend to show their partiality for one approach or another depending on their familiarity or ignorance, even sometimes displaying their disapproval or disdain for approaches outside their own preferred one, there should be little doubt that all the three approaches are helpful, perhaps even essential, to understand turbulence. There are few whose repertoire of expertise covers statistical, structural and dynamical turbulence. John Lumley is one such exception, having had made profound contributions in all three areas (e.g. Lumley 1970; 1980; Aubry *et al.* 1988).

1. Why Vortex Dynamics?

The scope of our research at the University of Houston has been narrower, focusing primarily on structural turbulence. Motivated by the desire that there is an order underlying the disorder of fluid turbulence, we, among others, have explored in turbulent shear flows large scale organized motions, which are now popularly called *coherent structures* (CS). We have defined a CS as a flow module with instantaneously phase-correlated vorticity (Hussain 1979, 1983). Our persistent interest in CS stems from the expectation that CS are the quasi-deterministic features of turbulence, and that they are the ones that do most of what turbulence does and thus may hold the key to explaining and controlling turbulence phenomena. While obviously defined so far

---

[+] Permanent Address: Southern Methodist University, Dallas, Texas

in a qualitative way, the CS concept has enabled us to unravel specific details of their geometrical features and dynamics in a variety of turbulent shear flows that we have studied experimentally and numerically, as well as to partly explain in mechanistic ways the role of turbulence in transports of heat, mass, and momentum, in chemical reaction and combustion, and in the generation of drag and aerodynamic noise. There is little doubt now that turbulence management can be achieved via the control of initiation, growth, interaction and decay of CS and that CS are crucial to turbulence control; i.e. *no CS, no turbulence control.*

We will not discuss here the variety of CS in different turbulent flows, their topology (by which we mean knottedness of the vortex lines of the coherent vorticity field), nor their dynamical significance, as these have been discussed previously by us and others in considerable detail (see, for example, Hussain 1983; Fiedler 1988).). We, however, note that our understanding of CS is still fairly qualitative. This should not be surprising because CS research is still in its infancy. A much clearer understanding of CS is required for obtaining fundamental explanation of turbulence phenomena and for incorporation of CS concepts in a viable turbulence theory or a predictive model. What is clearly lacking is a mathematical framework for CS.

The overwhelming majority of thousands of papers written on CS involve flow visualization, as if showing the mere presence of an eddying pattern of motion is a challenge in itself. However, our own work has focused mainly on quantitative data, and we have reiterated that flow visualization can not only be confusing but also can be grossly misleading. Even in the case of unity Schmidt number (Sc), scalar marker concentration matches vorticity intensity only in 2D flows; as we will show here, even in a totally axisymmetric flow, with $Sc \equiv 1$ and with identical initial distribution of scalar and vorticity characterizing a structure, contours of constant scalar concentration deviate significantly from contours of vorticity magnitude level. This should not at all seem surprising because, in a 3D flow, vorticity undergoes self-augmentation by stretching; markers are not subject to this effect. Consequently, markers are typically depleted from regions of vorticity intensification but are accumulated in regions of vortex compression or lowered vorticity, thus obscuring dynamically significant/strong flow events from visualization. We have previously demonstrated and reiterated the limitations of flow visualization, and will elucidate this point further here.

Our definition of CS in terms of vorticity has had its roots in a deeper motivation. Pending the development of a mathematical definition as well as of evolutionary equation for CS, it was our expectation that as vortical modules in turbulent flows, CS evolution and interaction can be rendered tractable via the well-developed theoretical framework of vortex dynamics. Of course, the definition, let alone the dynamics, of a vortex in a turbulent flow is a nontrivial challenge that requires further investigation (subject of a detailed study in our group). Thus, while vortex studies are of fundamental hydrodynamic interest in their own right, care is necessary to be judicious in proper selection of the problem as well as in the extrapolation of vortex dynamics results to interpret turbulence dynamics or even in establishing the relevance of the results to turbulence. We have carried out a number of vortex studies with a view to understanding CS interactions. In the following, we briefly review two studies which shed new insight on the physics of cascade and large scale/fine scale interaction in turbulent shear flows.

2. Vortex Reconnection

This study has been motivated by our early conjecture that vortex reconnection (also called crosslinking or cut-and-connect) plays a key role in cascade and mixing as well as in the production of turbulence, helicity and aerodynamics noise. In addition, reconnection of vortex lines happens continuously in a viscous 3D, particularly turbulent, flow. Thus, this study is relevant both to CS interactions at large scales and to fine-scale viscous interactions, depending on the Reynolds number of interacting vortices.

In spite of its extensive study, there is yet no clear definition of vortex reconnection. Vortex reconnection is a local change of vortex line topology in the neighborhood of points of zero vorticity. To clearly understand this idea consider a region of space where two nearly antiparallel vortices are in close proximity (Fig. 1). In such a region we can partition the set of all vortex lines into three types: non-reconnected lines (e.g. A_nA_n' and B_nB_n'), reconnected lines (e.g. A_rB_r) and separating lines (e.g. AO). Associated with vortex lines we have a physically relevant measure, namely circulation. The set of separating lines has measure zero, *that is no* circulation. Thus in cross section 1 of vortex A the total circulation Γ consists of two contributions, namely the circulation Γ_n of the non-reconnected lines, and the circulation Γ_r of the reconnected lines. Obviously $\Gamma = \Gamma_n + \Gamma_r$ by Gauss's theorem. We may now define *reconnection as the process which decreases Γ_n and increases Γ_r .*

Some remarks are appropriate at this point.

a) Vortex lines are lines to which the vorticity field is everywhere tangent. They should not be confused with vortex filaments, which are a mathematical model of thin vortex tubes. Strictly speaking, filaments (lines with circulation) do not exist in a classical fluid.
b) Occasionally it is believed that vortex lines are only meaningful in an inviscid fluid, as vortex lines are generally not material lines in a viscous fluid. Presumably this belief is rooted in a failure to distinguish between vortex filaments and vortex lines. Much of the classical 3D vortex dynamics deals with thin vortex tubes in an inviscid fluid and models them as material vortex filaments. Of course this branch of vortex dynamics does not apply to viscous fluids.
c) Our definition of vortex reconnection does not build on any direct or implicit assumption concerning a connection between mass transport and vorticity transport. For example we do not assume that vortex lines are material lines. However, it is sometimes useful to connect the two anyway. Namely, in a problem like reconnection the viscous effects are not equally important everywhere in the physical space; in some regions the viscous effect are very important (e.g. in the reconnection region where the vorticity gradients are large) while in other regions the viscous effects are only of secondary importance and can thus be ignored to a first approximation . Such an approach can be helpful as it allows us to assign an approximate time history to the vortex lines by following the lines from an essentially inviscid flow region into a highly viscous domain. This way using the inviscid flow region, where vortex lines are material lines gives the time-parameterization in the viscous domain. This idea has been used extensively to explain vortex reconnection by us (Melander & Hussain 1988, referenced as MH).

In the case of incompressible flow, vortex reconnection is possible only by annihilation of vorticity by viscous cross-diffusion (MH). In a compressible flow, nonbarotropic or other mechanisms of vorticity generation can produce new vorticity and hence apparent reconnection. Here we address reconnection of incompressible antiparallel vortices of equal circulation. The intricate details of the reconnection process will require instantaneous measurement of the 3D vorticity field in a laboratory experiment. Such is beyond the current measurement technology. Pending the development of 3D holographic velocimetry, now under development at UH, we have investigated the reconnection phenomenon by direct numerical simulation of the Navier-Stokes equation in a 64^3 periodic box employing spectral methods.

The initial condition consists of two vortices (of Reynolds number $\Gamma/\nu = 1000$) with 'compact Gaussian' support and with a small sinusoidal perturbation of the vortex axes to guarantee that the two vortices continue to press against each other in order to sustain viscous cross-diffusion and hence reconnection. Fig. 2 shows the time evolution of the reconnection process in the left column ($|\omega|$ level shown at 30% of the initial peak vorticity) and the corresponding scalar of one vortex in the right column (at a level of 5% of the initial peak). The initial scalar concentration was set equal to initial $|\omega|$ and the Schmidt number Sc = 1. Note that an iso-vorticity surface should not be confused with a vorticity surface which is tangential to vorticity vectors. A vortex line can cross an iso-vorticity or $|\omega|$ = constant surface.

A few observations can be made from the simulation. The scalar (even at a very low level) fails to faithfully mark the vortical fluid because stretching depletes marker away from locations of peak $|\omega|$. The reconnection consists of three distinct phases. (I) *Inviscid induction* when vortex cores approach as they deform from the initial circular cross section, form a contact zone with a dipole cross-section and then flatten and stretch to form a head-tail structure. (II) Vorticity annihilation at the contact zone as a result of viscous cross-diffusion causing crosslinking of vortex lines which orient orthogonally to the initial direction of vortex lines, and accumulate into two bundles (the *bridges*). (III) As the bridges pull apart by self induction they induce a downwash (an induced flow opposite to the direction of motion of the initial dipole) causing a reversal of the curvature of initial vortices. The remnants of the initial two vortices at the contact zone now pull apart from each other as a result of reversed self induction caused by the curvature reversal while being stretched as *threads* by the bridges. This arrests further reconnection and causes the threads to survive for a long time.

We show that bridging is the crucial feature of reconnection (see MH for detailed explanation). The threading, which so far has not been observed in flow visualization is an unavoidable consequence of reconnection. When a long time has elapsed after the two bridges have moved sufficiently apart by self induction, the dipole formed by the two threads can propel forward (in the direction of motion of the initial dipole) thus reversing *again* the curvature of the thread dipole, and then pushing the two threads against each other. This then initiates another reconnection, hence forming new bridgelets. While simulation to show the second stage of reconnection will require very long computation time and will be prohibitively expensive, there should be little doubt that such successive reconnections have to happen. This is what we propose as a *new cascade mechanism*.

The formation of the head-tail shape of the dipole itself is an interesting aspect of mixing and cascade. Vortical fluid from the two vortices in the dipole are peeled out forming two relatively thinner tails of opposite vorticity. Just like the heads, the tails also annihilate each other's circulation by viscous cross-diffusion. In addition, because the circulation in the heads is much higher than in the tails, the heads move ahead by mutual induction, leaving the tails behind (Fig. 3). The tails then undergo uneventful viscous decay while mixing their fluids. The heads eventually survive as the threads.

These processes produce increased branching of the initial vortex tubes, increase the vorticity surface much more than would happen by viscous diffusion and decrease the length scales of the vortices (e.g. by formation of threads). These are examples of cascade and mixing. Thus the ingredients of cascade and mixing associated with reconnection are: (i) bridging, (ii) head-tail formation with sheetlike tail, (iii) separation of head from tail, (iv) formation of threading resulting from heads and (v) successive reconnection of the threads. Thus, vortex reconnection, apart from being a hydrodynamic curiosity and an example of topological transformation, has profound implications in viscous turbulence phenomena. Further details of the reconnection phenomenon are available for antiparallel vortices (MH), and collision of vortex rings (Kida *et al.* 1989).

3. Core Dynamics of nonuniform vortices and Large scale/fine scale interactions

In spite of the usual assumptions made by theorists - of course, forced by analytical and conceptual expediency - there is no distinct scale separation in most turbulent flows, which consist of a continuum of scales from the largest determined by the flow geometry to the smallest at the Kolmogorov scale. Suppose we concede scale separation as a hypothetical possibility or consider interaction of large-scale CS and fine scales in the absence of intermediate scales; we have asserted, contrary to the conventional wisdom that these two disparate scales are decoupled (a centerpiece of the Kolmogorov theory and the local isotropy hypothesis), that these scales are indeed strongly coupled and that local isotropy is an unlikely abstraction (Hussain 1983). We address this question here by studying the interaction between large and fine scale turbulence. To clearly focus on the mechanism we study the highly idealized situation of a well-defined vortex (as a prototypical CS) superimposed with fine-scale turbulence. In order to avoid the additional complication of curvature-induced self induction, we consider a straight vortex. Our vortex has a sinusoidally varying core size. The vortex is then by itself dynamically evolving and not merely diffusing. The dynamics then represents some aspects of CS internal dynamics. Also, to focus on the coupling alone, there is initially a distinct wavenumber separation (i.e. spectral gap) between the CS and the fine-scale turbulence, the latter consisting of vorticity fluctuations in the dissipation range. This study was carried out by direct numerical simulation of the Navier-Stokes equations in a 128^3 box using 3D (alias-free) spectral method with fourth-order time stepping and periodic boundary conditions.

An important question we wanted to settle before the coupling question was addressed is the nature of core dynamics in a nonuniform vortex. As a result of interaction with turbulence, axisymmetric waves can be excited which involve axial flow with significant core dynamics. Such axial flow can be induced by nearby nonparallel vortices, as well as in situations of nonuniform core diameter or when the vorticity peak is not centered at the vortex core. All these situations occur rather frequently in flows with CS and also when two or more vortices are near each other; the reconnection of antiparallel vortices is just one example of many (see MH). Nonuniform core flow occurs in the bridges because of the presence of the orthogonal threads (this is not unlike the situation of ribs inducing axial flows in the cores of rolls in a mixing layer). When we consider the bridges, the cross-section at the dividing plane has a crescent shape with the peak vorticity not located at the centroid of the bridge cross-section while away from the symmetry plane the vortices have circular cross-section. This produces coiling of the vortex lines and hence pumping of fluid towards the reconnection region, and thus providing an invisid mechanism that tends to make the vorticity distribution uniform in the bridges.

The effect of nonuniform core and thus core dynamics can be significant in both self induction and local interaction. Many theoretical studies of organized structures in turbulent flows (such as hairpins in turbulent boundary layers and homogeneous shear flows) have modeled CS in such flows as being of uniform core diameter and have presumed core dynamics to be of insignificant effect.

Vortex core dynamics is also of considerable importance in computation and modeling of turbulence using the fairly popular and attractive vortex methods. In these approaches, the self-induced axial flow inside filaments is usually ignored (Lundgren & Ashurst (1989) is an exception). While Moore & Saffman (1972) postulates that if the vortex cores exhibited axial variation then an axial flow would immediately result so as to make the vortex cores essentially uniform on a short timescale. However, it is well known that vortices support waves which travel along the axis. These waves can be on the vortex core, or bending waves on the vortex axis, or both. We find that viscous effect is the primary damping mechanism for axisymmetric waves on the vortex core, thereby indicating that the assumption of uniform vortex cores is problematic, at best.

In this study we address the simplest case of a rectilinear laminar vortex with sinusoidal core diameter first before considering the coupling with turbulence. The core dynamics and axial flow in the vortex can be understood well in terms of coupling of the swirling flow with the meridional flow. Considering (u,v,w) to be the velocities in the cylindrical coordinates (r,θ,z) if we temporarily ignore viscous effects, we have, from Kelvin's theorem, that

$$\frac{D}{Dt}(rv) = 0\,,$$

where

$$\frac{D}{Dt} = \frac{\partial}{\partial t} + u\frac{\partial}{\partial r} + w\frac{\partial}{\partial z}$$

represents derivative following a particle in the meridional flow. Thus rv (circulation of a circular material circuit centered on the vortex axis) is convected with the meridional flow. On the other hand,

$$\frac{D}{Dt}(\omega_\theta / r) = r^{-4}\frac{\partial}{\partial z}(rv)^2$$

show the direct coupling between swirl and meridional flow, being strongest along the axis of the vortex. Although there is no azimuthal vorticity at $t = 0$, it is immediately created by the coupling term causing significant axial expansions and contractions of the vortex tube. It is this coupling that enables ω_θ/r to be transported as waves without corresponding mass transport. Fig. 4 shows a composite sketch of the rectilinear laminar vortex showing meridional plane motion (streamlines), axial and azimuthal velocity profiles and twisted vortex lines. Corresponding to the axial variation of the vortex core size, there is an axial variation in the swirling motion around the axis. As a result, the vortex lines which are initially sinusoidal lines in the meridional plane start twisting. The helical twists propagate from the ends ($z=0, 2l$) to $z = l$ while the vortex core expands at ends but shrinks in the middle. This is followed by a reversed coiling process, and so on. The sequence is demonstrated very well in Fig. 5 by vortex lines starting from an arbitrary radial rake. Since vortex lines are not material lines no explicit time history is implied among the successive panels. However, the relative coiling and uncoiling of vortex lines are very clear.

One might superficially view the evolution as a damped periodic wave motion, with the damping provided by viscous effect. This view is, however, only partially correct, because the vortex lines are torsion free only initially. We have an interesting result concerning this nearly damped periodic motion. Namely, the frequency (consider it to be roughly constant in time) as well as the amplitude of the oscillations depends strongly on the Reynolds number. The frequency increases significantly with Re and it appears that the frequency reaches a constant and finite value in the high Reynolds number limit. This highlights the difference between local vortex dynamics, where local self induction is of the utmost importance, and classical vortex dynamics (filaments) where interaction takes place over distances which are large compared to the core size. In the latter case only the circulation of the vortex (Reynolds number independent) and its position is of importance. For internal vortex dynamics, on the contrary the vorticity magnitudes (sensitive to Reynolds number effects) are of critical significance. Interestingly, this observation has relevance for vortex reconnection as well. Because, as we have pointed out on numerous occasions the reconnection time decreases with increasing Reynolds number, and we have conjectured that the reconnection time becomes constant in the high Reynolds number limit. This is of course counter-intuitive and requires some explanation. First of all we have a special definition of the reconnection time, namely the time where ($d\Gamma_r/dt$) reaches its first local maximum. This definition is necessary because of the reconnection cascade (mentioned earlier). While it is commonly believed that all circulation is transferred from Γ_n to Γ_r in one reconnection, we have conjectured that the circulation transfer occurs in bursts, the reconnection cascade being the sequence of all bursts. At low to moderate Re most of the circulation transfer occurs in the first burst (e.g. Fig. 2). In the high Reynolds number limit we have conjectured that the circulation transferred in each burst is infinitesimal, while the duration for each burst remains finite. In our view this is the only sensible way the high Reynolds number limit can approach the Euler equations, where no circulation is transferred at all. Thus $\Gamma_r(t, Re)$ should converge to zero as Re tends to infinity with the convergence being nonuniform in t. The reason why this scenario is possible is the very fact that local self-induction plays a critical role in reconnection and is highly sensitive to Reynolds number effects.

By considering dissipation associated with the different flows, we notice that while dissipation due to swirl is dominant at all times, the dissipation due to meridional flow becomes most significant during the formation of the low enstrophy bubble (see Melander & Hussain 1991 for details).

By marking this simple flow situation with a scalar distribution equalling $|\omega|$ at t=0, we see that the scalar remains spatially clumped even when vorticity has relaxed to a uniform state (Fig. 6). This again shows that even at Sc=1 and even in an axisymmetric flow, visualization can provide grossly misleading information. The difference between vorticity and scalar transports are specially interesting in the light of recent popularity of Lagrangian vortex methods, where vortex particles or sticks are transported with the flow as a passive scalar. Our simulation clearly illustrates the difference between mass and vorticity transports.

4. Helical Wave Decomposition

While the helicity density is not a very useful quantity because it is not Galilean invariant, we have found that *complex helical wave decomposition* (Moses 1971; Lesieur 1972) is a particularly powerful analytical tool for quantitative understanding of evolution of CS, their mutual interactions, coupling with fine-scale turbulence and particularly their breakdown and reorganization. If one decomposes a velocity field into complex helical components (i.e. in terms of the eigenmodes of the curl operator which are exact solutions of the Euler equation) one obtains the components as Beltrami flows. That is, if λ is the eigenvalue of the curl operator, then λ, which is real, characterizes the nature of the eigenmodes: potential if $\lambda = 0$, right-handed if $\lambda > 0$, and left-handed if $\lambda < 0$. We feel that this is highly promising in analyzing coherent structure dynamics, their mutual interactions and coupling with fine-scale turbulence. We define the right (left) handed component of a vector field as the sum of all right (left) hand eigenmodes. The right and left handed components of the flow field are unique and are Galilean invariant.

When both the vortex (a prototypical CS) and small scales are present initially we observe an interesting coupling. Although the small scales in the absence of the vortex are isotropic and homogeneous, the vortex destroys the isotropy by first aligning the small scale vorticity in the swirl direction of the vortex, thereby giving the small scale vorticity preferred directions, namely, aligning with or against the direction of swirl induced by the vortex. Consequently, the dynamics of the small scales becomes like 2D-turbulence, apart from the continual stretching by the shearing of the vortex's swirling motion. Pairing of small scales oriented in roughly the same direction (with or against the swirl of the vortex) takes place, whereby the small scales organize into increasingly larger vortical structures, see Fig. 7 (the process is roughly similar to the growth of the boundary layer on an impulsively rotated rod). We observe this organization no matter how the initial random small scales are specified; whether or not there is an initial spectral gap does not make any difference.

We have introduced the spectral gap so as to demonstrate that the growth of incoherent scales is due to vorticity dynamics, rather than vorticity diffusion. In the absence of the spectral gap one could argue that the very smallest vortical scales are erased by diffusion (and dissipation) leaving only the intermediate scales. This would give a false impression of dynamical small scale organization. With the presence of the spectral gap there are no random intermediate scales, thereby ruling out this diffusive mechanism. To make doubly sure that pairing or merger of small scale vortices is in fact the operating mechanism, we have tracked the small scale vortical structures in physical space and clearly observed them to pair. This small scale organization is in fact the *inverse cascade in 3D* , which has long been sought in vain. The reason why it has not been found before is two-fold. First, theoretical approaches have assumed local isotropy. Second, numerical simulation of turbulent flows with CS have had too many of these CS to also allow the presence of substantially smaller scales within the available resolution constraints of numerical simulations. One has to keep in mind that the difference between the length of the large and small vortical structures is not very big; clearly this will improve with future generations of supercomputers.

While the large scale CS (the vortex) has a profound influence on the random small scales, namely the organization of these into increasingly large scales (Fig. 8) the large scales are also affected in the process. There is the effect of an enhanced or turbulent diffusion, but the large scales can also be affected in a way which can not be modeled as an enhanced diffusion. Namely, the small scales may be sufficiently energetic so that they excite bending waves (non-axisymmetric) on the vortex, thus changing the nature of the CS (the vortex) internal dynamics.

It has been conjectured (e.g. Moffatt 1970; Lesieur 1972) that an inverse cascade should be related to vortical structures exhibiting high helicity. Indeed the small scales exhibit high helicity, which is not surprising as they align with the swirling flow induced by the vortex. However, we have not been able to make much use of the helicity densities investigating the present simulations in spite of our best efforts. Fortunately, the mathematical framework (the complex helical wave decomposition) is associated with helicity concepts. We find that the small scales polarize in the left and right handed components during the organization. During pairing, the predominantly right handed small vortical scales merge with themselves; likewise also for the left handed structures. Moreover, the decomposition separates waves which move in opposite directions on the CS.

We have found that the presence of a CS in a turbulent flow has a profound influence on the turbulence in its immediate vicinity. There is a local inverse cascade, which in turn affects the dynamics of the CS.

This research was funded by Office of Naval Research Grant N00014-89-J-1361. The simulation was performed on the CRAY YMP at NAS.

References

[1] Aubry, N., Holmes, P., Lumley, J. L. & Stone, E. 1988 *J.F.M.* **192,** 115

[2] Berge, P., Pomeau, Y. & Vidal, C. 1984 *Order Within Chaos,* Hermann, Paris

[3] Brown, G. L. & Roshko, A. 1974, *J.F.M.* **64,** 775

[4] Crow, S. C. & Champagne, F. H. 1971 *J.F.M.* **48,** 547

[5] Fiedler, H. E. 1988 *Progr. Aerosp. Sci* **25,** 231

[6] Hussain, A.K.M.F. 1979, *Symp. Turb.* p 207

[7] Hussain, A.K.M.F. 1983 *P. Fluids* **26,** 2816

[8] Kida, S., Takaoka, M. & Hussain, F. 1989 *P. Fluids* 630

[9] Kline, S. J., Reynolds, W. C., Schraub, F. A. & Runstadler, P. W. 1967, *J.F.M.* **30**, 741

[10] Kolmogorov, A. N. 1941, *Dokl. Akad. Nauk. SSSR,* **31**, 538

[11] Lesieur, M. 1972 *Turbulence,* Observatoire de Nice

[12] Lumley, J. L. 1970 *Stochastic Tools in Turbulence, Academic Pr.*

[13] Lumley, J. L. 1980 in *Transition and Turbulence* (ed. R. E. Meyer) Academic Pr., p. 215

[14] Lundgren, T.S. & Ashurst, W. T. 1989 *J.F.M.* **200**, 283

[15] Melander, M. V. & Hussain, F. 1988 CTR Report - S88, 259

[16] Melander, M. V. & Hussain, F. 1991 (in preparation)

[17] Moffatt, H. K. 1970 *J.F.M.* **44,** 705

[18] Moore, D. W. & Saffman, P. G. 1972 *Phil. Trans. Roy. Soc. Lond. A.* **272**, 403

[19] Moses, H. E. 1971 *SIAM J. Appl. Math* **21**, 114

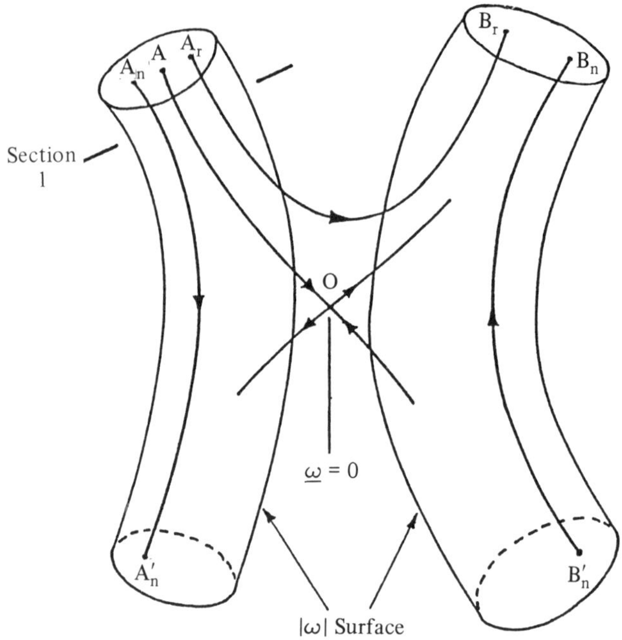

Figure 1. Definition of reconnection.

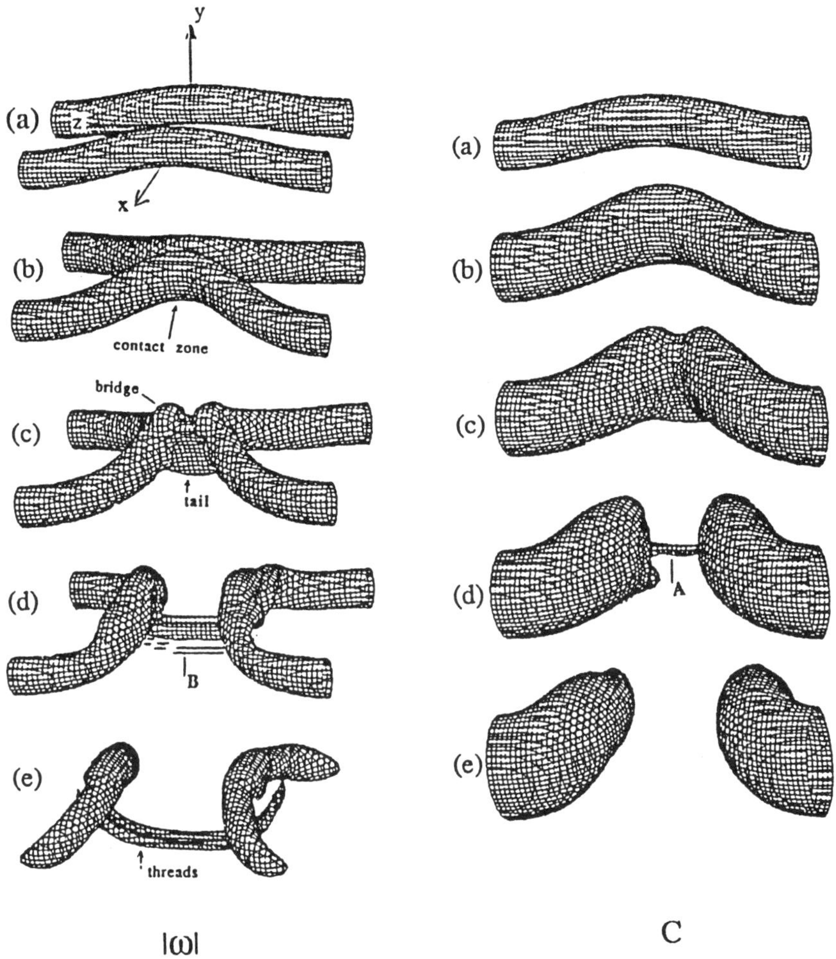

Figure 2. Time evolution of |ω| and scalar surfaces from top to bottom.

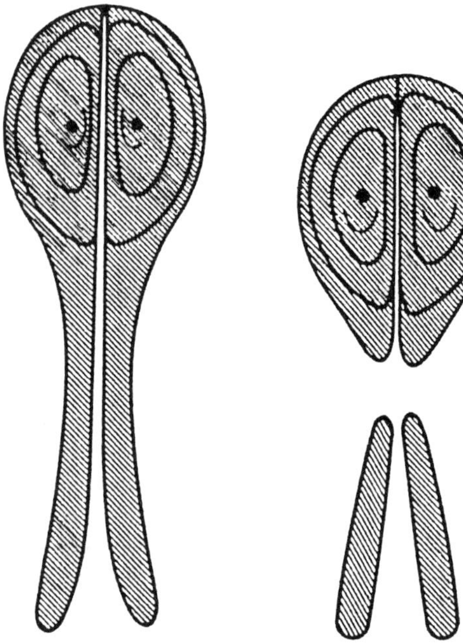

(a) (b)

Figure 3. (a) Head-tail formation; (b) separation of head from tail.

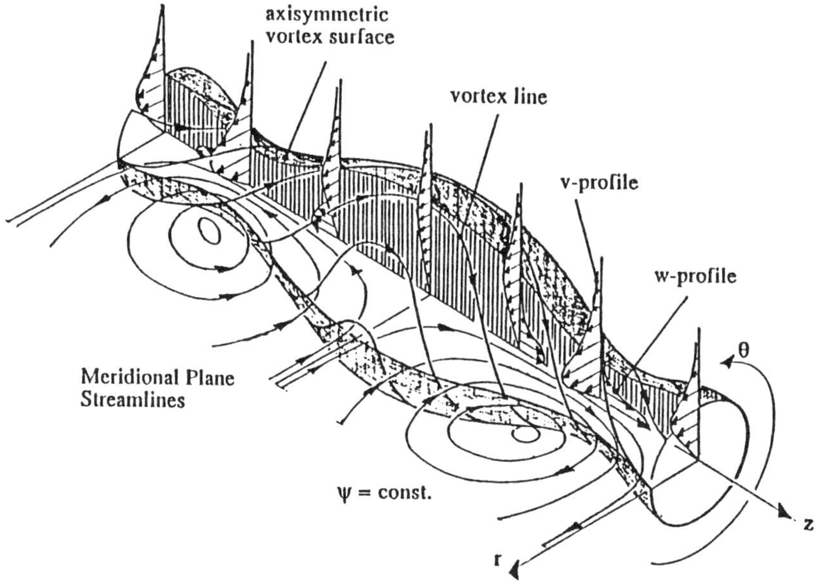

Figure 4. A schematic of coupling between swirl and meridional flow in a straight nonuniform vortex.

(a)
(b)
(c)
(d)
(e)
(f)
(g)
(h)

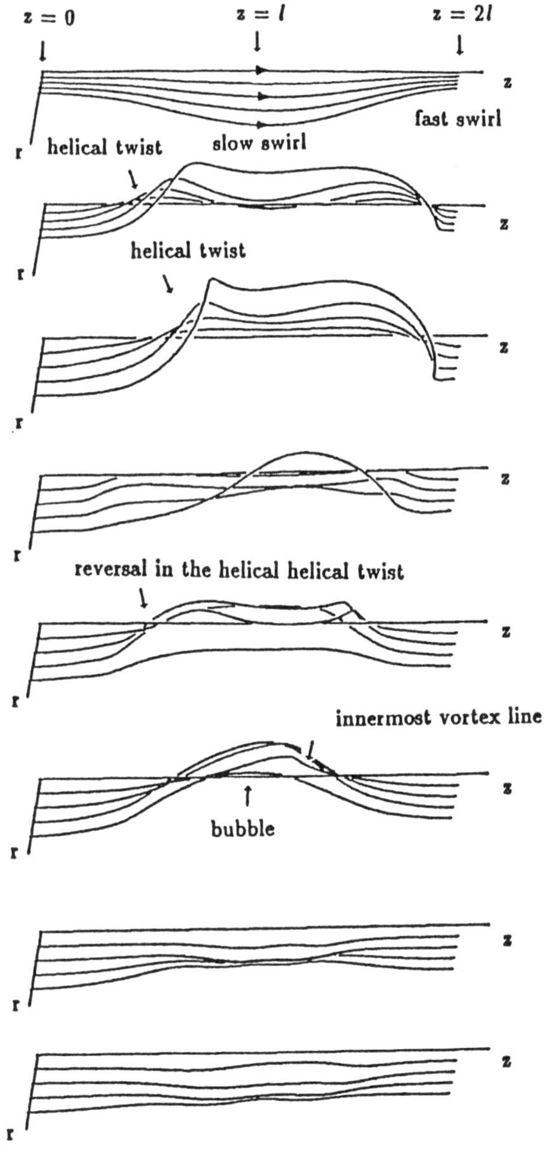

Figure 5. Vortex lines as a function of time.

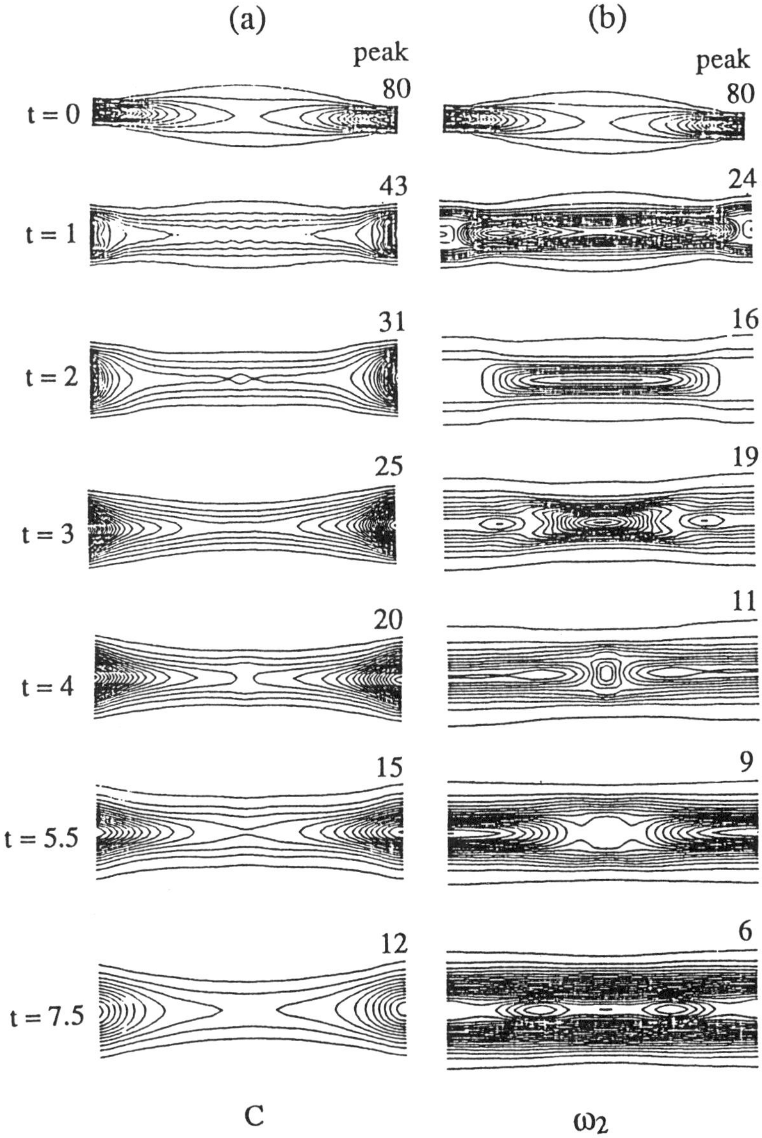

Figure 6. Contours of: (a) scalar; (b) vorticity.

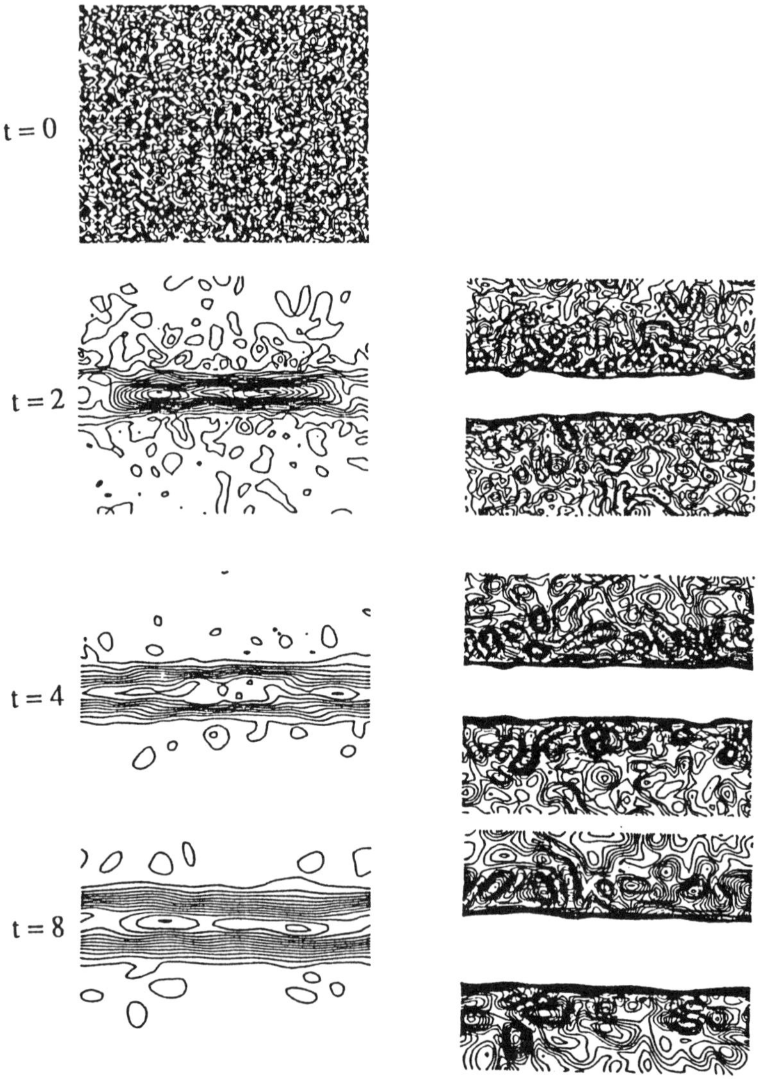

Figure 7. Time evolution of coupling between the vortex and fine scales. Shown are panels of $|\omega|$ in a plane through the axis of the unperturbed vortex. On the left, a full range of contour levels are displayed. On the right, only contour levels in the lower 10% are shown.

Figure 8. Evolution of the initially isotropic turbulence under the influence of CS. Shown are the positive and negative regions of the ω - component normal to a plane through the unperturbed vortex.

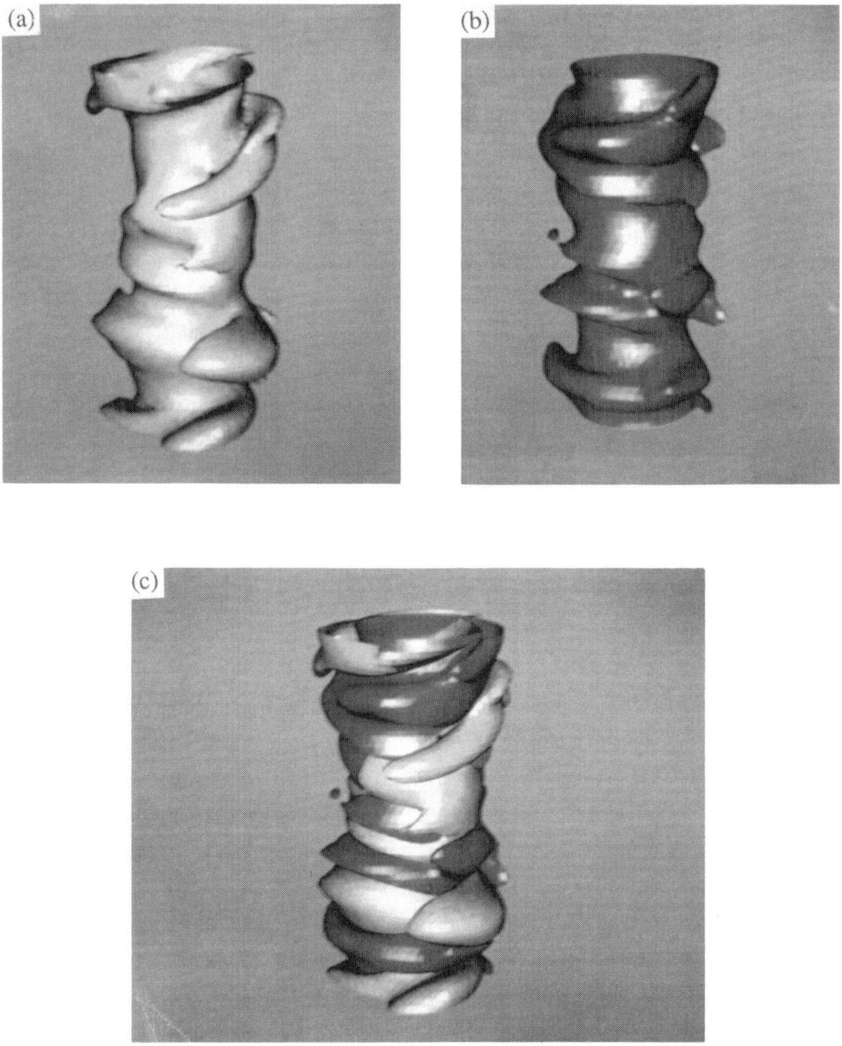

Figure 9. Spatial separation of left-handed and right-handed turbulence organized by the CS: (a) left-handed; (b) right-handed; (c) superposed.

HETEROCLINIC CYCLES, EXPONENTIAL TAILS, AND INTERMITTENCY IN TURBULENCE PRODUCTION

Philip Holmes[1] *and Emily Stone*[2]

Departments of Theoretical &
Applied Mechanics and Mathematics,
Cornell University,
Ithaca, NY 14853 USA

and

Department of Applied Mathematics
and Theoretical Physics,
Cambridge University,
Cambridge CB3 9EW ENGLAND

ABSTRACT

It has recently become clear that physical symmetries inherent in many fluid flows lead to interesting global dynamics in the phase spaces of the resulting governing equations. Examples include continuous and discrete translations in spanwise and streamwise direction in boundary layers, spanwise reflections in the same and rotations in circular jets and pipe flows. The phase space structures typically include heteroclinic cycles: orbits which link a set of unstable equilibria or other limit sets in a closed loop. Solutions attracted to such an object correspond to fluid flow regimes in which the flow remains relatively quiescent for "long" periods, punctuated by short, violent events in which many modes become active. Small random or deterministic perturbations of such cycles produce skewed distributions of inter-event durations with exponential tails remarkably similar to those observed for boundary

[1] Supported by AFOSR 89–0226A, Wall Layers and NSF 88-14553.
[2] Supported by SERC (U.K.) Nonlinear Systems Initiative.

layer bursts and in certain other flows. We describe the mechanism in physical and phase space and review some of the evidence in support of the view that heteroclinic cycles play a significant role in intermittent turbulence production.

1. Introduction

In this note we outline recent results which suggest that intermittent turbulence production in certain flows corresponds to orbits which cycle irregularly among a set of unstable solutions in the phase space of the dynamical system generated by the Navier Stokes equations. We focus on the bursting phenomenon in the wall region of the turbulent boundary layer. (Corino and Brodkey [1969], Kline [1978], Blackwelder [1989]). In such a layer, coherent, unsteady, streaky structures, containing significant streamwise vorticity, occupy much of the flow region "on average". Viewed as steady states, these flows are linearly unstable and, indeed, experiments (cf. Kline et.al. [1967]) and direct numerical simulations (cf. Moin and Moser [1989]) reveal that quasi-steady periods are punctuated by relatively rapid *bursts* in which fluid having low streamwise velocity erupts from the wall, followed by *sweeps* in which high speed fluid "replenishes" the wall region. These events occur irregularly in space and time, but with a characteristic distribution. Figure 1 shows examples of temporal distributions from a wide range of flows, the Reynolds number based on momentum thickness varying from 420 in (c) to 14,000 in (b). The key point for what follows is that the tails of these distributions are fitted well by exponentials: see (b) and (c).

Rather than modelling these phenomenon as random (Poisson) sequences (cf. Bogard and Tiederman [1986]), the "dynamical systems" approach to the boundary layer developed by Aubry et.al. [1988, 1989, 1990], cf. Berkooz et.al. Holmes [1990] suggests a different, deterministic and more natural origin. Specifically, the spatial *symmetries* inherent in (fully developed) flow over an infinite flat plate or in a circular pipe, such as invariance under spanwise translation and reflection, lead to global structures of recurrent solutions in the (finite dimensional) phase spaces of ordinary differential equations obtained by Galerkin projection of the Navier Stokes equation. Aubry et.al. use bases of empirical eigenfunctions obtained via the proper orthogonal decomposition (Lumley [1970]), but this observation is more general (Holmes [1991]). Briefly, the spanwise symmetries correspond to rotational and reflectional invariance with respect to the spanwise (Fourier) modes used in the velocity field representation and it can be proved that differential systems equivariant under such a group (0(2)) possess *structurally stable heteroclinic cycles* over open sets of parameter values.

In the simplest case (Armbruster et.al. [1988, 1989]), these cycles are orbits connecting diametrically opposite points on a circle of unstable, saddle-type equilibria corresponding to linearly unstable "laminar" flows occupying arbitrary spanwise locations relative to some fixed reference frame. More generally, the unstable limit sets can be periodic, quasiperiodic or more complex chaotic motions; see Aubry and Sanghi [1989, 1990] for example or Nicoalenko and She [1990 a,b]. See Figure 2. The important feature is that they (and all their images under the group orbit) occupy a well defined region of phase space and so correspond to flows with particular spatial structures. Similiarly, the manifolds of heteroclinic cycles connecting them are spatially well defined. These structures therefore provide an attractive analogue, in phase space, of the coherent structures observed in physical space and the associated cycles of turbulent energy production.

Solutions attracted to such a cycle spend periods of increasing lengths lingering near unstable limit sets, punctuated by violent transits. Reconstruction of velocity fields during such events (Aubry et.al. [1988], Aubry and Sanghi [1989, 1990]) reveals broad similarity with the phenomena reported in experiments and simulations, even down to a characteristic phase (lateral) shift of the streaks after an event (Kline [1967, p.55]). However, in the "perfectly symmetric", unperturbed case, the event spacing increases without bound and so the bursts die out. Holmes [1991] contains an introductory review of these ideas, with many references to relevant work in both fluid mechanics and the analysis of model evolution equations.

2. Perturbation of Heteroclinic Cycles

In modelling the wall region, Aubry et.al [1988] considered flow in a domain doubly periodic in the spanwise and streamwise directions, but bounded at fixed (small) distance X_2 normal to the wall ($y^+ = 40$, specifically). This "virtual boundary" is permeable: the empirical eigenfunctions do not decay until one reaches the free stream. This leads, in the projection process, to a boundary pressure term of the type

$$p(x_2)\phi(x_2)\Big|_{x_2=X_2} . \tag{1}$$

More generally, an additive perturbation, which may or may not break all the "perfect" symmetries, might arise from weak coupling to other "modes" in the underlying flow.

While small, such a term has a marked effect on solutions passing near a saddle point or other unstable set. As Stone and Holmes [1989, 1990] showed, it creates a characteristic distribution of passage times which is strongly skewed and has an exponential tail, the exponent being the largest unstable eigenvalue λ_u (in the simple case of a saddle point). They further showed that the mean passage time for an individual saddle, T_m, scales with the perturbation amplitude ε and eigenvalue λ_u as

$$T_m \sim \frac{1}{\lambda_u}|ln(\varepsilon)|. \tag{2}$$

Thus, distributions of relatively violent events—passages near the heteroclinic cycles—having exponential tails, arise rather naturally. It may even be possible to relate the exponent apparant in data such as those of Figure 1 with growth rates of linearly unstable modes.

To do their analysis Stone and Holmes assumed excitation by additive uncorrelated Weiner processes (white noise) and ignored the effects of perturbations during the rapid "global" transits of the heteroclinic cycles where the perturbed vector field dominates.

More recently, Stone and Holmes [1991] have shown that, at least in the case of saddle points, perturbation by additive colored noise or by certain deterministic, periodic terms yields essentially identical distributions.

Figure 3a. shows a comparison between the theory of Stone and Holmes [1990] and numerical simulation of the 10 dimensional boundary layer model of Aubry et.al. [1988] perturbed by Wiener processes. Figure 3b illustrates that the scaling relation (2) is remarkably good. Note that, since each cycle involves a pair of equilibria, the relation (2) must be imposed twice and the corresponding probability density obtained by convolution (Stone and Holmes [1991]).

It is important to recognize that, while the events are "catalyzed", and their distributions evoked, by perturbations which may be regarded as random, the structures of these events and the quiescent periods which accompany them are dominated by the structurally stable deterministic dynamics of the underlying symmetric dynamical system. This dynamics has clear signatures both in phase space (Figure 2) and in physical space (Aubry et al, Figure 19, 20; Aubry and Sanghi [1990], Figures 2,4). Morever, the energy source of the instability is deterministic (and, in the boundary layer, resides in the wall region) and the symmetric heteroclinic cycles provide a natural mechanism leading to the continual reformation of linearly unstable roll-like structures. This latter point seems to be new: in the "kinematic" descriptions of instability leading to bursting (eg. Kline et.al. [1967], Figure 19) the process leading to sweeps and subsequent recreation of rolls (or, more modestly, streaks) is left vague. There is no reason, a priori, why solutions should continually return to *unstable* limit sets in phase space. Heteroclinic cycles, induced by physical symmetries, provide a very natural mechanism. In this context it is interesting to note that Moffatt [1985, 1986] has pointed out the possible relevance of "unstable fixed points" in Euler flows to turbulence production mechanisms.

3. Conclusions

In this brief note we have outlined what we believe to be a rather general "explanation" of intermittent turbulence production in flows dominated, at the larger scales, by coherent structures. Symmetries in a "perfect" underlying flow lead to bundles of recurrent solutions which circulate near a heteroclinic attractor in the phase space of the dynamical system which models the flow. Small imperfections or perturbations, random or deterministic, act as catalysts which promote the rapid heteroclinic transits or events and establish a characteristic event distribution with an exponential tail. As Narashimha [1990] has observed, such distributions appear to be prevalent in many turbulent flows (transitional as well as full developed, cf. Mullin and Darbyshire [1989]).

References

[1] D.Armbruster, J. Guckenheimer, and P. Holmes [1988] *Physica 29D*, 257–282. Heteroclinic cycles and modulated travelling waves in systems with 0(2) symmetry.

[2] D. Armbruster, J. Guckenheimer, and P. Holmes [1989] *S.I.A.M. J. Appl. Math. 49*, 676–691. Kuramoto-Sivashinsky dynamics on the center-unstable manifold.

[3] N. Aubry, P. Holmes, J. Lumley and E. Stone [1988] *J. Fluid Mech. 1988*, 115–173. The dynamics of coherent structures in the wall region of a turbulent boundary layer.

[4] N. Aubry and S. Sanghi [1989] Streamwise and spanwise dynamics of the turbulent wall layer, in *Forum on Chaotic Flow* (ed. Ghia) ASME, New York.

[5] N. Aubry and S. Sanghi [1990] Bifurcation and bursting of streaks in the turbulent wall layer. In Turbulence 89: *Organized Structure and Turbulence in Fluid Mechanics*, ed. M. Lesieur and O. Métais, Kluwer Academic Publishers.

[6] G.Berkooz, P.J. Holmes and J.L. Lumley [1990a] (submitted) Intermittent dynamics in simple models of the turbulent wall layer.

[7] R.F. Blackwelder [1989] *AIAA paper # 89–1009, AIAA 2nd Shear Flow Conference.* Some ideas on the control of near wall eddies.

[8] D.G. Bogard and W.G. Tiederman [1986] *J. Fluid Mech. 162*, 389–413. Burst detection with single point velocity measurements.

[9] E.R. Corino and R.S. Brodkey [1969] *J. Fluid Mech. 37*, 1–30. A visual investigation of the wall region in turbulent flow.

[10] P. Holmes [1990] In *Whither Turbulence? Turbulence at the Crossroads*, (ed. John Lumley) Springer Lecture Notes in Physics #357, 195–249, Springer-Verlag, New York. Can dynamical systems approach turbulence?

[11] P. Holmes [1991] To appear in *Proc. IMA Workshop on Dynamical Systems and Turbulence*, (ed. G. Sell et.al) Symmetries, heteroclinic cycles and intermittency in fluid flow.

[12] H.T. Kim, S.J. Kline and W.C. Reynolds [1971] *J. Fluid Mech. 50*, 133–160. The production of turbulence near a smooth wall in a turbulent boundary layer.

[13] S.J. Kline [1967] In *Fluid Mech. of Internal Flow* (ed. G. Sovran), Elsevier, 27–68. Observed structural features in turbulent and transitional boundary layers.

[14] S.J. Kline [1978] In *Coherent Structure of Turbulent Boundary Layers, Proc. AFOSR/Lehigh* (eds. C.R. Smith and D.E. Abbott), 1–26. The role of visualization in the study of the turbulent boundary layer.

[15] S.J. Kline, W.C. Reynolds, F.A. Schraub and P.W. Rundstatler [1967] *J. Fluid Mech. 30* 741–773. The structure of turbulent boundary layers.

[16] J.L. Lumley [1970] *Stochastic Tools in Turbulence.* Academic Press, New York.

[17] H.K. Moffatt [1985] *J. Fluid Mech. 159*, 359–378. Magnetostatic equilibria and analogous Euler flows of arbitrarily complex geometry. Part I, Fundamentals.

[18] H.K. Moffatt [1986] *J. Fluid Mech. 166*, 359–378. Magnetostatic equilibria and analogous Euler flows of arbitrarily complex geometry. Part II, Stability considerations.

[19] P. Moin and R.D. Moser [1989] *J. Fluid Mech. 200*, 471–509. Characteristic-eddy decomposition of turbulence in a channel.

[20] T. Mullin and A.G. Darbyshire [1989] *Europhys. Lett.*, (submitted). Intermittency in a rotating annular flow.

[21] R. Narasimha [1990] *Whither Turbulence? Turbulence at the Crossroads*, (ed. John Lumley) Springer Lecture Notes in Physics #357, 13–48, Springer-Verlag, New York. The utility and drawbacks of traditional approaches.

[22] R. Narashimha and S.V. Kailas [1990] *Atmospheric Environment 24A* (in press). Turbulent bursts in the atmosphere.

[23] B. Nicolaenko and Z.S. She [1990a] Temporal intermittency and turbulence production in the Kolmogorov flow. In *Topological Dynamics of Turbulence*, Cambridge Univeristy Press.

[24] B. Nicolaenko and Z.S. She [1990b] Symmetry-breaking homoclinic chaos in Kolmogorov flows. (preprint, Arizona State University).

[25] K.R. Sreenivasan, A. Prabhu and R. Narasimha [1983] *J.Fluid Mech. 137*, 251–272. Zero crossing in turbulent signals.

[26] E. Stone and P. Holmes [1989] *Physica 37D*, 20–32. Noise induced intermittency in a model of a turbulent boundary layer.

[27] E. Stone and P. Holmes [1990] *S.I.A.M. J. Appl. Math. 50* 3, 726–743. Random perturbations of heteroclinic cycles.

[28] E. Stone and P. Holmes [1991] (submitted) Unstable fixed points, heteroclinic cycles and exponential tails in turbulence production.

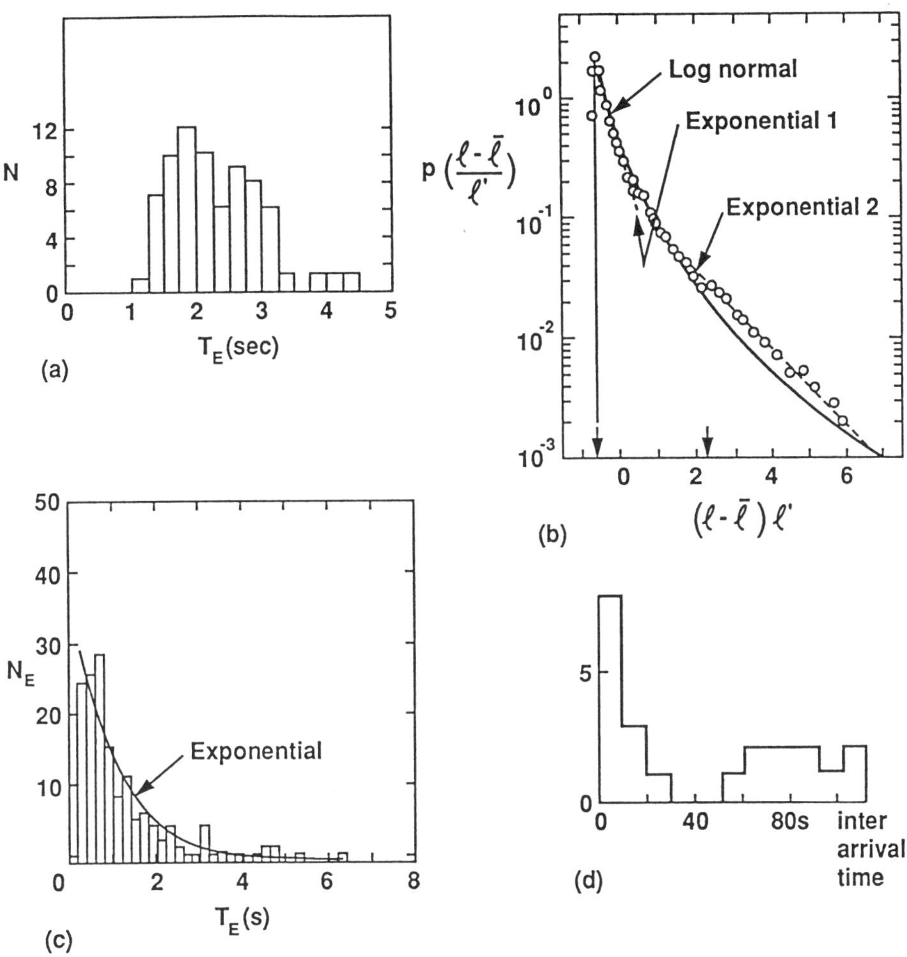

Figure 1.

Probability distribution of intervals between bursts in turbulent boundary layers. (a) from Kim et.al. [1971]; (b) from Sreenivasan et.al. [1983]; (c) from Bogard and Tiederman [1986]; (d) from Narashimha and Kailas [1990]. (d) is the atmospheric boundary layer, the others are laboratory flows. In (b) the "events" are zero crossings, in the others they are crossings of a preset $-uv$ threshold. See the references cited for details.

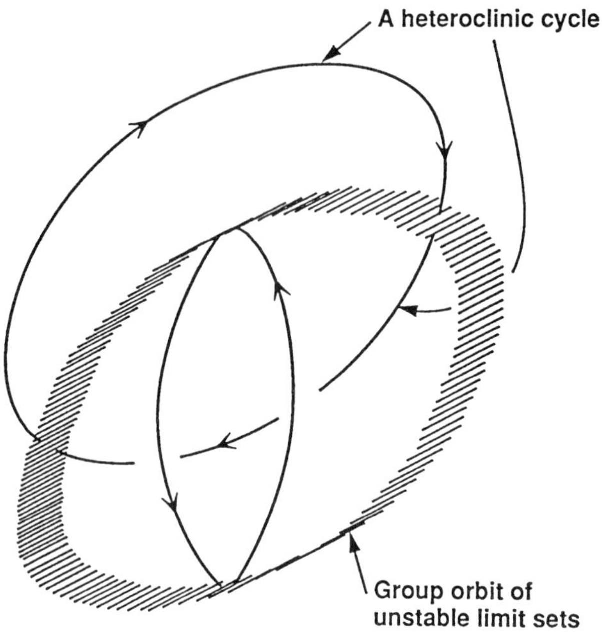

Figure 2.

Heteroclinic cycles in phase space: a schematic view: only two cycles show. See Armbruster et.al. [1989, 1990] and Aubry et.al [1988] for specific examples.

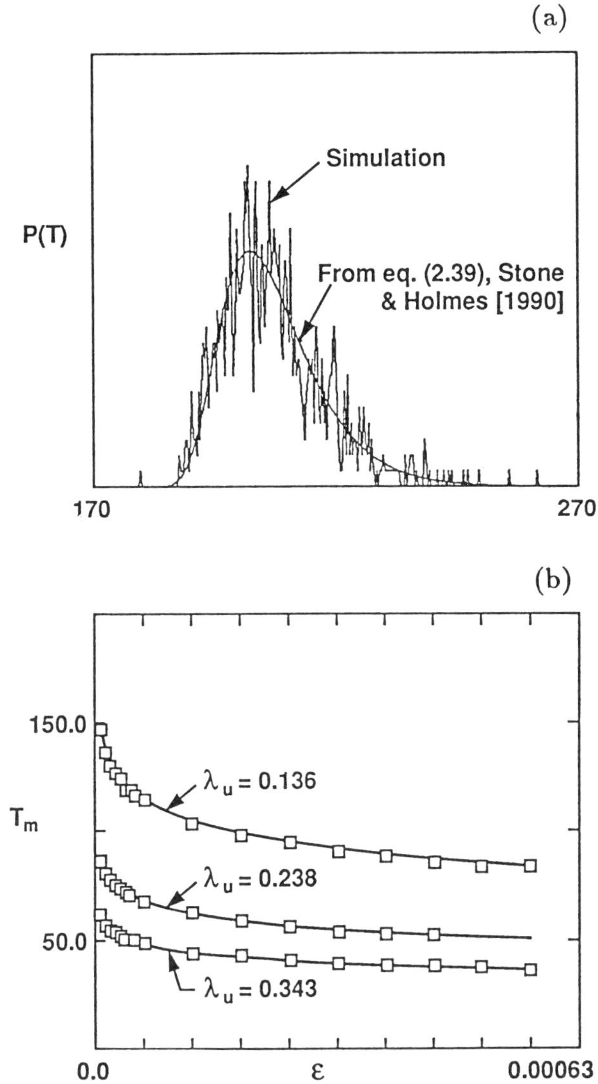

Figure 3(a)
Probability distribution of intervals between heteroclinic excursions in the model of Aubry et.al. [1988], excited by uncorrelated Wiener processes of RMS level 10^{-6}. Heisenberg viscosity $\alpha = 2.2$, other parameters as in Aubry et.al. (b) Effects of perturbation size on mean passage time. Solid lines show (fits to) theory, boxes indicate simulation.

MODELS FOR THE STRUCTURE AND DYNAMICS OF NEAR WALL TURBULENCE

*Sanjeev Sanghi and Nadine Aubry**

Benjamin Levich Institute and Department of Mechanical Engineering
City College of the City University of New York
New York, NY 10031

ABSTRACT

The dynamics of near wall turbulence is characterized by intermittent bursts during which fine scale turbulence is produced. Bursting events have been observed in low dimensional dynamical models built from empirical eigen functions of the proper orthogonal decomposition (Lumley 1967). Severely truncated models have been derived and studied by Aubry *et al.* (1988) which did not account for streamwise variations in an explicit way so that all structures were infinitely long. In the present work, we are investigating the persistency of intermittent behavior in higher dimensional models. First, the new truncations account for streamwise variations. Intermittent behavior persists in *all* the models, but can be of different nature. In particular, the non zero streamwise modes become excited during the bursts so that infinitely long streamwise rolls burst in the streamwise direction into smaller scales. In all cases, intermittency seems to be due to homoclinic cycles connecting hyperbolic fixed points or complicated (apparently chaotic) limit sets. In the latter case, the elongated rolls are no longer steady between the bursts but randomly active, as they should be. The influence of another proper orthogonal mode (i.e. smaller structures in the normal direction) is also investigated: bursts still occur in the same way, the second mode participating most of the time. Finally, a coherent structure study is performed in these models in order to connect our findings to those of other researchers. Most structures experimentally observed are clearly identified, i.e. streaks, streamwise rolls, horseshoe vortical structures and shear layers.

---

* This work was partially supported by a NSF/PYI award MSS89-57462 and by a Petroleum Research grant ACS-PRF#23009-G7 which are gratefully acknowledged. Computer resources were provided by the City University of New York, University Computer Center.

1. Introduction

Despite the contribution of many researchers, the mechanism of turbulence production in near wall turbulence is still poorly understood. It seems, however, well established that a large fraction of the production occurs during violent explosions (or 'bursts') of otherwise quiescent streamwise streaks (see for example Cantwell 1981). These streaks have been first visualized by Kline *et al.* (1967) using hydrogen bubbles and they have been associated with streamwise rolls (Bakewell and Lumley 1967, Blackwelder and Eckelman 1979, Hogenes and Hanratty 1982, Kim 1985, Adrian *et al.* 1987, Moin and Moser 1989, Guezennec 1989). The structure of near wall turbulence has received considerable interest and significant progress has been recently made with the generation of highly detailed numerical simulation data (Moin and Kim 1982, Spalart 1988). The complexity of the data, however, has made their interpretation extremely difficult in terms of coherent structures and their dynamics so that the formation of the streaks together with their dynamics are not well described. There is a need for simplified numerical simulations and models. Jimenez and Moin (1989) carried out an extremely reduced simulation in which the spanwise periodic length is about 100 wall units (i.e. the average distance between two streaks) so that one streak can be isolated and studied in detail. The striking result is that this coarse simulation generates approximately right first and second order statistics.

One of the most detailed studies of the structure of the wall region is that of Robinson *et al.* (1990) who analyzed Spalart's numerical data. The different instabilities, however, are rather absent there. The generally accepted idea is 'a violent small-scale secondary instability on a large-scale primary disturbance' (Landahl 1975). Streak formation as an instability mechanism is the object of many investigations and is not addressed here (see Jang *et al.* (1986)). Instead, we are interested in the streak breaking or 'secondary instability' and the streak reformation as a cyclic sequence of events (Kline *et al.*1967). One postulates that streak breaking is generated by an inflectional instability of the streamwise velocity profile observed in visualization experiments (Kline *et al* 1967) and numerical data (Moin and Kim, 1982). Jimenez *et al.* (1988) propose a mechanism based on the ejection of shear layer (spanwise vorticity) in 2D flows.

Another approach is that of Aubry *et al.* (1988) who derived a set of ten ordinary differential equations (10-D model) for the infinitely long streamwise rolls close to a wall. The rolls, extracted from experimental two point correlations and identified with the first proper orthogonal (or Kharhunen-Loève) eigenmode, exhibit intermittency whose nature is clearly identified: it is due to the presence of a heteroclinic cycle connecting hyperbolic fixed points. The solution spends a relatively long time near a fixed point where it stays quasi-steady, oscillates in a growing oscillatory motion and suddenly bursts through a brief jump along a heteroclinic orbit to another fixed point (only translated in the cross-stream direction). In physical space, streamwise rolls, quiescent for a long time, start to oscillate, suddenly burst into other structures in the cross-stream direction and get reformed. This strongly intermittent dynamics repeats itself an infinite number of times. The influence of noise of any kind has been clearly identified: it triggers the eruption event earlier and the time between two events, which otherwise keeps increasing, reaches an equilibrium. Stone and Holmes (1989) have shown that the presence of noise is responsible for exponential tails in the distributions of violent events due to saddle-saddle connections. Exponential tails have been experimentally observed for the distribution of bursting events (Kim *et al.* 1971 and Bogard and Tiederman 1986). Since the fluctuating pressure term at the upper boundary has such an effect which does not affect the qualitative behavior of intermittent solutions (see Aubry *et al.* 1988), it is omitted in the rest of this paper.

2. The Model

Our model,which accounts for streamwise variations, is an extension of the 10-D model of Aubry *et al.* (1988). Since the flow is approximately homogeneous in the streamwise and spanwise directions, we use a discrete Fourier expansion in these directions (assuming periodic boundary conditions) and proper orthogonal modes in the wall normal direction:

$$u(x,y,z,t) = \frac{1}{\sqrt{L_1 L_3}} \sum_{l,k,n} a^n_{l,k}(t)\, e^{2\pi i(lx+ky)}\, \varphi^n_{l,k}(y) \qquad (1)$$

We then project the Navier Stokes equations onto these modes to obtain sets of ODEs. The technique is thus a spectral method in which the basis modes are optimal in the sense that the resulting truncated system of ordinary differential equations captures the maximum amount of kinetic energy among all possible truncations of the same order (in the normal direction). These modes are provided by the proper orthogonal decomposition theorem of probability theory, which has been proposed by Lumley for identification of coherent structures in a turbulent flow. The method has obvious advantages over decompositions where the modes are known a priori, however, it requires the two point correlation tensor as an input for a three dimensional decomposition. Proper orthogonal modes have been or are being extracted from numerical or experimental data for turbulent flows by a fast growing number of researchers (Bakewell and Lumley, Payne and Lumley, Glauser *et al.*, Moin and Moser, Sirovich *et al.*, Glezer *et al.*, Ciliberto and Nicolaenko) and the dynamical systems approach has been recently applied by Sirovich, Deane *et al.*, Glauser *et al.* for different flow configurations. A symmetric space/time version of the proper orthogonal decomposition has been recently proposed by Aubry *et al.* (1990) for flows which undergo spatio-temporal bifurcations and eventually exhibit complexity in space and time (for references and details see Aubry in this volume).

Here, we are concentrating on the dynamical systems approach where only a few spatial degrees of freedom are considered. The flow studied is the same as in the 10-D model, that is the wall region: $0 \leq y^+ \leq 40$ ($^+$ denotes wall variable normalization with kinematic viscosity ν and friction velocity u_τ) of a pipe flow of Reynolds number 8750 based on the centerline velocity and the diameter of the pipe. The spatial two point correlation tensor and the empirical modes have been experimentally obtained by Herzog (1986).

In this paper, our first concern is the influence of streamwise variations, which were not explicitly included in the 10-D model, although equation coupling permitted transfer of energy among the various velocity components (see Berkooz *et al.*, 1990). While it is clear from the 10-D model that streamwise modes are not necessary for destabilizing the rolls, they undoubtedly participate in the overall dynamics and statistics.

Then, we add a second normal mode which was also neglected in the 10-D model. We thus account for normal mode interactions. In some sense, we partially address the general issue which systematically emerges from this type of truncated model, namely the robustness of the dynamical mechanism (here the intermittency) to the order of truncation.

3. Properties of the ODEs and specific sets of equations

The properties of the sets of ODEs can be found in Aubry and Sanghi (1989, 1990a,b). Let us only mention here the symmetry group: $O(2) \times SO(2)$ due to spanwise translation and reflection and streamwise translation. This symmetry group is inherited from the infinite dimensional system (i.e. the Navier Stokes equation subject to the boundary conditions at the wall) and that individual solutions do not necessarily satisfy these symmetries. This implies for example that a steady solution which has non zero streamwise and spanwise wave numbers is not unique but there is a torus of such solutions, all images of each other by the symmetry group. For a detailed description of invariant subspaces, see Aubry and Sanghi (1990b). Note that the symmetry group in the 10-D model was the subgroup $O(2)$.

Addition of a single non zero streamwise Fourier mode (and its opposite) in the 10 dimensional system leads to 32 dimensions. Addition of two modes gives 54 dimensions. Aubry (1987) showed that addition of two streamwise scales or more gave velocity variances and Reynolds shear stress approximating fairly well the original experimental data (Herzog, 1986). Improvement can be then achieved by addition of more proper orthogonal modes (which is not considered here). Although more systems have been derived and numerically integrated, we are reporting here in detail results obtained with 32 dimensional dynamical systems in a periodic cell of streamwise length $L_1^+=667$ (see Aubry and Sanghi 1990a for a detailed description of a 54 dimensional system including 2 non zero streamwise modes). We are also describing typical solutions obtained from a 64 dimensional system including two normal modes and one non zero streamwise mode.

We now outline the behavior of typical solutions of a 32 dimensional model (including 6 spanwise and 2 streamwise Fourier modes, 1 normal mode) as the loss of energy toward the unresolved scales is reduced (this loss is adjusted by a parameter α which is of order 1).

When losses of energy are excessively large compared to reality (α is larger than 1.53), the trivial solution is the global attractor. For $1.52 \leq \alpha \leq 1.51$ we observe a non zero fixed point in the ($l = 0$, even k) subspace (see equation 1 for the notation). As α decreases from 1.50 to 1.22, the solution becomes intermittent in the zero streamwise mode subspace. This intermittency is similar to window I of intermittency present in the 10-D model and is rather simple. For $1.22 \leq \alpha \leq 1.06$, the solution is steady again and is a combination of the modes (0,2) and (0,4) only. Another window (II) of intermittency appears for $1.05 \leq \alpha \leq 0.88$. This intermittency is of the same type as that of window II of intermittency in the 10-D model. It occurs in the zero streamwise mode subspace, the non zero streamwise modes decaying very fast. As in the window II of intermittency of the 10-D model, the (0,2)/(0,4) mixed modes are destabilized in the (0,1)/(0,3)/(0,5) directions. This solution encounters bursts involving growing oscillatory motions of the spanwise even and odd modes (at constant modal phases), corresponding to cross-stream oscillations of the streamwise rolls followed by a sudden breakup. Until this point, the sequence and nature of the solutions have been very similar to those observed in the 10-D model (see Aubry *et al.* (1988) for time series), indicating that the zero streamwise mode subspace is attracting for the whole system. In these solutions, the non zero streamwise modes are inactive after a very short transient during which they decay to zero. As α decreases more, the invariant zero streamwise subspace becomes unstable. For $0.80 \leq \alpha \leq 0.89$, we observe a third window of intermittency. Between the bursts, the solution is the same as in the first and second windows, almost steady since it stays in a neighborhood of one equilibrium in the zero streamwise/even spanwise subspace. However, it changes significantly during the bursts since all the modes then participate in the dynamics. Different projections of the phase space clearly show presence of a heteroclinic cycle connecting fixed points in the (0,2)/(0,4) subspace (see Aubry and Sanghi, 1989). For $0.80 \leq \alpha \leq 0.50$, the solution is attracted to the zero streamwise subspace where it becomes much more complex, apparently chaotic. The heteroclinic cycle starts travelling and then becomes unstable while the non zero streamwise modes collapse. For lower values of α, the ($l = 0$) subspace is unstable again and a fourth window of intermittency appears. In the intermittent behaviors previously described, the solution between the bursts was always quasi steady due to the proximity of a fixed point. The new solution is different in nature: it is very disorganized, apparently chaotic in between the bursts in the zero streamwise subspace and the degree of complexity increases even more during the bursts due to the sudden excitement of the non zero streamwise modes of high frequency motions. The time series for the coefficients is shown in figure 1.

Vectorial representations of the velocity field in cross-sections of the flow near a fixed point solution represent two pairs of symmetric (with respect to a plane (x_1,x_2)) streamwise rolls (figure 2a). In the first three windows of intermittency the field is in a quasi-steady configuration like this before and after the burst; it gets animated during the explosive event. More precisely, rolls vary in amplitude and position in a quasi-random way but a striking feature is that structures always appear in pairs: The reflection symmetry is not broken during the events. This type of saddle-saddle connection due to the symmetry subgroup O(2) is now well understood (Armbruster *et al.* 1988). Aubry and Sanghi (1990a) performed a linear analysis around the fixed points and showed that the second intermittency window occurs through a subcritical Hopf bifurcation and that the unstable manifold of the fixed point at the instability onset is two dimensional and lies in a space spanned by zero streamwise/odd spanwise modes. Then the zero streamwise/even spanwise modes get activated by non linear (quadratic) interactions which explains the time delay of the excitation of these modes. These results were also verified by looking at the POD (in the homogeneous direction) of the numerical solution.

In the fourth window of intermittency, things are rather different. Bursts occur in the streamwise direction only: streamwise modes which stay zero during quiescent periods suddenly explode at very high frequency and calm down again. The evolution of the flow cross-section displayed in Figure 3 shows that the solution is very disorganized and asymmetric even in between the bursts (figure 3). The intermittency scenario is thus different here than in previous windows. Projections of the phase space onto streamwise components strongly resembles the projection of the phase space in the other intermittency windows onto spanwise components. The new picture is thus similar to the previous one: We have hyperbolic limit sets connected by manifolds. We speculate that the cause of the intermittency is the presence of an attracting homoclinic cycle (in a generalized sense) which connects hyperbolic, apparently chaotic, limit sets. The saddle(s) is(are) no longer a fixed point but the complicated limit set of the zero streamwise subspace, the attractor of the solution when the initial condition lies in the zero streamwise subspace. For an initial condition outside of any invariant subspace, the solution leaves the ten dimensional zero streamwise subspace, starts a brief excursion in the full 32 dimensional phase space and comes back to the zero streamwise subspace. This suggests that the symmetry subgroup SO(2) is now responsible for intermittency. To our knowledge, this scenario has not been theoretically understood.

Similar solutions have been observed in 54 dimensional systems , the second non zero streamwise mode being a slave of the first one, gets excited or dies by quadratic interactions. We should finally mention that the presence of higher order streamwise modes does not change the previous picture: Higher order streamwise modes get activated with the first non zero one. This explains generation of fine scale turbulence during bursting, a phenomenon absent in a zero streamwise mode model.

We now consider the system with two proper orthogonal modes, one non-zero streamwise Fourier mode and five non-zero spanwise Fourier modes. This leads to a 64 dimensional dynamical system. The numerical integration leads to the following results. (Note that the coefficients a have three indices in this case; $a_{l,k}^{n}$ where n denotes the normal mode and as before l and k the streamwise and spanwise modes, respectively.)

$\alpha \geq 1.9$	trivial solution $a_{l,k}^{n} = \mathbf{0}$, $\forall(n,l,k)$ is the global attractor
$1.8 \geq \alpha \geq 1.7$	stable fixed point $a_{0,2}^{n}, a_{0,4}^{n} \neq 0$, $\forall n$, other modes are zero
$1.6 \geq \alpha \geq 0.9$	stable fixed point $a_{0,k}^{n} \neq 0$, $\forall$ (n,k), other modes are zero
$0.8 \geq \alpha \geq 0.6$	Intermitttency in the full subspace. fixed point in the $a_{0,k}^{n}$ subspace excited by the activation of $a_{1,k}^{n}$ and $a_{-1,k}^{n}$ modes, $\forall$ (n,l,k)
$0.5 \geq \alpha \geq 0.4$	complex solution where all the modes are active
$0.3 \geq \alpha \geq 0.2$	intermittency in the full space similar to IV window in the 32-D model. Complex behavior in the $a_{0,k}^{n}$ space for all times, $a_{1,k}^{n}$ and $a_{-1,k}^{n}$ activated only during bursts, $\forall$(n,l,k) $a_{0,k} \neq \mathbf{0}$, $\forall k$
$0.2 \geq \alpha$	complex solution in the full space

Table 1. Qualitative dynamical behavior in the 64 dimensional system.

We observe similar results when we consider a 64 dimensional dynamical system derived from the numerical data for the channel flow (Moin and Moser, 1989). The basic point we want to emphasize is the presence of intermittency in all the cases we have considered. Of particular interest is the occurrence of the fourth window of intermittency in all the cases where we allow for streamwise variations.

4. Physical Interpretation

We now concentrate on the kinematics and dynamics of coherent structures as they appear in our model. Structures of different types have been been extensively described as they have been observed in many visualization experiments as well as in direct numerical simulations. However, the connection between these structures is still not well understood and highly questioned. We first show that all structures which have been previously reported are present in our low dimensional models. Moreover, due to the low dimensionality of our system, connections between structures become rather clear.

We now follow the world-wide survey of Kline and Robinson (1989) who reported various types of structures gathered from many contributions. We discuss here only the first five categories as the others seem to have been observed by a few researchers only or are not relevant to the present work (outer structures, for example).

Low speed Streaks in the region $0 < y^+ < 7$ are perhaps the most common structure reported in the literature. They consist of streamwise, elongated filaments of low speed, outward moving fluid close to the wall. They have a cross-stream spacing and length of about 100 and 1000 wall units, respectively. Moreover, Smith and Metzler (1983) have shown that the probability distribution function of the streak spacing around the mean value of 100 wall units is log normal. For simplicity, we first consider the velocity field associated with the fixed point of our models which appears at large values of the α parameter. This fixed point, already present in the 10-D model, represents a pair of symmetric, counter-rotating rolls. In order to localize the streaks, we have plotted the (spatial) mean and instantaneous streamwise velocities at a specific distance from the wall as a function of spanwise coordinate (figure 4). The narrow region in between the rolls is slow moving and has normal velocity outward oriented, corresponding to streaks. Their spanwise width is about 30 wall units and they are separated by a distance of approximately 130 wall units. These values are consistent with experimental evidence. Note that this picture of streaks lying in between rolls is not new. Blackwelder and Eckelman (1979) used conditional averaging to show that pairs of counter-rotating streamwise vortices frequently occur in the wall region. The defect region of a pair thus extracted corresponds to a low speed streak.

Ejections of low-speed fluid outward from the wall in the region $10 < y^+ < 40$ We quote Kline and Robinson (1990): 'the low speed streaks seen by marking fluid essentially at the wall, migrate slowly outward as they move downstream, and at some downstream distance, turn sharply and move far more rapidly away from the wall in ejection-motions, called "lifting". So Kline and Robinson distinguish between 'low speed streak' in the sublayer and 'lifted low-speed streaks'. It is clear from the picture of the rolls obtained in our models (figure 2a) that strong ejections of slow speed fluid occur in the region $10 < y^+ < 40$ where the normal velocity component is the strongest. Note that both the streaks and the lifted streaks originate from the rolls and are placed one above the other. We attribute the weakness of the normal velocity component near the wall to its quadratic tangency (which is because of the continuity equation).

Sweeps of high-speed fluid toward the wall. Although flow visualizations do not clearly show wallward sweeps of high speed fluid, their presence can be inferred by the continuity equation (for references see Kline and Robinson, 1989). In our model, the region outside of the rolls has negative normal velocity and excess of streamwise velocity compared to the mean. These have all characteristics of sweeps. In particular, the inward motion, more spread out in the spanwise direction than the outer motion of the lifted streaks, is not as strong. This of course satisfies the continuity equation and explains the difficulty in marking sweeps in experiments. Note that the picture of rolls, when looked at from the point of view of sweeps and streaks, is one in which elongated regions of high and low streamwise momentum alternate. This is akin to the description given by Gupta *et al.* (1971).

Vortical structures. While the roll shaped structures which we observe in the cross-stream plane are indicative of streamwise vorticity, they may not imply streamwise vortices, i.e. vortices in which the vorticity is only streamwise and the vortex lines are aligned in the streamwise direction. Although streamwise component of the vorticity has been measured, other components have been found to be non zero as well. On another hand, visualization experiments have shown the presence of horseshoe (sometimes also called hairpin) or arch (half horseshoes) vortical structures rather than purely streamwise vortices. In particular, the presence of normal vorticity has been observed in spiralling motions. Assumptions have been made that streamwise vortices near the wall are the legs of the horseshoes. We now use the steady solution of our model to describe the vorticity lines (the vorticity considered here is the instantaneous one). Let us consider one pair of rolls in Figure.2a We first note that the vorticity due to the mean velocity is only spanwise and that the spanwise gradient of streamwise velocity is negative in the left roll, positive in the right one. If we neglect streamwise variations in a first approximation, normal vorticity is then negative in the left roll and positive in the right one. In both regions, spanwise vorticity, which essentially originates from the mean velocity gradient, is negative. This simple argument implies that the vorticity lines corresponding to a pair of counter-rotating rolls has a horseshoe shape with its head lying in between the rolls. Vorticity vectors have been plotted in Figure 2b. However, this horseshoe is inclined upstream.

Internal shear layers are defined as local regions where the shear rate exceeds its mean value. They can be due to either normal or spanwise gradients of streamwise velocity. Jimenez *et al* (1988). used contours of spanwise vorticity in numerically simulated channel flow to establish that the near wall region is dominated by intense three dimensional shear layers. Blackwelder and Swearingen (1990) also found three dimensional shear layers and compare the situation to Gortler vortices next to a concave wall. The presence of rolls in our model leads to both horizontal and vertical shear layers. While the former comes from the juxtaposition of slow and fast moving fluid (sweeps and streaks) in the spanwise direction (see figures 2a and 4), the latter originates in the presence of the change of streamwise velocity as a function of distance from the wall: a clear inflection point occurs in the sweep region as the fluid gets accelerated as it approaches the wall, making the streamwise velocity profile more blunt; in the streak region fluid has a momentum defect and changes in the velocity profile are less clear.

Note that we have described until now the structures according to the steady solution emerging out of our model, which is simply a pair of reflection symmetric rolls. When the flow gains some dynamics, as of course it does in real turbulence, observation of the temporal evolution of velocity vectors in a cross section of the flow shows that the reflection symmetry is often broken (except in the earlier windows of intermittency). Then structures are very similar to those previously described, the main difference being that the rolls appear isolated or in asymmetric pairs (figure 3). In this case, rolls are not symmetric but they are still accompanied by sweeps, streaks and three dimensional shear layers. Perhaps, the most significant change is that inclined vorticity lines which become half horseshoes or hook like structures in the case of one roll, asymmetric horseshoes in the case of two asymmetric rolls. Then the 1 to 1 relation of structures which has been established in the case of symmetric rolls is still valid.

This picture of near wall region structures, however, is not complete since it does not explain finite streamwise structures as observed in numerical simulations. For this, we exploit our more realistic solution which does depend on time and displays streamwise variations. For this purpose, we can take a complex (probably chaotic) solution in our 32D model with non zero streamwise modes or an intermittent solution in the fourth window of intermittency as previously described. Figure 5 shows fluctuating streamwise velocity contours resembling those obtained by other researchers (for example Stretch *et al.*,1990). Spanwise vorticity contours which have strong similarity with the picture obtained by Jimenez *et al.*(1988) are presented in figure 6. In figure 7 we plot vortex lines for a box corresponding to the top part of the box (dotted streaks) in figure 5. We show a perspective view of various vortex lines which start at the same spanwise and normal locations (but at different streamwise locations). As we have seen above, the slow moving streaks lie in between the rolls and the vorticity lines have the shape of a horseshoe. This pattern is very similar to the one reported by other workers. See, for example, Kim and Moin (1986).

Finally, comparison of the dynamics itself with other works seems difficult. Our model seems to provide the basic qualitative features described in the literature, more specifically oscillations and fine scale generation during bursting events.

References

[1] Adrian, R. J., Moin, P. and Moser R. D. 1987 *Stochastic estimation of conditional eddies in turbulent channel flow.* In Proc. of the 1987 Summer Program of Center for Turbulence Research (NASA Ames/Stanford Univ., Stanford, CA 1987)

[2] Armbruster, D., Guckenheimer, J. and Holmes, P. 1988 *Heteroclinic cycles and modulated travelling waves in systems with 0(2) symmetry.* Physica 29D, p. 257.

[3] Aubry. N. 1987 *A dynamical system / coherent structure approach to the fully developed turbulent wall layer.* Ph. D. thesis, Cornell University.

[4] Aubry, N., Holmes, P., Lumley, J.L., & Stone, E. 1988 *The dynamics of coherent structures in the wall region of a turbulent boundary layer.* J. Fluid Mech. 192, p. 115.

[5] Aubry, N. & Sanghi, S. 1989 *Streamwise and spanwise dynamics in the turbulent wall layer.* In Forum on Chaotic Flow, ed. K.N. Ghia. New York: ASME.

[6] Aubry and Sanghi, S. 1990a *Bifurcations and bursting of streaks in the turbulent wall layer.* In Turbulence and Coherent Structures, eds. M. Lesieur and O. Métais, Kluwer academic publishers.

[7] Aubry, N. and Sanghi, S. 1990b *Mode interaction models for near wall turbulence. Kinematics and dynamics.* Preprint.

[8] Aubry, N., Guyonnet, R. and Lima, R. 1990 Spatio-temporal analysis of complex signals: Theory and applications. Preprint.

[9] Bakewell, P & Lumley, J.L. 1967 *Viscous sublayer and adjacent wall region in turbulent pipe flow.* Phys. Fluids 10, p. 1880.

[10] Berkooz, G., Holmes, P.J. and Lumley, J.L. 1990 *Intermittent dynamics in simple models of the turbulent boundary layer* Preprint

[11] Blackwelder, R. F. & Eckelman, H. 1979 *Streamwise vortices associated with the bursting phenomenon.* J. Fluid Mech. 94 (3), p. 577.

[12] Blackwelder, R. F.and Swearingen, J. D.1990 *The role of inflectional velocity profiles in wall bounded flows* In Near Wall Turbulence, eds. S. J. Kline *et al.*, Washington DC: Hemisphere

[13] D. G. Bogard and W. G. Tiederman, (1986) *Burst detection with single point velocity measurements.* J. Fluid Mech. 162, p. 389.

[14] Cantwell, B.J. 1981 *Organized motion in turbulent flow.* Ann. Rev. Fluid Mech. 13, p. 457.

[15] Corino, E.R. & Brodkey, R.S. 1969 *A visual investigation of the wall region in turbulent flow.* J. Fluid Mech. 37, p. 1.

[16] Guezennec, Y. G. 1989 *Stochastic estimation of coherent structures in turbulent boundary layers.* Phys. Fluids A 1 (6), p. 1054.

[17] Gupta, A. K., Laufer,J. and Kaplan,R. E.1971 *Spatial structure in the viscous sublayer* J. Fluid Mech. 50, p. 493

[18] Herzog, S. 1986 *The large scale structure in the near-wall region of turbulent pipe flow.* Ph.D. thesis, Cornell University.

[19] Hogenes, J.H.A. & Hanratty, T. J. 1982 *The use of multiple wall probes to identify coherent flow patterns in the viscous wall region.* J. Fluid Mech. 95, p. 655.

[20] Jang, P. S., Benney, D. J. & Gran, R. L. 1986 *On the origin of streamwise vortices in a turbulent boundary layer.* J. Fluid Mech. 169, p. 109.

[21] Jimenez, J., Moin, P. , Moser, R.& Keefe, L. 1988 *Ejection mechanisms in the sublayer of a turbulent channel.* Phys. Fluids 31 (6), p. 1311.

[22] Jimenez, J.& Moin, P. 1989 *A minimal turbulent channel.* American Physical Society/Division of Fluid Mechanics, 42nd annual meeting, DA3. Palo Alto, CA.

[23] Kim, J. 1985 *Turbulence structures associated with the bursting event.* Phys. Fluids 28, p. 52.

[24] Kim, H. T., Kline, S. J. &Reynolds, W. C. 1971. *The production of turbulence near a smooth wall in a turbulent boundary layer* J. Fluid Mech. 50, p. 133

[25] Kim, J.& Moin,.P. 1986 *The structure of vorticity field in turbulent channel flow. Part 2. Study of ensemble-averaged fields* .J Fluid Mech. 162, p. 339

[26] Kline, S.J., Reynolds, W.C., Schraub, F.A. & Runstadler, P.W. 1967 *The structure of turbulent boundary layers.* J. Fluid Mech. 30, p. 741.

[27] Kline, S.J. and Robinson, S.K.1990 *Quasi-coherent structures in the turbulent boundary layer: Part I. Status report on a community-wide summary of the data.* In Near Wall Turbulence, eds. S. J. Kline *et al.*, Washington DC: Hemisphere.

[28] Landahl, M. 1975 *Wave breakdown and turbulence.* SIAM J. Appl. Math. 28 (4), p.735.

[29] Lumley, J.L. 1967 *The structure of inhomogeneous turbulent flows.* In Atmospheric Turbulence and Radio Wave Propagation (ed. A.M. Yaglom & V.I. Tatarski), p 166 Moscow: Nauka.

[30] Moin, P. & Kim, J. 1982 *Numerical investigation of turbulent channel flow.* J. Fluid Mech. 118, p. 341.

[31] Moin, P. & Moser, R.D. 1989 *Characteristic-eddy decomposition of turbulence in a channel*, J. Fluid Mech. 200, p. 471.

[32] Robinson, S.K., Kline, S.J. and Spalart, P.R. 1990 *Quasi-coherent structures in the turbulent boundary layer: Part II. Verification and new information from a numerically simulated flat-plate layer.* In Near Wall Turbulence, eds. S. J. Kline *et al.*, Washington DC: Hemisphere.

[33] Smith, C.R. and Metzler, S. P. 1983 *The characteristics of low-speed streaks in the near-wall region of a turbulent boundary layer* J. Fluid Mech. 129, p. 27

[34] Spalart, P. R. 1988 *Direct simulation of a turbulent boundary layer up to* $R_t = 1410$. J. Fluid Mech. 187, p 61.

[35] Stone, E. and Holmes, P. 1989 *Noise induced intermittency in a model of a turbulent boundary layer.* Physica 37D, p. 20.

[36] Stretch, D., Kim J. and Britter, R. 1990 *A conceptual model for the structure of turbulent channel flow* , in Turbulent boundary layer structure workshop, NASA Langley, Hampton, 28-30 August.

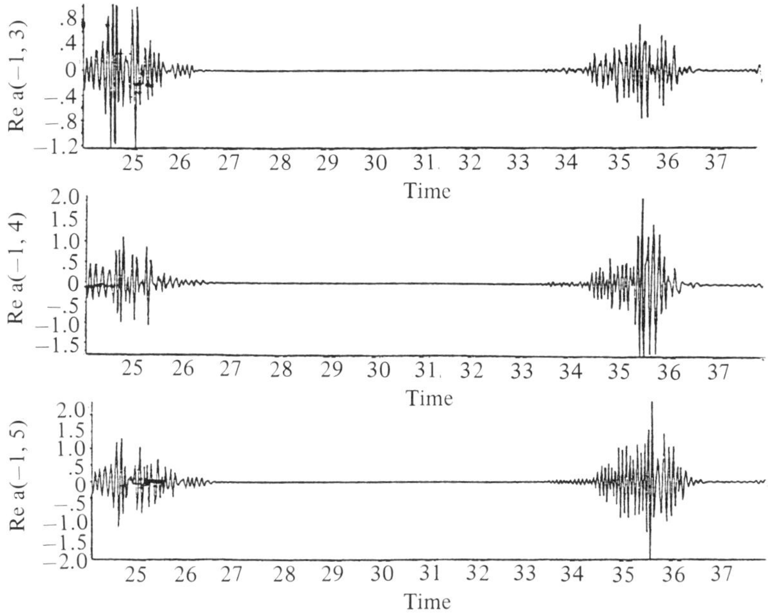

Figure 1. Typical solution in the window IV of intermittency of the 32 dimensional system. Time series of the real parts of a_{01}, a_{02}, a_{03}, a_{04},a_{05},a_{-11},a_{-12},a_{-13},a_{-14},and a_{-15} for $\alpha = 0.2$.

a)

b)

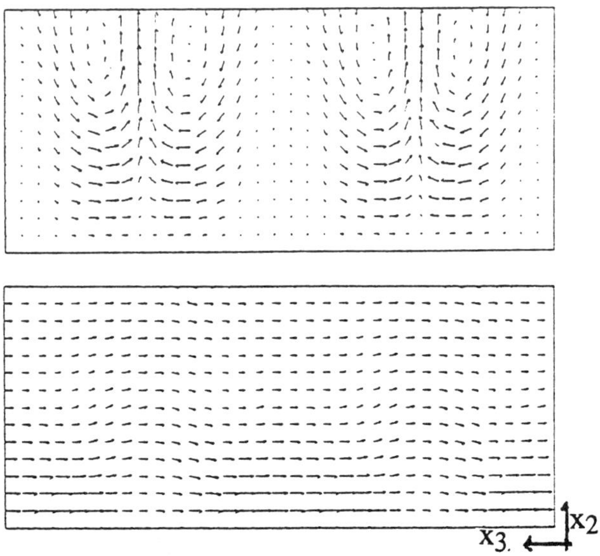

Figure 2. Vector representation of a) u_2-u_3 b) ω_2-ω_3 in a (x2-x3) plane during a fixed point region of the solution for $\alpha = 0.87$. The spanwise and the normal dimensions of the box are 333 and 40 wall units.

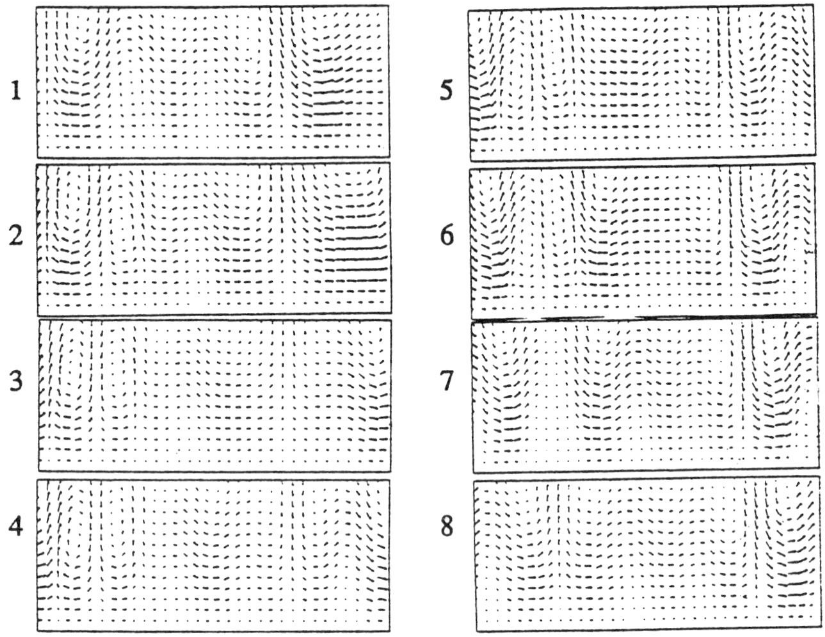

Figure 3. Time evolution of vector representation of u_2-u_3 in an (x_2-x_3) plane during the "quiescent" period of the IV window of intermittency for the 32 dimensional system ($\alpha = 0.2$). For the dimensions and orientation of the box see figure 2.

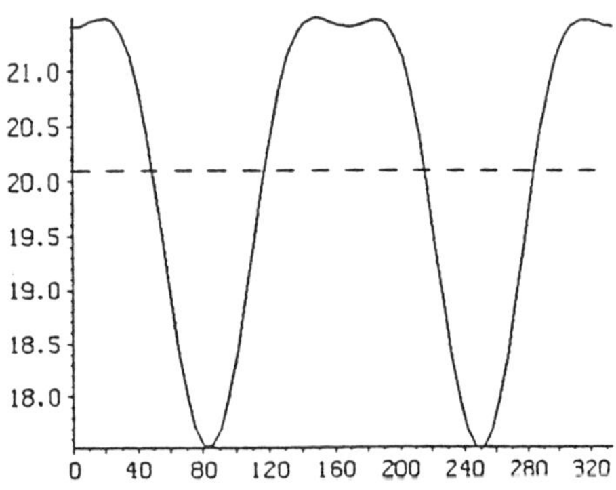

Figure 4. Distribution of streamwise velocity U, mean - - - - and instantaneous —, as a function of x_3 at $y^+ = 28$ and $x^+ = 0$. The time and α are the same as that for figure 2.

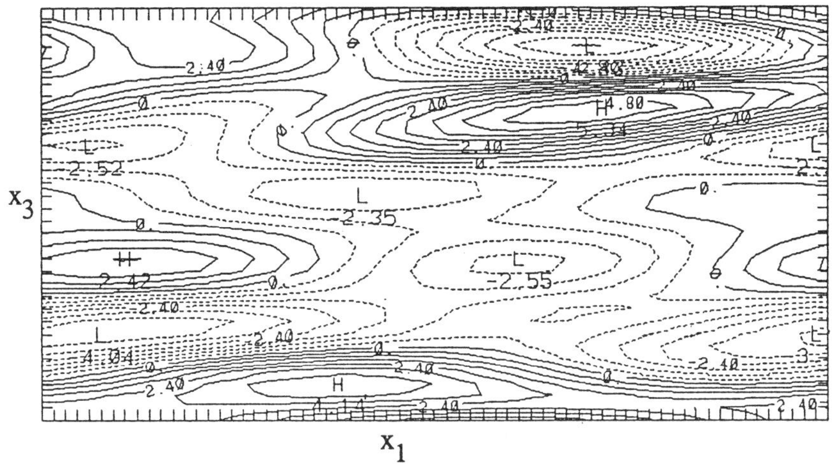

Figure 5. Contours of u in the (x_1-x_3) plane, located at $x_2^+ = 28$, for a solution in the IV window of intermittency ($\alpha = 0.2$) when all the modes are activated. The sreamwise and spanwise dimensions of the box are 666 and 333 wall units, respectively.

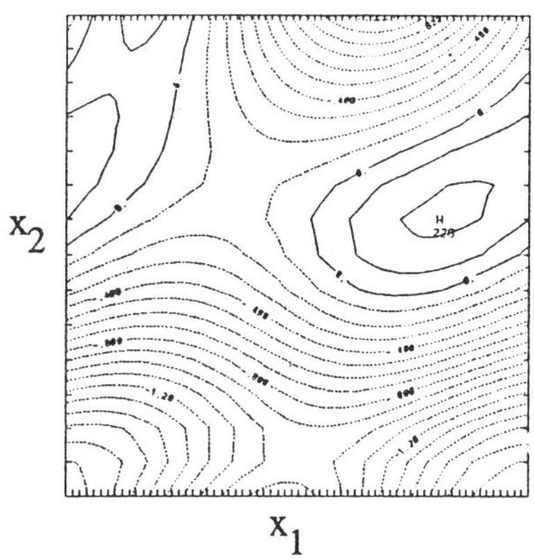

Figure 6. Contours of ω_3 in the (x_1-x_2) plane, located at $x_3^+ = 250$, for a solution in the IV window of intermittency ($\alpha = 0.2$) when all the modes are activated. The sreamwise and normal dimensions of the box are 666 and 40 wall units, respectively (note the high distortion of the box shape).

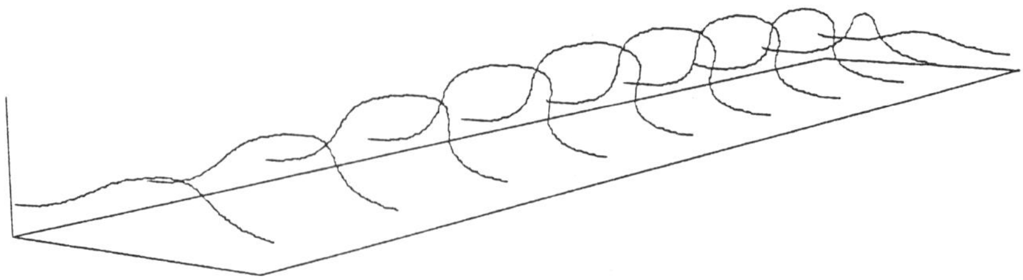

Figure 7. Vortex lines displaying the vortical structure during the IV window of intermittency when all the modes are excited. This is the perspective view of a three dimensional field. The box shown in the figure represents a portion of the lower wall of the channel. The mean flow direction is from lower left to the upper right. The streamwise, spanwise and normal dimensions of the box are 450, 100 and 40 units respectively.

THE STREAMWISE EVOLUTION OF COHERENT STRUCTURES IN THE AXISYMMETRIC JET MIXING LAYER[1]

Mark Glauser and Xiaowei Zheng
Clarkson University
Potsdam, N.Y. USA 13699
and
William K. George
University at Buffalo/SUNY
Buffalo, N.Y. USA 14260

ABSTRACT

The dynamical equations governing the streamwise evolution of coherent structures in the axisymmetric jet shear layer are derived. The method utilized is similar to that used by Aubry et al (1988) and Glauser et al (1989). This method consists of performing a Galerkin projection, using the basis functions obtained from application of the proper orthogonal decomposition (POD), onto the Navier Stokes equations. In this study, however, the so called random coefficients are written as a function of the streamwise direction, and not time as in the previous studies. This results in a boundary value problem and not an initial value problem as was the case in the afore mentioned work. This type of an approach is important from an experimentalist's point of view. The two-point correlation tensor (needed for application of the POD) can be measured much easier at one streamwise location as a function of time difference rather than at many streamwise separations. The basic idea is to measure the two-point correlation at one streamwise location and infer the evolution of the coherent structures in the streamwise direction from the dynamical equations. These equations and some ideas on how to solve them numerically will be discussed.

[1]This work was supported by NSF under grant number MSM-8808872 and Pratt and Whitney/UTC. These sources of support are gratefully acknowledged.

1. Introduction

In recent years, two separate developments have altered the basic statistical framework of turbulence. There is an abundance of experimental evidence implying the existence of coherent structures and recent applications of dynamical systems theory to turbulence suggest that such flows reside on relatively low dimensional manifolds or attractors (v. Aubry et al 1988 and Bartuccelli et al 1989).

Aubry et al (1988) were the first to link low dimensional chaotic dynamics to a turbulent open flow system. They used the proper orthogonal decomposition (POD) to provide basis functions to obtain low dimensional sets of ordinary differential equations from the Navier–Stokes equations. Used in conjunction with Galerkin projection, the POD yields an optimal set of basis functions in the sense that the resulting truncated system of ordinary differential equations captures the maximum amount of kinetic energy among all possible truncations of the same order. Sirovich and Rodriguez (1987) have shown, for the Ginzburg–Landau system, that these basis functions are fairly robust and can be used over a wide range of the bifurcation parameter. Chambers et al. (1988) have seen similar trends for Burgers equation. The basis functions that Aubry et al (1988) used were those obtained experimentally by Herzog (1986) for the near wall region of a pipe flow. The solutions to the equations that they derived exhibited intermittent behavior, and were then analyzed using dynamical systems theory. The results to date show consistency between the behavior of these equations and events seen in experimental work.

The axisymmetric jet is also a good candidate for a similar approach because the series converges quickly. This was demonstrated by Glauser et al (1987), where the instantaneous signals were almost completely reconstructed with only 3 terms from the expansion. By assuming the flow to be approximately homogeneous in the streamwise direction, Glauser et al (1989) and Zheng and Glauser (1990) were able to use a similar approach (where the so called random coefficients are written as a function of time) and examined the sequence by which the streamwise wavenumbers and azimuthal mode numbers retained in the model contribute in time. Their initial results indicate that there is a transfer of energy between certain streamwise wavenumbers and azimuthal modes 4, 5 and 6, consistent with the mechanism for turbulence production suggested by Glauser and George (1987) for the jet shear layer.

In the study reported here the so called random coefficients, for which the equations are written, are functions of the streamwise direction and not time as in the previous studies. This results in a boundary value problem and not an initial value problem as was the case in the afore-mentioned work. This type of approach overcomes the need for assuming the flow to be homogeneous in the streamwise direction, and thus more closely models its spatially developing character. Also it more closely reflects the experiments (from which the basis functions are taken) since the two-point correlation tensor (needed for application of the POD) can be measured much more easily at one streamwise location as a function of time difference rather than at the many streamwise separations necessary to decompose a developing flow.

The axisymmetric jet is stationary so the appropriate decomposition in time is the Fourier transform (v. George 1988). In the azimuthal direction the flow is periodic, hence the discrete Fourier modes are appropriate. In the radial direction the flow is strongly inhomogeneous so that the eigenfunctions obtained from applying the POD are utilized. Galerkin projection is then used in conjunction with the POD to obtain a truncated system of ordinary differential equations. The modes neglected in the truncation are accounted for by a Heisenberg model as was done in the temporal studies of Aubry et al (1988) and Glauser et al (1989). Finally, the evolution of the coherent structures in the streamwise direction is inferred from the resulting dynamical equations. This work appears to be the first which utilizes low dimensional dynamics in conjunction with Galerkin projection and the POD to examine the *spatial* dynamics of coherent structures in a high Reynolds number axisymmetric jet free shear layer.

2. Proper Orthogonal Decomposition

In 1967 Lumley suggested that the coherent structure should be that structure which has the largest mean square projection on the velocity field. This process involves maximizing the mean square energy via the calculus of variations and leads to the following integral eigenvalue problem.

$$\lambda \phi_i(\vec{x}) = \int R_{ij}(\vec{x}, \vec{x}')\phi_j(\vec{x}')d\vec{x}' \tag{1}$$

The symmetric kernel of this Fredholm integral equation is the two–point correlation tenser R_{ij} defined by

$$R_{ij}(\vec{x}, \vec{x}') = \overline{u_i(\vec{x})u_j(\vec{x}')}, \tag{2}$$

where $\vec{\phi}$ is the candidate structure and $\vec{x}$ and $\vec{x}'$ represent different spatial points in the inhomogeneous directions and different times if the flow is non–stationary.

From the Hilbert–Schmidt theory it can be shown that the solution of a Fredholm integral equation of the first kind for a symmetric kernel and a finite energy domain (i.e., statistically inhomogeneous) is a discrete set, hence equation (1) can be written as

$$\lambda^n \phi_i^n(\vec{x}) = \int R_{ij}(\vec{x}, \vec{x}')\phi_j^n(\vec{x}')d\vec{x}' \tag{3}$$

The eigenfunctions of the Fredholm equation are orthogonal over the interval and

$$\int \phi_i^n(\vec{x})\phi_i^m(\vec{x})d\vec{x} = \delta_{nm} \tag{4}$$

for normalized eigenfunctions. The eigenvalues of the Fredholm equation with a real symmetric kernel are all real and uncorrelated

$$\overline{a^n a^m} = \lambda^n \delta_{nm} \tag{5}$$

and the fluctuating random field $\tilde{u_i}$ can be reconstructed from the eigenfunctions in the following way

$$\tilde{u_i}(\vec{x}) = \sum_{n=0}^{\infty} a^n \phi_i^n(\vec{x}). \tag{6}$$

The random coefficients are calculated from

$$a^n = \int \tilde{u_i}(\vec{x}) \phi_i^{n*}(\vec{x}) d(\vec{x}) \tag{7}$$

where the ϕ_i^n are the eigenfunctions obtained from equation (3). The turbulent kinetic energy is the sum over n of the eigenvalues λ^n, and each structure makes an independent contribution to the kinetic energy and Reynolds stress.

If the random field is homogeneous or periodic in one or more directions or stationary in time, the eigenfunctions become Fourier modes, so that the POD reduces to the harmonic orthogonal decomposition in these directions. In this study we will treat the jet flow as periodic in the azimuthal direction and stationary in time. We will not transform over the streamwise direction in this case because we are writing the random coefficients as a function of the streamwise variable z. Under these conditions the spectral tensor may be defined by

$$S_{ij}(r, r', m, f, \overline{z}) = \int R_{ij}(r, r', \theta, \tau, \overline{z}) e^{-i(2\pi f + m\theta)} d\tau d\theta \tag{8}$$

where r and r' represent different spatial locations in the radial direction (the strongly inhomogeneous direction), τ and θ are the separations in time and the azimuthal direction respectively, f is the frequency, m is the azimuthal mode number and $\overline{z}$ denotes the streamwise location where the correlation tensor is measured. Equation (3) now becomes

$$\lambda^n(m, f) \phi_i^n(r, m, f, \overline{z})$$

$$= \int_\Omega S_{ij}(r, r', m, f, \overline{z}) \phi_j^n(r', m, f, \overline{z}) r' dr'. \tag{9}$$

This equation can be solved numerically using the measured values of $S_{ij}(r, r', m, f, \overline{z})$ obtained by Glauser and George (1987) which were obtained at 3 diameters downstream in the jet shear layer. It should be noted that in this case the eigenvalues and eigenfunctions are now a function of azimuthal mode number m and frequency f.

3. The Dynamical Equations

The momentum equations for an incompressible flow in cylindrical coordinates are

$$\rho(\frac{Du_r}{Dt} - \frac{u_\theta^2}{r}) = -\frac{\partial p}{\partial r} + \mu(\nabla^2 u_r - \frac{u_r}{r^2} - \frac{2}{r^2}\frac{\partial u_\theta}{\partial \theta}) + F_r \quad (10)$$

$$\rho(\frac{Du_\theta}{Dt} + \frac{u_\theta u_r}{r}) = -\frac{1}{r}\frac{\partial p}{\partial \theta} + \mu(\nabla^2 u_\theta - \frac{u_\theta}{r^2} + \frac{2}{r^2}\frac{\partial u_r}{\partial \theta}) + F_\theta \quad (11)$$

$$\rho(\frac{Du_z}{Dt}) = -\frac{\partial p}{\partial z} + \mu \nabla^2 u_z + F_z \quad (12)$$

where

$$\frac{D}{Dt} = \frac{\partial}{\partial t} + u_r \frac{\partial}{\partial r} + \frac{u_\theta}{r}\frac{\partial}{\partial \theta} + u_z \frac{\partial}{\partial z}$$

and

$$\nabla^2 = \frac{\partial^2}{\partial r^2} + \frac{1}{r}\frac{\partial}{\partial r} + \frac{1}{r^2}\frac{\partial^2}{\partial \theta^2} + \frac{\partial^2}{\partial z^2}$$

We assume that the ρ and μ are constant and $F_r = F_\theta = F_z = 0$.

By decomposing the instantaneous dependent variables into a fluctuation and an average over the azimuthal direction and time, the equation for the fluctuating velocity can be obtained as

$$\frac{\partial}{\partial t}\tilde{u}_i + M\tilde{u}_i + (-\frac{\tilde{u}_\theta^2}{r} + \overline{U}_z \frac{\partial \tilde{u}_r}{\partial z} + \overline{U}_r \frac{\partial \tilde{u}_r}{\partial r} + \tilde{u}_r \frac{\partial \overline{U}_r}{\partial r})\delta_{ir}$$

$$+(\frac{\tilde{u}_r \tilde{u}_\theta}{r} + \overline{U}_z \frac{\partial \tilde{u}_\theta}{\partial z} + \overline{U}_r \frac{\partial \tilde{u}_\theta}{\partial r})\delta_{i\theta}$$

$$+(\overline{U}_z \frac{\partial \tilde{u}_z}{\partial z} + \overline{U}_r \frac{\partial \tilde{u}_z}{\partial r} + \tilde{u}_r \frac{\partial \overline{U}_z}{\partial r} + \tilde{u}_z \frac{\partial \overline{U}_z}{\partial z})\delta_{iz}$$

$$-\overline{M\tilde{u}_i} + \frac{\overline{\tilde{u}_\theta^2}}{r}\delta_{ir} + \frac{\overline{\tilde{u}_r\tilde{u}_\theta}}{r}\delta_{i\theta}$$

$$= -\frac{r(\delta_{ir}+\delta_{iz})+\delta_{i\theta}}{r\rho}\frac{\partial\tilde{p}}{\partial i}$$

$$+\nu[\nabla^2\tilde{u}_i - (\frac{\tilde{u}_r}{r^2}+\frac{2}{r^2}\frac{\partial\tilde{u}_\theta}{\partial\theta})\delta_{ir} + (-\frac{\tilde{u}_\theta}{r^2}+\frac{2}{r^2}\frac{\partial\tilde{u}_r}{\partial\theta})\delta_{i\theta}] \qquad (13)$$

where $M = \tilde{u}_r\frac{\partial}{\partial r} + \frac{\tilde{u}_\theta}{r}\frac{\partial}{\partial\theta} + \tilde{u}_z\frac{\partial}{\partial z}$, $i = r, \theta, z$, δ is the Kronecker delta, $\tilde{p}$ is the fluctuating pressure, $\tilde{u}_r$, $\tilde{u}_\theta$ and $\tilde{u}_z$ are the fluctuating velocities in the radial, azimuthal, and streamwise directions respectively, and $\overline{U}_z$ and $\overline{U}_r$ are the mean streamwise and radial velocity respectively. These mean velocities must also satisfy the averaged continuity equation. It should be noted that the terms involving $\overline{U}_r$ are kept in the *spatially* evolving problem considered here whereas Zheng and Glauser (1990) neglected these terms. In the high Reynolds number jet studied here we can neglect the viscous terms in equation (13) and this will be done in the following section when the Galerkin projection is performed.

We seek a relationship which relates the mean velocities $\overline{U}_z$ and $\overline{U}_r$ to the Reynold's stress since the actual measured mean velocity profiles will be incorrect for the truncated system of equations to be studied. The equation which relates these (to second order) can be derived from the z momentum equation as

$$\frac{1}{r}\frac{\partial}{\partial r}(r\overline{\tilde{u}_r\tilde{u}_z}) = -(\overline{U}_z\frac{\partial}{\partial z}\overline{U}_z + \overline{U}_r\frac{\partial}{\partial r}\overline{U}_z) \qquad (14)$$

where the mean pressure term has been eliminated from the streamwise equation by integrating the averaged radial equation (v. Tennekes and Lumley 1972). Equation (14) indicates that the Reynold's stress is balanced by the mean convection and provides communication between the mean flow and the Reynold's stress for the *evolving* shear layer. This result is quite different than that of Zheng and Glauser (1990) who treat the flow as homogeneous in z, resulting in the Reynold's stress being balanced by the viscous terms. This modifies the character of the ODEs as will be seen in the next section.

4. Galerkin Projection

The Galerkin method is well known and has been used extensively to study turbulence and the instability of various fluid flows (v. Lin et al (1987) and references therein). The essential idea of the method is to expand the dependent variables in terms of a finite series of independent basis functions. The basis functions form a complete basis for the relevant class of functions, and they satisfy the relevant boundary conditions. In this work we use a Galerkin projection in conjunction with the POD (to supply the basis functions) which minimizes the error due to the truncation and yields a set of ordinary differential equations for the coefficients (v. equation 7).

The Galerkin projection is performed on the Fourier Transform (over time and in the azimuthal direction) of the Navier Stokes equations so that is is useful to define the following equations:

$$\tilde{u}_i(z,\theta,r,t) = \sum_{m=-\infty}^{\infty} \int_{-\infty}^{\infty} e^{-2\pi j(tf+m\theta)} \hat{u}_i(f,m,r,z)df \qquad (15)$$

$$\hat{u}_i(f,m,r,z) = \int_{-\pi}^{\pi} \int_{-\infty}^{\infty} e^{2\pi j(tf+m\theta)} \tilde{u}_i(z,\theta,r,t)dtd\theta \qquad (16)$$

We expand $\hat{u}_i(f,m,r,z)$ in terms of the coefficients and eigenfunctions in the following manner

$$\hat{u}_i(f,m,r,z) = \sum_n a^n_{f,m}(z)\phi^n_{i,f,m}(r). \qquad (17)$$

Note the z dependence of the random coefficients. We then substitute these equations into the Fourier Transform of equation (13) *after eliminating the viscous terms* and perform the Galerkin projection. This can be written as

$$(\phi^l, N) = \int_0^R N_i(r,z)\phi_i^{l*}(r)dr = 0 \qquad (18)$$

where $N_i(t, z)$ represents the Fourier Transform of the Navier–Stokes equations (i.e. equation 13 with the viscous terms neglected).

Finally, after performing the above and utilizing the orthogonality conditions (v. equation 4) we obtain the following set of coupled ordinary differential equations for the coefficients:

$$\begin{aligned}
0 &= \sum_n a^n_{f,m}[-2\pi j f \delta_{nl}] \\
&+ \sum_{f'm'}\sum_{pq} a^p_{f',m'}\{a^q_{f-f',m-m'} \int_0^R [\phi^p_{r,f',m'} \frac{d}{dr}\phi^q_{i,f-f',m-m'} \\
&\quad + \frac{-j(m-m')}{r}\phi^p_{\theta,f',m'}\phi^q_{i,f-f',m-m'} - (\frac{1}{r}\phi^p_{\theta,f',m'} \\
&\quad \phi^q_{\theta,f-f',m-m'})\delta_{ir} + (\frac{1}{r}\phi^p_{r,f',m'}\phi^q_{\theta,f-f',m-m'})\delta_{i\theta}] \cdot \phi^{l*}_{i,f,m} dr \\
&\quad + b^q_{f-f',m-m'} \int_0^R [\phi^p_{z,f',m'}\phi^q_{i,f-f',m-m'}] \cdot \phi^{l*}_{i,f,m} dr\} \\
&+ MeanFlowFluctuationInteractionTerms \\
&- \int_0^R [\frac{r(\delta_{ir}+\delta_{iz}) + \delta_{i\theta}}{\rho r}\frac{\partial \hat{p}_{f,m}}{\partial i} \cdot \phi^{l*}_{i,f,m}] dr \qquad (19)
\end{aligned}$$

where

$$b^l_{f-f',m-m'} = \frac{da^l_{f-f',m-m'}}{dz}. \qquad (20)$$

In our ordinary differential equation

$$0 = Bl + Cq_1 + Dq_2 + MeanFlowTerms$$
$$+ PressureTerm \qquad (21)$$

there are 5 parts. The first term on the right hand side (RHS) is a linear term that comes from the time derivative of the Navier–Stokes equations. The second and third terms on the RHS are quadratic terms that are a consequence of the fluctuation–fluctuation interactions and exhibits the energy transfer between the different eigenfunctions(from the POD) and the Fourier modes and B, C, and D, are all matrices. The fourth term, not shown here due to lack of space, is a result of the mean flow fluctuation terms in equation (13). It should be noted that the third and fourth terms contain the spatial derivative of our coefficient a with respect to z. The final term is the pressure term which will vanish if the integration covers the whole domain. In the jet shear layer we cover most of the domain in our integration (unlike Aubry et al 1988 who studied the near wall region and not the whole domain of the turbulent boundary layer), hence we will neglect this pressure term. This equation, unlike the temporal equations derived by Glauser et al (1989), does not have a cubic term. This is a direct result of how we are handling the relationship between the mean flow and the Reynold's stress for the *evolving* shear layer. Instead of eliminating the mean flow terms from the system of ODEs as was done by Glauser et al (1989), here we propose to solve equation (14), along with the mean continuity equation, simultaneously with the system of ODEs. Perhaps it would be easier to decompose the flow initially by using the mean and fluctuating components instead of only the fluctuating. However, looking at the fluctuating part by itself, is more instructive for now since it highlights the differences from Glauser et al (1989) and Aubry et al (1988).

An interesting point to note is that the b terms (i.e., $\frac{da^l_{f,m}}{dz}$) can be eliminated from equations (19) through (21) by utilizing the equation of continuity for the fluctuations, which, in terms of the random coefficients can be written as

$$\frac{da^l_{f,m}}{dz}\phi^l_{z,f,m} = -a^l_{f,m}\left[\frac{d}{dr}\phi^l_{r,f,m} + \frac{1}{r}\phi^l_{r,f,m} + \frac{im}{r}\phi^l_{\theta,f,m}\right]. \tag{22}$$

This reduces the ODEs to a system of coupled algebraic equations which is an intriguing result.

5. The Energy Transfer Model

When we truncate at some cutoff point we need to account for the energy transfer between the included and neglected modes. The effect of the neglected modes will be accounted for by utilizing a Heisenberg model (v. Aubry et al 1988). The assumption is that the neglected modes withdraw energy from the modes that our kept, as if a certain turbulence viscosity were present.

The equation for ν_T, our turbulence viscosity, can be shown to be (v. Zheng 1991)

$$\nu_T = \frac{\alpha \sum_{f,m,n} \lambda^n_{f,m}}{R \sum_{f,m,n} \lambda^n_{f,m} \overline{S}} \tag{23}$$

where

$$\overline{S} = \overline{\int_0^R [\frac{d\phi^n_{i,f,m}}{dr}\frac{d\phi^{n*}_{i,f,m}}{dr} - \frac{1}{r^2} 4\pi^2 m^2 \phi^n_{i,f,m}\phi^{n*}_{i,f,m}]dr - 4\pi^2 f^2\}}$$

α is a dimensionless parameter and the sums are over the first neglected modes. We now substitute ν_T into our ordinary differential equations. The equations then have the form

$$0 = B_1 l_1 + \nu_T B_2 l_2 + C q_1 + D q_2$$

$$+ MeanValueFluctuationTerms \tag{24}$$

where l_1 is the linear term from the time derivative of the Navier–Stokes equations as was shown earlier in equation (21) and l_2 is a linear term introduced because of the assumption of the neglected modes acting as a viscosity on the modes that are kept.

The effect of the energy transfer model is to introduce the parameter α which becomes the bifurcation parameter in our system of ODEs. The larger α the more energy that the neglected modes take from our system so that the system should be stable. As α decreases less energy is extracted so that we expect our system to become unstable.

6. Truncations and Discussion

A truncated version of equation (19) must be developed. A basic guide in the selection of which terms to keep in the model is to retain a minimum number of terms but yet keep as much energy in the system as is necessary to retain the essential dynamics of the flow phenomena (v. Aubry et al 1988 and Glauser et al 1989). Figure 1 shows the dominant eigenvalue obtained from the application of the POD in the jet shear layer at 3 diameters downstream, plotted versus frequency and azimuthal mode number. This plot indicates that there is an exchange of energy between the various frequencies and azimuthal mode numbers, and in particular, that there is a maximum energy ridge (remember that the eigenvalues are energy integrated across the jet shear layer) between the two peaks for azimuthal modes 0 and 1 and the peak at azimuthal mode 5. The peak in the frequency direction for azimuthal mode 0 corresponds to the preferred frequency of the jet. The peak in the azimuthal mode number direction is approximately at mode 5. A possible combination of frequencies and azimuthal modes that could be used as an initial truncation are shown in figure 2. These were selected to try and capture the maximum energy ridge shown in figure 1. As an initial step, only the dominant eigenfunction will be necessary in the inhomogeneous direction. This corresponds to setting $l = n = p = q = r = 1$ in equation (19). This is justified because the dominant eigenvalue is significantly larger than the next smallest eigenvalue and in fact, typically contains 50 percent or more of the energy. All of the higher azimuthal modes exhibit this same dominance of the first eigenvalue (v. Glauser and George 1987). This particular truncation will result in a set of 18 complex or 36 real, ordinary differential equations.

The integrations needed to obtain the coefficient matrices in the truncated version of equation (19) can be obtained using the trapezoidal rule. A fourth order Runge Kutta scheme could then be used to integrate the truncated version of equation (19) with the necessary boundary conditions. As a first step in the selection of the boundary conditions, we can argue that at $z = 0$ the value of our coefficient a is zero, since the fluctuations are small at this position.

7. Conclusions

The dynamical equations have been derived for the spatially evolving random coefficients in the axisymmetric jet shear layer. An initial truncation has been proposed which would involve a system of 36 ODEs. Instead of eliminating the mean flow terms from the system of ODEs as was done by Glauser et al (1989), here we propose to solve for the mean quantities, $\overline{U}_z$ and $\overline{U}_r$, simultaneously with the system of ODEs. Hence, the resulting ODEs do not exhibit the cubic character as described by Aubry et al (1988) and Glauser et al (1989).

An interesting extension of this work would be to write the random coefficients as a function of time and the streamwise direction (i.e., $a = a(z,t)$). This would result in a system of partial differential equations in z and t. This type of an approach would allow for the temporal and spatial problems to be studied simultaneously.

ACKNOWLEDGMENTS

The authors would like to thank Professors S.P. Lin, W.R.C. Philips and C.R. Doering from the Institute for Nonlinear Studies at Clarkson for their many helpful discussions.

REFERENCES

Aubry, N., Holmes, P., Lumley, J.L. and Stone, E. (1988) The Dynamics of Coherent Structures in the Wall Region of a Turbulent Boundary Layer, *JFM vol.192, pp.115-173.*

Bartuccelli, M., Constantin, P., Doering, C.R., Gibbon, J.D., and Gisselfalt, M. (1989), Hard Turbulence in a Finite Dimensional Dynamical System?, *Physics Letters A–Nonlinear Science*

Chambers, D.H., Adrian, R.J., Moin, P. Stewart, D.S., and Sung, H.J., (1988) *Phys. of Fluids* **31** p.2573.

George, W.K. (1988), Insight into the Dynamics of Coherent Structures from a Proper Orthogonal Decomposition, Proceedings Symposium on Near Wall Turbulence, Dubrovnik, Yugoslavia, May 16-20.

Glauser, M.N. and George, W.K. (1987) An Orthogonal Decomposition of the Axisymmetric Jet Mixing Layer Utilizing Cross-Wire Velocity Measurements. *Proc. 6th Symp. Turbulent Shear Flows*, Toulouse, France, Sept. 7-9.

Glauser, M.N., Leib, S.J. and George, W.K. (1987) Coherent Structures in the Axisymmetric Jet Mixing Layer. *Turbulent Shear Flows 5, Springer-Verlag.*

Glauser, M.N., Zheng, X. and Doering, C.R. (1989) The Dynamics of Organized Structures in the Axisymmetric Jet Mixing Layer, *Turbulence and coherent structures, M. Lesieur and O. Metais, eds., Kluwer academic publishers.*

Herzog, S. (1986) The Large Scale Structure in the Near-Wall Region of Turbulent Pipe Flow. PH.D. Dissertation, Cornell University.

Lin, S.P., Chen, H.B., and Tobak, M.(1987) Instability of A Non-parallel Flow,*Proceedings of 1st International Conference of Fluid Mechanics, Beijing.*

Lumley, J.L. (1967) The Structure of Inhomogeneous Turbulent Flows, In *Atm. Turb. and Radio Wave. Prop. (Yaglom and Tatarsky, ed.) Nauka, Moscow, 166-1,8.*

Sirovich, L. and Rodriguez, J.D. (1987) Coherent Structures and Chaos: A Model Problem,*Physics Letters A, Volume 120, Number 5.*

Tennekes, H. and Lumley, J.L. (1972), A First Course In Turbulence, *MIT Press.*

Zheng, X. and Glauser, M. N.(1990), A Low Dimensional Description of the Axisymmetric Jet Mixing Layer, ASME Computers in Engineering, Vol. 2 pp. 121-127.

Zheng, X. (1991), A Low Dimensional Description of the Axisymmetric Jet Mixing Layer, Ph.D. Thesis, Dept of Mechanical and Aeronautical Engineering, Clarkson University.

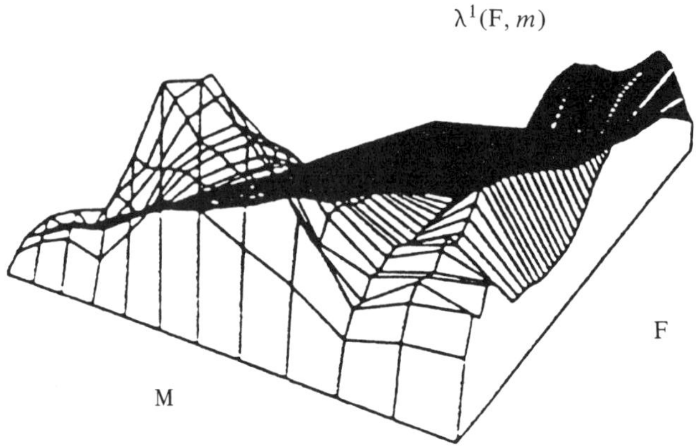

Figure 1: Dominant eigenvalue obtained from the application of the POD in the jet shear layer at 3 diameters downstream.

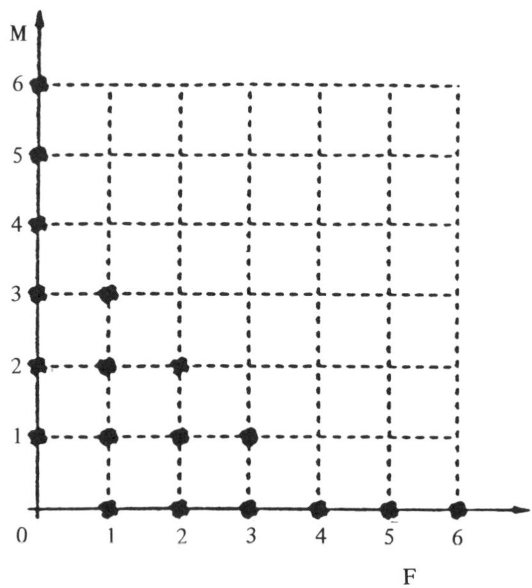

Figure 2: Possible combination of frequencies and azimuthal modes that could be used as an initial truncation.

ON THE COMPLEXITY OF TURBULENCE NEAR A WALL

Parviz Moin

Dept. of Mechanical Engineering
Stanford University Stanford, CA 94305,
& NASA-Ames Research Center

ABSTRACT

Some measures of the intrinsic complexity of the near wall turbulence are reviewed. The number of modes required in an "optimal" eigenfunction expansion is compared with the dimension obtained from the calculation of Liapunov exponents. These measures are of the same order, but they are very large. It is argued that the basic building block element of the near wall turbulence can be isolated in a small region of space (minimal flow unit). When the size of the domain is taken into account, the dimension becomes more manageable.

1. Application of Lumley's Orthogonal Decomposition

In this section, the complexity of the flow is measured by considering the number of terms required in an eigenfunction expansion of the velocity field. Functional decomposition into a sum of mutually orthogonal eigenfunctions of the two-point correlation tensor is optimal in the sense that the mean square value of the function is recovered with the fewest terms in the expansion. This decomposition is known as the Karhunen-Loeve or Lumley's expansion. The eigenfunctions were proposed by Lumley [1] for the representation of characteristic organized structures in turbulent flows. Moin & Moser [2] applied Lumley's Decomposition to turbulent channel flow using the direct numerical simulation database of Kim, Moin & Moser [3]. Fourier expansions were used in the directions of flow homogeneity, x and z. The decomposition can be performed for any sub-domain by simply changing the limits of the integral in the Fredholm eigenvalue problem, and the corresponding eigenfunctions in that sub-domain can be found.

This procedure allows for separate measures of complexity in different flow regimes such as the near wall region or the core of the channel.

It can be shown that turbulent kinetic energy in the entire computational box is given by

$$< u_i u_i > = \sum_n \sum_{k_x} \sum_{k_z} \lambda^{(n)}(k_x, k_z) \tag{1}$$

where, $\lambda^{(n)}(k_x, k_z)$ is the eigenvalue corresponding to the nth K-L eigenfunction and Fourier waves, k_x and k_z. The eigenvalues are arranged in descending order with $\lambda^{(1)}$ as the largest eigenvalue. The convergence of the K-L expansion is normally expressed in terms of the number of modes n required to recover a certain (high) percentage of the turbulent kinetic energy after summation is taken over *all Fourier modes*, k_x and k_z. Indeed, Moin & Moser found that in the domain extending from the wall to the centerline, the first term carries 50% of the kinetic energy. However, when only the near wall region is considered ($y^+ \leq 40$), 76% of the energy is recovered with only one term in the expansion. For the domain of the same size but away from the wall ($140 \leq y^+ \leq 180$), the first term contains 57% of the energy. The strong inhomogeneity near the wall imposes a length scale on the flow which apparently leads to more organization. Thus, from the energy considerations *the flow near the wall is less complex than the more homogeneous flow away from the wall.* Another result that is not predicted by the theory but is obtained from the computations is that the method is also very efficient in recovering the Reynolds shear stress, $< u_1 u_2 >$.

A more appropriate measure of the complexity should include the total number of modes including the Fourier modes in the homogeneous directions. Examimation of $\lambda^{(n)}(k_1, k_3)$ surface shows that to recover 80% of the kinetic energy in the near wall domain with two K-L modes in the wall-normal direction, 15 complex modes (with $k_x, k_z > 0$) are required in each of the spanwise z- and streamwise x-directions. Noting that there are two independent velocity components (the third being determined from the continuity), the total number of degrees of freedom in the near wall region is estimated to be

$$D_L = 3600.$$

This estimate is significantly less than approximately 10^6 degrees of freedom used in the direct simulations of Kim, Moin & Moser [3] in the same region (for 100% energy recovery). The dimension estimate for the domain extending from the wall to the centerline is about four times larger.

2. Dimension from the Spectrum of Lyapunov Exponents

Keefe *et al.* [4] estimated the dimension of the strange attractor underlying fully developed turbulent channel flow at the very low Reynolds number of 1780 (based on the centerline velocity and channel half-width). Direct numerical simulations were carried out with the resolution ranging from $16 \times 33 \times 8$ to $32 \times 33 \times 32$ spectral modes (6.8×10^4 degrees of freedom). The Lyapunov spectrum was calculated for a direct application of Kaplan-Yorke formula for dimension. For this Reynolds number the dimension was estimated at

$$D_\lambda = 780.$$

This value should probably be viewed as a lower estimate because at the highest computational resolution the full spectrum necessary for application of Kaplan-Yorke formula could not be obtained due to computational constraints.

To compare different estimates of the dimension we must take into account the difference in Reynolds numbers. The Reynolds number for the calculations of Keefe *et al.* was lower than that of Moser & Moin by the factor of 1.8. Using the $Re^{9/4}$ scaling for the dimension, D_λ for the higher Reynolds number calculation is estimated to be

$$D_\lambda = 2900.$$

This estimate is for the entire channel. The dimension for one side of the channel should be about half of this value, and for the near wall region, we should divide by another factor of four (see §1). Thus, for the near wall region the dimension is estimated to be

$$D_{\lambda-w} = 363.$$

The dimension estimates, D_L and D_λ, differ significantly. Part of the discrepancy is due to different methodologies used. However, the sizes of the computational domains in the homogeneous directions, x and z, are very different in the two computations which can affect measures of the complexity. This issue will be taken up in the next section.

3. The Minimal Flow

Recently, Jimenez & Moin [5] have shown that turbulent channel flow can be sustained in a relatively small computational box. When the computational box is reduced beyond a critical size, the flow becomes laminar. The critical box size was found to scale with the wall variables, and it was approximately 100 wall units in the spanwise z-direction and between 200 and 300 wall-units in the streamwise direction. In the near wall region, the mean velocity profile and turbulence intensities were in good agreement with the experimental data. However, away from the wall the agreement was only fair, presumably because the larger eddies that exist in the outer region could not be accommodated in the computational box. The structure of the minimal unit was investigated. Generally, it consists of one or a pair of streamwise vortices, a shear layer and a wall-layer streak. Apparently, the non-linear interaction between these structures in the presence of the wall leads to a stochastic field with the correct second order statistics. The minimal flow unit is considered as the building block element of near wall turbulence because it reproduces the essential turbulence statistics in the wall region. Clearly, two-point statistics in the minimal flow would not be in agreement with the experiments or 'natural' channel simulations. However, one may speculate that two-point correlations (at large separations) can be reproduced by an appropriate spatial sprinkling of the minimal flow units and may not be of any dynamical significance.

Due to significantly lower computational cost, the minimal flow can be integrated for large times, and has found several applications. It is used to study the dynamics of the near wall structures in two- and three-dimensional boundary layers and in simulation and analysis of turbulent flow over riblets.

The size of the minimal flow unit in planes parallel to the walls is about 100×250 square wall units. This is 68 times smaller than 1.7×10^6 square wall units used in computations of Moin & Moser and 10 times smaller than 2.5×10^5 units used in computations of Keefe *et al.*. There is evidence that the dimension of the attractor underlying fully developed periodic solutions of the Ginzburg-Landau equation increases linearly with the spatial periodic size [6]. Using the minimal flow unit as the yardstick and dividing the computed dimensions by the ratios of the areas of the respective computational domains, the minimal flow unit results in the following normalized dimensions

$$D_{L-w-N} = 53 \qquad \text{and} \qquad D_{\lambda-N} = 36.$$

The estimates are of the same order of magnitude.

4. Conclusions

We have used the dimension as a measure of the intrinsic complexity of turbulence. Our estimates indicate that the dimension of the essential dynamical element of low Reynolds number wall-turbulence is in the tens rather that thousands. Although a turbulent boundary layer is still more complicated than Taylor Couette flow or Benard convection, the prospects for dynamical systems analysis is brighter in the context of the minimal flow unit.

Acknowledgements

I am grateful to Bob Moser, Laurence Keefe, John Vastano, and Javier Jimenez for stimulating discussions.

References

[1] Lumley, J.L. 1967 The structure of inhomogeneous turbulent flows. In *Atmospheric Turbulence and Radio Wave Propagation* (ed. A.M. Yaglom & V.I. Tatarsky), p. 166. Moscow: Nauka.

[2] Moin, P. and Moser, R.D. 1989 Characteristic-eddy decomposition of turbulence in a channel. *J. Fluid Mech.* **200**, 471-509.

[3] Kim, J., Moin, P. & Moser, R. 1987 Turbulence in fully developed channel flow at low Reynolds number. *J. Fluid Mech.* **177**, 133-166.

[4] Keefe, L., Moin, P. & Kim, J. 1987 The dimensions of an attractor in turbulent Poiseuille flow. *Bull. of the Am. Phys. Soc.* **32**, 10, 2030.

[5] Jimenez, J. & Moin, P. 1990 The minimal flow unit in near wall turbulence. *CTR Manuscript 105* Ctr. for Turbulence Research. Also to appear in *J. Fluid Mech.*

[6] Ghidaglia, J.M. & Heroń, B. 1987 *Physica* **28D**, 282.

OBSERVATIONS ON THE PROPER ORTHOGONAL DECOMPOSITION

*Gal Berkooz**

Cornell University
Ithaca, NY 14853

ABSTRACT

The Proper Orthogonal Decomposition (P.O.D.), also known as the Karhunen-Loeve expansion, is a procedure for decomposing a stochastic field in an L^2 optimal sense. It is used in diverse disciplines from image processing to turbulence. Recently the P.O.D. is receiving much attention as a tool for studying dynamics of systems in infinite dimensional space. This paper reviews the mathematical fundamentals of this theory. Also included are results on the span of the eigenfunction basis, a geometric corollary due to Chebyshev's inequality and a relation between the P.O.D. symmetry and ergodicity.

1. Introduction

The Proper Orthogonal Decomposition, also known as the Karhunen-Loeve decomposition, is a procedure by which a random vector field $\mathbf{u}(\mathbf{x}, t)$ is represented as

$$\mathbf{u}(\mathbf{x}, t) = \sum_{k=1}^{\infty} a_i(t)\, \psi_i(\mathbf{x}) \tag{1}$$

where $a_i(t)$ are time dependent coefficients and $\psi_i(\mathbf{x})$ are deterministic time independent functions. Throughout it is assumed that $\mathbf{u}(\mathbf{x}, t)$ is a field on some domain Ω and that $\mathbf{u}(\mathbf{x}, t)$ is almost always with respect to t in $L^2(\Omega)$.

The Proper Orthogonal Decomposition (P.O.D.) produces the "optimal" representation in an "average" sense. Both these notions, of "optimality" and "averaging" will be discussed in the sequel. The utility of the P.O.D. comes from the fact that the eigenfunctions $\{\psi_i(\mathbf{x})\}_{i=1}^{\infty}$ may be obtained by averages of the random field. Once these eigenfunctions and their associated average energy content is known, possibilities for

* Center for Applied Math. 305 Sage Hall, Cornell University, Ithaca, NY 14853

modeling and deeper understanding of the random phenomena become apparent.

According to Lumley (1970) the P.O.D. was suggested independently by several scientists (e.g. Kosambi (1943), Loeve (1945) Karhunen (1946) Pougachev (1953) and Obukhov (1954)). In the context of coherent structures in turbulent flows it was suggested by Lumley (1967). It seems that the most common application today is in communications and signal processing, to maximize throughput of communications channels. See Andrews et al. (1967), Algazi et al. (1969). The P.O.D. has been applied to the extraction of coherent structures in turbulent flows cf. Bakewell and Lumley (1967), Herzog (1986), Moin and Moser (1989). Recently it has been applied to the study of the dynamics of turbulent systems. cf. Aubry et al. (1988), Sirovich (1989), and to obtain upper bounds on the dissipative spectrum of turbulence by Foias et al. (1990).

This work is motivated by applications of the P.O.D. to the understanding of dynamics in turbulent systems, such applications have become the subject of recent interest. Some of the material included here is known and included for the sake of completeness, and some are recent observations. The organization is as follows: we start by defining the different averaging operations possible and their relations to the P.O.D. We then review the motivation for the empirical eigenvalue approach, identify the span of the eigenfunctions and show the optimality of the expansion. We then discuss the cases when the random field is homogeneous and there are symmetries in the problem. We show a necessary condition for ergodicity of complex systems which may be verified by examining the eigenspaces. We finish by showing an application of Chebyshev's inequality to arrive at bounds for the location of the dynamics or attractors in phase space, in a probabilistic sense.

For simplicity we will assume that $\mathbf{u}(\mathbf{x}, t)$ is a one dimensional vector field. The dimension of the domain to which x belongs is unimportant. The general vector case requires nothing more than additional notation. The subject of averaging of chaotic or turbulent systems is fundamental to our discussion. To the experimentalist the notion of ensemble average is very useful. The rationale is as follows. If one performs the same experiment starting from the same initial condition (up to the resolution of the measurements) a large number of times such that a quantity of interest (say $\mathbf{u}(\mathbf{x}, t)$ after a given time t_0) has a well defined average, with convergence as one would expect from the Law of Large Numbers, then one calls this "the ensemble average" of the quantity of interest. The above may be translated to precise mathematical language, using stochastic processes and their transition probabilities. For our purpose this is not worth the technical effort involved. We will simply state our assumption regarding ensemble averages (denoted by $\langle\bullet\rangle$).

<u>Ansatz</u>: If $P = P(\mathbf{u}(\mathbf{x}, t))$ is an operation defined on a random field and L is some linear operator of interest on the space which includes the range of P then $\langle LP(\mathbf{u}(\mathbf{x}, t))\rangle = L\langle P(\mathbf{u}(\mathbf{x}, t))\rangle$, assuming both sides exist.

We give an example showing that this ansatz holds for very general systems. Although stated in a general form, for the sake of generality, our primary concern is for commutation of the averaging operation with spatial or phase space averaging. Assume $(L^2(\Omega), \mathfrak{B}, \mu, T)$ is a measure theoretic dynamical system on $L^2(\Omega)$, i.e. $T : L^2(\Omega) \to L^2(\Omega)$ is a measurable map with respect to the σ-algebra $\mathfrak{B}$ defined by the topology on $L^2(\Omega)$ and it preserves the measure μ, i.e.

$$\mu(T^{-1}(A)) = \mu(A) \tag{2}$$

where $T^{-1}(A) = \{\varphi \in L^2(\Omega) \mid T(\varphi) \in A\}$.

In this case if one assigns (discrete time).

$$u(x, t) = T^t \mathbf{u}(\mathbf{x}, 0) \tag{3}$$

and assumes that the measure on the ensemble of initial conditions is μ, (i.e. $<\cdot>$ – averages are the same as μ – averages) then one has an explicit expression for the ensemble averages

$$<P(u(x, t))> = \int_{L^2(\Omega)} P(T^t u)\, d\mu(u) = \int_{L^2(\Omega)} P(u) d\mu(u). \tag{4}$$

If $P : L^2(\Omega) \to H$ in an operation from $L^2(\Omega)$ into a Hilbert space H and, $L : H \to S$ is a linear operation from H to another Hilbert space S in order for us to have

$$L \int_{L^2(\Omega)} P(u)\, d\mu(u) = \int_{L^2(\Omega)} LP(u)\, d\mu(u)\ , \tag{5}$$

then it is sufficient to assume that $< \| u(x, t) \| > < \infty$, and that is P a bounded operator on supp$\{\mu\}$, and S bounded on the range of P.

Of specific interest is the case when one can replace the ensemble with a time average. This happens when the dynamical system is ergodic. See Cornfeld Fomin & Sinai (1982). Also note that one can use the time averages of a given experiment as an averaging operator. In this case the measure to be produced is the measure on the ergodic component containing the initial condition. In the treatment given below we will think primarily of time as the ensemble index. This is done for simplicity of notation.

2. The Karhunen-Loeve or Proper Orthogonal Decomposition

2.1 The basic procedure

The K-L or P.O.D. provides a means of producing a basis to linearly decompose a random signal. We start with $\tilde{u}(x, t)$, the random signal. Denote $U = <\tilde{u}(x, t)>$ and u the fluctuation around the mean, i.e. $\tilde{u}(x, t) = U + u$. We start by a motivation for the P.O.D. presented by Lumley (1970).

Given the random real valued signal u(x, t) we ask which element φ in $L^2(\Omega)$ is best correlated with u, i.e. which φ maximizes

$$\frac{\langle (u, \varphi)^2 \rangle}{\|\varphi\|^2} \tag{6}$$

where $(\cdot,\cdot)$ is the inner product in $L^2(\Omega)$ and $\|\cdot\|$ is the norm. Note that this correlation is slightly different than the common notion of a correlation coefficient and represents a similarity of shape. But due to our assumption on $\langle\bullet\rangle$, we have

$$\langle (u,\varphi)^2 \rangle = \left\langle \int_\Omega u(x)\bullet\varphi^*(x)\,dx \int_\Omega u^*(x')\bullet\varphi(x')dx' \right\rangle =$$

$$\int_\Omega \left(\int_\Omega \langle u(x)u^*(x')\rangle \bullet\varphi(x')dx' \right)\bullet\varphi^*(x)dx \tag{7}$$

Denote $\langle u(x)\, u^*(x')\rangle = \mathbb{R}(x, x')$ the so called "two point correlation tensor". $\mathbb{R}$ will also denote the linear operator on $L^2(\Omega)$ of which $\mathbb{R}(x, x')$ is the kernel (see Appendix A). We see that our problem is equivalent to

$$\max \frac{(\mathbb{R}\varphi,\varphi)}{\|\varphi\|^2} \tag{8}$$

The last expression shows that φ will be a solution to an eigenvalue problem. But first let us see what assumptions we must impose on u(x, t) to make our discussion rigorous.

The first detail is a change of the order of integration for

$$\int_\Omega u(x)\bullet\varphi^*(x)dx \int_\Omega u(x')\bullet\varphi(x')dx' = \int_{\Omega x\Omega} ((u(x)u^*(x'))\bullet\varphi(x'))\bullet\varphi^*(x)$$

dx'dx.

for this we assume that $\int_{\Omega x\Omega} u(x)u^*(x')dxdx' < \infty$. We will make a stronger assumption, namely

$$\left| \int_{\Omega x\Omega} u(x)u^*(x)\,dxdx' \right| < C_1 < \infty \text{ a.e w.r.t } \langle\bullet\rangle \tag{9a}$$

In addition we will assume

$$\text{ess sup } \| u(x, t))\|^2 < C_2 < \infty \text{ with respect to } \langle\bullet\rangle \tag{9b}$$

This will assure us that we can interchange the order of integration with the averaging operation, hence (7) is meaningful. It also assures us that $\mathbb{R}$ has a square integrable kernel, as per Appendix A. In addition $\mathbb{R}$ is a non negative operator. Hence we get that the solution to (8) is given by solving

$$\mathbb{R}\,\varphi_1 = \mu_1 \varphi_1$$

where μ_1 is the largest eigenvalue of $\mathbb{R}$. The eigenfunction φ_1 will be the desired solution.

One can proceed in the following manner, define

$$u_1 = u - (u, \varphi_1)\varphi_1 \tag{10}$$

and $\mathbb{R}_1(x, x') = \langle u_1(x)\, u_1^*(x')\rangle$ and compute μ_2 and φ_2 as solutions to the variational problem (8) for $\mathbb{R}_1$ and so on. φ_2 will also be an eigenfunction of $\mathbb{R}$, with eigenvalue μ_2, thus $\mu_1 \geq \mu_2$. Under the assumption (9), using (A5) from Appendix A we get that

$$\sum_{i=1}^{\infty} \mu_i^2 < \infty . \tag{11}$$

and we recover all the non-zero eigenvalues of $\mathbb{R}$. The orthonormal set $\{\varphi_i\}_{i=1}^{\infty}$ is the basis used for the P.O.D.

2.2 Span of the eigenfunction basis

The first step in understanding what can be done with the basis $\{\varphi_i \mid i = 1 \ldots \infty\}$ is characterizing the set $S = \{ \sum_{\mu_i \neq 0} a_i \varphi_i \mid \Sigma a_i^2 < \infty, \mu_i > 0\}$, which is the closure of the span of $\{\varphi_i \mid i = 1\ldots\infty, \ \mu_i > 0\}$ in $L^2(\Omega)$. Equality will be interpreted as a.e in Ω. This question is important to prevent misconceptions regarding the completeness of this basis.

Proposition: $u(x, t) \in S$ a.e. with respect to $\langle\bullet\rangle$.

Proof: Given (9b) we know that for a.e.w.r.t $\langle\bullet\rangle$ we have

$\sum_i (u(x,t), \varphi_i) < \infty$. Let us consider $u_s(x, t) = \sum_i (u(x, t), \varphi_i)\varphi_i$.

We have $u_s(x, t) \in S$. We need to show that

$$\langle \| u(x, t) - u_s(x, t) \|^2 \rangle = 0 \tag{12}$$

We have, from the real valueness of $u(x, t)$.

$$\langle \| u(x, t) - u_s(x, t)\|^2\rangle = \langle(u-u_s, u-u_s)\rangle = \langle(u,u) - 2(u, u_s) + (u_s, u_s)\rangle \tag{13}$$

$$<(u, u)> = < \int_\Omega u(x, t)\, u^*(x, t)\, dx> = \int_\Omega \mathbb{R}(x, x)\, dx . \qquad (14)$$

$$<-2\,(u, u_S)> = -2 < \int_\Omega u(x, t) \left(\sum_i (u^*(x, t), \varphi_i^*)\, \varphi_i^* \right) =$$

$$= -2 \int_\Omega \sum_i \left(\int_\Omega <u(x, t)\, u^*(x', t)>\, \varphi_i(x')dx' \right) \varphi_i^*(x)\, dx$$

$$= -2 \int_\Omega \left(\sum_i \mathbb{R}\, \varphi_i\right) \varphi_i^* = -2 \int_\Omega \sum_i \mu_i\, \varphi_i(x) \varphi_i^*(x)\, dx = -2 \int_\Omega \mathbb{R}(x, x)\, dx . \qquad (15)$$

Observe that (15) is true because of Mercers theorem (Appendix A) for the uniform convergence of (4).
A similar calculation shows that

$$<(u_S, u_S)> = \int_\Omega \mathbb{R}(x, x)\, dx \qquad (16)$$

Putting (13), (14), (15), (16) together we obtain the desired conclusion. ♦

Throughout the remainder of this section we will assume u(x, t) is continuous in t. This is done to avoid the cumbersome treatment for the measurable case.

The proposition above implies that for any sequence $\{t_i \mid i = 1,\dots, \infty\}$ and $\{b_i \mid i=1,\dots, \infty\}$ if the series

$$\sum_{i=1}^{\infty} b_i u(x,t_i) \qquad (17)$$

converges in $L^2(\Omega)$ then it is in S. The following proposition together with the last comment will characterize S.

<u>Proposition</u>: If $\theta \in S$, then there exist sequences $\{t_i \mid i=1,\dots, \infty\}$ and $\{b_i \mid i=1,\dots, \infty\}$ such that

$$\theta(x) = \sum_{i-1}^{\infty} b_i\, u(x, t_i).$$

Before we prove this we will need the following.

<u>Lemma</u>: For $\theta \in L^2(\Omega)$ we have $(\theta, u(x, t)) = 0$ a.e w.r.t. $<\bullet>$ iff $(\theta, \varphi_i) = 0$ for every i such that $\mu_i > 0$.

Proof: Assume that $(\theta, \varphi_i) = 0$ for every i such that $\mu_i > 0$. For a.e t we can write

$u(x, t) = \Sigma\, b_i(t)\, \varphi_i(x)$ where $\mu_i > 0$ (previous proposition). Therefore a.e.w.r.t. t $(\theta, u(x, t)) = 0$.

Assume now, $(\theta, u(x, t)) = 0$ for a.e t then

$$\int_\Omega \theta^*(x') \bullet u(x', t)\, dx' = 0 \quad \text{for a.e. t}$$

$$\Rightarrow \text{for a.e. t} \quad u(x, t) \int_\Omega u^*(x', t) \bullet \theta(x')\, dx' = 0$$

$$\Rightarrow \int_\Omega \langle u(x, t)\, u^*(x', t)\rangle \bullet \theta(x')\, dx' = 0$$

$\Rightarrow \mathbb{R}\, \theta = 0$

$\Rightarrow (\theta, \varphi_i) = 0$ for every i such that $\mu_i > 0$. ♦

Proof of proposition: Denote

$S' = \{\Sigma\, b_i\, u(x, t_i) \mid \text{the series converges in } L^2\}$.

We have shown that $S' \subset S$, we want to show $S' \supset S$. Let $v \in S$ then $v = w + v'$ where $v' \in S'$ and $w \in S'^{\perp} \cap S$. We have that $(w, u(x, t)) = 0$ for a.e. t, by the definition of w.

$\Rightarrow$ From the lemma $(w, \varphi_i) = 0$ for every i such that $\mu_i > 0$

$\Rightarrow w = 0$ ♦

We thus have a complete characterization of the span of the eigenfunctions. The span is exactly the span of all the realizations of u(x, t). a.e. w.r.t time. This result has further consequences to be discussed in the sequel,

From this we see that the sequence need not be complete. It is complete only if one includes the kernel of the operator $\mathbb{R}$, that is, all the (generalized) eigenfunctions with zero eigenvalues. See also Appendix A. Of course if $\mathbb{R}$ is a positive (rather then non negative) operator the kernel is trivial and in that case one would get a complete basis.

In many applications one can argue on physical grounds that the realizations u(x, t) do not span the whole space. Thus the basis is not complete. This result highlights a strong property of the P.O.D. It *apriori* limits the space studied to the smallest linear subspace that is sufficient to describe the observed phenomena.

2.3 Optimality of the P.O.D. basis, and a physical interpretation

Given a signal u(x, t) and a decomposition of it (a.e. in time) with respect to an orthonormal basis $\{\psi_i(x) \mid i=1,\dots,\infty\}$.

$$u(x,t) = \sum_i b_i(t)\, \psi_i(x) \tag{18}$$

one makes the following observations. If the $\psi_i(x)$ are dimensionless objects such that the inner product $\int_\Omega \psi_i(x)\, \psi_j^*(x)\, dx = \delta_{ij}$ is normalized appropriately to give dimensionless δ_{ij}, then the coefficients $b_i(t)$ carry the dimension of the quantity u.

If we think of u(x, t) as a velocity, the average kinetic energy per unit mass over the experiment is given by

$$< \int_\Omega u(x,t) \bullet u^*(x,t)\, dx> = <\sum_{ij} b_i(t)\, b_j^*(t) \int_\Omega \psi_i(x) \bullet \psi_j^*(x)\, dx>$$

$$= \sum_i <b_i(t)\, b_i^*(t)> .$$

This fits well with the notion of "average kinetic energy in the i-th mode" being given by $<b_i(t)\, b_i^*(t)>$.

The following proposition establishes what is called the optimality of the P.O.D. or Karhunen-Loeve decomposition.

<u>Proposition</u>: Let u(x, t) be a random signal in $L^2(\Omega)$ for a.e. t. Let $\{\varphi_i, \mu_i \mid i=1, \dots, \infty, \mu_i \geq \mu_{i-1} > 0\}$ be the P.O.D. orthonormal set with the associated eigenvalues. Let (equality a.e.)

$$u(x,t) = \sum_i a_i(t)\, \varphi_i(x) \tag{19}$$

be the decomposition w.r.t to this basis. Let $\{\psi_i(x) \mid i=1, \dots, \infty\}$ be an arbitrary orthonormal set such that

$$u(x,t) = \sum_i b_i(t)\, \psi_i(x).$$

Then the following holds,

1. $<a_i(t)\, a_j^*(t)> = \delta_{ij}\, \mu_i$ i.e. the P.O.D. random coefficients are uncorrelated

2. For every n

$$\sum_{i=1}^{n} \langle a_i(t)\, a_i^*(t)\rangle = \sum_{i=1}^{n} \mu_i \geq \sum_{i=1}^{n} \langle b_i(t) b_i^*(t)\rangle. \qquad (20)$$

Proof: The first assertment comes from the theorem on the representation of $\mathbb{R}(x, x')$.

$$\mathbb{R}(x, x') = \langle u(x, t)\, u^*(x', t)\rangle = \langle \sum_i a_i(t)\, \varphi_i(x) \sum_j a_j^*(t)\, \varphi_j^*(x)\rangle$$

$$= \sum_{i,j} \langle a_i(t)\, a_j^*(t)\rangle\, \varphi(x) \varphi_j^*(x')$$

But we know that

$$\mathbb{R}(x, x') = \sum_i \mu_i \varphi_i(x) \varphi_i^*(x')$$

therefore, since $\varphi_i(x)\, \varphi_j^*(x')$ are an orthonormal family in $L^2[\Omega \times \Omega]$, we have that

$$\langle a_i(t) a_j^*(t)\rangle = \delta_{ij} \mu_i.$$

The second assertion comes from the lemma in Appendix A. Let $\psi_1(x)$, ..., $\psi_n(x)$ be n arbitrary orthonormal vectors in $L^2(\Omega)$, let us complete them to obtain an orthonormal basis. Let Q be the projection on Span $\{\psi_1, ..., \psi_n\}$. We can express the kernel $\mathbb{R}$ in terms of $\{\psi_i | i=1, ..., \infty\}$.

$$\mathbb{R}(x, x') = \langle \sum_i b_i(t)\, \psi_i(x) \sum_j b_j^*(t) \psi_j^*(x')\rangle = \sum_{i,j} \langle b_i b_j^*\rangle\, \psi_i \psi_j^* \qquad (21)$$

One can then write $\mathbb{R}$ in operator matrix notation as

$$\begin{bmatrix} \langle b_1 b_1^*\rangle & \langle b_1 b_2^*\rangle & \langle b_1 b_3^*\rangle & \cdots\cdots\cdots \\ \langle b_2 b_1^*\rangle & \langle b_2 b_2^*\rangle & \cdots\cdots\cdots & \\ \langle b_3 b_1^*\rangle & & & \end{bmatrix}$$

and the projector Q may be written as

$$
n \rightarrow \begin{bmatrix} 1 & 0 & 0 & \\ 0 & 1 & : & O \\ 0 & .. & 1 & \\ & O & & O \end{bmatrix}
$$

The product $A_0 Q$ yields

$$
\begin{bmatrix} <b_1 b_1^*> & <b_1 b_2^*> & ---- & <b_1 b_n^*> & 0---- & 0---- \\ <b_2 b_1^*> & <b_2 b_2^*> & ---- & <b_2 b_n^*> & 0---- & 0 \\ <b_n b_1^*> & <b_n b_2^*> & ---- & <b_n b_n^*> & 0---- & 0 \\ 0 & 0 & & 0 & 0 & \end{bmatrix}
$$

The lemma in Appendix A tells us that

$$
\sum_{i=1}^{n} \mu_i \geq \mathrm{Tr}(A_0 Q) = \sum_{i=1}^{n} <b_i b_i^*> \qquad \blacklozenge
$$

The last proposition is the basis for the claim that the P.O.D. or Karhunen-Loeve is optimal for modelling or reconstructing a signal u(x, t). It implies that among all linear decompositions this is the most compact in the sense that for a given number of modes the projection on the subspace used for modelling will contain the most possible kinetic energy. In addition the time series of the coefficients $a_i(t)$ are linearly uncorrelated.

2.4 Symmetric Attractors and Invariant Measures

The purpose of this section is to see how *apriori* knowledge about the system which produces u(x, t) manifests itself in the Karhunen-Loeve eigenfunctions and to investigate what the K-L basis tells us about attractors or ergodicity. As an immediate result of the characterization of S found in 2.2 we may state the following:

Claim: If P is a closed linear property of u(x, t) (i.e. all $u \in L^2(\Omega)$ with that property form a closed linear subspace). Then the eigenfunctions of K-L decomposition will have the same property. The converse is also true.

This takes care of a whole list of properties such as periodicity, Neumann boundary conditions, being odd (i.e. odd subspace in a Fourier sense), incompressibility for fluid flows, etc. Another observation regards what happens when a finite number of eigenfunctions suffices for the representation. In this case the "dimension" (by any sensible definition) of the system is finite, and bounded by the number of non zero eigenvalues. The converse is not true, as shown in an example (Berkooz 1990). We next want to describe a particular kind of symmetry. We say that the two point correlation tensor $\mathbb{R}(x, x')$ is homogeneous if $\mathbb{R}(x, x') = f(x-x')$ i.e. $\mathbb{R}$ depends only on the difference of the two coordinates. Assuming Ω is bounded and u periodic, we may develop f in a Fourier series,

$$f(x-x') = \Sigma C_k \, e^{2\pi i k \bullet (x-x')}$$

One can then solve the eigenvalue problem via the representation

$$\mathbb{R}(x, x') = \Sigma C_n \, e^{2\pi i n \bullet x'} e^{-2\pi i n \bullet x'} ,$$

which implies that $e^{2\pi i n \bullet x'}$ are eigenfunctions with eigenvalues C_n. This observation is especially useful in systems where the domain Ω is of higher dimension. For example if $\Omega \subset \mathbb{R}^2$, we then have

$$\mathbb{R}\ (\mathbf{x}, \mathbf{x}') = \mathbb{R}\ (x, y, x', y')$$

If the x-direction is homogeneous, the problem of finding eigenfunctions in a 2-D domain is decoupled into a set of 1-D problems by writing

$$\mathbb{R} = \mathbb{R}\ (x-x', y, y')$$

and performing the same procedure as above, yielding a 1-D eigenvalue problem for every Fourier wave number. Examples of such applications can be found Herzog '86, Moin & Moser '89 and also see Lumley '70. In the context of coherent structures in the turbulent boundary layer this observation leads to structures which are not localized in space, unlike the events observed. Preliminary results of work done by Berkooz, Elezgaray and Holmes 1990 shows that a wavelet decomposition in the homogeneous direction may be feasible.

We now go back to the context of a measure theoretical dynamical system discussed in the introduction. If a system has certain symmetries, these symmetries will show up as symmetries of the invariant measure. However, starting from a given initial condition it is indeed possible that the solution will not explore <u>all</u> the states associated with the symmetry group. If such a discrepancy exists it enables us to say that the system is not ergodic. We are motivated by the following simple observation. Assume a 2-dimensional map is invariant under the symmetry $x \to y$, $y \to x$, with two distinct regions of attractors (see [6,7]). Obviously the invariant measure concentrated on the two attractors is not ergodic. A typical time average will reproduce just one of the pieces of the attractor. There are no established ways for numerically representing and studying invariant measures, even those defined by time averages. However, if the empirical eigenfunctions have less symmetry than the problem as a whole, this provides an important hint. Formally we have:

<u>Proposition</u>: Let S_t be a statistically stationary solution semi group defined for $t \geq 0$ on $L^2(\Omega)$ and μ the invariant measure associated with it. Let G be a linear symmetry group for S_t. i.e. G is a group such that every

member $g \in G$ is a linear isomorphism of $L^2(\Omega)$, such that the following diagram commutes

$$\begin{array}{ccc} u(0) & \xrightarrow{\quad S_t \quad} & u(t) \\ g \Big| & & \Big| g \\ \tilde{u}(0) & \xrightarrow[\quad S_t \quad]{} & \tilde{u}(t) \end{array}$$

Then a necessary condition for μ to be ergodic is: For almost every experiment each of the finite dimensional eigenspaces corresponding to a given eigenvalue is invariant under G.

<u>Proof</u>: By the ergodicity of μ the two point correlation tensor

$$A(x_1, x_2) = \langle u(x_1, t)\, u^*(x_2, t)\rangle. \tag{22}$$

is independent of the initial condition u(x, 0) for a.e. initial condition. We know we can write

$$A(x_1, x_2) = \sum_{i=1}^{\infty} \mu_i\, \varphi_i(x_1)\, \varphi_i^*(x_2). \tag{23}$$

Denote $N_\mu = \text{Span}\,\{\varphi_i \mid \mu_i = \mu\}$.

We want to show that if S_t is ergodic than, for every $g \in G$ $gN_\mu = N_\mu$. It is enough to show $gN_\mu \subset N_\mu$ for every g, since then $g^{-1}N_\mu \subset N_\mu$ and $N_\mu \subset gN_\mu$, showing that $gN_\mu = N_\mu$.

let $\{u(x, t) \mid t \in [0, \infty)\}$ be almost any experiment, then

$$u(x, t) = \sum_{i=1}^{\infty} a_i(t)\, \varphi_i(x)\,, \tag{24}$$

where $\quad \langle a_i a_i^* \rangle = \mu_i\,.$

Thus, for every $g \in G$,

$$\tilde{u}(x, t) = g\,(u(x, t)) = \sum_{i=1}^{\infty} a_i(t)\, g\,(\varphi_i(x)).$$

is also an experiment .

From the independence of the representation of A on the initial condition we have that

$$A(x_1, x_2) = \Sigma\, \mu_i\, g(\varphi_i(x))\,(x_1)\, g(\varphi_i(x))\,(x_2)^*$$

$$\Rightarrow g(\varphi_i(x)) \in N_{\mu_i}.$$ ♦

Note that in a physical situation one might have a measure on the set of possible initial conditions which differs from the invariant measure. This technical point does not change the above results.

The way one would go about checking this condition in an experiment would be:

1. Perform the experiment and measure $A(x_1, x_2)$.
2. Decompose using P.O.D. $A(x_1, x_2) = \Sigma\, \mu_i \varphi_i(x_1)\, \varphi_i^*(x_2)$.
3. Denote $N_i = \mathrm{Span}\,\{\varphi_j \mid \mu_j = \mu_i\}$.
4. Check that for every $\varphi_j \in N_i$ and every $g \in G$, $g(\varphi_j) \in N_i$.

Recently Aubry, Lian and Titi (1990) performed numerical integrations of the Kuramuto - Shivashinksy equation and computed the P.O.D. basis. By using the results above, and the calculations cited one can conclude empirically that for certain values of the bifurcation parameter the system is not ergodic. In the study of Aubry et al. 1990 taking the images of the basis computed based on one experiment under the symmetry group yields a better description of the overall dynamics (expressed by the attractors). However, one should be cautious as there are examples (Berkooz 1990) where the partition into ergodic components is finer than the partition into symmetric components, in this case the image of the basis obtained by one experiment under the symmetry group will not suffice.

2.5 A Geometric Result from the P.O.D.

Let $\{\varphi_i \mid i=1,\dots,\infty\ \mu_i > 0\}$ be the P.O.D. associated with $u(x, t) \in L^2(\Omega)$. One can see that the random coefficients $a_i(t)$ are actually the variances of the random variable $(u(x, t), \varphi_i)$. It is then possible to give a probabilistic - geometric picture of the location of the dynamics in phase space using Chebyshev's inequality. This is a more general approach to the type of result in Foias et al. (1990), who essentially reproduced the proof of Chebyshev's inequality in a specific case. Recall Chebyshev's inequality from basic probability theory. (See Feller (1957).)

Theorem: Let $\mathbf{X}$ be a vector valued random variable, with mean $\langle \mathbf{X} \rangle$ and variance $\sigma^2 = \mathrm{var}(\mathbf{X}) = \langle |\mathbf{X} - \langle \mathbf{X} \rangle|^2 \rangle$. Then for any $\varepsilon > 0$

$$\langle 1_{|\mathbf{X}-\langle \mathbf{X}\rangle|}(\varepsilon) \rangle = P\{\, |\mathbf{X} - \langle \mathbf{X} \rangle| \geq \varepsilon \,\} \leq \frac{\sigma^2}{\varepsilon^2}$$

$$\text{where} \quad 1_s(\varepsilon) = \begin{cases} 1 & S \geq \varepsilon \\ 0 & S < \varepsilon . \end{cases}$$

This is a formal expression for the notion that the variance gives an indication for the variation from the mean. In our case we define

$$S_n(\varepsilon) = \{ u \in L^2(\Omega) \mid \sum_{m=n+1}^{\infty} |(u, \varphi_m)|^2 < \varepsilon \}.$$

$$W_n(\varepsilon) = L_2(\Omega) / S_n(\varepsilon).$$

This is a slab of thickness 2ε around the finite dimensional space span $\{\varphi_1, .., \varphi_n\}$. We want to use Chebyshev's inequality to estimate what fraction of the time is spent in $S_n(\varepsilon)$. Denote by $\mathbf{X}_n$ the vector valued random variable

$$X_n(t) = ((u, \varphi_m)) |_{m=n+1}^{\infty}$$

then we have

$$<\mathbf{X}_n> = \mathbf{0} \quad \text{and} \quad \sigma^2(\mathbf{X}_n) = \sum_{m=n+1}^{\infty} \mu_m$$

and by the Chebyshev inequality.

$$P\{u \in W_n(\varepsilon)\} = P\{ |X_n| \geq \varepsilon \} \leq \frac{\sum_{m=n+1}^{\infty} \mu_m}{\varepsilon^2}$$

Note that this is meaningless for a fixed m and $\varepsilon \to 0$. The way to get something meaningful is to take a sequence $\varepsilon_n \to 0$ such that

$$\frac{\sum_{m=n+1}^{\infty} \mu_m}{\varepsilon_n^2} \to 0$$

in other words, the ε_n are chosen such that their squares go to zero slower than the decay of the norm in the residual modes. This will give a series of slabs with thickness going to zero and the probability of being those slabs goes to one.

The problem is now shifted to computing the rate of decay of the residual $\sum_{m=n+1}^{\infty} \mu_m$. There is analytical evidence that suggests that this residual decays at least exponentially fast for turbulent flows when the P.O.D. basis is used. See Foias et al. (1990). Also results of numerical computations of Moin and Moser (1989) are consistent with such a rate. This enables us to take a series $\varepsilon_n^2 \to 0$, with a slightly smaller exponent.

The result will be a series of slabs with thickness going exponentially to zero, and the probability of being in that slab goes exponentially to 1. This gives rise to a picture where an attractor is very thin, even though possibly infinite dimensional. Thus the essentials of the dynamics may be controlled by a finite number of modes.

3. Summary

Given u(x, t), a random signal in $L^2(\Omega)$ we have shown what information can be obtained by the P.O.D. or Karhunen-Loeve decomposition. The question of the span of the eigenfunction basis is settled, showing that this is not a complete basis in general. The optimality of the eigenfunction expansion is established by using results from linear operator theory. We find the relation between ergodicity of invariant measures in symmetric cases and invariance of the P.O.D. eigenspaces of particular interest. This presents a possibility of examining the fundamental ergodic hypotheses employed regularly in turbulence, and suggests possible experiments. However, one must conclude that the primary utility of the P.O.D. is for ergodic systems. The slight improvement offered in cases the ergodic components coincide with the symmetric components of the problem is not a general remedy. A possible way around this problem is the use of ensemble averages which one believes will sample all ergodic components. The geometric interpretation given through the Chebyshev inequality helps strengthen notions of finite dimensional subspaces containing the "dominant" behaviour of systems in infinite dimensions.

However, one should face up to the fact that the P.O.D. is a linear theory, and that it deals with average properties. One can give examples (see Berkooz 1990) that show that the P.O.D. cannot distinguish between degrees of complexity in certain systems, or even between deterministic and purely stochastic systems. Thus in general one has to employ more information to understand the phenomena studied.

It seems that a reasonable role for the P.O.D. is in models of chaotic systems where the goal is to give models which give "average" best performance. See Holmes et al. (1990). However, a rigorous theory for such an approach has yet to be put forward.

Appendix A - Some useful facts from Functional Analysis.

The purpose of this section is to summarize results from classical functional analysis which are necessary for understanding the mathematical foundations of the Proper Orthogonal or Karhunen-Loeve Decomposition. See Riesz and Sz Nagy (1955) and Teman (1988).

Our interest is in the space $L^2(\Omega)$, where for simplicity Ω is the interval [a, b]. For any two real valued functions f, g: $[a, b] \to \mathbb{R}$ denote $(f,g) = \int_a^b fg\,dx$ and $\| f \|^2 = (f, f)$.

Given a function K(x, y) one can formally define a linear transformation K on $L^2(\Omega)$

$$(Kf)(x) = \int_a^b K(x, y)\, f(y)\, dy \tag{A1}$$

If

$$| K | = \int_a^b \int_a^b | K(x, y)) |^2\, dxy\, dy < \infty , \tag{A2}$$

then, K is a well defined transformation. We are specifically interested in transformations generated by a Hermitian kernel,

$$A(x, y) = A(y, x)^*$$

where $\bullet^*$ is the complex conjugate. The following theorems are due to Hilbert and Schmidt and are of interest for us. (see Reisz and Sz-Nagy (1955) pp 242). The first theorem deals with decomposing arbitrary functions using the eigenfunctions of A (the operator defined by the kernel A). The second deals with a representation for A itself.

<u>Theorem</u>: If the kernel A(x,y) is not zero almost everywhere the transformation A has at least one eigenvalue different from 0, and each of its eigenvalues is of finite multiplicity. There is an orthonormal sequence (finite or infinite) of eigenfunctions $\varphi_i(x)$ of A, corresponding to the eigenvalues $\mu \neq 0$. Every function $g \in L^2$ admits a decomposition, convergent in the L^2 sense.

$$g(x) = h(x) + \sum_i (g, \varphi_i)\, \varphi_i (x) \tag{A3}$$

where h(x) is a function such that Ah(x) = 0.

<u>Theorem</u>: Every function A(x, y) which is symmetric and square summable can be developed, in the L^2 sense, into the series

$$A(x,y) = \sum_i \mu_i\, \varphi_i(x)\, \varphi_i(y)^* \tag{A4}$$

where $\{ \varphi_i(x)\}$ denotes the orthonormal sequence of eigenfunctions, and $\{ \mu_i \}$ the corresponding eigenvalues, of A.

In particular we have $|A|^2 = \sum_i \mu_i^2$. (A5)

For the purpose of approximation and calculation the following result of Mercer is relevant (Riesz - Sz-Nagy (1955) pp 245), in the case of continuous functions.

Theorem: If the transformation A generated by the continuous symmetric kernel is non negative,that is $(Af, f) \geq 0$ for all f , or, equivalently, if all the eigenvalues $\mu_i \neq 0$ are positive, the development of A, is uniformly convergent.

The following lemma will establish the optimality of the proper orthogonal or Karhunen Loeve decomposition. See Riesz-Sz Nagy 1955 pp. 244 or Teman 1988 pp 260.

Let $\psi_1, \ldots, \psi_m$ be m orthonormal elements in $L^2(\Omega)$. Denote Q the projection defined by them. Then:

Lemma: For any operator bounded, self adjoint A , we have

$$\text{Trace}\,(A_0 Q) \leq \sum_{i=1}^{m} \mu_i \qquad \text{(A6)}$$

where μ_i is the i-th eigenvalue of A in descending order, $\mu_i \geq \mu_{i-1}$.

Also recall that for operators A of the type we are interested in, the dimension of each eigenspace corresponding to a non zero eigenvalue is finite, by Fredholm Theory, Riesz - Sz Nagy 1955.

ACKNOWLEDGEMENTS

The author expresses gratitude to P. Holmes and J.L. Lumley for their support and interest and useful suggestions and discussions on the topic and to E. Titi and J. Elezgaray for their comments. This work was supported in part by the NASA Langley Research Center under the program Hypersonic Turbulence Modeling, grant no.NAG-11-954. and AFOSR under the program Structure and Control of the Wall Region in a Turbulent Boundary Layer, grant no. AFOSR-89-0226.

References

[1] Algazi, V.R. and Sakrison, D.J. (1969). On the optimality of the Karhunen-Loeve expansion, IEEE Trans. Inform. Theory **15** 319-321.

[2] Andrews, C.A., Davies, J.M. and Schwarz, G.R. (1967). Adaptive data compression, Proc. IEEE **55** 267-277.

[3] Aubry, N., Holmes, P., Lumley, J.L. & Stone, E. (1988). The dynamics of coherent structures in the wall region of a turbulent boundary layer. J. Fluid Mech. **192** 115-173.

[4] Aubry, N., Lian, W.-Y., Titi, E.S. The P.O.D. in physical problems with symmetry in preparation.

[5] Bakewell, P. and Lumley, J.L. (1967). Viscous sublayer and adjacent wall region in turbulent pipe flow. Phys. Fluids **10** 1880-1889.

[6] Berkooz, G. (1990). In Preparation, Preprint.

[7] Berkooz,G., Elezgaray, J., Holmes, P. 1990. In preparation.

[8] Cornfeld, I.P., Fomin, S.V., Sinai, Ya. G. Ergodic Theory. Springer 1982.

[9] Feller, W. An introduction to probability theory and its applications. 1957 Wiley.

[10] Foias, C., Manley, O. and Sirovich, L. 1990 Empirical and Stokes eigenfunctions and the far dissipative turbulent spectrum. Phys. Fluids A **2** (3) pages.

[11] Herzog, S. 1986. The large scale structure in the near wall region of turbulent pipe flow. Ph.D. Thesis, Cornell University, Ithaca, NY 14853.

[12] Holmes, P., Berkooz, G. and Lumley, J.L. 1990. On the unreasonable effectiveness of empirical eigenfunction in turbulence. Proc. of the Kyoto Congress of Mathematicians.

[13] Karhunen, K. 1946. Zur Spektral theorie stochastischer prozesse, Ann. Acad. Sci. Fennicae Ser. A**1**, 34.

[14] Knight, B.W. and Sirovich, L. 1986. The eigenfunction problem in higher dimensions. Exact Results. Proc. Natl. Acad. Sci. USA **83** pp 527-530.

[15] Kosambi, D.D. 1943. Statistics in function space, J. Indian Math. Soc. **7**, 76-88.

[16] Loeve, M. 1945. Functions aleatoire de second ordre, Compte Rend. Acad. Sci. (Paris) **220**.

[17] Lumley, J.L. 1967. The structure of inhomogeneous turbulent flows. In Atmospheric Turbulence and Radio Wave Propagation (ed. A.M. Yaglom & V.I. Tatorski) pp 166-178, Moscow: Nauka.

[18] Lumley, J.L. 1970. Stochastic Tools in Turbulence. Academic Press.

[19] Moin, P. and Moser, R.D. 1989. Characteristic-eddy decomposition of turbulence in a channel. J. Fluid Mech. vol. 200 pp. 471-506.

[20] Obukhov, A.M. (1954). Statistical description of continuous fields, Trudy Geophys. Int. Aked. Nauk. SSSR, Ser. No. 24 3-42 (in Russian).
[21] Pougachev, V.S. 1953. General theory of the correlations of random functions, Izv. Akad. Nauk. SSSR, Ser. Mat. **17** 401-402 (in Russian).
[22] Riesz, F. Nagy, Sz. B. Functional Analysis (Ungar, New York, 1955).
[23] Sirovich, L., 1989. Chaotic dynamics of coherent structures. Physica D 37, 1989, 126-145.
[24] Teman, R. 1988. Infinite dimensional dynamical systems in mechanics and physics, Springer.

LUMLEY'S PODT DEFINITION OF LARGE EDDIES AND A TRIO OF NUMERICAL PROCEDURES

Fred R. Payne [1]

The University of Texas at Arlington
Arlington, Texas 76019

ABSTRACT

Lumley's 1967 Moscow paper provided, for the first time, a completely rational definition of the physically-useful term "Large Eddy," popular for a half-century. The numerical procedures are based upon his results: 1) PODT (Proper Orthogonal Decomposition Theorem) which extracts the Large Eddy structure of stochastic processes from physical or computer simulation two-point covariances, and 2) LEIM (Large-Eddy Interaction-Model) is a predictive scheme for the dynamical Large Eddies based upon higher order turbulence modelling. Earlier Lumley work (1964) forms the basis for the final member of the triad of numerical procedures: this predicts the global neutral modes of turbulence which have surprising agreement with both structural eigenmodes and those obtained from the dynamical equations. The ultimate goal of improved engineering design tools for turbulence may be near at hand, partly due to the power and storage of "super-" microcomputer workstations finally becoming adequate for the demanding numerics of these procedures.

1. Origins: PODT, OSL and LEIM Methodologies and Histories of Their Applications, 1962-1990

1962-63: Lumley's first use of Townsend's 1956 monograph (which can be titled "The Care and Feeding of Big Eddies") sensitized him and students (Bakewell, Payne, Wyngaard, others) to the imprecise "guessing game" (Payne, 1966, page 5-6) method for constructing large eddy structures based upon experimental two-point double velocity correlations. Townsend (1956), Grant (1959) and others "guessed" an analytic form and then checked that the proper sign is produced in each octant. Such a procedure offers little hope of providing uniqueness: results are bounded only by human imagination.

1963-64: Lumley first introduced the PODT methodology and formal results to his courses. All the essentials were now formulated; even the use of "generalized functions" in his 1970 book only added formal elegance and ease; there he notes that concepts like PODT were occurring

[1] Supported, in part, by NASA Ames Grants NSG-2077 and NCC 2-386 and by Vice-President for Research, UTA, 1972-89.

to various workers at this time. Lumley's treatment seems the most elegant and complete.

Basic Ansatz: Given a mixed Hilbert space of stochastic and determinate functions, seek a deterministic set of orthonormal functions which recover a maximum of statistical information from the measured [or computed] stochastic vector field [also works for scalar or tensors of any order]. Results are strong: 1) Variational approach yields a system of homogeneous Fredholm integral equations and, hence, eigenvalue problems. Lumley interpreted the dominant eigenmodes as the "big Eddy" which Payne (1966), Payne and Lumley (1967) later verified for the wake: 2) Homogeneity and stationarity allow analytic solution for the eigenmodes as simply harmonic with dense eigenvalue spectra; non-homogeneity, however, yielded discrete eigen-spectra and modes requiring computer numerics; 3) Both the stochastic vector and its double correlation are expandable in terms of extracted eigen-modes; such expansions are optimal in the sense of containing maximum information at each n-partial sum. [This, via a simple induction argument, guarantees uniqueness to within a set of measure zero.]; 4) The discrete eigenvalues form an ordered set with finite total energy. Lumley's construction so they gave the mean square energy for each mode was verified by Payne, 1966. This ensures statistical realizability.

$$\int_{\Omega} R_{ij}(.,.') \, \phi_j (.') \, d(.') = \lambda \, \phi_i (.)$$

$$\nabla \cdot \underline{\phi} (.) = 0$$

$$u_i (\underline{x}, t) = \sum_{n=1}^{\infty} a^n (t) \, \phi_i^n (\underline{x})$$

$$R_{ij} (.,.') = \sum_{n=1}^{\infty} \lambda^n \, \phi_i^n (\underline{x}) \, \phi_j^n (\underline{x}')$$

$$\sum_{n=1}^{\infty} \lambda^n < \infty; \quad \lambda^{(1)} \geq, \lambda^{(2)} \geq, \lambda^{(3)} \geq, \lambda^{(4)} \geq \; ....$$

1964: Lumley spent a summer at Stanford and further refined PODT.

1965-66: Payne extracted 15 complex-pair eigenvectors from Grant's 1959 data in the far, 2-D wake of a circular cylinder. About 60 curves and 2-D plots of data were fit (Generalized Gram-Charlier, Payne, AIAAJ 1969) to permit analytic fourier Transforms to 7x7 grids in k-space. Of interest is that Grant found two distinct structures necessary to match his data; even then, he admits to inability to incorporate or match certain other data. Payne's PODT extraction used all Grant's data and gave a single structure which blended both Grant's models (vortex pair and "reentrant" jet). Only problem was the tiny computer (8 K-words) which used most of the 5000 man-hours of effort.

**For the first time, a Big Eddy structure was produced by totally rational, objective procedures. (Some engineering judgment was used during the inverse Fourier Transforms.)
**Only the dominant mode was interpreted; the second mode was judged to contain 50% error (experiment, numeric, ...) and not worthy of analysis.

1966: After the initial PODT success with Grant's data, Lumley suggested that Bakewell use PODT on his pipe flow data. Payne's codes were converted to 1-D and showed that PODT could extract one physically meaningful mode; the others were slightly negative, and thus non-physical, indicating data inaccuracies as above.

1967-68: Lumley's ORL memo, 1964, applied a second extremum principle to the "Orr-Energy" (1907) method of stability analysis as applied to turbulent flow; this remains unpublished due to Serrin, 1959, for laminar flow. (Joseph has many papers, c. 1965-80.) After extremizing a representative value of the eddy viscosity, the Orr equations become the linear Orr-Lumley-Serrin (OLS):

$$S_{ik}\, u_k = -P_{,i} + (\nu_e\, u_{i,k})_k \quad \text{(Newtonian fluid)}$$

$$u_{i,i} = 0 \quad \text{(incompressibility)}$$

OLS results (Payne, 1968; 1978) agreed well qualitatively with PODT extractions from the wake (Payne, 1966) and even quantitatively for many wave-number pairs. This justifies Townsend's conjecture of energetic equilibrium because PODT and OLS eigenvalues and eigenvectors have different physical meanings (Table 1):

Table 1. Meanings of Eigenvalues

	Eigenvalue	**Eigenvectors**
PODT	Mean Square Energy of Eigenmode	Dominant Structure
OLS	Stability Parameter	Dominant Neutral Mode

(1974-80: Under NASA Ames Grants (1975-80): Lemmerman (1976), Lemmerman-Payne, (1977) did a second full PODT decomposition (on merged flat-plate data sets: Grant 1959 and Tritton 1967). Hong (1978) did a 2-D stability analysis of the flat-plate via OLS:

Wadia (1979), Wadia and Payne (1981) did the second full OLS; results for boundary-layers compared well to Lemmerman's PODT extractions, 1976.

1983: George used PODT on round jet data. S. K. Hong formulated his LEIM (Large-Eddy Interaction-Model) based upon another section of Lumley's Moscow 1966 paper, namely, a scheme for computing large eddies from dynamical equations. Hong amalgamated Lumely's infinite series for the eddy velocity field into a sum of two terms: the first term can be a sum of several PODT modes and carries some fixed percentage of the mean square energy, typically 60-80% for design purposes. The second term is a "lump" of the rest of the turbulent field and is modelled by a small-eddy viscosity.

1985-87: Hong and Payne further refined LEIM as applied to the, now, classic channel flow; Hong formulated codes for mixing layer and boundary-layer but these have not yet been exercised.

1988-89: Two significant developments: 1) a relatively new integral solver, applied to the same turbulent channel flow of eight coupled, nonlinear LEIM equations with 28 distinct derivatives allowed reduction of the usual "Crocco" iteration in "psuedo-time" and, thus, a 4-D problem to one of 3-D via avoiding Fourier Transform and using planar elliptic sweeping; the same problem in 1988 upon CRAY-XMP/48 required 10 CPU hours; in 1989 CPU time of 20-60 minutes was required upon VAX 11/780. This is a CPU savings of about 1/300. 2) Ahn (1989) did a third LEIM computation, predicting the structure of the 2-D turbulent wake. Of interest is that one parameter, the skewness S, has the same value as for the channel. This is encouraging for design purposes.

1988: Bob Moser, NASA Rep #100092, did the first(?) PODT analysis of digitally simulated data; of course, PODT cares not the origin of numbers fed it. Flow was the much-computed straight-wall channel and the vorticity modes were also computed for the first time. To relate vorticity and velocity eigenmodes is not simple; one vorticity mode = the infinite sum of all derivatives of the velocity modes.

1988-89: M-S Kim repeated the flat-plate PODT work of Lemmerman but, by use of supercomputer (CRAY-XMP/24) evaded, for the first time, the necessity to perform Fourier transform for homogeneous/stationary directions and thereby did all calculations in "real-" space. From Payne, 1966, use of Fourier transform meant that one had to make some approximation to enable the inverse transform (due to phase ambiguity of complex eigenvalue problems). As example of the size of Kim's problem: initially 4.6 billion words of RAM were required; use of symmetries and other simplifications reduced this to 40 million words, within CRAY-SSD capacity; the eigenvalue problem has now grown to 20,000x20,000.

2. PODT, OLS, LEIM Algols

A. PODT:

1) Required Data Pool:
   * Massive sets of two-point statistical correlations
   * Major Computer Simulations or Physical Experiments
2) Computation requirements:
   * High dimension Fredholm IE eigenvalue problem: Payne, 1966, of 49x30x30=44, 100; Kim, 1989, 20Kx20K = 40 Meg (No Fourier-Transform)
   * If Fourier Transform used then must make a phase approximation
   * About equal work with or without Fourier Transform; inverse transform will be in error.

B. OLS implemented 1968, 1978, 1979:

1) Requires Mean Velocity Strain Rates, and eddy-viscosity model:

$$S_{ij}\, u_j = -P_{,i} + (\nu_e,\, u_{i,\,k})_k$$

2) Curl of OLS and Green's function => eigenvalue problem of the form:

$$u_i = \lambda \int K_{ij}\, u_j\, dy$$

3) Lumley (1970) iterative procedure efficiently extracts the largest value: deflate for others:

$$\operatorname*{Limit}_{k\to\infty} \frac{u_k}{u_{k-1}} \to \lambda^{(1)},\ \forall_y$$

4) Unlike PODT, results are dynamic and global

C. LEIM is also dynamics dominated:

$$u_i + \nabla \bullet (u_i\, \underline{u}) + S_{ij}\, u_j = -P_{,i} + \nu \nabla^2 u_i$$

$$u_{i,i} = 0$$

1) Mean Velocity required
2) A Pre-set % of KE is required to well-pose the problem:

$$\langle u_i\, u_j \rangle = A\, \phi_i\, \phi_j + \nu_{se} \frac{\partial U_i}{\partial x_j}$$

where A = % KE selected for Big Eddy:
ν_{se} models all the rest.

3) Must search for skewness, S, and "small-" eddy viscosity to match % of KE.

3. Summary of Applications: All 3-D Except Where Noted

Table 2. Applications of the Triad of Procedures

	Boundary Layer	Wake	Channel	Misc.
PODT	1976	1966	1988	1966 Pipe (1-D) 1983 Jet
OLS	1978 (2-D) 1979	1968	---	---
LEIM	---	1989	1987 1989	---

Author's experience includes 50% of PODT and all of OLS and LEIM. Hence, he should be able to estimate accurately relative human and computer times required for the three methods.

Table 3. Relative Time Required by the Algols.

	Human	Computer	Total
PODT	5	10	15
OLS	3	5	8
LEIM	10	5	15

Here, "10" for a human, "starting from scratch," should be somewhat less than a year's work. For computer, "10" will be 20-30 hours CRAY-XMP/48 CPU time including debugging.

Table 4. Comparison of theThree Algols.

	Eigenvalue	Eigenvector
PODT	Mean Square Kinetic Energy	Structurally Dominant Mode
OLS	Stability Parameter	Dynamically Dominant Neutral Mode
LEIM	Sum of Low Portion of PODT Series	Combined Dynamic Structure for 60-80% of Turbulent KE

Table 5. Usage of Eddy Viscosity.

	Eddy Viscosity Use
PODT	Not Applicable
OLS	Stability Parameter
LEIM	Model of the "Small Eddies"

4. PODT Results (3), 1966-1989

Due to time/space limitations, the choice is made to provide a diverse menu of PODT, OLS and LEIM results over the past quarter-century rather than a detailed exposition of a single flow or method. Hopefully, this will provide a "flavor" of these approaches due totally to the impetus provided by Lumley, 1964-1965.

See Figure 1 for the Wake (1966), Figure 2 for the Boundary-layer (1976), Figure 3 for the Boundary-layer (1989) redone without Fourier Transform.

5. OLS Results (2), 1978-1979

See Figure 4 for the Wake (1968) neutral mode.

6. LEIM Results (2), 1985-87, 1989

See Figure 5 for some channel flow results.

7. Summary

Perhaps a sketch will display the interactions of these three procedures more clearly:

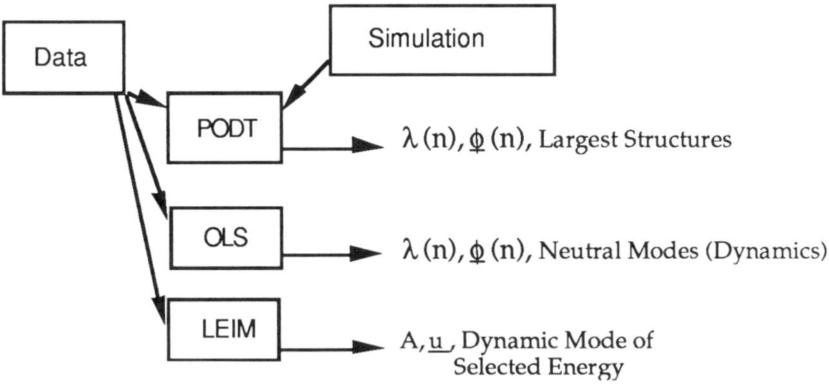

The Goal: To improve engineering design procedures for machinery having turbulent flow modes. LEIM appears to offer the most direct aid here; all three, somewhat complementary procedures are of value in the search for better understanding of turbulence.

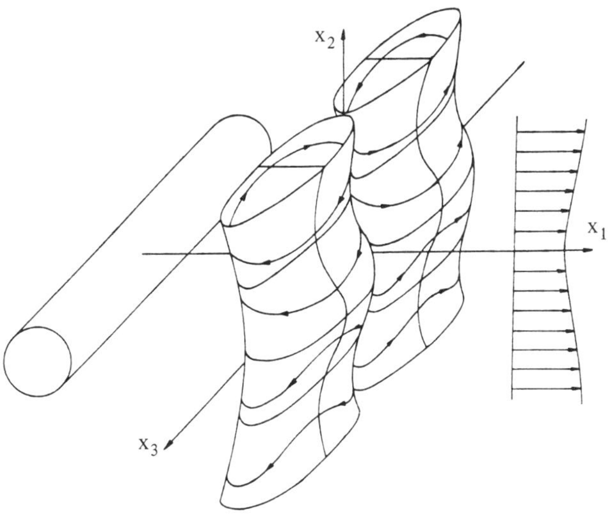

Figure 1. Artist's conception of the vortex pair (not to scale). Curved lines with arrows on surface are streamlines: note tip of planes of circulation normal to positive strain rate. (Payne and Lumley, 1967).

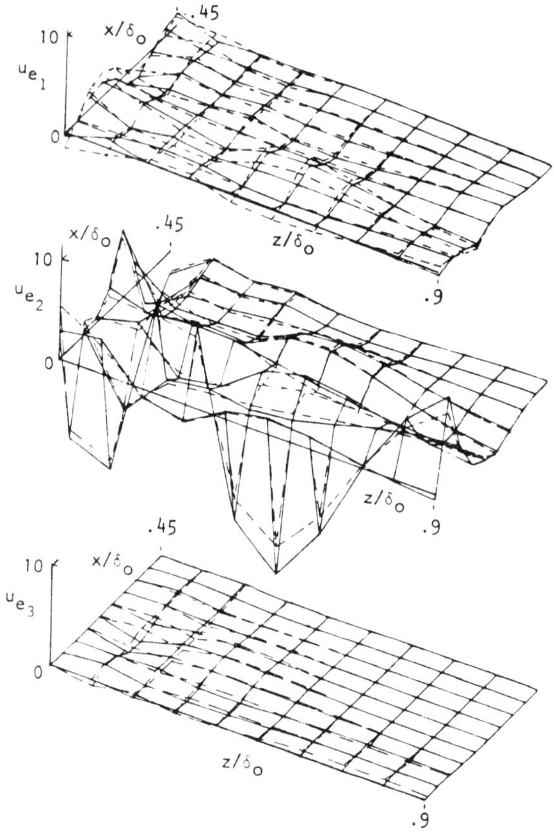

Figure 2 Velocity Contours at $y/\delta_0 = .5$
——————— First Mode, ------------ First Three Modes
(Lemmerman and Payne, 1977)

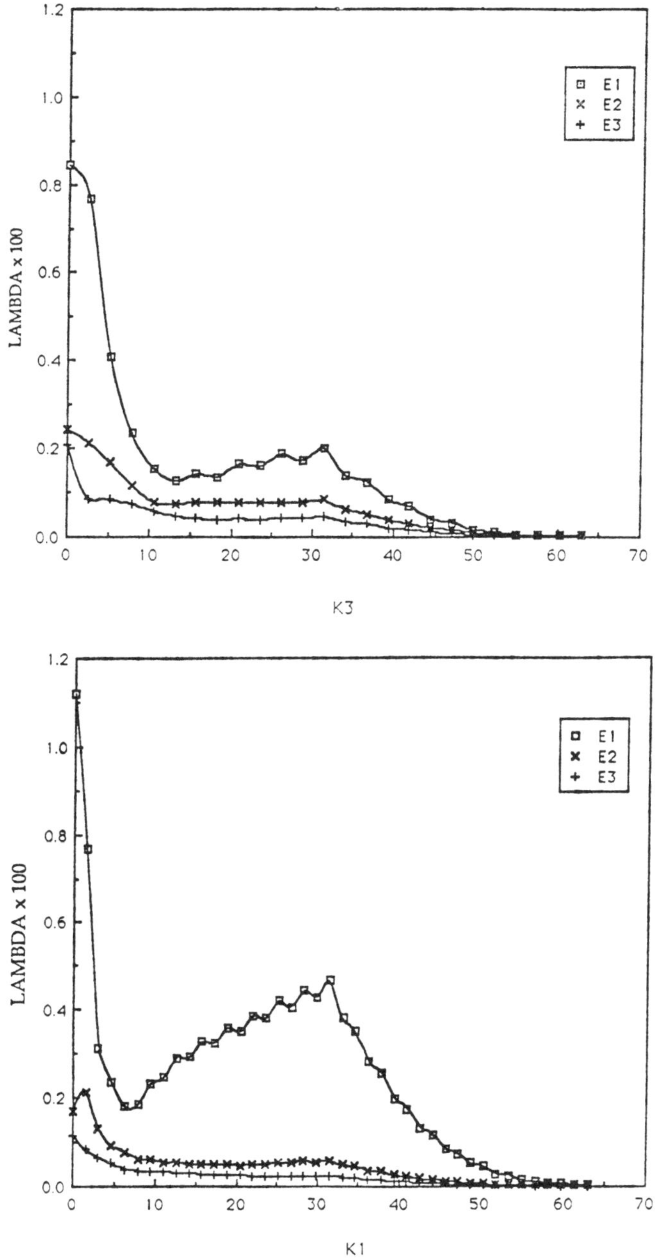

Figure 3a & 3b. First three eigenmode variation over spanwise wavenumber, k3, at a) k1 = 1.57 and b) k1 = 2.618. E1/2/3 denote first, second and third eigenvalues respectively. (Kim, 1989).

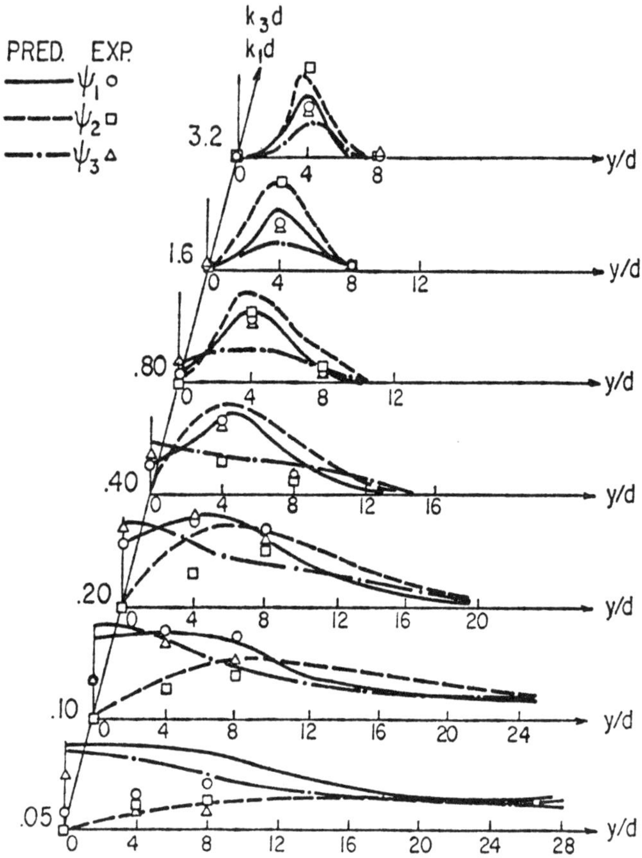

Figure 4. Predicted (OLS) eigenfunctions in Fourier space. (Payne, 1968)

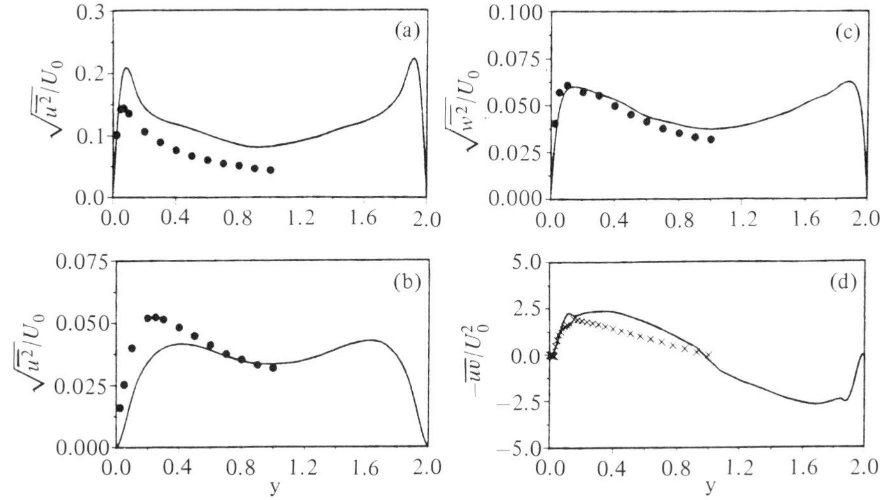

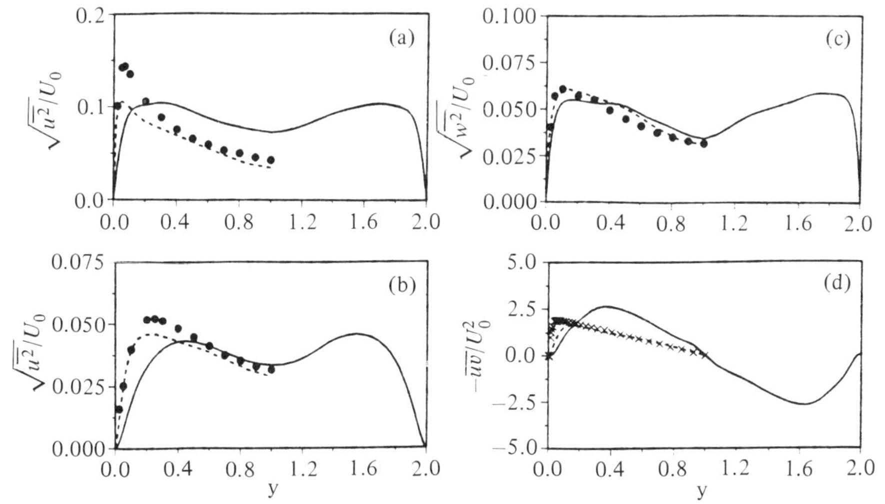

Figure 5a & 5b. (From Hong & Payne, 1987) Turbulence intensities and shear-stress distributions for a) Re = 3850: Points (•) from Kreplin and Eckelmann, PHYSICS OF FLUIDS, 1979; b) Re = 13,800: Dashed lines from LES (Moin & Kim, JOURNAL OF FLUID MECHANICS, 1982).

Acknowledgement

John L. Lumley's influence has been extensive: experimental techniques, physical and mathematical insights and high standards of rigor and elegance. May he have a second set of birthdays as productive as his first.

References

Ahn, C-S, 1989, PhD Dissertation, UTA

Bakewell, H.P., 1966, PhD Dissertation, Penn State

Grant, H.L., 1959, "The Large Eddies of Turbulent Motion," p. 149, J. FLUID MECH.

Hong, S. K., 1978, MSAE Thesis, UTA

Hong, S. K., 1983, PhD Dissertation, Purdue University.

Hong, S. K. and Payne, F. R., 1987, "Development of Large Eddy Interaction Model for Inhomogeneous Turbulent Flows," AIAA Paper No. 87-1248, 19th Fluids and Plasma.

Hong, S. K. and Payne, F. R., 1987, "Dynamics of Large Eddies in Turbulent Channel Flow," Math Dept. Rep. No. 246 to NASA Ames under NCC 2-386.

Kim, M-S, 1989, PhD Dissertation, UTA

Leib, S. J., Glauser, M. N. and George, W.K., BULL. AMERICAN PHYSICAL SOC., p. 1362, 1983.

Lemmerman, L. A., 1976, PhD Dissertation, UTA

Lemmerman, L. A. and Payne, F., 1977, "Extracted Large Eddy Structure of a Turbulent Boundary-Layer," AIAA Paper No. 77-717.

Lumley, John L., 1964, Internal Memo,ORL, Penn State

Lumley, John L., 1967, "The Structure of Inhomogeneous Turbulence Flows," in ATMOSPHERIC AND RADIO WAVE PROPAGATION, Nauka, Moscow.

Lumley, John L., 1970, STOCHASTIC TOOLS IN TURBULENCE, Academic Press, NY.

Moser, Robert D., 1988, "Statistical Analysis of Near-Wall Structures in Turbulent Channel Flow," NASA Technical Memo 100093, May.

Payne, F. R., 1966, PhD Dissertation, The Pennsylvania State University.

Payne, F. R. & Lumley, J. L., 1967, "Large Eddy Structure of the Turbulent Wake Behind a Circular Cylinder," PHYSICS OF FLUIDS, Sept. 1967, p. S194-S196.

Payne, F. R.,, 1968, "Predicted Large Eddy Structure of a Turbulent Wake," Report to USN/ONR, June.

Payne, F. R., 1978, "The OLS Method of Non-linear Stability Analysis of Turbulence in Newtonian Fluids," p. 417-437, NONLINEAR EQUATIONS IN ABSTRACT SPACES, ed. V. Lakshmikantham, Academic Press.

Townsend, A. A., 1956, THE STRUCTURE OF TURBULENT SHEAR FLOW, The University Press, Cambridge, England

Wadia, A. R., 1979, PhD Dissertation, UTA

Wadia, A. R. and Payne, F. R., 1981, "Prediction of the Unstable Modes of the Turbulent Boundary Layer," p.505-516, Proc. 2nd Int'l NUMERICAL METHODS IN LAMINAR AND TURBULENT FLOW, Pineridge Press.

PROPAGATING STRUCTURES IN WALL-BOUNDED TURBULENT FLOWS*

L. Sirovich
Center for Fluid Mechanics, Turbulence and Computation and
Division of Applied Mathematics
Brown University
Providence, RI 02912

K.S. Ball
Department of Mechanical Engineering
The University of Texas at Austin
Austin, TX 78712

R.A. Handler
Laboratory for Computational Physics and Fluid Dynamics
Naval Research Laboratory
Washington, DC 20375-5000

ABSTRACT

The Karhunen-Loeve (K-L) procedure for generating the *empirical eigenfunctions* is used to analyze two turbulent channel flow simulations of different geometry and origins. In one instance the Reynolds number Re_τ based on wall shear velocity, u_τ, and channel halfwidth, δ, is 80 and in the second instance $Re_\tau = 125$. The latter case showed a well defined log-layer, while this was absent in the former case. According to accepted convention, the turbulence is said to be *continuous* when $Re_\tau = 80$, and fully developed when $Re_\tau = 125$. In both instances the empirical eigenfunctions reveal the presence of propagating plane wave structures. Thus the contrasts in the two cases is of some importance since they virtually eliminate the possibility that the waves are artifacts.

In addition to the propagating structures the K-L analysis reveals a complimentary set of modes (non-propagating) which can be identified with the streaks that are found in wall bounded turbulence. The *energy* contained in the propagating structures is relatively small compared to that contained in the remaining streaky modes. Nevertheless the waves are shown to play an essential role in the local production of turbulence via *bursting* and *sweeping*.

The interaction of the propagating modes with the streaky modes is found to form the most significant portion of the local Reynolds stress, both in regard to magnitude and to time course.

The *group velocity* of the propagating modes is found to have a value which lies close to the mean flow velocity at the locus of maximal average Reynolds stress. Another feature of the analysis is that the most energetic and effective propagating modes lie in a direction which is in the range of 60 to 70 degrees from the streamwise direction. This is remarkably close to the range of directions of secondary instabilities that account for the transition process from laminar stability. Other analogies with critical layer mechanisms of transitional flows have also been found.

---

*The full text of this paper is to appear in *Theoretical & Computational Fluid Dynamics.*

ON THE HIDDEN BEAUTY OF THE PROPER ORTHOGONAL DECOMPOSITION*

Nadine Aubry [1]

Benjamin Levich Institute and Department of Mechanical Engineering
City College of the City University of New York
New York, NY 10031

ABSTRACT

The proper orthogonal decomposition theorem (Loève (1955)) of probability theory has been proposed and adapted by Lumley (1967) for detecting spatial coherent patterns in turbulent flows. More specifically, the decomposition extracts deterministic functions from second order statistics of a random field and converges optimally fast in the quadratic mean (i.e. in energy). Aubry *et al.* (1990) showed that the technique can be made completely deterministic in the sense that it can be applied to (deterministic) spatially and temporally evolving flows. The remarkable property of this deterministic decomposition is not only in its optimal convergence (as emphasized previously in the literature) but also in its space/time symmetry which permits access to the spatio-temporal dynamics. We discuss here the implications of this property for fluid mechanics studies. The flow is decomposed into both spatial and temporal orthogonal modes which are coupled: each space component is associated with a time component partner. The latter is the time evolution of the former and the former is the spatial configuration of the latter. This generalizes the notion of spatial structures to that of temporal structures both of which can be followed through the various instabilities that the flow undergoes as the Reynolds number increases (in the first flow instability stages as well as in the fully developed turbulence). It also provides a nonlinear dynamics tool for spatio-temporal dynamical systems and can be used for bifurcation detection and analysis as well as dimension and degree of complexity estimates. Simple examples are given to illustrate the theory. In particular, we describe a Hopf bifurcation occurring in time showing how the system complexifies spatially and temporally complexifies. Finally, we describe how the tool presented in this paper is ideal for analysis and compaction of experimental data obtained by simultaneous measurements at multiple locations such as laser scanning techniques, image processing of visualization and particle image velocimetry.

---

[1] This work was supported by a NSF/PYI award MSS89-57462, and partially by a NATO grant #900265 which are gratefully acknowledged.

References

[1] Aubry, N., Guyonnet, R. and Lima, R. 1990 Spatio-temporal analysis of complex signals: theory and applications. *J. Stat. Phys.*, submitted.

[2] Loève, M. 1955 *Probability Theory*. Van Nostrand, New York.

[3] Lumley, J.L. 1967. The structure of inhomogeneous turbulent flows. In: Atmospheric turbulence and radio wave propagation, A.M. Yaglom and V.I. Tatarski:, eds.: 166-178. Moscow: Nauka.

---

*The full text of this paper is to appear in *Theoretical & Computational Fluid Dynamics*.

PART 3

COMPRESSIBLE TURBULENCE

MACH NUMBER EFFECTS ON FREE AND WALL TURBULENT STRUCTURES IN LIGHT OF INSTABILITY FLOW INTERACTIONS

Mark V. Morkovin[1]

Professor Emeritus, Illinois Institute of Technology

ABSTRACT

Our perception of fluid instabilities and turbulence is largely based on properties of incompressible vorticity distributions ω and the Biot-Savart Law of "instantaneous" induction at distance. In high-speed shear layers it is the distribution of local angular momentum $\rho\omega$ that matters; modified interaction-induction between vortical elements is likely to take place only within relative Mach cones of influence, with acoustic time delay. This and the large mean density stratifications influence strongly instability vortex roll-ups, vortex mergings, and large-scale turbulent structures. Consequences of these observations on transition to turbulence and turbulence energetics are discussed in physical terms. Free shear layers and wall layers may require distinct M modelling.

1. Introduction

This paper compares the behavior of incompressible and high-speed shear layers which are inhomogeneous, primarily in the y direction, perpendicular to the nearly parallel mean flow. It focuses on the energetic coherent structures as manifestations of ongoing nonlinear instabilities in laminar and turbulent shear flows. At low speeds there is a significant difference between the geometries of observed energetic eddies in wall layers and in free shear layers, as illustrated in the canonical boundary layer and the canonical mixing layer, respectively. Flow interactions between and within the highly three-dimensional wall structures are much less sensitive to increases in the Mach number ΔM across the layer than are

[1]This work was partially supported under NASA Contract NAS1-18240, while the author was a consultant to High Technology Corporation, Hampton, VA, 23666.

the interactions in the quasi-two-dimensional energetic structures in mixing layers. (This could be anticipated from the sweepback principle for high-speed flows.)

In studies of homogeneous compressible flows, the amplitude of the fluctuating Mach number M_t is often assumed as a parameter, and the focus is on the interactions between the divergence-free (rotational) $\boldsymbol{v}_r$ and irrotational $\boldsymbol{v}_i$ components of the velocity field, their development and decay. Here we ask: how can fluctuating energy be extracted from the mean-flow energy? What mechanisms set the levels of M_t for various shear layers as ΔM increases?

2. Low-speed mixing layers

Coherent structures build up initially by a quasi-two-dimensional slide-and-roll mechanism into discrete vorticity concentrations, exceeding in size the original thickness δ of the layer. Spiral-like advection of preexistent mean-flow vortices Ω_z and of tongues of external irrotational fluid into neighboring large eddies leaves the stretched, thinned layer between them (the braid region) depleted of ω_z. This convoluted state corresponds to the nonlinear saturation of the 2D linearly unstable motions arising in layers with an extremum of mean vorticity $\Omega_z(y)$, see the survey by Ho and Huerre (1984). (When the density varies across the layer, the instability condition requires an extremum of the local angular momentum $\rho(y)\Omega_z(y)$ at the so-called generalized inflection point even when the $U(y)$ profile is convex.) Strong secondary instabilities cause successive vortices to roll around each other and merge, thereby entraining further fluid and thickening the layer, on the average. This quasi-two-dimensional inverse cascade to larger scales continues even when the layer becomes turbulent; the eddies become unequal and the amalgamations irregular, occasionally involving three eddies, Roshko (1976, Fig. 4). The Navier-Stokes (NS) spatially growing simulations of Sandham and Reynolds (1989) leave no doubt about the underlying instability mechanisms of the energetics. When they excited the disturbances at the origin of a laminar layer at a single sinusoidal frequency, but with a random-walk type of small changes in the phase $\Delta\phi$ at $x = 0$ at each time step, the statistics of the growing layer mimicked those of a turbulent layer even though the simulation contained no ω_x or ω_y fluctuations.

When there is some streamwise vorticity present in the thinning layer between the rolling-up eddies, this vorticity is tilted, stretched and amplified in each successive roll-up process. Metcalfe et al. (1987) introduced randomized spanwise perturbations into their NS simulation of a temporally growing mixing layer and demonstrated the rapid growth of smaller-scale irregular streamwise vortex pairs, the "braids." These are part of the downward cascade of energy transfer to smaller scales of classical 3D turbulence. The quasi 2D largest-scale eddies extract energy from the mean flow via roll-ups and mergings and simultaneously mediate the energy input into the streamwise roller-pairs at smaller scales. The similarity between the laser-sheet/fluorescent dye visualizations of braid cross sections in a water turbulent mixing layer in Bernal and Roshko (1986) and the computer shapes of Metcalfe et al. (1987, Fig. 21) provides convincing evidence as to the mechanism.

When the shear $\partial U / \partial y$ is constant and homogeneous to infinity, there is no extremum of Ω_z and no roll-up. However, when the shear is strong enough (dimensionlessly close to that in a viscous sublayer at a wall), instantaneous "streaky structures" elongated in the U direction appear in NS simulations, Lee, Kim and Moin (1990). Because of the high shear, small cross-stream perturbations v, w give rise to large perturbations u. The structures are probably related to families of lightly damped eigenmodes of linearized elongated viscous vortices that arise under irregular excitation. The structures resemble the experimentally observed streakiness under turbulent wall layers, where the boundary conditions maintain the high shear.

3. Supersonic mixing layers and the communication problem

Detailed modern experiments of Papamoschou and Roshko (1988), Papamoschou (1989), Samimy and Elliott (1990), and Elliott et al. (1990) document that turbulent mixing layers grow increasingly less and extract less energy from the mean flow as ΔM across the layers increases. Because of the changes in flow behavior outlined in the Abstract, it is no surprise that linear instability theory and limited analyses of secondary instability indicate the same trends with ΔM. In free shear layers, the slide-and-roll and the merging mechanisms (even though ultimately nonlinear) do not require threshold amplitudes for initiation and vigorous growth at low speeds so that linear theory is directly relevant to the input into the turbulent flow.

In building up instability eigenfunctions we need some upstream—downstream influence and especially communication across the shear layer. Mack (e.g., 1984) formulated one such clear cut criterion for the linear slide-and-roll instability of a supersonic boundary layer. At a fixed x consider the point y_s at the generalized inflection location (which travels at the phase speed c_{ph}) and the point y_1 immediately above it, at the edge of the layer. When fluid elements at these points can no longer influence each other, i.e., when $\Delta M = M_c$ between these exceeds unity, the slide-and-roll mechanism becomes inoperative! Papamoschou (1990) explored some consequences of this communicability problem. It seems very likely that interactions within supersonic turbulent configurations are subject to similar constraints of communicability.

In a mixing layer, there is a point y_2 at the edge of the slower stream U_2 with which fluid at y_s may continue to communicate subsonically until much higher ΔM and initiate an "**inhibited**" **roll-up**, corresponding to a different analytical solution discovered by Gropengiesser in 1968. Results of Sandham (1989), Ragab and Sheen (1990) and at the Univ. of Washington, (D. S. Eberhardt, private communication) strongly suggest that the inhibited slide-and-role leads to weaker flatter vortices in which the entrainment comes predominantly from the stream side with respect to which the amplifying wave travels subsonically. In a high-speed mixing layer, Jackson and Grosch (1989), among others, have shown that past the Mach number for which the slide-and-roll subsonic mechanism ceases, there are two very much less amplified linear modes, classified as supersonic/subsonic and subsonic/supersonic depending upon whether the wave travels supersonically or subsonically with respect to the upper, faster stream U_1. At still higher ΔM the modes travel supersonically with respect to both streams and are likely to generate very little mixing. The definition of the convective Mach number M_c as $|U_i - c_{ph}|/a_i$ for these linear modes is unambiguous (c_{ph} = phase velocity, $i = 1,2$). Since nonlinear simulations suggest that linear c_{ph} changes little through nonlinear saturation, true $M_c > 1$ for experimental mixing layers should depend upon what actual mixture of unstable modes is present in the layer for which the results are analyzed.

According to the sweepback principle, instability waves at non-zero angle θ with respect to the z axis should and do amplify more rapidly than the two-dimensional mode as ΔM exceeds unity. The energy input into the primary and secondary instabilities and into the resulting turbulence will thus depend upon the 3D mixture of rotational and irrotational forcing disturbances near the origin of the mixing layer and their effective receptivity at this location. Note that Papamoschou and Roshko's turbulent boundary layers had small pockets of separation at the finite-angle trailing wedge which generated visible, moderate, unsteady shock waves. Such quasi 2D pressure waves inject nonnegligible quasi 2D (albeit irregular in time) vorticity waves into the mixing layer, as it forms. A supersonic mixing layer originating at a skew trailing edge would probably amplify preferentially the corresponding skew mode for which the effective M_c would then be $M_{c2D} \cdot \cos\theta$. This mixing layer would grow less at low speeds, but continue mixing more efficiently at higher ΔM. Would this mixing layer ultimately approach the same "developed asymptotic turbulent state" (if it exists) as that with $\theta = 0$ at the same ΔM? However, the most important mixing in supersonic layers of technological interest should take place near their origin so that their performance is likely to be input-conditioned and even partly manipulatable.

4. Multi-faceted compressibility effects

We have seen compressibility effects enter in the communicability problem as well as in the generation of energy drain through shocklets when vortices roll up at high ΔM (one shocklet per instability roll-up). Both effects are related to M_c. At incompressible speeds the isotropic instantaneous communicability stems from the singular solutions of the Poisson equation associated with the Biot-Savart Law, which ceases to apply strictly as soon as $M > 0$. According to Baskin (1980), the Biot-Savart Law survives only in its linearized form and even then we must include delay times because the "induction" proceeds at the speeds of sound, i.e., within Mach cones of influence (pseudo Mach cones of Lagerstrom when viscosity and heat conductivity are included). The symmetries of the differential equation systems definitely change at supersonic speeds. Yet, there has been practically no discussion of the consequences the strong anisotropy in the direction of the mean flow on the various assumptions of symmetry and isotropy in theories of supersonic turbulence!

It is clear that this is the main reason for the demise of the two-dimensional slide-and-roll and the merging instabilities at high ΔM of mixing layers as the inviscid numerical simulations of Soestrino (1990) and Soestrino et al. (1988) indicate. Section 2 described how at low speeds the energy extraction from the mean flow is completely dominated by the slide-and-roll and the continuing merging mechanisms. The average ω_z eddy structures have aspect ratios of 4 to 5 with occasional spanwise dislocations. The lesson from the above simulations is that past $\Delta M \sim 3$ such quasi 2D structures will not build up. The low-speed input mechanisms may possibly remain operative for highly skewed structures, but the net energy extraction from the mean flow will be drastically different. In fact, any estimates concerning the expected levels of M_t based on low-speed experiences are likely to be excessive.

Another cornerstone of incompressible turbulence is the 3D vorticity growth through the nonlinear mechanism of tilting and stretching of ω. When the density ρ varies, the curl of the momentum equations can be reduced to the form:

$$\frac{D\omega/\rho}{Dt} = \frac{\omega}{\rho} \bullet \nabla V - \frac{1}{\rho} \nabla\left(\frac{1}{\rho}\right) \times \nabla p + \text{viscous terms} \;, \qquad (1)$$

where D/Dt indicates the material derivative. The terms involving ω/ρ show that the tilt-and-stretch mechanism persists, but now applies to vorticity per unit mass! The $1/\rho$ factor appears as a simultaneous consequence of the conservation of angular momentum and of the conservation of mass:

$$-\nabla \bullet V = \frac{1}{\rho}\frac{D\rho}{Dt} \qquad (2)$$

This density factor explains why even the v and w components of free-stream turbulence attenuate in stretching of turbulent ω_x through a subsonic-supersonic nozzle past M of 3, contrary to incompressible behavior in contractions of wind tunnels. The effect of ρ is much less spectacular in flows with mild $\partial\rho/\partial x$ as encountered in most shear layers. Nevertheless, the tilt-and-stretch mechanisms fuels the 3D braid instability of Section 2 when the roll-ups and mergings are sufficiently strong. It is also present in compressible wall layers when a near-wall hairpin vorticity element is lifted and stretched by the shear.

The second term on the right of (1) represents the barotropic source of vorticity when pressure and density isosurfaces do not coincide. The resultant pressure force on a fluid element then does not pass through its center of mass, creating a local torque and new vorticity. This term was compared informatively to other expanded terms in (1), in particular $\omega_z \nabla \cdot \boldsymbol{V}$ in the nonlinear simulation of unstable 2D compressible mixing layers by Lele (1989) and Sandham and Reynolds (1990). They illustrate the non-tilt-and-stretch effects of varying density on vorticity. Both effects are present in supersonic turbulence.

The intrinsic compressible term in (2), the dilatation, looks deceivingly simple, but contains no dimensionless parameter of the flow! Thompson (1972, Eq. 3.46) uses all the other equations to express the dilation most instructively in terms of eight dimensionless flow parameters, including M and an unsteadiness parameter in linear and nonlinear "sound," M^2 in convective accelerations, M^2/Re in effects associated with viscous forces in the direction of the flow, and a specific-volume thermodynamic parameter (multiplied by M^2/Re in dissipation effects and by $1/\mathrm{PrRe}$ in heat conduction effects). Since supersonic turbulence involves all these listed effects through nonvanishing dilatation, understanding of their relative importance may require asymptotic comparisons as the various parameters dominate. Special difficulties arise with the asymptotics in M and Re. This has a bearing on the interactions between the irrotational and solenoidal components of the velocity field in various regimes.

5. Instability in supersonic laminar boundary layers

The density at the wall of an adiabatic laminar boundary layer decreases as M increases so that the angular momentum distribution $\rho\Omega_z$ develops a maximum farther away from the wall. The associated generalized inflection point would make the layer increasingly more unstable if it were not for the stabilization effect caused by decreased communicability between elements described earlier for the compressible mixing layer. The net result is that the 2D compressible vorticity waves (TS waves at $M = 0$) almost cease to amplify near M of 1.7. Just as for the mixing layer, they are replaced as most amplified modes by waves swept back 55°-60°, see Mack (1984).

However, a new two-dimensional unstable "acoustic" or "Mack" mode (nonexistent in mixing layers) becomes most unstable for M exceeding about 4 as verified in detail experimentally by Kendall (1975). It is believed to be the primary instability which leads to transition to turbulence in low-disturbance experiments, although no associated secondary instability modes have been identified. A still higher Mack mode with roughly three times the frequency may become dominant for $M > 8$ or so. The Mack family of unstable so-called acoustic modes is yet another manifestation of compressibility: acoustic wave lengths, too long at low M, can now match the wavelengths of vorticity disturbances. When properly phase-tuned the combination can amplify through the interaction. Let us note parenthetically that similar enhancing interaction between pressure and vorticity structures in supersonic turbulent boundary layers is highly unlikely: the irregularity of turbulence would spoil the required fine phase tuning between the irrotational and vortical motions. Note that this family of instabilities is absent in the asymmetric mixing layers, but it can be present in free symmetric shear layers, i.e., supersonic jets and wakes, see detailed studies of Mack (1990) and Tam and Hu (1989a). In the interaction, the pressure waves in these modes reflect internally back and forth from "relative sonic layers" (relative to c_{ph}) and are mostly subsonic with respect to the free stream, without radiation losses. These modes ensue from tight coupling between the acoustic, vorticity, and entropy modes of Kovasznay, brought about by the commensurability of their wavelengths and the non-zero lateral gradients of mean U, ρ, and T. Somehow, the trapped acoustic waves energize the vorticity and entropy waves.

Bypasses to transition (i.e., roads to turbulence unexplained by known instability mechanisms) including turbulent wedges caused by roughness have been widely observed in supersonic laminar boundary layers often associated with forced ω_x motion. (Bypasses in mixing layers for $\Delta M > 4$ probably exist but have not been documented carefully enough experimentally.)

6. M-variation in turbulent boundary layers

In low-speed mixing layers, the linearizable primary slide-and-roll instabilities and the secondary merging instabilities are directly related to the mechanisms and shapes of the observed coherent turbulent structures. However, in turbulent boundary layers neither the large coherent eddies in the outer part of the layer nor the coherent near-wake structures associated with "bursting" processes are directly connected to any clear instability mechanisms, Sreenivasan (1988). The mean velocity profile is stable to linear perturbation, Landahl (1967). The excess of the near-wall generated fluctuation energy diffuses outward to invigorate the outer turbulence, which would otherwise slowly decay with distance (as it does for relaminarized inner layers). It is widely believed that both sets of coherent structures are a manifestation of (loosely coupled) three-dimensional nonlinear instabilities operating above a finite threshold of disturbances, Robinson (1990).

In a significant contrast to the quasi 2D structures in turbulent mixing layers, the structures in the boundary layers are narrow. As documented by Kovasznay et al. (1970) the outer structures are inclined, about 3δ long and one δ wide, with the downstream head protruding slightly beyond the average δ height. They convect downstream at 80-90% of the free-stream speed. The wall structures are narrower still, the universal spanwise wave length of the streaky structures (similar to those described in the last paragraph of Section 2) being about 100 in viscous wall units. Robinson (1990) compares the many kinematic and dynamic features observed by various experimenters (via different imperfect techniques) with the features gleaned painstakingly from a numerical NS simulation with Spalart's boundary-layer code. (Initially the code was operated at Re_θ of 670 where the inner and outer behavior is not clearly separated.) What comes across is the diversity of the features, inner intermittency in x rather than in t, and the narrowness of the structures, some vortical with single orientation, inclined spanwise at 5-10° to the x axis. Comparable NS simulation for supersonic turbulent boundary layers is a decade or two away.

Not having a theory for the wall turbulent instabilities nor a simulation extendable to supersonic speeds we have no direct way of judging the influence of M on structures and fluctuation levels. However, from the discussions in Sections 3 and 4 we can surmise whether the low speed structures can continue to interact internally as M increases. The sweep-back principle, in particular, suggests that the narrow outer layer structures should not be very sensitive to M, although there might be 3D shocklet formations for $M > 5$ or so. This sonic drain, compared to the total fluctuation energy production may not be very significant, but the pressure fields might be distorted near the compressed back of the eddies, and their shape and inclination in y modified. Smits et al (1990) and Spina et al. (1990) provide space-time correlations and other evidence (also unpublished visualization via Raman excitation and laser-induced electronic fluorescence) of large coherent structure at $M = 2.9$. At any height the supersonic space-time correlations are somewhat foreshortened in the x direction and the structures seem to curve towards the vertical at higher y. However, in contrast to the drastic changes of the mixing layer structures at ΔM of 3, these boundary layer structures seem to remain functionally similar to those at low speeds.

No instrumentation has yet resolved the wall structures at supersonic speeds. But the narrow, 80-85° swept back structures should be very nearly M insensitive, even if they occasionally protrude beyond the $M = 1$ height of the mean layer. We have no way of judging how energetic the fluctuations might be. Near the adiabatic wall, temperatures become high, but the subsonic communicability should remain, possibly similar to the processes in strongly heated low-speed layers. (Numerical and experimental studies of the latter are somewhat easier and perhaps desirable.)

7. Discussion and conclusions

In my 1961 pre-coherent-structures paper (out of which Bradshaw extracted inequalities which should really be called the Morkovin-Bradshaw hypothesis — see Bradshaw 1977) , I organized hot-wire findings and limited analysis of supersonic turbulent boundary layers up to $M \sim 4.7$ into a seemingly consistent viewpoint. Conceptually it followed the classical asymptotic division into an outer inviscid layer and an inner vigorous viscous layer which interacted and merged smoothly in an overlap layer. Emphasis was placed on the adiabatic heating near the wall which grows as $(\gamma - 1)M^2$ and causes the layer to be increasingly stratified. In the schematic diagram explicit sources of mechanical mixing energy were drawn in both the inner and outer layer which churned and mixed the thermal-density layer. The inference was that to the first order the fluctuating velocity products such as uv needed to be weighted with mean density at each height to be comparable. To the then prevailing accuracy of the experiments, the weighted fluctuation products were indeed similar across the layer up to the highest $M = 4.67$ adiabatic wall experiments. For this supersonic regime the incompressible behavior was modified primarily by the stratification of the layer, not through an intrinsic dynamic compressibility effect.

The recent experimental and computational information for the simpler mixing layer and the concern for required communicability between elements in interaction processes, which are the essence of turbulence, call for an extension of the 1961 viewpoint. The supply of the mechanical fluctuating energy was shown to decrease rapidly with M in the mixing layer. For other supersonic shear layers a similar question arises: how much energy can a given family of instability modes (perhaps growing from linearizable amplitudes into groups of large coherent eddies) extract from a given flow as a function of M? And how much true mixing does it entail? The contrast between the (limited) information on free shear layers and turbulent boundary layers appears traceable to the differences in geometry of the coherent eddies, the presumably most amplified structures. As we have seen, two-dimensional, purely vortical modes (see Note 1) become extremely inefficient extractors, unquestionably because of the communication problem. In contrast, the performance of nearly streamwise vortices or pairs of vortices should be very insensitive to M; communication will persist though with some delay. Other geometrical structures will fall somewhere in between, the narrower, the less M-sensitive they are likely to be.

The 1990 view reaffirms the importance of the stratification of mean density and its consequences. The communicability problem requires attention to the M-dependence of the extraction of fluctuating energy from the mean flow, which goes hand in hand with amplification and spreading rates. The Papamoschou-Roshko experiment documents the decreased extraction rate even for mixing layers which grow from initially turbulent boundary layers. Their correlation parameter M_c is in fact a communicability parameter for a single (!) fixed-mode geometry identifiable through laminar instability theory. Normalization of amplification rates and thickness growth rates with respect to the corresponding quantities at $M = 0$, hopefully disposes of the contributions due to the stratification effects.

Amplification modes in supersonic laminar boundary layers also decreases with M_1 but at present the effect of stratification is not separable from the communicability factor. However, these primary modes bear no relation to the presumably nonlinearly unstable structures in turbulent boundary layers at any speeds. In the absence of any experimental evidence, a rationale might propose that the "burst"-associated wall layer structures could be considered independent of M up to about 8 or 10. The large outer structures and the inner-outer coupling are likely to become less efficient in building fluctuations for $M > 4\text{-}5$. However the communicability of these structures is likely to decrease slowly because of their narrowness. If in a given design drag and high heat transfer are not desirable, the assumption of constant effectiveness would yield conservative estimates.

Nonadiabatic walls would bring about different density stratifications. Again, assuming with good probability that the wall layer becomes turbulent via unknown secondary and tertiary instabilities or bypass modes, the problem centers on relating the actual high-speed turbulent structures to those at $M = 0$ which have no such density stratification. For this we have no guidance at present. Similarly, we have no guidance for incorporating the effects discussed in this paper into turbulence modeling for industrial purposes. The contrast between the behavior of mixing layers and boundary layers, say near Kistler's M of 4.67, suggests that there is a need.

Note 1 On so-called acoustic modes

Most amplified Mack's higher modes remain two-dimensional at all Mach numbers, because reflection of skew waves from a wall or from a "relative" sonic surface is less efficient. Because of evidence (not fully conclusive) that the second-mode growth precedes transition to turbulence in laminar boundary layers past M of 5 or so, and a few visual indications of apparent vortices near the generalized inflection point, it has been assumed that the pressure-vorticity interactions somehow lead to vortical motion and turbulence. Similar conclusions pertain to the higher modes in a supersonic axisymmetric jet, Tam and Hu (1989a, Section 9). It would be desirable to analyze the energetics and the scalar tracer mixing in these "inward reflecting" modes, linearly and nonlinearly a'la Sandham and Reynolds (1990), in order to clarify the mechanisms outlined in Section 5. How do the energetic and mixing features of the Mack modes differ from the slide-and-roll modes and from the "externally reflected" amplifying modes in the case of mixing layers, discovered by Tam and Hu (1989b) and corroborated by Greenough et al. (1989) and others?

Because of its asymmetry, a supersonic mixing layer does not possess intrinsic internally reflecting sound waves (Mack, 1990) but radiates outwards. However, when surrounded by a duct, the outward neutral acoustic radiation is reflected back upon the layer and gives rise to new duct-dependent unstable modes. The issue is whether these new modes contribute to mixing and lead to turbulence or represent just somewhat enhanced resonant sound fields. As noted earlier these amplifying modes are mono-chromatic and finely phase tuned; they are unlikely to have counterparts in jittered turbulent structures.

References

Baskin, V. E., 1980, *Induction of an arbitrary vortex filament in a gas*, Fluid Mech, Soviet Research, **9**, p. 51.

Bernal, L. P. and Roshko, A., 1986, *Streamwise vortex structure in plane mixing layers*, J. Fluid Mech, **170**, p. 499.

Bradshaw, P., 1977, *Compressible turbulent shear layers*, Annual Review of Fluid Mech., p. 33.

Elliott, G. S. and Samimy M., 1990, *Compressibility effects in free shear layers*, Phys. Fluids A, **2**, p. 1231.

Elliott, G. S., Samimy, M., and Reeder, M. F., 1990, *Pressure based real-time measurements in compressible free shear layers*, AIAA Paper 90-1980.

Greenough, J. A., Riley, J. J., Soestrino, M., Eberhardt, D. S., 1989, *The effects of walls on a compressible mixing layer*, AIAA Paper 90-0372.

Gropengiesser, H., 1968, *On the stability of free shear layers in compressible flows* (in German), PhD thesis, Tech. Univ. of Berlin, also NASA transl. NASA TT F-12786, Feb. 1970.

Ho, C.-M. and Huerre, P., 1984, *Perturbed free shear layers*, Review of Fluid Mechanics, **16**, p. 365.

Jackson, T. L. and Grosch, C. E., 1989, *Inviscid spatial stability of a compressible mixing layer*, J. Fluid Mech, **208**, p. 609.

Kendall, J. M., 1975, *Wind tunnel experiments relating to supersonic and hypersonic boundary-layer transition*, AIAA J., **13** p. 290.

Kovasznay, L. S. G., Kibens, V., and Blackwelder, R. F., 1970, *Large-scale motion in the intermittent region of a turbulent boundary layer*, J. Fluid Mech, **41**, p. 283.

Landahl, M. T., 1967, *A wave-guide model for turbulent shear flow*, J. Fluid Mech, **29**, p. 441.

Lee, M. J., Kim, J., and Moin, P., 1990, *Structure of turbulence at high shear rate*, J. Fluid Mech, **216**, p. 561.

Lele, S., 1989, *Direct numerical simulation of compressible shear flows*, AIAA Paper 89-0374.

Mack, L. M., 1984, *Boundary-layer linear stability theory*, Special Course on Stability and Transition on Laminar Flow, AGARD Report No. 709, 3-1.

Mack, L. M., 1990, *On the inviscid acoustic-mode instability of supersonic shear flows. Part 1. Two-dimensional waves*, to appear in Theory and Comput. Fluid Dyn., Springer Journal.

Metcalfc, R. W., Orszag, S. A., Brachet, M. E., Menon S., and Riley, J. J., 1987, *Secondary instability of a temporally growing mixing layer*, J. Fluid Mech, **184**, p. 207.

Morkovin, M. V., 1961, *Effects of compressibility on turbulent flows*, in Mecanique de la Turbulence, A. Favre, ed. CNRS, Paris, p. 368. Also Gordon and Breach, 1964.

Morkovin, M. V., 1987, *Transition at hypersonic speeds*, ICASE Interim Report 1.

Papamoschou, D., 1989, *Structure of the compressible turbulent shear layer*, AIAA Paper 89-0126.

Papamoschou, D., 1990, *Communication paths in the compressible shear layer*, AIAA Paper 90-0155.

Papamoschou, D. and Roshko, A., 1988, *The compressible turbulent shear layer: an experimental study*, J. Fluid Mech, **197**, p. 453.

Ragab, S. A. and Sheen, S., 1990, *Numerical simulation of a compressible mixing layer*, AIAA Paper 90-1669.

Robinson, S. K., 1990, *A perspective on coherent structures and conceptual models for turbulent boundary-layer physics*, AIAA Paper 90-1638.

Roshko, A., 1976, *Structure of turbulent shear flows: a new look*, AIAA J., **14**, p. 1347.

Samimy, M. and Elliott, G. S., 1990, *Effects of compressibility on the characteristics of free shear layers*, AIAA J., **28**, p. 439.

Sandham, N. D., 1989, *A numerical investigation of the compressible mixing layer*, PhD Thesis, Mech. Engr., Stanford Univ.

Sandham, N. D. and Reynolds, W. C., 1989, *Some inlet-plane effects on the numerically simulated spatially-developing mixing layers*, Proc. Turbulent Shear Flows Symposium No. 6, Springer Verlag.

Sandham, N. D. and Reynolds, W. C., 1990, *Compressible mixing layer: linear theory and direct simulation*, AIAA J., **28**, p. 618.

Sigalla, L., Eberhardt, D. S., Greenough, J. A., and Riley, J. J., 1990, *Numerical simulation of confined, spatially developing mixing layers: comparison to the temporal shear layer*, AIAA Paper 90-1462.

Smits, A. J., Spina, E. F., Alving, A. E., Smith, R. W., Fernando, E. M., and Donovan, J. F., 1990, *A comparison of the turbulence structure of subsonic and supersonic turbulent boundary layers*, to appear in Phys. Fluids A.

Soestrino, M., 1990, *Numerical simulations of instabilities in compressible mixing layers*, PhD thesis, Dept. Aero. & Astro., Univ of Washington, Seattle.

Soestrino, M., Eberhardt, D. S., Riley, J. J., and McMurtry, P. A., 1988, *A study of inviscid supersonic mixing layers using a second-order TVD scheme*, AIAA Paper 88-3676 and AIAA J., **27**.

Spina, E. F., Donovan, J. F., and Smits, A. J., 1990, *On the structure of supersonic turbulent boundary layers*, to appear in J. Fluid Mech.

Sreenivasan, K. R., 1988, *A unified view of the origin and morphology of the turbulent boundary layer*, 26 p. in Proc. Turbulence Management and Relaminarization, Liepmann, H. W. and Narasimha, R., editors, Springer Verlag.

Tam, Ch.K.W. and Hu, F. Q., 1989a, *On the three families of instability waves of high-speed jets*, J. Fluid Mech, **201**, p. 447.

Tam, Ch,K.W. and Hu, F. Q., 1989b, *The instability and acoustic wave modes of supersonic mixing layers inside a rectangular channel*, J. Fluid Mech, **203**, p. 51.

Thompson, P. A., 1972, *Compressible Fluid Dynamics*, McGraw-Hill, NY.

TOWARD A CONSTITUTIVE RELATION IN COMPRESSIBLE TURBULENCE

Otto Zeman
Center for Turbulence Research
Stanford University
Stanford, CA 94305-3030

ABSTRACT

In high-speed turbulent flows questions arise as to what extent compressibility affects turbulence structures, the Reynolds stress-mean velocity (constitutive) relation, and the accompanying processes of heat transfer and mixing. This paper addresses the problem of compressibility effects on the constitutive relation when the r.m.s. fluctuating Mach number is sufficiently high to cause significant coupling between the compressive and solenoidal modes. Emphasis is given to the analysis of homogeneous shear turbulence. It is found that turbulence model equations possess a compressible selfsimilar solution characterized by invariant r.m.s. Mach number and turbulence time scale (normalized by mean shear). The turbulence second-order quantities grow exponentially but at a much slower rate than in the incompressible regime.

1. Introduction

In his pioneering paper "Toward a turbulent constitutive relation", Lumley (1970) attempted to formulate in a rational way the problem of constitutive relation, or a closure problem, in a homogeneous incompressible turbulent fluid. During the following 20 years, much progress has been made in the development of approximate turbulence closures for a variety of flows but the general problem of the pointwise stress-strain relation in a turbulent fluid remains far from being resolved. The turbulent fluid is neither materially indifferent nor deterministic and, as often argued on the basis of coherent structures in turbulent flows, it may also have a long memory. Now, with the plethora of turbulence closure models available on the market it is instructive to reread Lumley's (1970) paper to view the modeling effort in proper perspective. With the recent revitalization of high speed flow research, compressibility presents a new set of challenging problems to turbulence researchers, and the purpose of this paper is to address a few of these problems. In particular we shall be concerned with the compressibility effects on the constitutive relation in shear-driven turbulence.

The measure of the compressibility effects on turbulence is the so-called r.m.s. fluctuating Mach number M_t based on an r.m.s. fluctuating speed and the sonic speed corresponding to the background (mean) temperature. The sonic speed can be considered as a new intrinsic velocity scale which limits the propagation of perturbations through the fluid and sets a limit, in a statistical sense, on the velocity fluctuations as well. As M_t of a given turbulent increases, the velocity fluctuations are accompanied by growing levels of density and pressure fluctuations and a growing degree of coupling between momentum and internal energy fields. Zeman (1990), (hereafter referred to as Z90) suggested that at a sufficiently large M_t the flow may generate shock-like structures or shocklets which entail high levels of dilatation dissipation. The existence of turbulent shocklets has been confirmed in direct numerical simulations (DNS) of decaying turbulence (Lee, Lele, and Moin 1990) and in shear driven turbulence (Blaisdell, 1990). In laboratory flows, the most prominent effect of compressibility on turbulence is the inhibition of shear layer growth rate with the (convective) Mach number M_c as observed in experiments of Papamoschou and Roshko (1988), Samimy and Elliott (1989) and others. In Z90 it was suggested that the growth rate inhibition is due to the (anomalous) dilatation dissipation (ϵ_d) caused by turbulent shocklets and a mathematical model was proposed to account for ϵ_d in Reynolds stress closure models.

In the following section we shall review the present state of modeling dilatation dissipation and pressure dilatation correlation; in Section 3 we shall discuss the physics and findings inferred from the DNS and modeling of homogeneous shear turbulence. Section 4 offers a simplified constitutive relation for compressible shear turbulence, and Section 5 concludes the paper.

2. Explicit compressibility terms

As shown in Z90, utilizing Helmholtz' decomposition and Favre averaging, one can identify two terms in the second order turbulence equations which explicitly depend on the dilatation velocity field. These are: 1) the dilatation dissipation ϵ_d formally defined as $\epsilon_d = 4/3\mu\overline{(u_{j,j})^2}/\overline{\rho}$, and 2) the pressure-dilatation correlation Π_d defined as $\Pi_d = \overline{p u_{j,j}}/\overline{\rho}$.

Dilatation dissipation: On the basis of a turbulent shocklet model, Z90 obtained the following expression for the shocklet contribution to ϵ_d:

$$\epsilon_d \propto \frac{q^3}{L}[(\frac{1+\gamma}{2})^2 \frac{1}{M_t^4} \int_1^\infty \left(\frac{m^2-1}{m}\right)^3 p(m)dm] \propto \frac{q^3}{L} F(M_t). \quad (1)$$

Here, $m = (u_j u_j)^{1/2}/a^*$ is the instantaneous Mach number based on the sonic temperature $T^* = (2/(\gamma+1))\widetilde{T}$, $p(m)$ is a probability distribution function (pdf) of m, and L is a turbulence length scale. The function $F(M_t)$ represents the expression in square brackets; its values are obtained by numerical integration over the pdf $p(m)$. We note that $m(\mathbf{x}, \mathbf{t})$ is a positive quantity and in shear turbulence it must be estimated from the DNS data; in the strictly isotropic Gaussian turbulence $p(m) \propto m^2 exp\{-m^2/2M_t^2\}$. As defined earlier, $M_t = q/a$ is the r.m.s. Mach number, $q^2 = \widetilde{u_j u_j}$ designates twice the Favre-averaged turbulent kinetic energy, and a is the sonic speed at the Favre-averaged temperature $\widetilde{T}$.

In 3D turbulence q^3/L is proportional to the solenoidal dissipation ϵ_s, so the total dissipation rate of turbulence kinetic energy (in 3D) can be expressed as

$$\epsilon_{tot} = \epsilon_s(1 + c_d F(M_t)). \quad (2)$$

The constant c_d and $F(M_t)$ must be evaluated from experiments or DNS data. In homogeneous shear, we inferred the approximation (used in the following section)

$$F(M_t) = 1 - exp\{-((M_t - 0.25)/.80)^2\}. \quad (3)$$

with $c_d = 0.75 - 1.0$ and $F(M_t) = 0$ when $M_t < 0.25$. This signifies the physical fact that shocklets cannot exist below certain value of M_t. In the simulations of Lee et al. (1990), the shocklets were observed only when M_t exceeded values of about 0.3. The recent DNS results indicate that in the absence of shocklets $\epsilon_d/\epsilon_s \propto M_t^2$, which is in agreement with the model proposed by Sarkar et al. (1989). This is a suprising result, suggesting that the compressive and solenoidal velocity spectra $E_d(\mathbf{k})$, $E_s(\mathbf{k})$ are selfsimilar. A more plausible explanation is that the partition between compressive and solenoidal components depends, at each scale, on the local Mach number i.e.

$$\frac{E_d(\mathbf{k})}{E_s(\mathbf{k})} \propto M^2(\mathbf{k}) \approx E_s(\mathbf{k})k/a^2. \quad (4)$$

It is then easy to show that the compressive-to-solenoidal energy ratio is $q_d^2/q_s^2 \propto M_t^2 Re_t^{-1/4}$. In the DNS turbulence, $Re_t \leq 100$, and it is thus plausible that at higher Reynolds numbers the M_t^2-contribution to ϵ_d would become negligibly small.

Pressure-dilatation correlation: In Zeman (1990a) and Zeman and Blaisdell (1990), it was shown that in decaying or shear-driven turbulence the pressure-dilatation term $\Pi_d = \overline{p\theta}/\overline{\rho}$ can be approximated as

$$\Pi_d = -\frac{1}{2}\frac{D\overline{p^2}}{Dt}\frac{1}{\overline{\rho}a^2}, \tag{5}$$

as in linear acoustics. The DNS data indicate that the two terms in (5) are virtually identical throughout the tested range $0 < M_t < 0.7$, even in the presence of shocklets (Blaisdell 1990). The physical meaning of (5) is best illustrated by utilizing (5) in the turbulence energy equation (Zeman 1990a):

$$\frac{D}{Dt}[\frac{1}{2}(q^2 + \overline{p^2}/\overline{\rho}^2 a^2)] = -(\epsilon_s + \epsilon_d). \tag{6}$$

The quantity in square brackets in (6) is the sum of kinetic and potential energies; hence, the Π_d term mediates energy transfer between the kinetic and potential (pressure variance) modes.

In view of (5), the closure for Π_d is reduced to determining $D\overline{p^2}/Dt$. A model for $D\overline{p^2}/Dt$ proposed by Zeman (1990a) is

$$\frac{D\overline{p^2}}{Dt} = -\frac{\overline{p^2} - p_e^2}{\tau_a}. \tag{7}$$

The equation states that in the absence of $\overline{p^2}$-producing mechanisms the pressure variance will relax to its equilibrium (equipartition) value (p_e^2) on the acoustic time scale $\tau_a \propto L/a \propto M_t q^2/\epsilon_s$. The equilibrium value p_e^2 is related to the kinetic energy q^2 and M_t through the relation

$$p_e^2/\overline{\rho}^2 q^2 a^2 = P_e(M_t) = \frac{\alpha M_t^2 + \beta M_t^4}{1 + \alpha M_t^2 + \beta M_t^4}, \tag{8}$$

where optimal values of the constants are $\alpha = 1.0$ $\beta = 1.0$. The validity of (7) is questionable in low Mach number decaying turbulence ($M_t \leq 0.1$) where the dilatational and solenoidal fields tend to be decoupled. However, (7) yields good agreement with the DNS of decaying turbulence when $M_t > 0.5$, and as shown later, the agreement is also good in shear-driven turbulence.

3. Shear-driven turbulence

Following Lumley (1970), we write a formal constitutive relation for Reynolds stresses $R_{ij} = \overline{u_i u_j}$ in an incompressible turbulent fluid with $Re_t >> 1$ as

$$\frac{DR_{ij}}{Dt} + C\frac{R_{ij}}{\tau} = \frac{C-2}{3}\epsilon_s \delta_{ij} + \mathbf{F}_{ij}\{U_{p,q}, R_{pq}, \epsilon_{pq}\}, \tag{9}$$

where $\tau = q^2/\epsilon_s$ is the (vortical) turbulence time scale, $\mathbf{F}_{ij}$ is a functional of its arguments, and C is a return-to-isotropy invariant (note that $q^2 = R_{jj}$). The simplest form of (9) is such that $\mathbf{F}_{ij}$ is bilinear in $U_{p,q}$ and R_{pq} and represents the combination of the so-called rapid-pressure and explicit (production) terms; C is a constant and the dissipation tensor ϵ_{pq} is taken to be isotropic so that $\epsilon_{pq} = \epsilon_s \delta_{pq}/3$.

Now, on the basis of the differential equation for fluctuating pressure in compressible turbulence, it can be argued that the form of the slow and rapid pressure terms in the Favre-averaged setting will retain their incompressible forms and that the compressibility modifications to (9) will appear as additive terms only. In the case of plane shear, these additive terms are the dilatation dissipation and pressure dilatation terms. As shown earlier, these compressive terms are coupled to the solenoidal field according to their model expressions, i.e., $\epsilon_d\{\epsilon_s, M_t\}$ and $\Pi_d\{q^2, \overline{p^2}, M_t\}$. In summary, we suggest that the first order effects of compressibility appear in the constitutive relation for $R_{ij} = \overline{\rho u_i u_j}/\overline{\rho} = \widetilde{u_i u_j}$ as additive isotropic terms or as modifying factors of the dissipation term ϵ_s. Hence, in the formal equation (9), only the first term on the RHS will be altered to

$$\frac{DR_{ij}}{Dt} + ... = \frac{C - 2(1 + (\epsilon_d - \Pi_d)/\epsilon_s)}{3}\epsilon_s \delta_{ij} + ... \tag{10}$$

With the mean shear $dU_1/dx_2 = S$, the actual set of equations needed to compute evolution of the (Favre-averaged) stress components R_{ij} are

$$1/2\frac{\partial q^2}{\partial t} = P_s - (\epsilon_s + \epsilon_d - \Pi_d), \tag{11}$$

$$\frac{\partial R_{12}}{\partial t} + C\frac{R_{12}}{\tau} = -0.4R_{22}S, \tag{12}$$

$$\frac{\partial R_{22}}{\partial t} + C\frac{R_{22}}{\tau} = 0.4P_s - \frac{C - 2(1 + (\epsilon_d - \Pi_d)/\epsilon_s)}{3}\epsilon_s, \tag{13}$$

$$\frac{\partial \epsilon_s}{\partial t} = -C_\epsilon \frac{(\epsilon_s - 0.75P_s)}{\tau}. \tag{14}$$

The above equations are based on the Z90 version of a compressible turbulence model. Here the return-to-isotropy constant is $C = 3.25$, $C_\epsilon = 3.7$, $\tau = q^2/\epsilon_s$, and $P_s = -R_{12}S$ is the rate of turbulence production by shear. The closure equations for the dilatation terms Π_d and ϵ_d are as outlined in Section 2.

As alluded to in Zeman and Blaisdell (1990) and shown in Figures 1 and 2, the model equations (11)-(14) possess two asymptotic solutions depending on the initial Mach number M_t. If M_t is identically zero, the solution agrees with the incompressible DNS and experimental data (Tavoularis 1985, Rogers et al. 1986). The kinetic energy and R_{ij} grow exponentially, i.e. $q^2 \propto R_{ij} \propto exp\{\lambda_o St\}$, and the turbulent time scale approaches a constant value dictated by the shear, i.e. $S\tau \approx 11.7$. If, on the other hand, M_t is initially finite, the model solutions asymptote to a different state given by a unique $M_t = M_{t\infty} \approx 0.62$ and $S\tau \approx 7.3$. The energy growth is again exponential but with a smaller growth parameter $\lambda < \lambda_o$. Both the model and DNS data suggest that before the asymptotic state is reached, $\lambda = \lambda(M_t)$ decreases with increasing M_t, although the DNS data exhibit rather different transient behavior due likely to low DNS Reynolds numbers. The model-computed evolutions of M_t shown in Figure 2 also include DNS data with the initial $M_{to} = 0.4$ (reproduced from Zeman and Blaisdell 1990). In order to match the initial behavior of the DNS-computed M_t, the model initial value of $S\tau$ was 3, rather than 5.9 used for the DNS run. As evident from Figure 2, the DNS data of M_t do not show any sign of converging toward an asymptote. However, it is inevitable that ultimately the exact solutions also reach a saturation state where velocity fluctuations are controlled by the sonic speed through shocklet generation and dissipation.

In order to verify whether the model equations may possess more asymptotic solutions, the model equations have been reduced to two ODE's for the (phase) variables $X = M_t$, and $Y = S\tau$

$$\frac{dX}{dt} = F_x(X,Y), \quad \frac{dY}{dt} = F_y(X,Y). \tag{15}$$

Figure 3 shows phase space trajectories $\{X(X_o,t), Y(Y_o,t)\}$ for different initial pairs (X_o, Y_o). It appears that for $X_o > 0$ there exists, indeed, only one attracting point, indicated by the cross-hatched circle on the right side of the diagram. Although if the initial Mach number is very small ($X_o << 0.01$) the convergence to the attracting point is very slow. This result, if physically correct, would considerably simplify the compressible constitutive relation and could be useful in parameterization of turbulence in real flows. In the following we shall attempt to show that the asymptotic solutions of the model equations do represent an actual physical state.

First, as mentioned earlier the DNS computed dilatation dissipation appears to follow the functional dependence $\epsilon_d/\epsilon_s \approx M_t^2$. Using the latter relation rather then (3) we found that the equation solutions approached almost the same asymptotic state as before. This signifies that the model results are virtually independent of the choice of the dissipation function $F(M_t)$ as long as ϵ_d/ϵ_s is a realistic (increasing) function of M_t. Second, the presure-dilatation Π_d plays, in the turbulence dynamics, an equally important role as ϵ_d. But its effects on turbulence are rather subtle, and it is essential that the relation (5) and the pressure relaxation model in (7) represent the correct physics. The evolution of modeled Π_d (normalized by initial ϵ_{so}) is compared with the DNS results of Blaisdell (1990) in Figure 4. Disregarding the (acoustic) fluctuations in the DNS data, the model agrees with the DNS data reasonably well including the initial transient. As turbulence approches equilibrium (for $St > 5$) the average Π_d becomes a negative constant fraction of the local solenoidal dissipation. The model predicts $\Pi_d/\epsilon_s \approx -0.07$, while the DNS yield a somewhat smaller value (-0.05). The agreement between the DNS and model is suprisingly good considering the fact that in the model Π_d depends on the subtle difference between the current value of $\overline{p^2}$ and the equilibrium value p_e^2 in the pressure relaxation model (7). Evidently, $\overline{p^2}$ always lags behind the equilibrium value which is continually growing with q^2 and M_t. Physically this means that $\Pi_d < 0$ mediates a transfer of kinetic-to-potential energy ($\overline{p^2}$) and in the kinetic energy balance (11) it represents a loss (of the same order as ϵ_d).

Another important aspect of the pressure relaxation model is its capability to predict the (normalized) pressure variance $\overline{p^2}/\overline{p}^2$ as a function of M_t as shown In Figure 5. Here, the cross-hatched area represents the spread of DNS data reproduced from Blaisdell (1990);

the model data (full circles) are based on the equilibrium function P_e in (8). Evidently the model-DNS agreement is more then satisfactory and if necessary, further improvement could be made by reducing the constant α in (8) to about 0.9. We note that since the acoustic time scale τ_a is small compared with the turbulence growth time scale, according to (7) the normalized values of $\overline{p^2}$ and p_e^2 are almost identical and therefore

$$\overline{p^2}/\overline{p}^2 \approx \gamma^2 M_t^2 P_e(M_t). \tag{16}$$

The DNS data appear to favor the M_t^2 -dependence in the equilibrium function P_e in (8).

In view of the the above arguments and comparison with the DNS results, there can be no doubt that the combined model for the pressure dilatation and pressure variance (in (7) and (8)) is physically sound and we may safely conclude that the set of turbulence model equations (11)-(14) represent the essential physics of compressible shear turbulence in quasi-equilibrium state.

4. The constitutive relation in the limit $M_t = M_{t\infty}$

On the basis of the existence of the asymptotic state discussed above it is now possible to formulate a simple version of the constitutive relation and to determine the second-order turbulence statistics. Using the approximation inferred in the previous section $\overline{p^2} \approx p_e^2$, the kinetic energy equation (11) reduces to

$$\frac{1}{q^2}\frac{dq^2}{d(St)} = \frac{2}{1+P_e(M_t)}(|r_{12}| - \frac{1+c_d F(M_t)}{S\mathcal{T}}). \tag{17}$$

Since the asymptotic solution is characterized by the invariant the normalized stresses $r_{ij} = R_{ij}/q^2$, Mach number ($M_{t\infty} = 0.63$), and $S\mathcal{T} = 7.3$, all second-order quantities q^2, R_{ij}, and the temperature $\widetilde{T}$ follow the same exponential growth $exp\{\lambda_\infty St\}$ with the growth parameter λ_∞ defined by the RHS of (17). Before the asymptotic state is reached, the turbulence growth rate is dependent on the initial M_t. Since P_e and F in (17) are increasing functions of M_t, the initial growth rate is lower for higher initial M_t in agreement with the DNS results. To complete the turbulence description, the equations (11)-(14) can be solved for the equilibrium values of $r_{ij} = R_{ij}/q^2$, yielding

$$r_{11} = 0.55,\ r_{22} = 0.225,\ r_{12} = -0.191. \tag{18}$$

Given the energy level q^2 at some time $t > S^{-1}$, equations (17), and (18) determine fully the asymptotic turbulence state.

5. Conclusions

We have shown that if the (anomalous) dilatation dissipation in shear-generated turbulence has a general functional form of (2), the modeled turbulence will approach an asymptotic state which is characterized by a constant (saturation) Mach number $M_{t\infty} \approx 0.62$ and a constant normalized time scale $S\mathcal{T} \approx 7.3$. The turbulence energy growth is exponential, but the growth rate is significantly reduced due to the elevated dissipation rate and the transfer of kinetic energy to the pressure fluctuations (potential energy). This energy transfer is mediated by the pressure-dilatation term (Π_d) and is sensitive to the particulars of the Π_d model. A comparison with DNS results indicates that the present (relaxation) model for $\overline{p^2}$ (and for Π_d is in excellent agreement with the DNS results.

From the physical viewpoint, it can be argued that the asymptotic selfsimilar state is maintained by strong (nonlinear) coupling among the solenoidal, compressive, and entropic modes and that the maximum (saturation) Mach number is controlled by the shocklet dissipation. We conclude that the existence of asymptotic selfsimilar state in compressible shear turbulence is a physical possibility and not an artifact of the turbulence closure model. It significantly simplifies the turbulent constitutive relation, and could be useful for turbulence parameterization in high-speed flows such as supersonic mixing layers. The asymptotic state and the Mach-number saturation limit have so far not been identified in DNS computations because of the lack of resolution of shock structures.

References

[1] Blaisdell, G. A., 1990, *Numerical simulations of compressible homogeneous turbulence*, Ph. D. Thesis, Mechanical Engineering Dept., Stanford University.

[2] Lee, S., Lele, S. and Moin P., 1990, *Eddy-shocklets in decaying compressible turbulence* , CTR Report No. 110, Stanford university (to be published in Phys. Fluids).

[3] Lumley, J. L., 1970, *Toward a turbulent constitutive relation*, J. Fluid Mech., **41**, 413.

[4] Papamoschou, D. and Roshko A., 1988, *The compressible turbulent shear layer: an experimental study*, J. Fluid Mech. **197**, 453.

[5] Rogers, M. M., Moin, P. and Reynolds, W. C., 1986, Report No. TF-25, Stanford University.

[6] Samimy, M. and Elliott, G. S, 1990, *The effects of compressibility on the characteristics of free shear layers*, AIAA J., **28**, 429.

[7] Sarkar, S., Erlebacher, G., Hussaini, M. Y. and Kreiss, H. O., 1989, ICASE Rep. No. 89-79.

[8] Tavoularis, S., 1985, Phys. Fluids **28**, 999.

[9] Zeman, O., 1990, *Dilatation dissipation: The concept and application in modeling compressible mixing layers*, Phys. Fluids, **A 2**, 178.

[10] Zeman, O., 1990a, *On the decay of compressible isotropic turbulence*, CTR Report No. 115, Stanford University (submitted to Phys. Fluids).

[11] Zeman, O. and Blaisdell G. A., 1990, *New physics and models for compressible turbulence*, in Advances of Turbulence 3, Springer-Verlag, 1991.

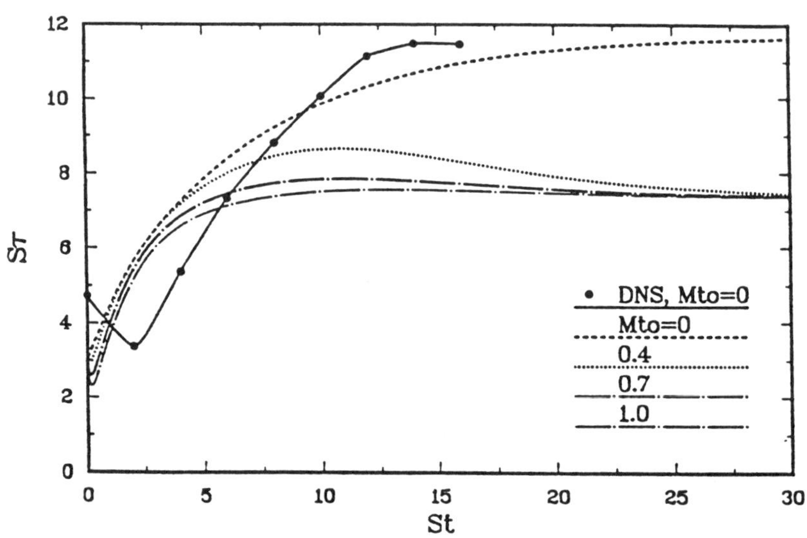

Figure 1. Evolution of normalized turbulent time scale; DNS data are from Rogers et al (1986).

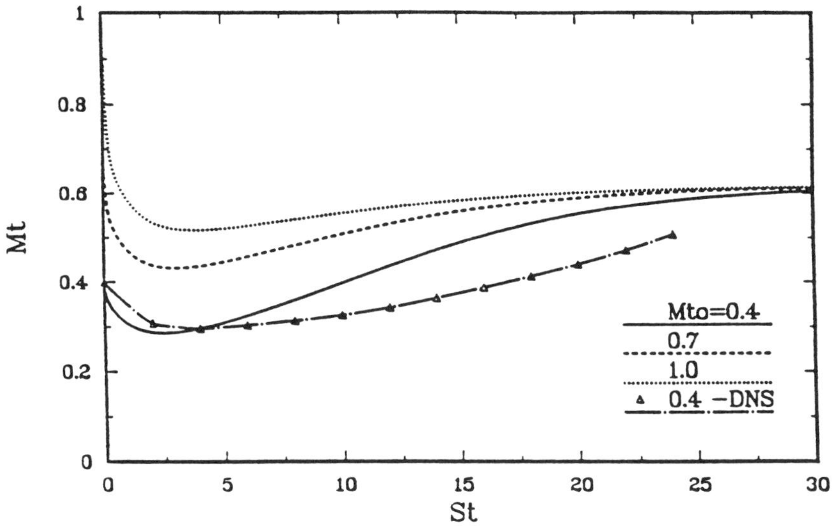

Figure 2. Evolution of the r.m.s. Mach number; DNS data are from Blaisdell (1990).

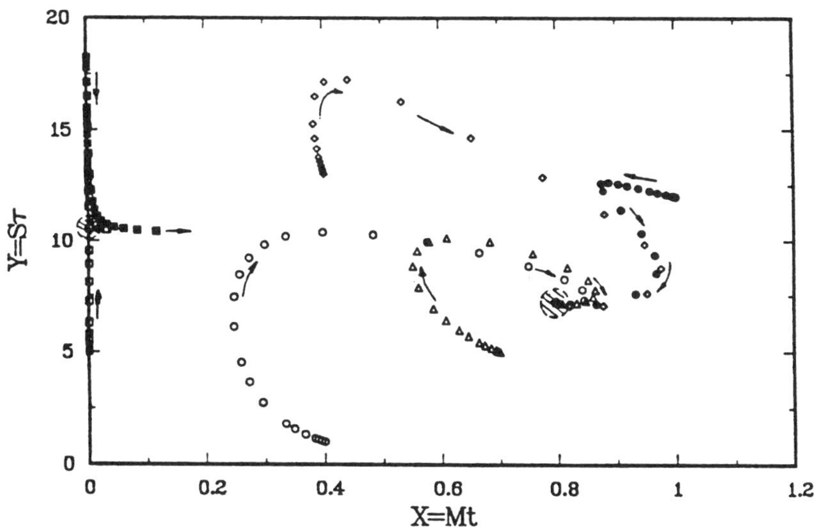

Figure 3. Phase-space trajectories of variables (M_t, $S\tau$) for different initial conditions.

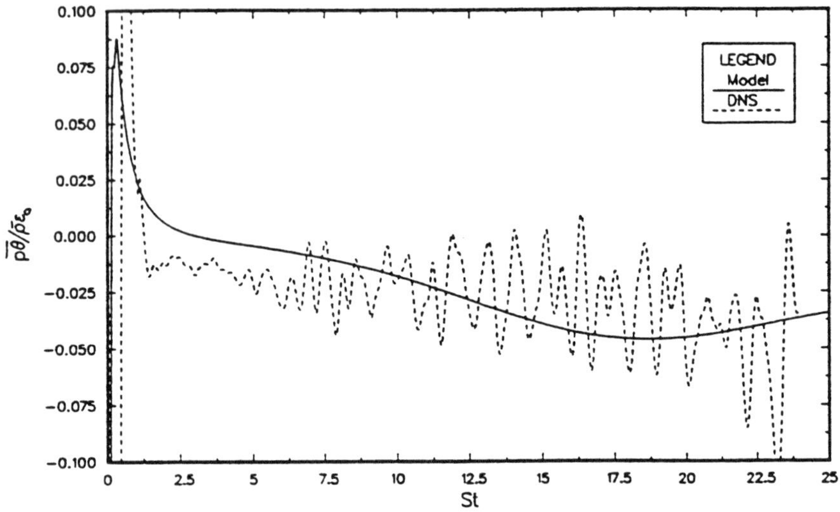

Figure 4. Evolution of the pressure-dilatation term ($\overline{p\theta}/\rho = \Pi_d$); DNS data are from Blaisdell (1990).

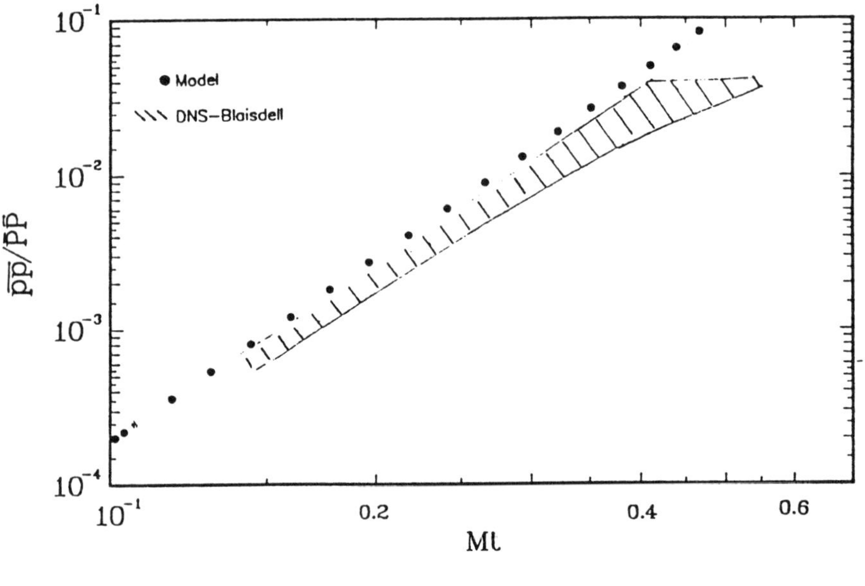

Figure 5. Equilibrium normalized pressure variance vs. M_t.

FLUID DYNAMICS AND NOISE EMISSION ASSOCIATED WITH SUPERSONIC JETS

John M. Seiner

NASA Langley Research Center
Hampton, VA 23665

Abstract

Methods have long been sought to find an efficient means for reduction of jet noise using either active or passive turbulence control measures. Progress in this area is limited by unclear understanding of the physical supersonic jet noise source mechanisms as they relate to the jet plume turbulence structure. These mechanisms have been extensively studied using round jets. This paper shows that jets with non-round jet exit geometry can provide beneficial noise reduction relative to round jets. Both the fluid dynamic structure and noise of several non-round jets are examined in the paper.

1. Introduction

Jet noise represents a formidable technological barrier to the future development of supersonic commercial aircraft. Environmental regulations require suppression methods that are capable of removing 99 percent of the emitted jet engine noise without incurring any significant performance penalties (Seiner and Krejsa 1989). This requirement represents a formidable challenge to the current level of understanding of aerodynamically generated sound by high speed turbulent shear layers.

The acoustic source mechanisms for high speed turbulent shear layers have been studied extensively for both hot and cold supersonic round jets including complex geometry suppressor nozzles (Seiner 1984). Theoretical concepts based on either the *Lighthill Acoustic Analogy Approach* (Ffowcs Williams 1984) or the *Linear Inviscid Instability Wave Model* (Tam 1987) have provided invaluable guidance to the development of empirical jet noise prediction codes (Zorumski 1982). Major differences between the two theoretical models exist in their need to represent turbulence in the jet shear layer. The linear inviscid spatial stability model only accounts for the growth and decay of a deterministic-like large scale structure through solution of the

Rayleigh equation. While the real physical process involves non-linear interaction of various turbulent scales, spatial stability models provide good comparison to round jet data and are easily implemented on a small mini-computer. The analogy approach is fully non-linear, requiring accurate representation of a wide range of turbulence scales. When available, this method will require the use of the latest computational technology.

Application of suppression methods invariably involves the use of non-round complex nozzle geometry for which little understanding exists. The purpose of this paper is to provide some account of present capabilities as they exist for round nozzle geometry and to discuss new measured flow and acoustic properties associated with non-round generic geometries. Both rectangular and elliptic supersonic jet exit geometries are shown to provide lower noise per unit jet mechanical power than round geometry.

2. Mach Wave Radiation

Two fundamental differences exist between low and high speed jet flows that have a profound effect on what is observed as the dominant sound production mechanism. These differences are present when turbulence is convected at supersonic speed relative to the ambient sound speed and/or shock waves are present in the flow. The second phenomenon of course requires the flow to be supersonic, but the first phenomenon also occurs with heated subsonic jets (i.e. $M_e = 0.85$; $T_j = 800°F$). Both mechanisms are known to emit intense noise and attain sufficient energy to be directly visualized by standard optical methods.

By way of illustration, consider the Figure 1 spark schlieren photograph of an overexpanded ($M_j = 1.5$) supersonic nozzle designed for Mach 2 and operating at a jet total temperature of 1370°K. This record, taken with a vertical knife edge and spark duration less than 0.1 μ-sec., captures a nearly instantaneous view of both the flow and near acoustic field. The pressure waves that appear to emanate from regions along the jet shear layer edge are produced by turbulence convecting supersonically relative to the ambient medium. This mechanism was first analytically described by Phillips (1960), who termed the mechanism eddy *Mach wave emission*. Physically the model considers the radiation to be generated in the same manner that ballistic waves are created. The data in Figure 1 reveals the presence of at least two families of waves. The first are high frequency waves located near the nozzle exit. The second are waves with lower frequency, whose axial spacing increases with increasing downstream distance. This latter result is expected, since turbulence length scales are proportional to the increasing mixing layer width.

The Mach wave emission angle, as defined in Figure 1, is determined from

$$\theta = \cos^{-1}(1/M_c) = \cos^{-1}(a_0/\alpha V_j) \tag{1}$$

where $\alpha = V_c/V_j$. Here V_j, V_c, and M_c are the fully expanded jet velocity, turbulence convection velocity and convective Mach number relative to ambient sound speed. For the hot jet example in Figure 1, $V_j = 1120$ m/sec. and $\alpha = 0.61$ based on Tam's (1990) analysis where $T_j/T_a = 4.66$. Equation1 predicts the Mach wave emission angle to be $\theta = 59°$, whereas the data in Figure 1 shows $\theta = 60°$. The sound field displayed in Figure 1 provides a realistic example of what occurs with full scale unsuppressed engine nozzles, like that used in the Concorde aircraft.

Within the framework of the acoustic analogy approach, Ffowcs Williams and Maidanik (1965) theoretically predict that acoustic energy emitted by the eddy Mach wave mechanism is

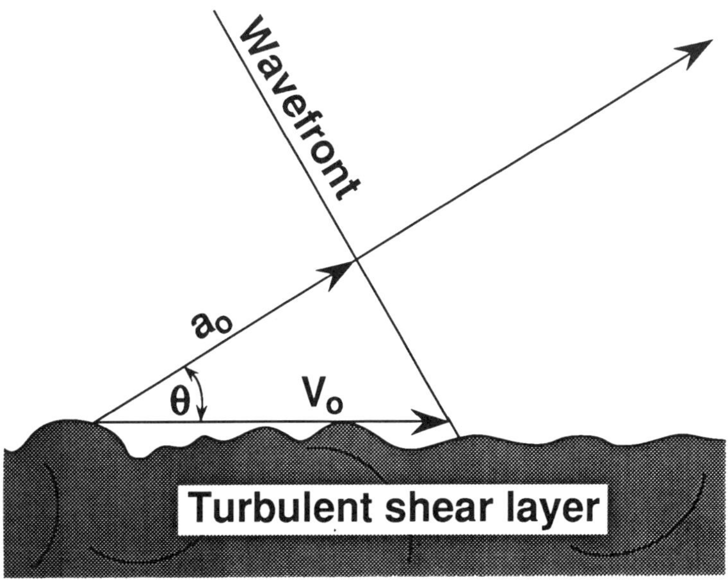

Fig. 1 a. Definition for Mach Wave Emission Angle, θ

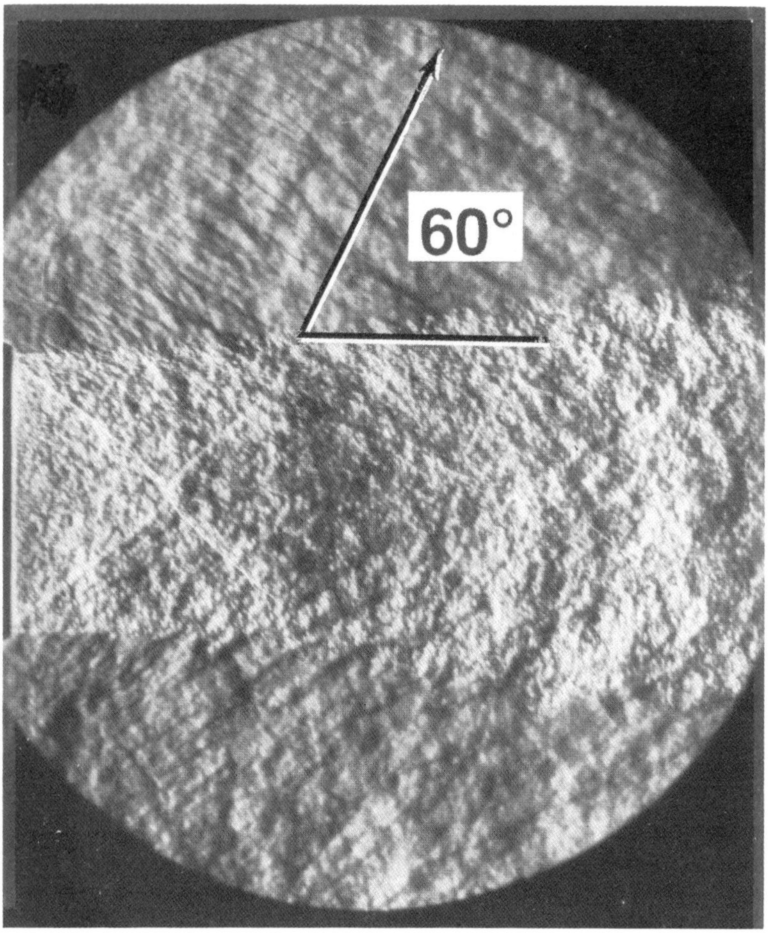

b. Spark Schlieren with Horizontal Knife Edge

Figure 1. Visualization of overexpanded Mach 2 jet at 1370°K.

proportional to M_c^3. This dependence on source strength is indicative of a high acoustic source efficiency. At supersonic speeds and in the direction of Mach wave emission, the fundamental quadrupole sources radiate as individual monopole sources with no net near field cancellation. Thus source efficiency is maximum, producing acoustic energy near 0.1 percent for cold jets and 1.0 percent for hot jets of the jet mechanical energy at narrow angles to the jet axis. At angles beyond the Mach cone, source efficiency drops to that associated with quadrupole emission, which is in the order of 0.01 percent of the jet mechanical power.

The circular arc acoustic array data of Figure 2 show the angular variation of overall acoustic pressure level measured on an arc with a 12 foot radius. The data is associated with operation of a round nozzle at its Mach 2 design point. The angle Ψ is measured relative to the nozzle inlet. The results in Figure 2 illustrate the substantial increase in acoustic energy at narrow angles to the jet axis due to Mach wave emission (between 130 to 180°). Predicted values for overall levels are also shown in Figure 2. These predicted values are derived from the empirical jet noise prediction scheme of the Stone module in the ANOPP code (Zorumski, 1982). Despite the general overall agreement with the data, the Mach wave mechanism is not being adequately represented by the empirical prediction procedure.

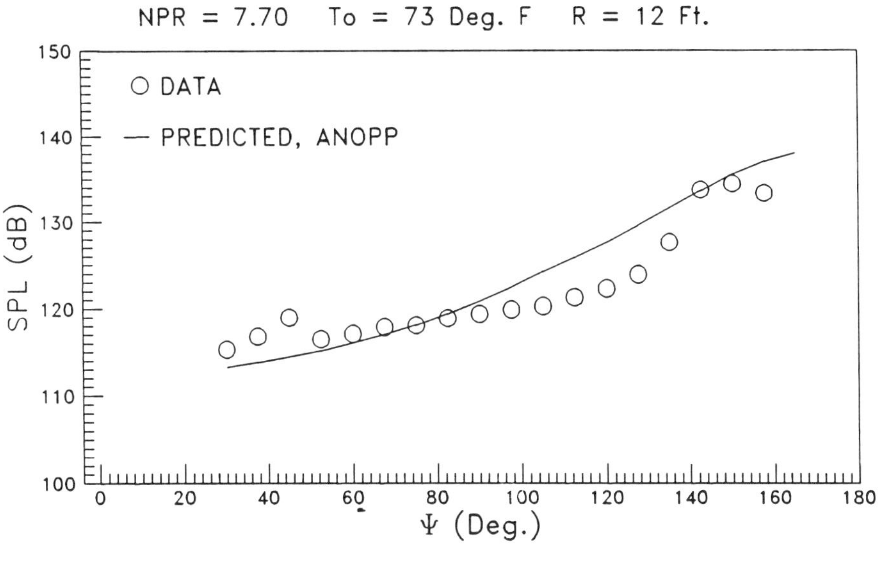

Figure 2. Directional behavior of Mach wave mechanism for a Mach 2 unheated jet.

The noise emitted at narrow angles to the jet axis produces a broadband spectrum with an energy peak at a Strouhal number, $S_t = fD/V_j$. Here f is circular frequency and D is the jet exit diameter. The Strouhal number decreases with increasing jet Mach number, but is independent of the jet Reynolds number as shown in Figure 3. The low Reynolds number narrow band spectral results obtained by McLaughlin, Morrison, and Troutt (1977) for a Mach 2.1 nozzle are compared to that obtained at high Reynolds number at NASA Langley using the same Mach 2.0 nozzle used to acquire the directional

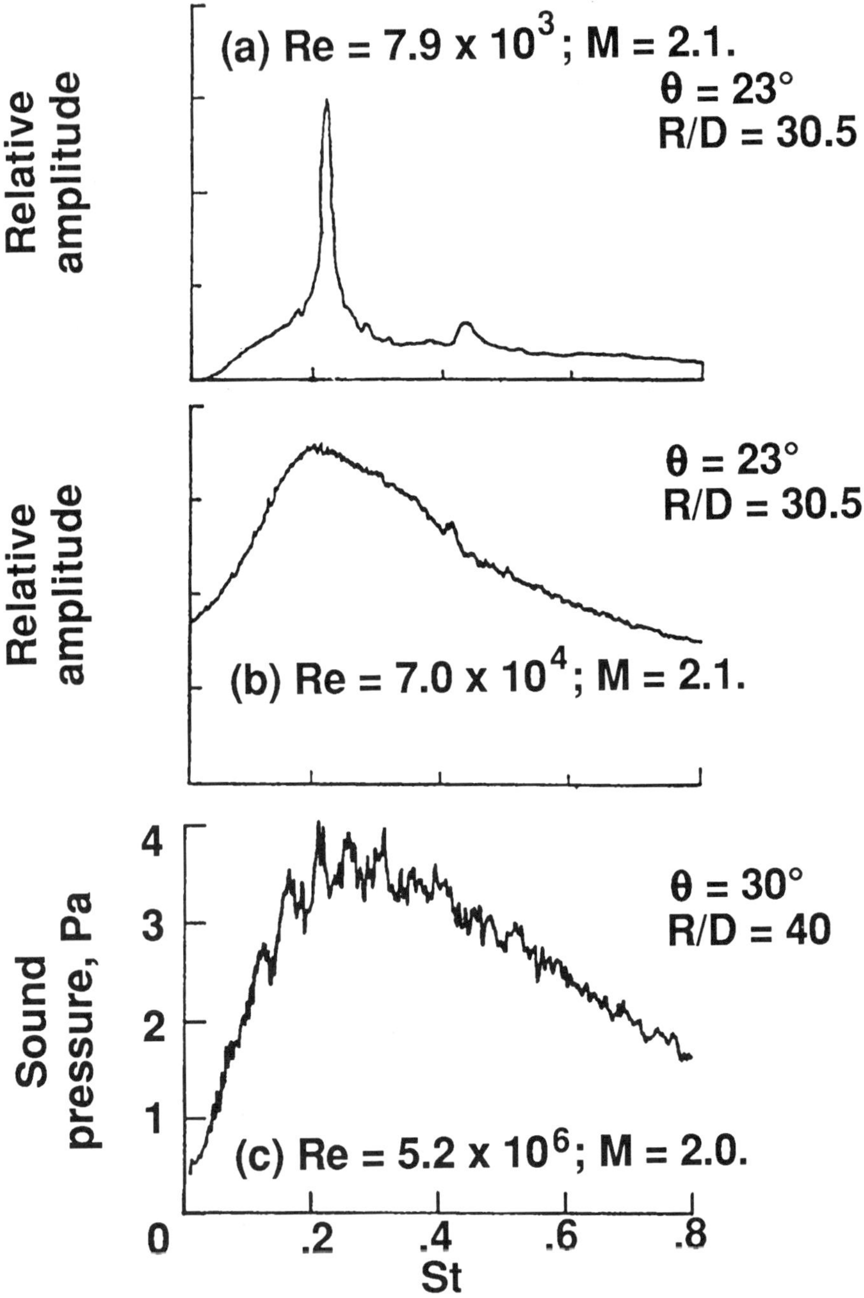

Figure 3. Dependence of acoustic spectrum on jet Reynolds number.

acoustic data described above. All nozzles are operated unheated at their design points. The $Re_d = 7900$ nozzle's flow is transitional, containing only one significant frequency contribution near $S_t = 0.2$. The spectrum width increases with jet Reynolds number, but the energy remains centered near $S_t = 0.2$.

3. Jet Noise Prediction

The low Reynolds number jet, with its wave-like flow field development, represents the cornerstone on which the spatial stability theory, first introduced by Michalke (1965) and later perfected by Tam and Burton (1984), has evolved. This analysis should work best when applied to the initially laminar low Reynolds number jet, and in particular for those undergoing artificial excitation at the preferred Strouhal number. Both the flow and acoustic data of McLaughlin et.al. (1977) were acquired with excitation of the initial jet shear layer. In the laminar region, consideration of a Reynolds stress model is unimportant and flow variables can be decomposed as follows,

$$u_i = \underset{\text{(Mean)}}{\overline{u_i}} + \underset{\text{(Turbulent)}}{u_i'} + \underset{\text{(Deterministic)}}{\tilde{u}_i} \tag{2}$$

Based on this decomposition, the linearized inviscid compressible equations for deterministic instability waves are to first order given by,

$$\tilde{u}_{i,t} + \overline{u}_j \tilde{u}_{i,j} + \tilde{u}_j \overline{u}_{i,j} = -(\overline{\rho})^{-1} P_{,i} \tag{3}$$

$$\tilde{P}_{,t} + \overline{u}_j \tilde{P}_{,j} + M^{-2} \tilde{u}_{j,j} + \gamma P \overline{u}_{j,j} = 0 \tag{4}$$

where any interaction with turbulent structure is considered second order, the mean velocity field is assumed given, and the mean density $\overline{\rho}$ is related to mean velocity $\overline{u_i}$ through Crocco's relation. The instability wave pressure is given by $\tilde{P}$. With a known mean flow, the equations are linear permitting complex wave analysis. Thus, according to Tam and Burton (1984), a fixed frequency wave of azimuthal order m can be expressed as

$$\tilde{P}(r,\Phi,x,t) = \int_{-\infty}^{\infty} \sum_{m=0}^{\infty} \delta_m(\varepsilon)\hat{P}_m(r,\varepsilon x)$$

$$X \exp\left\{ i \left[\int_0^x \alpha_m(\varepsilon x)dx + m\phi - \omega t \right] \right\} d\omega \tag{5}$$

where r, ϕ, x are cylindrical coordinates. In Equation 5, the fixed frequency wave ω is real, the wavenumber α_m is complex, and ε is a parameter that permits slow axial growth of the wave. The eigenfunctions $\hat{P}_m(r,\varepsilon x)$ in Equation 5 satisfy the compressible Rayleigh equation which is obtained from combination of Equations 3 and 4. Substitution of Equation 5 into the Rayleigh equation produces the following equation for a fixed frequency wave:

$$\frac{d^2\hat{P}_m}{dr^2} + \left[\frac{1}{r} - \frac{1}{\bar{\rho}}\frac{d}{dr} + \frac{2\alpha_m}{\omega - \bar{u}\alpha_m}\frac{d\bar{u}}{dr} \right] \frac{d\hat{P}_m}{dr}$$

$$+ \left[\frac{(\omega - \bar{u}\alpha_m)}{\bar{a}^2} - \frac{m^2}{r^2} - \alpha_m^2 \right] \hat{P}_m = 0 \tag{6}$$

The functions $\delta_m(\varepsilon)$ in eq. 5 are used to match the eigenfunctions $\hat{P}_m(r,\varepsilon x)$ to a far field solution using the method of matched asymptotic expansions between an inner solution of the Rayleigh problem discussed above and an outer solution that satisfies the radiation boundary condition.

The near field acoustic radiation for the unheated Mach 2 jet operating fully expanded was computed using the LSNOIS code (Morris and Tam, 1979) for several fixed frequency waves. This code is based on the spatial stability analysis discussed above. Calculations were performed for both varicose m=0 and helical m=1 azimuthal modes using the half-Gaussian mean velocity profile,

$$\bar{u}/\bar{u}_c = \begin{cases} 1 & r \leq h \\ \exp\left(-\ln 2\left(\eta^2\right)\right) & r \geq h \end{cases} \tag{7}$$

where u_c is the centerline velocity and η is a non-dimensional radial coordinate given by,

$$\eta = (r - h) / b \tag{8}$$

where b is the velocity half width of the mixing layer and h is the radius of the potential core. Designating $r_{.5}$ as the radial distance to where the velocity is half the centerline value, the velocity half width becomes $b = r_{.5} - h$. Both h and $r_{.5}$ were determined from the measured mean velocity profiles shown in Figure 4 which cover a range from $0.9 \leq X/D \leq 33$. A comparison between the predicted and measured 1/3 octave near field acoustic pressure for the $S_t = 0.2$ frequency wave is shown by the contour maps of Figure 5 for the m = 1 helical mode. The

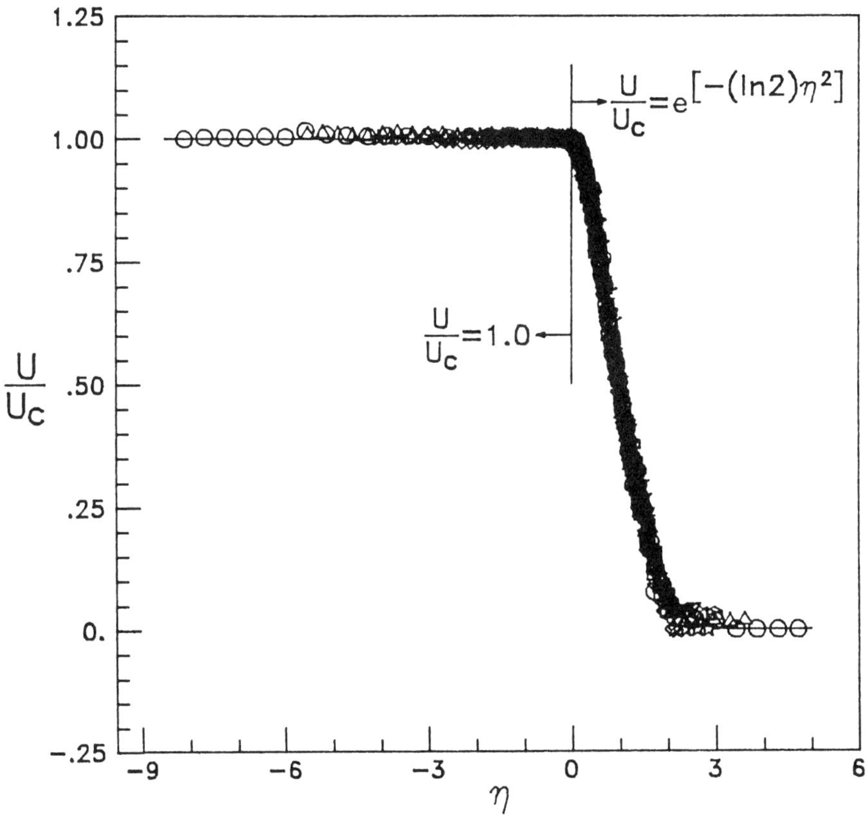

Figure 4. Normalized mean axial velocity profiles.

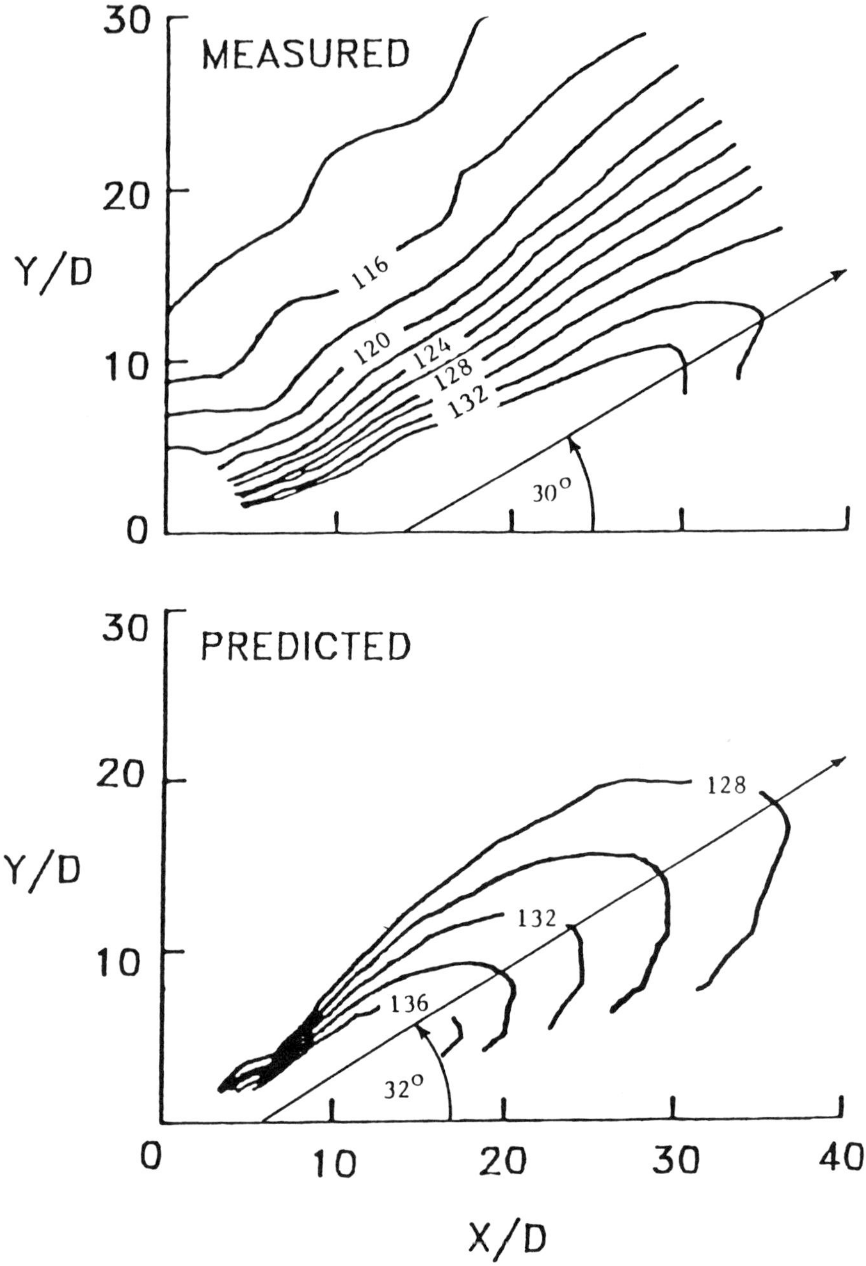

Figure 5. Predicted and measured nearfield pressure contours for the $S_t = 0.2$ component using the LSNOIS code.

calculations for the m = 0 varicose mode indicated a much smaller amplified wave. The initial wave amplitude is adjusted to enable scaling of the computed results to measured overall amplitude levels. In general there is good qualitative agreement between the measured and computed contours. The axial location for sound emission for the $S_t = 0.2$ frequency differs between measurement and prediction. The measured contours indicate that intense sound emission originates from axial locations further downstream then predicted.

4. Shock Generated Noise

Improperly expanded nozzles, a likely event for complex nozzle geometries, produce shock noise that dominates acoustic emission in the jet forward quadrant ($0 \leq \psi \leq 90$). A comprehensive review of the literature on this subject can be found in Seiner (1984). For purposes of this paper only a few aspects of this source mechanism need be discussed.

Shock noise is produced by the interaction of turbulence as it is convected through shock waves in the jet plume. The interaction gives rise to a broadband frequency spectrum with an energy peak that is Doppler dependent. When the shock cell spacing is near an integer multiple of the wavelength for the most highly amplified fixed frequency wave of the shear layer, the shock cells will oscillate at that frequency producing what is known as jet screech (Seiner and Yu, 1984). This mechanism requires feedback of acoustically generated noise to the nozzle exit to be maintained.

An illustrative narrow band shock noise spectrum is shown in Figure 6 at the peak emission angle for shock generated noise. This example is associated with the Mach 2 nozzle previously discussed, but operating unheated and overexpanded at a fully expanded Mach number $M_j = 1.5$. The advantage of using a nozzle designed for shock free flow is shown in Figure 7. As is evident, shock noise contributions quickly surpass those of jet noise in the forward arc with only modest departure from the nozzle design point A. The shock noise intensity is found to be proportional to $I_s \approx (M_j^2 - M_d^2)^2$, which is a measure of the strength of the oblique shocks in the plume. The shock noise intensity continues to follow this proportionally until formation of a strong shock (i.e. Mach disc) in the plume at points C and B.

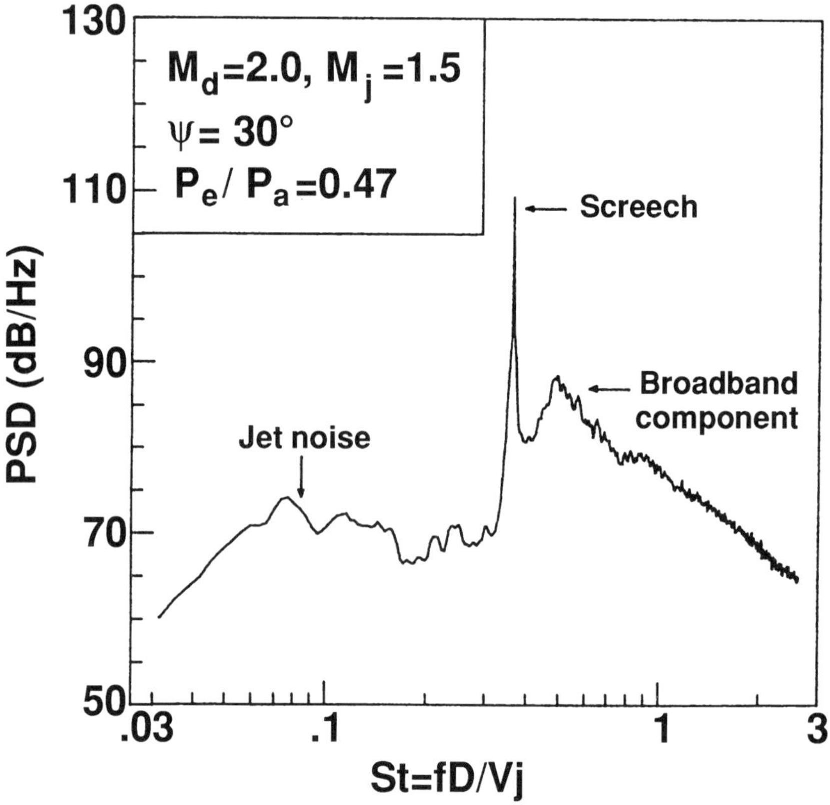

Figure 6. Typical narrowband spectrum illustrating shock noise features.

Figure 8 is a spark schlieren photograph that, in addition to illustrating Mach wave emission, also shows the locations for emission of broadband shock noise. From Figure 8, it can be observed that broadband shock noise is emitted from the supersonic plume at those locations where oblique shocks terminate at the sonic line. The shock noise appears as a set of concentric rings centered at these shock locations and superimposed with the Mach wave emission. In this example jet screech is not present. Jet screech is inhibited by the presence of the large Mach disc (I.e. normal or shock) evident in the Figure 8 photograph. Evidently the stability characteristics of the plume are altered by the presence of mixed subsonic pockets along the jet centerline.

The screech frequency can be used to investigate the spatial modes of stability of the jet shear layer. This can be

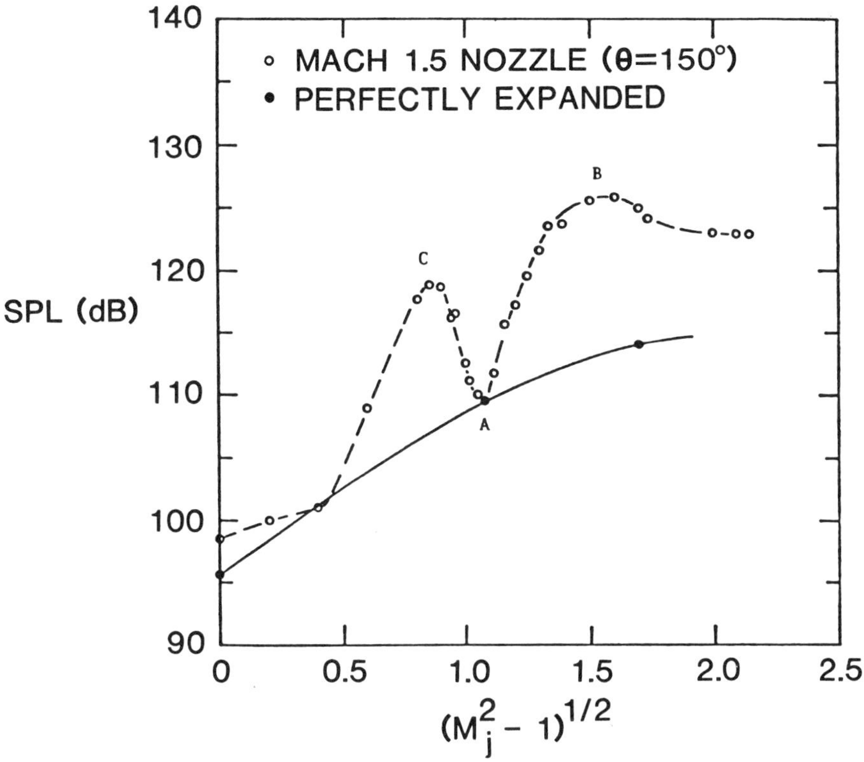

Figure 7. Advantage of shock-free nozzle design.

accomplished by using a nano-second duration sub-millimeter phase fired spark-gap source in a schlieren optics assembly. The screech frequency, like that shown in Figure 6, is used to provide the phase reference for firing the spark-gap at the screech frequency or a sub-multiple. Thus only those turbulent structures that are phase coherent with the screech frequency will be visualized. The phase averaged schlieren records (32 averages) of Figure 9 show representative examples for the varicose (9a) and helical (9b) azimuthal modes. These flow visualization results clearly show the presence of large scale wave-like turbulent structures. The shock cell spacing associated with the helical instability is nearly one half that of the large scale wave. Shock cells are seen to follow the large scale wave.

The helical instability for round jets can be found to be purely spinning or in a flapping about some temporary plane of reference. The flapping plane can become fixed by nozzle geometry as shown by the twin jet phase averaged schlieren

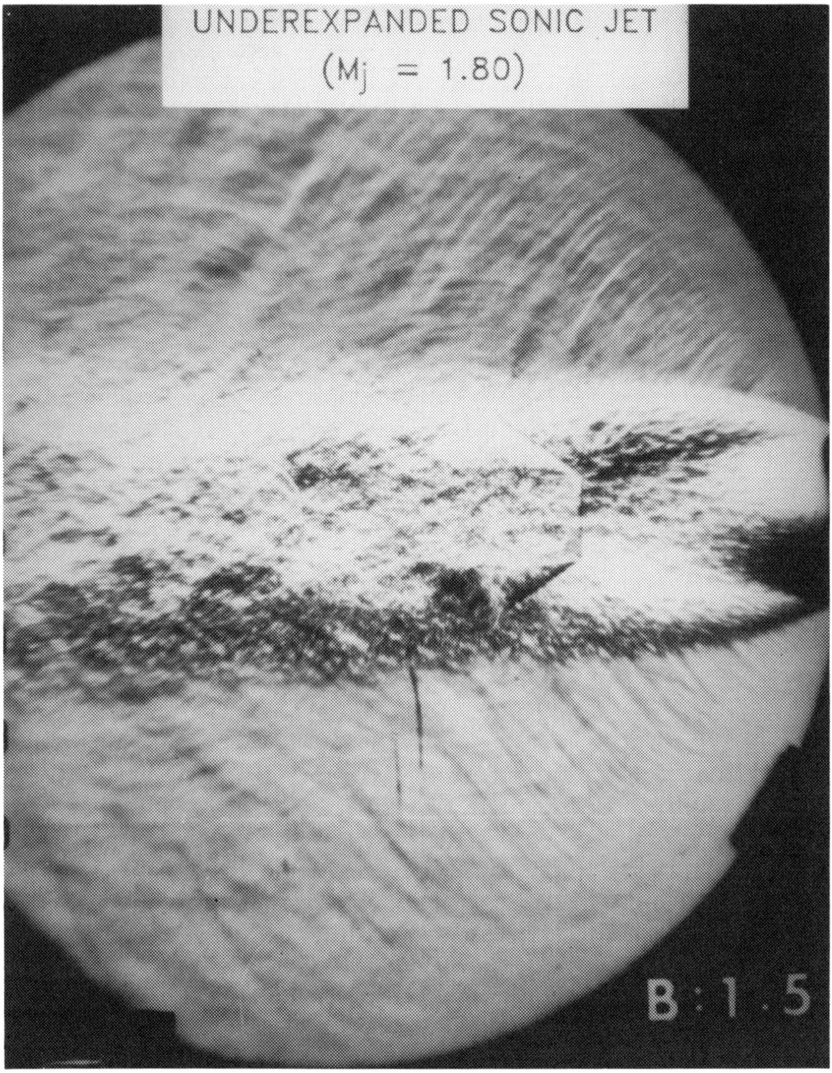

Figure 8. Spark Schlieren of underexpanded jet illustrating broadband shock noise and Mach wave emission.

data of Seiner, Manning, and Ponton (1988). Their results show that, for closely spaced nozzles, the helical structure of the jet is coupled producing intense sound emission and large growth to each jet's mixing layer. A representative example is shown in the Figure 10 phase averaged schlieren record.

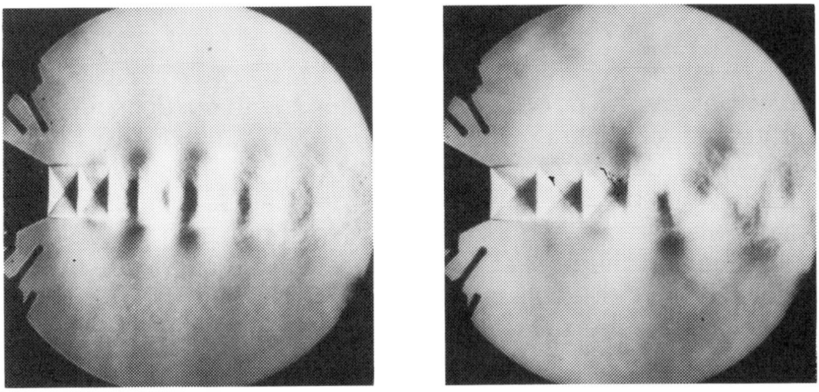

a. $M_j = 1.11$ Varicose Instability b. $M_j = 1.22$ Helical Instability

Figure 9. Phase averaged Schlieren of underexpanded jet from sonic convergent nozzle.

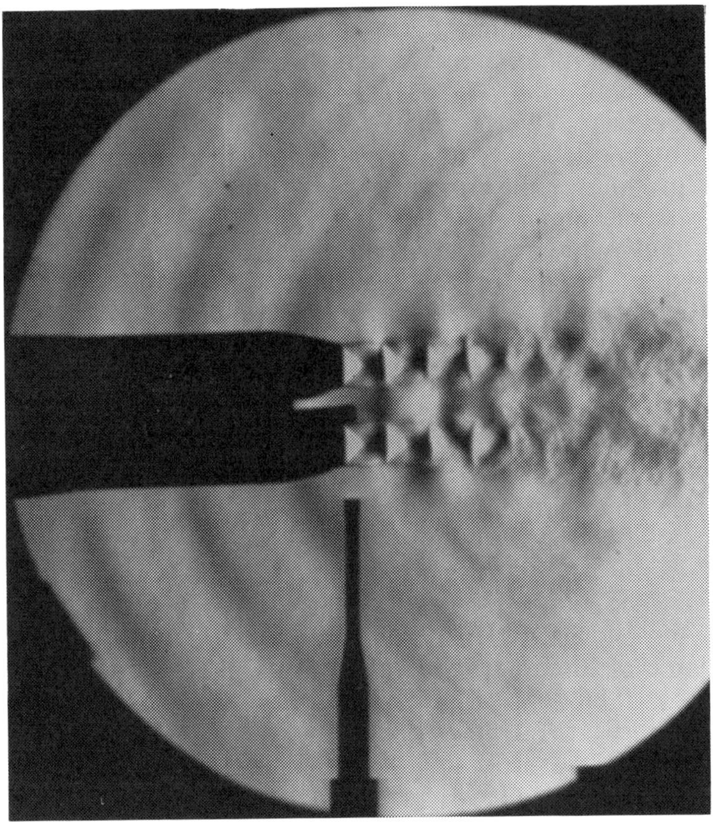

Figure 10. Phase averaged Schlieren of underexpanded twin sonic nozzles illustrating plume coupling.

5. Rectangular Nozzles

In the above, we have seen that Mach wave emission and shock generated noise are the dominant acoustic sources of supersonic jet noise. Shock noise can be minimized through appropriate design of nozzle geometry. For round laboratory type nozzles, this is of course straight forward as illustrated by the results in Figure 7. However, the geometric tolerance requirements resulting from such a design procedure cannot be transferred to full scale aircraft. As discussed by Seiner and Kresja (1989), the axisymmetric plug nozzle with porous walls exhibits good shock noise control even when constrained by full scale design procedures. This nozzle also reduces noise radiated to the jet rear arc including Mach wave emission. The level of noise reduction achieved is nevertheless well below needed requirements.

Reduction of Mach wave emission represents the most serious challenge to the successful design of a suppressor nozzle. This is because it is unclear how to accomplish this without incurring substantial installed performance losses or adding additional unknown noise sources. Shock generated noise reduction procedures, in contrast, present formidable yet clear objectives.

The reduction of Mach wave emission certainly involves finding methods that reduce the region in a jet plume with supersonically convecting turbulence. In general this may be thought synonymous with finding methods that produce a more rapid mixing of the turbulent jet shear layer with the surrounding medium.

Figure 11 provides a comparison of the difference in mixing capabilities between a round convergent and aspect ratio 4 convergent rectangular nozzle. Here the centerline velocity profiles are normalized by the jet exit velocity. The data indicates that a rectangular nozzle can mix much faster with the surrounding medium than a round nozzle. In Figure 11, the axial coordinate is normalized by an area equivalent diameter for each nozzle. Both the round and rectangular nozzle have equivalent exit area. The increase in mixing with the rectangular nozzle can possibly be explained by the vortex dynamics associated with the asymmetric distortions produced by three dimensional jet mixing as discussed, for example, by Ho and Gutmark (1987).

Based on the flow results of Figure 11, far field noise measurements were obtained for a series of supersonic round and rectangular nozzles with various exit design Mach numbers and aspect ratios. The nozzles were operated unheated at their

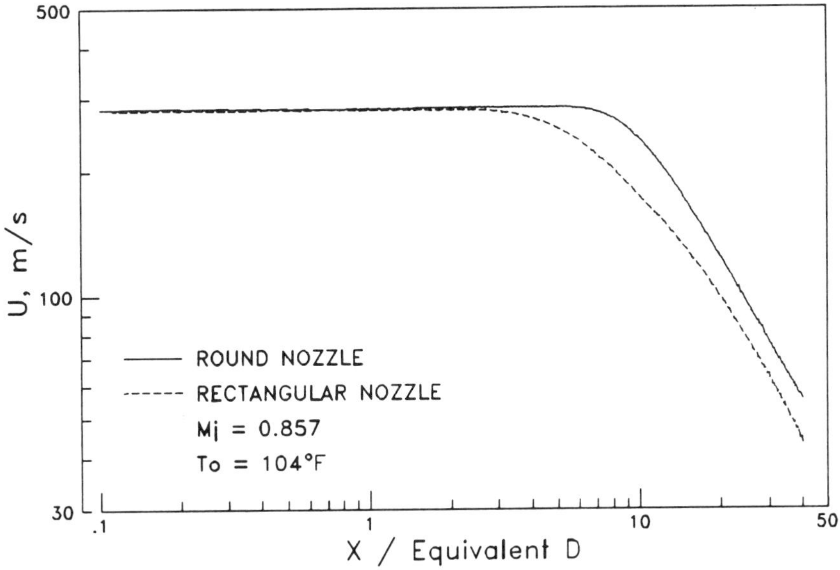

Figure 11. Comparison of centerline velocity decay between round and rectangular nozzle flow fields.

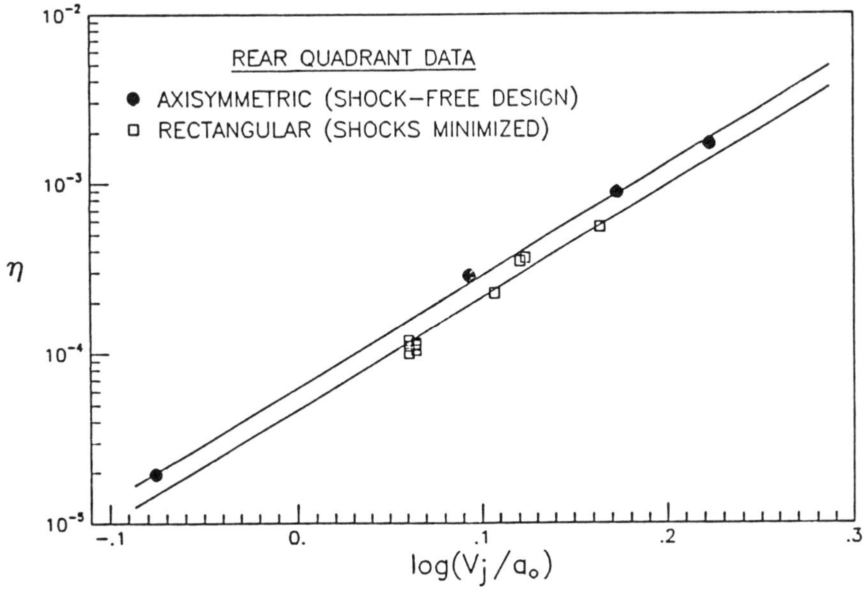

Figure 12. Acoustic efficiency of round and rectangular nozzles.

design point minimizing the influence of shock generated noise. The rear quadrant (i.e. region for Mach wave emission) noise power for each nozzle was determined by integrating the measured sound field over the surface of an imaginary hemisphere surrounding the rear quadrant of the jet. Performance data was obtained for each nozzle and the total mechanical energy of the jet flow was determined.

From these measurements, the acoustic efficiency, which is the ratio of the acoustic to jet mechanical power, was determined for each nozzle. The results are presented in Figure 12. The acoustic efficiency is shown to increase with jet velocity, but a lower overall efficiency can be associated with the rectangular nozzles. This result is most encouraging considering that the rectangular nozzles used in this study also produce shock noise contributions in the rear arc. The round nozzles have no detectable shock noise in this quadrant.

6. Elliptic Supersonic Nozzle

Based on the above supersonic rectangular nozzle result, the low subsonic elliptic nozzle studies of Ho and Gutmark (1985) grew in significance. In their work, they found that a low aspect ratio elliptic nozzle could entrain three times the jet's own mass flow. At the same time, recent developments in the spatial stability analysis of elliptic supersonic nozzles by Morris (1988) and Koshigoe and Tubis (1987) provided encouragement to study both the flow and sound fields of these elliptic jets. Since spatial stability theory is only valid for jets that are shock-free, it was at first necessary to develop a procedure for designing a shock free elliptic nozzle. Seiner, Baty, and Kumar (1990) have managed to develop such a procedure. In their work, a Mach 1.52 aspect ratio elliptic nozzle with exit aspect ratio of 2 was designed, constructed, and found to achieve a shock-free plume at its fully expanded design point. The nozzle was constructed with a major radius of 2.540 cm., producing an area equivalent diameter of 3.592 cm.

The importance of achieving a successful design procedure for shock-free elliptic nozzles is apparent from the phase averaged schlieren data of Figure 13. From these optical records, comparison can be made of the difference in jet spread rate between fully expanded and overexpanded operation. At the fully expanded condition, the major and minor axis mixing layers appear to experience similar growth rates. At the overexpanded condition, the jet's minor axis growth rate is

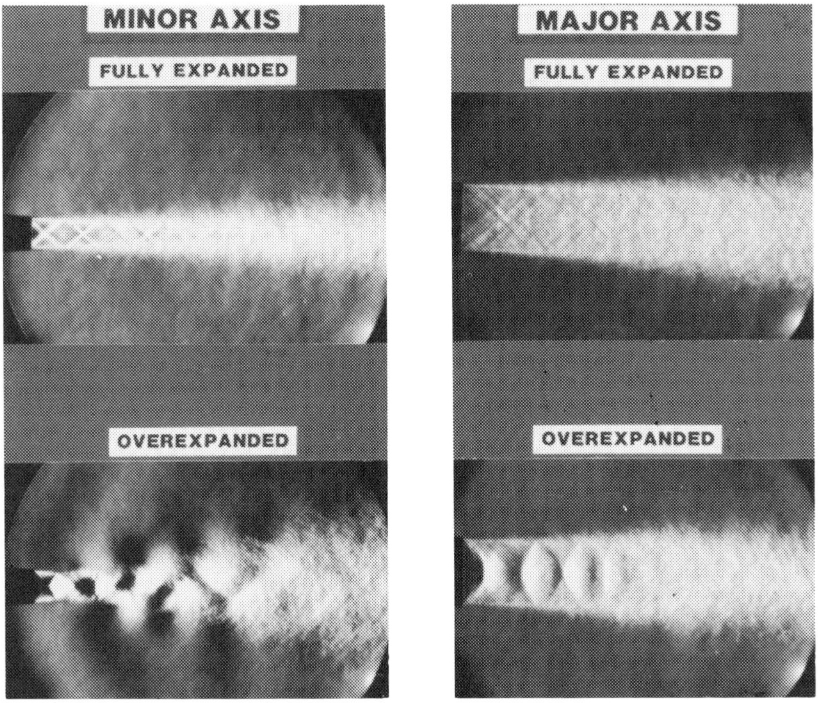

Figure 13. Effect of shocks on evolving elliptic jet turbulent flow structure.

greatly enhanced, whereas the major axis growth rate may even decrease. These results clearly signify that with shocks in the plume, large modifications to the growth of the large scale structure will occur. Since growth rate modifications are dependent on the operating nozzle pressure ratio, no rational basis would exist for conducting a stability analysis of these three dimensional plumes. The design of a shock-free elliptic nozzle is essential.

The mixing effectiveness of the supersonic elliptic nozzle is shown in Figure 14. The data shows that the elliptic jet's centerline axial velocity is nearly equivalent to that for a subsonic round $M_j = 0.857$ jet. In this comparison, the $M_d = 1.52$ elliptic nozzle is operating fully expanded at $M_j = 1.52$.

The supersonic elliptic nozzle's mixing layer has a larger growth rate than the round jet, for according to the empirical formula of Nagamatsu and Horvay (1970), a round Mach 1.52 nozzle would have a potential core length of 7.8 nozzle diameters. The data of Figure 14 indicate that the elliptic jet's potential core length is 5.2 equivalent jet diameters. Thus

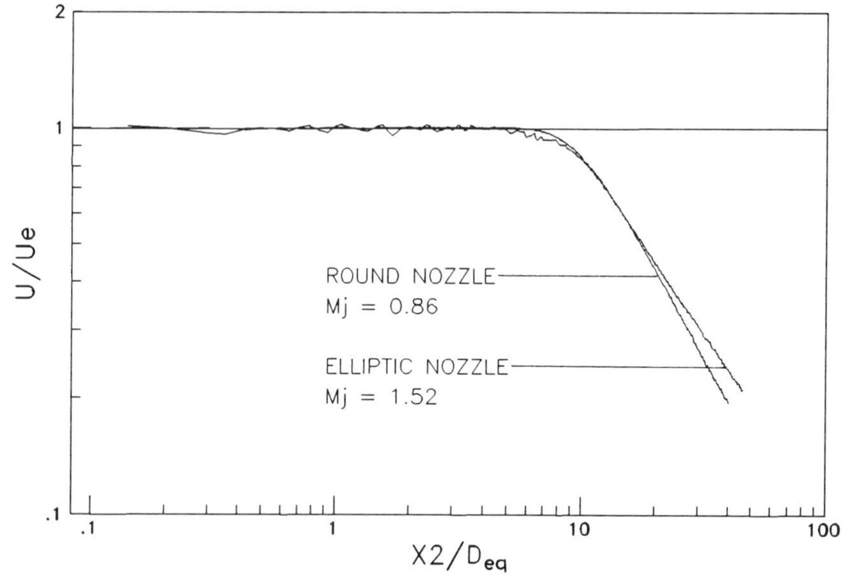

Figure 14. Comparison of centerline velocity decay between round and elliptic nozzle flow fields.

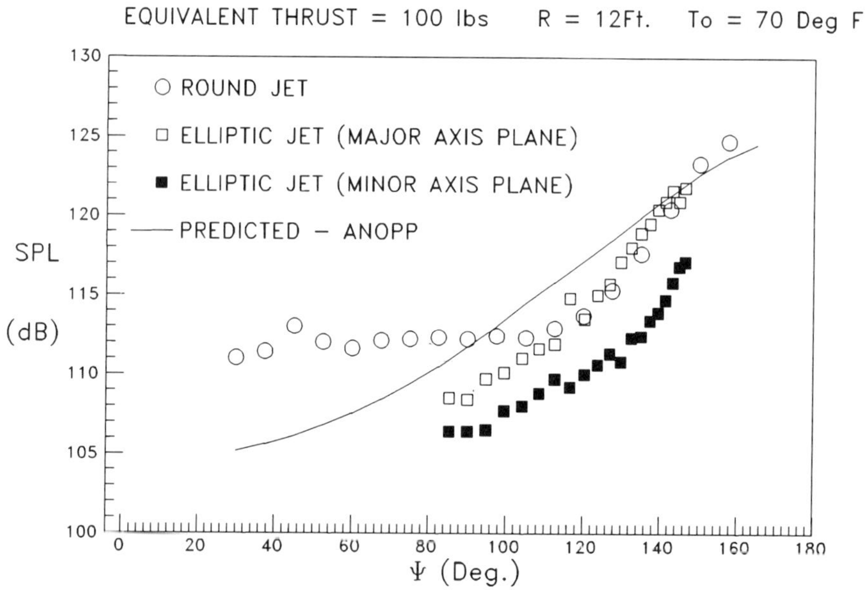

Figure 15. Comparison of round and elliptic Mach 1.5 nozzle acoustic far fields.

compared to a round jet, the axial extent of the supersonically convecting turbulence should be reduced relative to a round nozzle.

Figure 15 shows a comparison of the far acoustic field data between the elliptic nozzle and an equivalent round supersonic nozzle. The round nozzle was designed to be fully expanded at $M_d = 1.5$. Both nozzles in this comparison have equivalent ideal thrust of 100 lbs. The data indicates that the elliptic nozzle in general produces less noise per unit thrust than round geometry nozzles. The noise reduction is greatest along the jet's minor axis plane. Even though the observed noise reduction in Figure 15 is modest, it is important to understand the nature of how it occurs through the enhanced growth rates of the shear layer.

The initial distribution of momentum thickness has been used in the past by Ho and Gutmark (1987) and Bridges and Hussain (1989) as a length scale for the initial turbulent shear layer. The azimuthal distribution of the momentum thickness around the periphery of the nozzle exit controls the initial role-up of shed vorticity and the eventual three dimensional distortion of the jet plume with downstream distance.

Mean radial velocity profiles were measured at various axial stations across both major and minor axis planes. This was accomplished using both total and supersonic pressure tubes. Following Crocco, the jet total temperature across the shear layer was assumed constant. The momentum thickness was then obtained for both the major and minor axis by numerical integration of the data according to

$$\theta = \int_0^\infty \frac{\rho V}{(\rho V)_0}\left(1 - \frac{V}{V_0}\right) dr \tag{9}$$

where in Equation 9, ρV is the axial mass flux and $(\rho V)_o$ is the centerline reference value. Figure 16 shows the results of applying Equation 9 to the velocity data to obtain the axial dependence of the momentum thickness. Both the momentum thickness θ and axial distance X are normalized by the equivalent area diameter.

The data in Figure 16 shows that, at the nozzle exit, the momentum thickness is nearly equal along the major and minor axis, and remains so almost to the end of the potential core. Recall from Figure 14 that the length of the potential core is 5.2 D_{eq}. Beyond the potential core region of the jet, the major axis

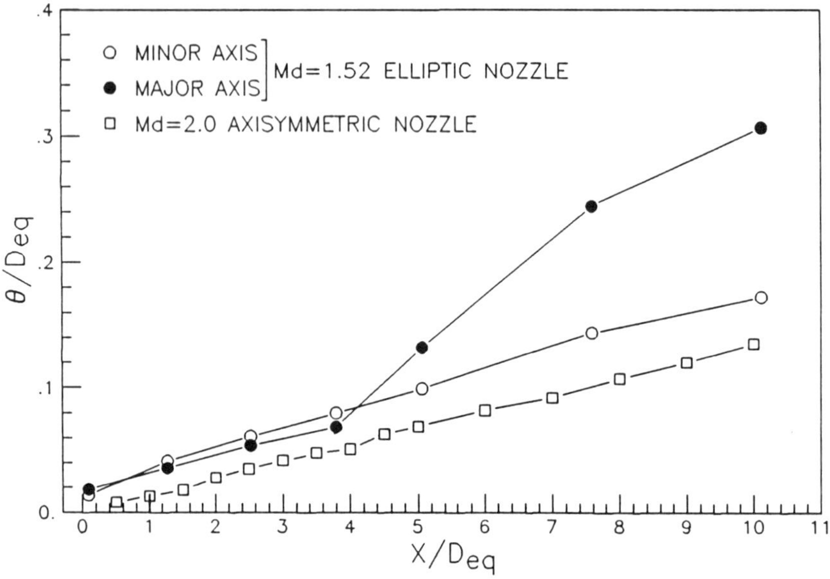

Figure 16. Growth of shear layer momentum thickness for round and elliptic nozzle flow fields.

momentum thickness increases much faster than the minor axis momentum thickness. This three dimensional distortion is unexpected based on the results reported by Ho and Gutmark (1985) for their low speed aspect ratio 2 elliptic nozzle. They reported a similar behavior for the momentum thickness in the potential core region, but, with their nozzle the minor axis momentum thickness increases at a much accelerated rate. As a means of reference, the axial variation of momentum thickness for a round supersonic nozzle is also shown in Figure 16. Even though the round nozzle is designed for Mach 2, there is general agreement with the behavior found along the elliptic jet's minor axis.

As an aid to the stability analysis by Baty and Morris (1989) for this nozzle, a 12 element near field microphone array was arranged around the nozzle exit. The microphones were placed along an elliptic cylindrical coordinate and spaced with equal arc distance as shown in Figure 17. With this array, it is possible to determine for a given frequency component the azimuthal content for the large scale waves. Since little distortion of the jet column occurs in the potential core region, it was assumed that noise radiated to the nozzle exit from upstream would lie along contours in the elliptic cylindrical coordinate system. The noise in a band around $S_t = .16$ was filtered and the cross correlation coefficient computed between microphone pairs using microphone 4 as reference. The resulting azimuthal distribution shown in Figure 18 was then Fourier analyzed into fundamental spatial modes. The table indicated on the figure provides a summary of the Fourier analysis. The component at

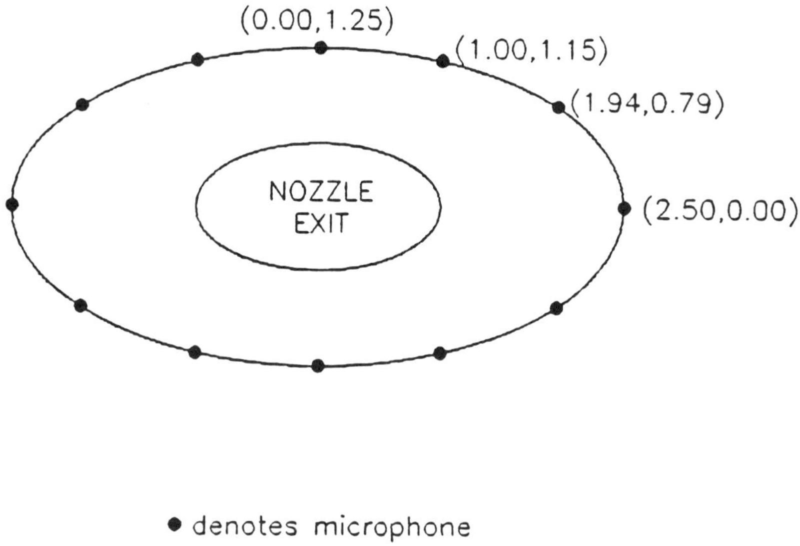

Figure 17. Microphone array along an elliptical cylindrical coordinate at nozzle exit.

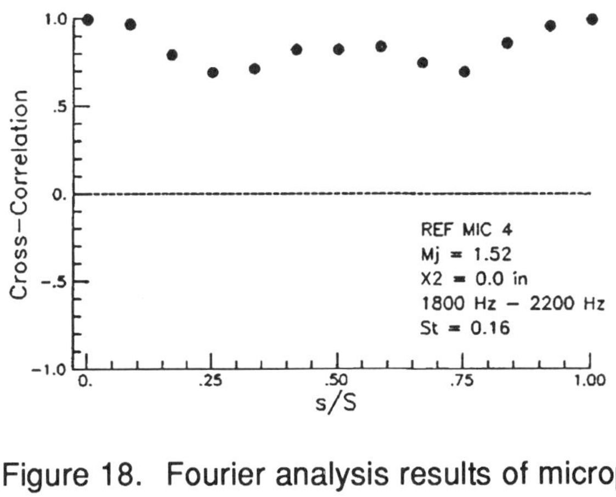

Mode Coefficients

A0 =	0.82408
A1 =	0.07940
A2 =	0.05667
A3 =	0.00100
A4 =	0.00467
A5 =	−0.01240
B1 =	0.02773
B2 =	0.09613
B3 =	−0.00383
B4 =	−0.02569
B5 =	−0.00806

Figure 18. Fourier analysis results of microphone array data.

$S_t = 0.16$, which can be identified with the peak energy radiated, is primarily a varicose mode as evidenced by the mode coefficient $A_0 = 0.824$ amplitude.

The results in Figure 16 suggest that near the end of the potential core the varicose mode becomes rapidly distorted generating significant jet noise. Recall in this region the major axis begins to grow at a much faster rate.

7. Discussion and Conclusions

In this paper, the two major acoustic source mechanisms for supersonic jet noise have been discussed. Mach wave emission, shown in the Figure 1 schlieren, dominates noise radiation at narrow angles to the jet axis. This source mechanism is associated with direct emission of sound by turbulent shear layer structures convecting supersonically relative to the ambient sound speed. The second acoustic source mechanism is associated with the generation of sound when turbulence is convected through shocks in the plume. This mechanism, shown in the Figure 8 spark schlieren and phase averaged schlieren of Figures 8, 9, and 10, can be reduced or eliminated through proper design of nozzle wall geometry. This represents a formidable challenge when applied to the design of full scale aircraft engines.

Large scale wave-like structures were found to be present in supersonic plumes. Only low order azimuthal modes, such as the varicose and first order helical modes, appear to dominate the flow. The compressible Rayleigh equation governs the linear spatial stability of these wave-like large scale turbulent structures. Solution of the Rayleigh problem for wave amplitude, along with application of the method of matched asymptotic expansions, is shown to provide a reasonable agreement between measured and predicted acoustic emission from a round supersonic jet, as shown by Figure 5.

Despite this success, major difficulties exist with application of spatial stability theory. First, application of the theory requires input of the mean jet flow field. For round unheated supersonic jets this information is readily available, but hot round jet data is scarce. The computed large scale wave amplitude is found to be extremely sensitive to the method used to represent the mean flow field. In the present application, the half-Gaussian is used to input mean velocity data, although the hyperbolic tangent profile is known to have smoother second derivative properties than the half-Gaussian.

The second difficulty with spatial stability theory is that it is not straight forward to extend the method to jets with arbitrarily shaped jet plume cross section. The acoustic data of this paper provide strong motivation to extend the method to jet nozzles with non-round exit geometry. Both rectangular and elliptic supersonic nozzles were found to produce less noise per unit mechanical energy than the round jet nozzle. The results of this paper indicate that the elliptic jet turbulent shear layer spreads faster than that from a round nozzle reducing the region for Mach wave emission and consequently noise. For the elliptic jet without plume shocks, the varicose mode was found to be the preferred spatial structure.

When shocks are present in the plume, a more likely event with non-round jet exit geometry, spatial stability theory is not strictly applicable. Only weak shock waves can be considered, and these can only provide perturbations to the flow that are of second order. The application of CFD to assist the empirical prediction of noise from shock containing jets by Seiner, Dash, and Wolf (1985) has proven useful, but new computational methods are needed to extend to the complex three dimensional shear layer of non-round jet nozzles.

In a recent paper, Dash, Sinha, York, and Lee (1990) discuss requirements necessary to apply the Reynolds Averaged Navier-Stokes equations in a time-asymptotic manner to solve these complex flow fields. The principle conclusion of their study is the need for the development of compressible turbulence models that are physically realizable. Such models, like that proposed by Sarkar, Erlebacher, Hussaini, and Kreiss (1989), are able to capture the growth rate of the high speed shear layer by accounting for compressible dissipation. The most important feature required of a model is that it be able to realistically account for the mechanics of shock turbulence interaction in the shear layer. Numerical accuracy is required in treating the reflection of shock waves back into the shear layer at the sonic line. Both the amplitude and frequency for shock related noise depend on the accuracy of this method.

Acknowledgements

The help of Michael Ponton of NASA/LaRC and Bernie Jansen of Lockheed E & S Company is greatly appreciated for their assistance in acquiring data for the elliptic and rectangular jets.

References

Baty, R. S. and Morris, P. J., 1989, Instability of Jets of Arbitrary Geometry, AIAA Paper 89-1796.

Bridges, J. E. and Hussain, A. K. M. F., 1987, Roles of Initial Condition and Vortex Pairing in Jet Noise, J. Sound and Vib., Vol. 117, pp. 289-311.

Dash, S. M., Sinha, N., York, B. J., and Lee, R. A., 1990, Progress in the Development of Advanced Computational models for the Analysis of Generalized Supersonic jet Flowfields, AIAA Paper No. 90-3915.

Ffowcs Williams, J. E., and Maidanik, G. 1965, The Mach Wave Field Radiated by Supersonic Turbulent Shear Flows, J. Fluid Mech., Vol. 21, Pt. 4, pp. 641- 657.

Ffowcs Williams, J. E. 1984, The Acoustic Analogy - Thirty Years On, IMA J. of Applied Math., Vol 32, pp. 113-124.

Ho, C. M. and Gutmark, E. 1987, Vortex Induction and Mass Entrainment in a Small Aspect Ratio Elliptic Jet, J. Fluid Mech., Vol 179, pp. 383-405.

Koshigoe, S. and Tubis, A. 1987, Wave Structure in Jets of Arbitrary Shape. II. Application of a Generalized Shooting Method to Linear Instability Analysis, Physics of Fluids, Vol. 30, pp. 1715- 1723.

McLaughlin, D. K., Morrison, G. L., and Troutt, T. R. 1977, Reynolds Number Dependence in Supersonic Jet Noise, AIAA J., Vol. 15, pp. 526-532.

Michalke, A. 1965, On Spatially Growing Disturbances In An Inviscid Shear Layer, J. Fluid Mech., Vol. 23, Pt. 3, pp. 521- 544.

Morris, P. J., and Tam, C. K. W. 1979, On the Radiation of Sound by the Instability Waves of a Compressible Axisymmetric Jet, Mechanics of Sound Generation In Flows, Springer -Verlag, pp. 55-61.

Morris, P. J. 1988, Instability of Elliptic Jets, AIAA J., Vol. 26, No. 2, pp 172-178.

Nagamatsu, H. and Horvay, G., 1970, Supersonic Jet Noise, AIAA Paper No. 70- 237.

Phillips, O. M. 1960, On the Generation of Sound by Supersonic Turbulent Shear Layers, J. Fluid Mech., Vol. 9, Pt. 1, pp. 1-28.

Sarkar, S., Erlebacher, G., Hussaini, M.Y., and Kreiss, H. O., 1989, The Analysis and Modelling of Dilatational Terms in Compressible Turbulence, ICASE Report No. 89-79.

Seiner, J. M. 1984, Advances in High Speed Jet Aeroacoustics, AIAA Paper No. 84- 2275.

Seiner, J. M., and Yu, J. C. 1984, Acoustic Near Field Properties Associated With Broadband Shock Noise, AIAA J., Vol. 22, No. 9, pp. 1207-1215.

Seiner, J. M., Dash, S. M., and Wolf, D. E., 1985, Analysis of Turbulent Under-expanded Jets, Part II: Shock Noise Features Using SCIPVIS, AIAA J., Vol. 23, pp. 669-677.

Seiner, J. M., Manning, J. C., and Ponton, M. K. 1988, Dynamic Pressure Loads Associated with Twin Supersonic Plume Resonance, AIAA J., Vol. 26, No. 8, pp. 954-960.

Seiner, J. M. and Kresja, E. A., 1989, Supersonic Jet Noise and the High Speed Civil Transport, AIAA Paper No. 89-2358.

Seiner, J. M., Baty, R. S., and Kumar, A. 1990, A Procedure for the Design of Elliptic Supersonic Shock-Free Nozzles, to appear.

Tam, C. K. W. and Burton, D. E. 1984, Sound Generated by Instability Waves of Supersonic Flows. Part 2. Axisymmetric Jets, J. Fluid Mech., Vol. 138, pp. 273- 295.

Tam, C. K. W. 1987, Stochastic Model Theory of Broadband Shock Associated Noise From Supersonic Jets, J. Sound and Vib., Vol. 116, pp. 265-302.

Tam, C. K. W. 1990, Broadband Shock-Associated Noise of Moderately Imperfectly Expanded Supersonic Jets, J. Sound and Vib., Vol. 140, Pt. 1, pp. 55- 71.

Zorumski, W. E. 1982, Aircraft Noise Prediction Program Theoretical Manual, NASA TM-83199.

FORMULATION OF THE STATISTICAL EQUATIONS OF TURBULENT FLOWS WITH VARIABLE DENSITY

Alexandre J.A. Favre [1]

Institut de Mécanique Statistique de la Turbulence, Marseille
Académie des Sciences, Paris

ABSTRACT

The statistical equations of turbulent flows were derived by Boussinesq and Reynolds, in cases where density ρ is constant, using a separation of each random quantity w into its average value $\overline{w}$ and a fluctuation w", so $\overline{w''} = 0$. For the general case when the density is variable, the same "Reynolds average" may be used, but the formulae are not simple and the physical meanings of the terms are not clear. In order to express the principles of physics with the fundamental quantities per unit volume ρw, such as mass ρ, momentum ρu_i, internal energy ρe, entropy ρs, Favre retains the separation into the average value $\overline{\rho w} = \overline{\rho}\,\tilde{w}$ and a fluctuation $\rho w'$, so $\overline{\rho w'} = 0$. The formulae are simpler and the physical meanings clearer. The main statistical equations are given.

1. Introduction

There has been a recent renewed interest in chaos and turbulence. The chaos usually considered is temporal, while the turbulence is spatiotemporal, and the chaos may be deterministic or disorderly. We propose to use an experimental criterion to characterize a spatiotemporal chaos [9] : the space-time correlations are going to zero when the separations in time and in space are going to infinity, so the statistical memory [8] has a limited duration time. As the disorder is characterized by conditional probabilities equal to zero, the correlations are also equal to zero. Thus the chaos in disorder has a memory duration time which is naught, while the deterministic chaos has a memory duration time which is limited but not null.

Therefore, the fully established turbulence is a space-time chaos, with vorticity. As the set of equations is closed, the turbulent flows are deterministic [9], and in fact the correlations are not null. In addition, the unpredictability of turbulence for long time delays is not due to disorder, but to its complexity.

---

[1] This work was supported in part by the Air Force Office of Scientific Research through the European Office of Aerospace Research, United States Air Force, under contract F 61052-67-C-0025.

2. Formulation methods for the equations

Where density is constant, Boussinesq and Reynolds derived the statistical equations of fluids in turbulent flows by averaging the Navier-Stokes equations, with a separation of each random quantity w into its mean value $\overline{w}$ and a fluctuation w" such that:

$$\text{"A"} \qquad w = \overline{w} + w'' \Rightarrow \overline{w''} = 0,\ \rho = \overline{\rho} + \rho',\ \overline{\rho w} = \overline{\rho}\,\overline{w} + \overline{\rho' w''}, \tag{1}$$

This is the "Reynolds average" method. When the flow is stationary, the average is operated in time, with the assumption of ergodicity. When generalizing to cases where the density is variable, even random, some authors write the equations by the same method "A", but this leads to complicated formulations and the physical meanings of the terms are not simple.

Another method has been employed by the meteorologists, Hesselberg [10], Van Mieghem [12], and Blackadar [2], wherein the velocity component u_i is separated into a mass-weighted average velocity $\tilde{u}_i$ and a fluctuation u'_i, which is no longer centered. This method has been extended by Favre [3-9] to fluids experiencing turbulent fluctuations of velocity, density, pressure p, internal energy e (or temperature θ), entropy s, concentration c in binary mixture, also viscosity, mass diffusivity D_α, heat conductivity λ, and specific heats c_v and c_p. Each turbulent transportable quantity ρw, which is associated with the density, is separated into its mean value $\overline{\rho w}$ and a centered fluctuation $\rho w'$ such that:

$$\text{"B"} \quad \rho w = \overline{\rho w} + \rho w' \ \Rightarrow\ \overline{\rho w'} = 0,\ \ \rho = \overline{\rho} + \rho' \ \Rightarrow\ \overline{\rho'} = 0,\ w = \tilde{w} + w',\ \ \tilde{w} = \frac{\overline{\rho w}}{\overline{\rho}} \tag{2}$$

We call this the "B average" method. When the flow is stationary, the average is operated in time; when the flow is not stationary, it is operated in set average. The resulting equations have simpler formulations, and the physical meanings of the terms are clearer [3-9], [1], [11].

Since the equations are expressions of the principles of physics, their formulations should start with the fundamental quantities of which the variations are described by these principles: the mass ρ per unit volume for the principle of conservation of mass, the momentum components ρu_i for the principle of dynamics (and not the velocity), the internal energy per unit volume ρe, or heat $c_v\rho\theta$, for the principle of conservation of energy (and not the temperature), the entropy per unit volume for the second principle of thermodynamics. All these quantities are summarized by ρw, where w is a transportable scalar, or a component of a vector.

It should also be noted that, by an algebraic operation, the equations of fluid mechanics may be tranformed into the Lagrange equations, which express the variational principle of least action. This principle is based on the momentum multiplied by a length, or the energy multiplied by a time. So method "B" based on the momentum and on the energy per unit volume for fluid mechanics fits well with the methods used for statistical mechanics, relativity, and quantum mechanics [7], [9].

The equations of fluid mechanics are obtained by expressing the balance of the fundamental transportable quantities, ρw per unit volume, in a reference volume (Ω) bounded by a fixed, closed surface S of which the outward normal has direction cosines denoted by l_k.

The balance in (Ω) is made by taking into account: the rate of variation of ρw in the volume (Ω) per unit time, the flux of w transported by the mass flow ρu_i (and not by the velocity) through the surface S, the rate of production or destruction of ρw per unit time in the volume (Ω) (i.e. $\mathcal{D}(\rho w)$ per unit volume),

$$\iiint_\Omega \frac{\partial}{\partial t}(\rho w)\,d\Omega + \iint_S \rho\, u_k l_k w\, d\sigma = \iiint_\Omega \mathcal{D}(\rho w)\,d\Omega \tag{3}$$

Transformation of the surface integral into a volume integral, such that the flux becomes the divergence of the flux by Stokes formula, gives

$$\iiint_\Omega \left[\frac{\partial}{\partial t}(\rho w) + \frac{\partial}{\partial x_k}(\rho u_k w) - \mathcal{D}(\rho w) \right] d\Omega = 0 \tag{4}$$

It should be noted that the equations of fluid mechanics are thus integro-differential, taking into account the interactions between the whole and the parts.

The theorem of integral null is used to write the equation in the elementary volume ($d\Omega$) and to express the equations at "one point" as the "local balance" of variations of ρw,

$$\frac{\partial}{\partial t}(\rho w) + \frac{\partial}{\partial x_k}(\rho u_k w) = \mathcal{D}(\rho w) \tag{5}$$

(i.e., the local variation in time, transport by the divergence of the mass flux, the rate of production or local destruction of ρw per unit time and volume).

The statistical balance of the mean variables is obtained by set averages of the terms of equation (5), and using the two methods of separation "A" (1) and "B" (2), one obtains [3-9],

"A" $$\frac{\partial}{\partial t}\left(\overline{\rho}\,\overline{w}+\overline{\rho' w''}\right)$$

$$+\frac{\partial}{\partial x_k}\left(\overline{\rho}\,\overline{w}\,\overline{u}_k+\overline{\rho}\,\overline{w''u''_k}+\overline{\rho' u''_k}\,\overline{w}+\overline{\rho' w''}\,\overline{u}_k+\overline{\rho' w'' u''_k}\right)=\overline{\mathcal{D}(\rho w)} \tag{6}$$

"B" $$\frac{\partial}{\partial t}\left(\overline{\rho}\,\tilde{w}\right)+\frac{\partial}{\partial x_k}\left(\overline{\rho}\,\tilde{w}\,\tilde{u}_k+\overline{\rho w' u'_k}\right)=\overline{\mathcal{D}(\rho w)} \tag{7}$$

The equations obtained by the "B average" will have a much simpler formulation, and their terms will have clearer physical meanings than those obtained by the "Reynolds average" method. In equation "A" the local variation in time is separated into two terms, and the transport term is separated into one term for the mean motion, and four terms for the diffusion by turbulence. In equation "B", the mean variations of ρw are separated into: one term for the local variation in time, one for the transport by the mean mass flow $\overline{\rho}\tilde{w}\tilde{u}_k$, the transport by the turbulent mass flow (which is the diffusion by turbulence and is given by the divergence of the turbulent flux), and the mean production or destruction per unit time of ρw.

Because of the non-linearity of transport terms, the average produces, even in "B", correlation terms which are new unknowns; thus the set of statistical equations is open. The statistical theories of turbulence can be used in developing closure models. It is possible to write the equations for the correlations, but correlations of a higher order appear, and so on. The process is thus divergent.

The equation of mass conservation is obtained by substituting in (5), (6) and (7), $\rho w = \rho$ and $\mathcal{D}(\rho) = 0$ ([2],[3]),

$$\frac{\partial\rho}{\partial t}+\frac{\partial}{\partial x_k}(\rho\, u_k)=0\,;\text{"A"}\ \frac{\partial\overline{\rho}}{\partial t}+\frac{\partial}{\partial x_k}\left(\overline{\rho}\,\overline{u}_k+\overline{\rho' u''_k}\right)=0\,;\text{"B"}\ \frac{\partial\overline{\rho}}{\partial t}+\frac{\partial}{\partial x_k}\left(\overline{\rho}\,\tilde{u}_k\right)=0 \tag{8}$$

In equation "A" using the Reynolds average, there is a term $\overline{\rho' u''_k}$ which is due to the mean mass flux by turbulence. Thus, the mean stream-surfaces are crossed by mean mass flux and are losing their physical meaning. Such terms are found in all the statistical equations of balance.

In contrast, in the statistical equations derived using the "B average", such terms are equal to zero, by definition. The mean stream-surfaces are tight on average. In a volume limited by a surface moving with the mean mass-weighted velocity, the total mean mass remains constant. In all the statistical equations of balance such terms are zero, by definition.

The "B average" applied to the transportable quantities simplifies the terms of the statistical equations and clarifies their physical meanings, consistent with the fundamental principles of physics. For entropy, we shall see that the mean mass-weighted average $\overline{\rho}\,\tilde{s}$ fulfills the second principle of thermodynamics because the fluctuations are equal to zero in mean value, while this is not the case for the Reynolds average method. The total enthalpy statistical equation is simplified by the "B average" [5], [6], [7], and it may be integrated along a mean mass-weighted stream-line in the zones where the gradients are not strong. In addition, the mean state equation for perfect gases with the "B average" is $\overline{p} = R\,\overline{\rho}\,\tilde{\theta}$, whereas the Reynolds average gives $\overline{p} = R\,(\overline{\rho}\;\overline{\theta} + \overline{\rho'\theta''})$. As for the experimental measurements, the hot-wire anemometers measure the momentum, ρu_i, components, but not the velocity, and the laser-doppler velocimeters measure the velocities of the particles transported by the mass-flow, not exactly the fluid velocities, so the type of average obtained is open to question [1].

Having examined the statistical equations written by the two methods, we have retained the "B mass-weighted average" [3-8]. We now see that it is preferable in cases in which the density fluctuations are large, for instance in cases of strong heat and mass transfer, supersonic flows, and flows with chemical reactions [1], [11], [6], and [7].

For derivation of the statistical equations of variances and of turbulent fluxes [6], [7], we use the following,

$$\frac{d}{dt}()=\frac{\partial}{\partial t}()+u_k\frac{\partial}{\partial x_k}(); \quad \frac{D}{Dt}()=\frac{\partial}{\partial t}()+\tilde{u}_k\frac{\partial}{\partial x_k}() \tag{9}$$

The mass conservation equations (8) and balance of ρw (5) and (7) become

$$\frac{d\rho}{dt}+\rho\frac{\partial u_k}{\partial x_k}=0; \quad \text{"B"}\ \frac{D\overline{\rho}}{Dt}+\overline{\rho}\frac{\partial \tilde{u}_k}{\partial x_k}=0 \tag{10}$$

$$\rho\frac{dw}{dt}=\mathcal{D}(\rho w); \quad \text{"B"}\ \overline{\rho}\frac{D\tilde{w}}{Dt}+\frac{\partial}{\partial x_k}(\overline{\rho\, w' u'_k})=\overline{\mathcal{D}(\rho w)} \tag{11}$$

Let us consider two random quantities ξ and η which obey the first transport equation (11) and which are separated into

$$\xi=\tilde{\xi}+\xi',\ \ \overline{\rho\xi'}=0,\ \overline{\rho\xi}=\overline{\rho}\,\tilde{\xi}\,;\ \eta=\tilde{\eta}+\eta',\ \overline{\rho\eta'}=0;\ \overline{\rho\eta}=\overline{\rho}\,\tilde{\eta} \tag{12}$$

The first equation (11) being applied to ξ and multiplied by η', then applied to η and multiplied by ξ', added, then averaged so that ([6], [7])

$$\frac{D}{Dt}\overline{\rho\,\xi'\eta'}+\overline{\rho\,\xi'\eta'}\frac{\partial\tilde{u}_k}{\partial x_k}+\frac{\partial}{\partial x_k}\overline{\rho\,u'_k\xi'\eta'}+\overline{\rho\,u'_k\eta'}\frac{\partial\tilde{\xi}}{\partial x_k}+\overline{\rho\,u'_k\xi'}\frac{\partial\tilde{\eta}}{\partial x_k}=$$
$$=\overline{\eta'\mathcal{D}(\rho\xi)}+\overline{\xi'\,\mathcal{D}(\rho\eta)} \tag{13}$$

In order to get the equation for the variances of ξ, we applied $\eta = \xi$ and $\mathcal{D}(\rho\eta) = \mathcal{D}(\rho\xi)$ in (13), giving ([6],[7])

$$\frac{D}{Dt}\overline{\rho\,\xi'\xi'}+\overline{\rho\,\xi'\xi'}\frac{\partial\tilde{u}_k}{\partial x_k}+\frac{\partial}{\partial x_k}\left(\overline{\rho\,u'_k\xi'\xi'}\right)+2\overline{\rho\,u'_k\xi'}\frac{\partial\tilde{\xi}}{\partial x_k}=2\overline{\xi'\mathcal{D}(\rho\,\xi)} \tag{14}$$

The statistical equations of the turbulent fluxes are obtained by substitution in (13) of $\eta = u_i$; and $\mathcal{D}(\rho\eta) = \mathcal{D}(\rho u_i)$. For the expression of $\mathcal{D}(\rho u_i)$, we shall use the classical equations of motion (17), with Cartesian coordinates rotating with the Earth, the Coriolis force components T_k (due to the angular velocity of the earth's rotation), and the gravity ρg_i. This gives ([6], [7]),

$$\frac{D}{Dt}\overline{\rho\,u'_i\xi'}+\overline{\rho\,u'_i\xi'}\frac{\partial\tilde{u}_k}{\partial x_k}+\overline{\rho\,u'_k\xi'}\frac{\partial\tilde{u}_i}{\partial x_k}+\overline{\rho\,u'_k u'_i}\frac{\partial\tilde{\xi}}{\partial x_k}$$
$$+\frac{\partial}{\partial x_k}\left(\overline{\rho\,u'_k u'_i\xi'}+\overline{\xi'\,p'}\,\delta_{ik}-\overline{\xi'\,f_{ik}}\right)=-2\,\varepsilon_{ikl}T_k\overline{\rho u_l\xi'}+\overline{p'\frac{\partial\xi'}{\partial x_i}} \tag{15}$$
$$-\overline{f_{ik}\frac{\partial\xi'}{\partial x_k}}+\overline{u'_i\mathcal{D}(\rho\xi)}-\overline{\xi'}\frac{\partial\bar{p}}{\partial x_i}$$

It will be seen that the last term means a variation of the flux proportional to the gradient of mean pressure-it contains the effects of gravity.

3. Main statistical equations of turbulent flows

In addition to equations (8), mass conservation may be applied to binary mixtures between gases or between liquids, when the contaminant of density ρ_α, has a small concentration, $c = \rho_\alpha/\rho = \tilde{c} + c'$, and $\overline{\rho c} = \overline{\rho}\,\tilde{c}$. With $w = c$ in (5), and using Fick's law for mass diffusion D_α, we obtain ([6], [7])

$$\frac{\partial}{\partial t}\left(\bar{\rho}\tilde{c}\right)+\frac{\partial}{\partial x_k}\left(\bar{\rho}\,\tilde{u}_k\tilde{c}+\overline{\rho\, u'_k c'}-\overline{\rho D_\alpha\frac{\partial c}{\partial x_k}}\right)=0 \tag{16}$$

The equations of motion are obtained from (7) by substitution of u_i = w and with the components of the forces for $\mathcal{D}(\rho u_i)$, giving ([3-7], [12])

$$\frac{\partial}{\partial t}\left(\rho u_i\right)+\frac{\partial}{\partial x_k}\left(\rho u_i u_k\right)=\rho\, g_i-2\varepsilon_{ikl}T_k\rho u_l-\frac{\partial p}{\partial x_i}+\frac{\partial f_{ik}}{\partial x_i}=\mathcal{D}(\rho u_i) \tag{17}$$

$$\frac{\partial}{\partial t}\left(\bar{\rho}\tilde{u}_i\right)+\frac{\partial}{\partial x_k}\left(\bar{\rho}\tilde{u}_i\tilde{u}_k\right)=\frac{\partial}{\partial x_k}\left(-\bar{p}\,\delta_{ik}+\bar{f}_{ik}-\overline{\rho u'_i u'_k}\right)+\bar{\rho}\, g_i-2\varepsilon_{ikl}T_k\bar{\rho}\,\tilde{u}_l \tag{18}$$

For each component of ρu_i per unit time and volume, these equations are the balance between: mean local variation in time, transport by mean mass flow, effect of the gradient of mean pressure, divergence of the mean molecular friction stresses f_{ik}, divergence of Reynolds stresses for mean transport by turbulent diffusion, mean gravity force component, and mean Coriolis force component.

The statistical equation for the kinetic energy of turbulence is obtained by substitution in (14) of $\xi = u_i$, and of $\mathcal{D}(\rho\xi) = \mathcal{D}(\rho u_i)$ (17), thus ([2], [3-7])

$$\frac{D}{Dt}\frac{1}{2}\overline{\rho\, u'_i u'_i}+\frac{1}{2}\overline{\rho\, u'_i u'_i}\frac{\partial\tilde{u}_k}{\partial x_k}=\frac{\partial}{\partial x_k}\left(-\overline{u'_i p'}\,\delta_{ik}+\overline{f_{ik}u'_i}-\frac{1}{2}\overline{\rho\, u'_k u'_i u'_i}\right)$$
$$+\overline{p'\frac{\partial u'_i}{\partial x_i}}-\overline{\rho\, u'_i u'_k}\frac{\partial\tilde{u}_i}{\partial x_k}+\frac{\overline{\rho' u'_i}}{\bar{\rho}}\frac{\partial\bar{p}}{\partial x_i}-\overline{f_{ik}\frac{\partial u'_i}{\partial x_k}} \tag{19}$$

The three last terms are: kinetic production of turbulence by exchange with the part of kinetic energy due to the mean motion, the enthalpic production of turbulence by exchange with enthalpic energy (29), and the dissipation $\overline{\phi'}$ into heat by the action of the molecular friction stress f_{ik} and the corresponding velocity fluctuations. For newtonian fluids

$$f_{ik}=-\frac{2}{3}\mu\frac{\partial u_l}{\partial x_l}\delta_{ik}+\mu\left(\frac{\partial u_i}{\partial x_j}+\frac{\partial u_j}{\partial x_i}\right),\quad \phi=f_{ik}\frac{\partial u_i}{\partial x_k}=2\mu\, n_{ik}n_{ik}-\frac{2}{3}\mu\frac{\partial u_l}{\partial x_l}\frac{\partial u_m}{\partial x_m},$$

where the rate of strain is:

$$n_{ik} = \frac{\partial u_i}{\partial x_k} + \frac{\partial u_k}{\partial u_i}.$$

We can write ([6], [7]) the statistical equation for the Reynolds stresses, by application of (15) with $\rho\xi = \rho u_i$ and $\mathcal{D}(\rho\xi) = \mathcal{D}(\rho u_i)$, giving

$$\frac{D}{Dt}\overline{\rho u'_i u'_j} + \overline{\rho u'_i u'_j}\frac{\partial \tilde{u}_k}{\partial x_k} + \frac{\partial}{\partial x_k}\left[\overline{\rho u'_k u'_i u'_j} - \overline{f_{ik} u'_j} - \overline{f_{ik} u'_j} + \overline{(u'_i \delta_{jk} + u'_j \delta_{ik}) p'}\right]$$

$$+ \overline{\rho u'_k u'_j}\frac{\partial \tilde{u}_i}{\partial x_k} + \overline{\rho u'_k u'_i}\frac{\partial \tilde{u}_j}{\partial x_k} = -2\varepsilon_{ikl} T_k \overline{\rho u'_l u'_j} - 2\varepsilon_{jkl} T_k \overline{\rho u'_l u'_i}$$

$$+ \overline{p'\left(\frac{\partial u'_j}{\partial x_i} + \frac{\partial u'_i}{\partial x_j}\right)} - \overline{f_{ik}\frac{\partial u'_j}{\partial x_k}} - \overline{f_{jk}\frac{\partial u'_i}{\partial x_k}} + \frac{\overline{\rho' u'_j}}{\bar{\rho}}\frac{\partial \bar{p}}{\partial x_i} + \frac{\overline{\rho' u'_i}}{\bar{\rho}}\frac{\partial \bar{p}}{\partial x_j} \tag{20}$$

The two last terms show that the variations of the Reynolds stresses may be due also to the gradient of mean pressure including gravity effects.

The principle of conservation of energy may be expressed by equation (5), with $\rho w = \rho e + 1/2\ \rho u_i u_i$ and $\mathcal{D}(\rho w) = \mathcal{D}(\rho e + 1/2\ \rho u_i u_i)$, by taking into account : the power of the gravity force $\rho g_i u_i$ (that of the Coriolis force is zero because it is orthogonal to the velocity), the divergence of the surface stresses power including the molecular friction stress power and that of the pressure, and the heat flux by conduction into the fluid, $-h_j$. The rate of absorption or radiation of heat may be added where appropriate. The equation reads

$$\frac{\partial}{\partial t}(\rho e + \frac{1}{2}\rho u_i u_i) + \frac{\partial}{\partial x_k}(\rho u_k e + \frac{1}{2}\rho u_k u_i u_i) = \rho g_i u_i + \frac{\partial}{\partial x_j}\left(f_{ij} u_i - p u_i \delta_{ij} - h_j\right) \tag{21}$$

It is well-known that by multiplying the equations of motion (17) by 1/2 u_i and adding, one obtains the kinetic energy equation, and by subtracting it from (21), one gets the classical equations of internal energy e and of enthalpy $i = e + p/\rho$

$$\frac{\partial}{\partial t}(\rho e) + \frac{\partial}{\partial x_k}(\rho e u_k) = \varphi - p\frac{\partial u_i}{\partial x_i} - \frac{\partial h_j}{\partial x_j} = \mathcal{D}(\rho e) \tag{22}$$

$$\frac{\partial}{\partial t}(\rho i) + \frac{\partial}{\partial x_k}(\rho i u_k) = \varphi + \frac{dp}{dt} - \frac{\partial h_j}{\partial x_j} = \mathcal{D}(\rho i) \tag{23}$$

With $\theta ds = de + pd\,(1/\rho)$, the equation of entropy is

$$\frac{\partial}{\partial t}(\rho s)+\frac{\partial}{\partial x_k}(\rho s u_k)=\frac{\varphi}{\theta}-\frac{1}{\theta}\frac{\partial h_j}{\partial x_j}=\mathcal{D}(\rho s) \tag{24}$$

In terms of temperature (22) and (23) read, with $de = c_v\, d\theta$ and $di = c_p\, d\theta$

$$\frac{\partial}{\partial t}(\rho\theta)+\frac{\partial}{\partial x_k}(\rho\theta u_k)=\frac{1}{c_v}\left[\varphi-p\frac{\partial u_i}{\partial x_i}-\frac{\partial h_j}{\partial x_j}\right]=\frac{1}{c_p}\left[\varphi+\frac{dp}{dt}-\frac{\partial h_j}{\partial x_j}\right]=\mathcal{D}(\rho\theta) \tag{25}$$

After (21), the total enthalpy $i_T = i + 1/2 u_i u_i$ is given by

$$\frac{\partial}{\partial t}(\rho\, i_T)+\frac{\partial}{\partial x_k}(\rho i_T u_k)=\rho\, g_i u_i+\frac{\partial}{\partial x_j}(f_{ij}u_i-h_j)+\frac{\partial p}{\partial t}=\mathcal{D}(\rho i_T) \tag{26}$$

It is known that for a stationary, non turbulent, adiabatic, non viscous fluid, the two last terms are zero, and the integration along a stream-line l gives:

$$i_T+G=i+\frac{1}{2}u_k u_k+G=e+\frac{p}{\rho}+\frac{1}{2}u_k u_k+g x_3=C_l \tag{27}$$

where C_l is constant along the stream-line l, x_3 is the altitude, and G is the geopotential $g_i = -\dfrac{\partial G}{\partial x_i}$.

We write ([6], [7]) the statistical equations of internal energy, enthalpy, and the same equations in terms of temperature (with the assumption that the specific heats c_v and c_p may be variable but not turbulent). With Fourier's relation, $-h_k = \lambda\dfrac{\partial q}{\partial x_k}$, these equations become

$$\frac{\partial}{\partial t}(\bar{\rho}\tilde{e})+\frac{\partial}{\partial x_k}\left(\bar{\rho}\tilde{e}\tilde{u}_k\right)=\bar{\varphi}-\bar{p}\frac{\partial \tilde{u}_i}{\partial x_i}-\overline{p\frac{\partial u'_i}{\partial x_i}}-\frac{\partial}{\partial x_k}\left(\bar{h}_k+\overline{\rho u'_k e'}\right) \tag{28}$$

$$\frac{\partial}{\partial t}(\bar{\rho}\,\tilde{i})+\frac{\partial}{\partial x_k}(\bar{\rho}\,\tilde{i}\,\tilde{u}_k)=\bar{\varphi}+\frac{\partial\bar{p}}{\partial t}+\tilde{u}_k\frac{\partial\bar{p}}{\partial x_k}-\frac{\overline{\rho' u'_k}}{\bar{\rho}}\frac{\partial\bar{p}}{\partial x_k}+\overline{u'_k\frac{\partial p'}{\partial x_k}}$$
$$-\frac{\partial}{\partial x_k}\left(\bar{h}_k+\overline{\rho u'_k i'}\right) \tag{29}$$

$$\frac{\partial}{\partial t}(\bar{\rho}\,\tilde{\theta})+\frac{\partial}{\partial x_k}(\bar{\rho}\,\tilde{\theta}\,\tilde{u}_k)+\frac{\partial}{\partial x_k}\overline{\rho' u'_k \theta'}=\frac{1}{\bar{c}_v}\left[\bar{\varphi}-\bar{p}\frac{\partial \tilde{u}_k}{\partial x_k}-\overline{p\frac{\partial u'_k}{\partial x_k}}+\frac{\partial}{\partial x_k}\overline{\left(\lambda\frac{\partial \theta}{\partial x_k}\right)}\right]$$

$$=\frac{1}{\bar{c}_p}\left[\bar{\varphi}+\frac{\partial \bar{p}}{\partial t}+\left(\tilde{u}_k-\frac{\overline{\rho' u'_k}}{\bar{\rho}}\right)\frac{\partial \bar{p}}{\partial x_k}+\overline{u'_k\frac{\partial p'}{\partial x_k}}+\frac{\partial}{\partial x_k}\overline{\left(\lambda\frac{\partial \theta}{\partial x_k}\right)}\right] \tag{30}$$

The terms in the statistical equation for the internal energy are: mean local rate of variation in time, transport of the mean internal energy by the mean motion, mean dissipation into heat by viscosity, power developed by the mean pressure in the mean variation of volume, mean power developed by pressure in the variation of volume due to turbulence, the divergence of heat flux by mean conduction, and diffusion by turbulence of internal energy.

Making $\xi = \theta$ and $\mathcal{D}(\rho\xi) = \mathcal{D}(\rho\theta)$ in (14) and (25), we obtain the statistical equations of the variations of $\overline{\rho\theta'^2}$ [6],[7]

$$\frac{D}{Dt}\overline{\rho\theta'^2}+\overline{\rho\theta'^2}\frac{\partial \tilde{u}_k}{\partial x_k}+\frac{\partial}{\partial x_k}\overline{\rho u'_k\theta'^2}+2\overline{\rho u'_k\theta'}\frac{\partial \tilde{\theta}}{\partial x_k}=$$

$$=\frac{2}{\bar{c}_v}\left[\overline{\theta'\varphi}-\overline{\theta' p\frac{\partial u_k}{\partial x_k}}+\frac{\partial}{\partial x_k}\overline{\left(\theta'\lambda\frac{\partial \theta}{\partial x_k}\right)}-\overline{\lambda\frac{\partial \theta'}{\partial x_k}}\frac{\partial \tilde{\theta}}{\partial x_k}-\overline{\lambda\frac{\partial \theta'}{\partial x_k}\frac{\partial \theta'}{\partial x_k}}\right] \tag{31}$$

$$=\frac{2}{\bar{c}_p}\left[\overline{\theta'\varphi}-\overline{\theta'\frac{dp'}{dt}}-\frac{\overline{\rho'\theta'}}{\bar{\rho}}\frac{\partial \bar{p}}{\partial t}+\overline{\theta' u_k}\frac{\partial \bar{p}}{\partial x_k}+\frac{\partial}{\partial x_k}\overline{\theta'\lambda\frac{\partial \theta}{\partial x_k}}-\overline{\lambda\frac{\partial \theta'}{\partial x_k}}\frac{\partial \tilde{\theta}}{\partial x_k}-\overline{\lambda\frac{\partial \theta'}{\partial x_k}\frac{\partial \theta'}{\partial x_k}}\right]$$

It will be noticed that the last term on the second (or third lines), resembling a "dissipation", is damping the fluctuations of $\overline{\rho\theta'^2}$.

Making $\xi = \theta$ and $\mathcal{D}(\rho\xi) = \mathcal{D}(\rho\theta)$ in (15) and (25) we obtain the statistical equations for the turbulent fluxes of heat and enthalpy ([6], [7])

$$\frac{D}{Dt}\overline{\rho u'_i\theta'}+\overline{\rho u'_i\theta'}\frac{\partial \tilde{u}_k}{\partial x_k}+\overline{\rho u'_k\theta'}\frac{\partial \tilde{u}_i}{\partial x_k}+\overline{\rho u'_k u'_i}\frac{\partial \tilde{\theta}}{\partial x_k}+\frac{\partial}{\partial x_k}\overline{\rho u'_k u'_i\theta'}$$

$$+\frac{\partial}{\partial x_k}\left(-\overline{\theta' f_{ik}}+\overline{\theta' p'}\delta_{ik}\right)+2\varepsilon_{ikl}T_k\overline{\rho u'_l\theta'}-\overline{p'\frac{\partial \theta'}{\partial x_i}}+\overline{f_{ik}\frac{\partial \theta'}{\partial x_k}}+\frac{\overline{\rho'\theta'}}{\bar{\rho}}\frac{\partial \bar{p}}{\partial x_i}$$

$$=\frac{1}{\overline{c_v}}\left[\overline{u'_i\varphi}-\overline{u'_i p\frac{\partial u_k}{\partial x_k}}+\overline{u'_i\frac{\partial}{\partial x_k}\left(\lambda\frac{\partial\theta}{\partial x_k}\right)}\right]$$

$$=\frac{1}{\overline{c_p}}\left[\overline{u'_i\varphi}-\overline{u'_i\frac{dp'}{dt}}-\frac{\overline{\rho' u'_i}}{\overline{\rho}}\frac{\partial\overline{p}}{\partial t}+\overline{u'_i u'_k}\frac{\partial\overline{p}}{\partial x_k}+\overline{u'_i\frac{\partial}{\partial x_k}\left(\lambda\frac{\partial\theta}{\partial x_k}\right)}\right] \tag{32}$$

We obtain the statistical equation of total enthalpy [5], [6], [7], that is,

$$i_T=\tilde{i}_T+i'_T\,;\quad \overline{\rho i_T}=\overline{\rho}\ \tilde{i}_T, \tag{33}$$

Let $\xi = i_T$ and $\mathcal{D}(\rho\xi)=\mathcal{D}(\rho i_T)$ in (11), with (26), the equation becomes

$$\overline{\rho}\frac{D}{Dt}\left(\tilde{i}_T+G\right)=\frac{\partial}{\partial x_k}\left(\overline{f_{ik}u_j}-\overline{h}_k-\overline{\rho\, i'_T u'_k}\right)+\frac{\partial\overline{p}}{\partial t} \tag{34}$$

When the flow is statistically stationary, and in the parts of the flow where the turbulent diffusion of energy is negligible (small gradients) - as the turbulent mass flow is zero by the "B average" - we may integrate along the mean mass-weighted stream-line L (or stream-tube), and have [6], [7] with C_L being constant,

$$\tilde{i}_T+G=\tilde{i}+\frac{1}{2}\tilde{u}_j\tilde{u}_j+\frac{1}{2}\frac{\overline{\rho u'_j u'_j}}{\overline{\rho}}+G=\tilde{e}+\frac{\overline{p}}{\overline{\rho}}+\frac{1}{2}\tilde{u}_j\tilde{u}_j+\frac{1}{2}\frac{\overline{\rho u'_j u'_j}}{\overline{\rho}}+g x_3=C_L \tag{35}$$

For the entropy $s=\tilde{s}+s'$, $\overline{\rho s}=\overline{\rho}\ \tilde{s}$ with $\xi = s$ and $\mathcal{D}(\rho\xi)=\mathcal{D}(\rho s)$ in (7) and (24), the statistical equation takes the following form [3-7]

$$\frac{\partial}{\partial t}\left(\overline{\rho}\,\tilde{s}\right)+\frac{\partial}{\partial x_k}\left(\overline{\rho}\,\tilde{u}_k\tilde{s}\right)=\frac{\partial}{\partial x_k}\left(\overline{\frac{\lambda}{\theta}\frac{\partial\theta}{\partial x_k}}-\overline{\rho u'_k s'}\right)+\overline{\frac{\lambda}{\theta^2}\frac{\partial\theta}{\partial x_k}\frac{\partial\theta}{\partial x_k}}+\left(\overline{\frac{\varphi}{\theta}}\right) \tag{36}$$

In a volume V limited by a surface Σ moving at the mean mass-weighted velocity $\tilde{u}_i$, the mean mass remains constant, and the balance of entropy in mean value takes the form ([5], [6], and [7])

$$\frac{D}{Dt}\iiint_V \bar{\rho}\tilde{s}\,dv - \iint_\Sigma \left(\overline{\frac{\lambda}{\theta}\frac{\partial\theta}{\partial x_k}} - \overline{\rho u'_k s'} \right) l_k\, d\sigma = \iiint_V \left[\left(\overline{\frac{\varphi}{\theta}} \right) + \overline{\frac{\lambda}{\theta^2}\frac{\partial\theta}{\partial x_k}\frac{\partial\theta}{\partial x_k}} \right] dv \geq 0 \tag{37}$$

Thus, $\bar{\rho}\,\tilde{s} = \overline{\rho s}$ satisfies the second principle of thermodynamics, because $\overline{\rho s'} = 0$. The rate of variation of $\overline{\rho s}$, in the volume V, is not less than that which corresponds to the mean flux of entropy through this surface Σ by conduction of heat and by turbulent diffusion of entropy. If this surface Σ coincides with a surface such that the double integral is zero, the mean value of the entropy inside V is not decreasing.

With ξ = s and $\mathcal{D}(\rho\xi) = \mathcal{D}(\rho s)$ in (14) and (24), we obtain ([6], [7]) the statistical equation of the variations of $\overline{\rho s'^2}$

$$\frac{D}{Dt}\overline{\rho s'^2} + \overline{\rho s'^2}\frac{\partial \tilde{u}_k}{\partial x_k} + \frac{\partial}{\partial x_k}\overline{\rho u'_k s'^2} + 2\overline{\rho u'_k s'}\frac{\partial \tilde{s}}{\partial x_k}$$
$$= 2\left(\overline{s'\frac{\varphi}{\theta}}\right) + 2\overline{\frac{s'}{\theta}\frac{\partial}{\partial x_k}\left(\lambda\frac{\partial\theta}{\partial x_k}\right)} \tag{38}$$

With ξ = s and $\mathcal{D}(\rho\xi) = \mathcal{D}(\rho s)$ in (15), and (24), we obtain the statistical equations of the turbulent fluxes of entropy ([6], [7])

$$\frac{D}{Dt}\overline{\rho u_i s'} + \overline{\rho u'_i s'}\frac{\partial \tilde{u}_k}{\partial x_k} + \overline{\rho u'_k s'}\frac{\partial \tilde{u}_i}{\partial x_k} + \overline{\rho u'_k u'_i}\frac{\partial \tilde{s}}{\partial x_k}$$
$$+ \frac{\partial}{\partial x_k}\left(\overline{\rho u'_k u'_i s'} - \overline{s' f_{ik}} + \overline{s'p'}\,\delta_{ik} - \overline{\frac{u'_i}{\theta}\lambda\frac{\partial\theta}{\partial x_k}} \right)$$
$$= -2\varepsilon_{ikl}T_k\overline{\rho u'_l s'} + \overline{p'\frac{\partial s'}{\partial x_i}} - \overline{f_{ik}\frac{\partial s'}{\partial x_k}} + \overline{\frac{\rho' s'}{\bar{\rho}}\frac{\partial \bar{p}}{\partial x_i}} + \overline{u'_i\frac{\varphi}{\theta}} - \overline{\lambda\frac{\partial\theta}{\partial x_k}\frac{\partial}{\partial x_k}\left(\frac{u'_i}{\theta}\right)} \tag{39}$$

It will be seen that it is possible to measure the fluctuations s' of entropy by the measurements of the fluctuations of temperature θ' [6], [7], because s' is, in many flows with moderate fluctuations, approximately equal to $c_p\,\theta'/\,\bar{\theta}$.

Alternative forms of the momentum equations (17) in terms of vorticity are of interest; it is obtained by taking the curl of the terms.
Bearing in mind the following definitions and relations:

$$\frac{\partial u_i}{\partial x_k} = n_{ik} + r_{ik} \text{ with } n_{ik} = \frac{1}{2}\left(\frac{\partial u_i}{\partial x_k} + \frac{\partial u_k}{\partial x_i}\right); r_{ik} = \frac{1}{2}\left(\frac{\partial u_i}{\partial x_k} - \frac{\partial u_k}{\partial x_i}\right)$$

(n_{ik} is the rate of strain, r_{ik} the rotation tensor, the vorticity ω_j is the vector of r_{ik})

$$\omega_j = \varepsilon_{jki}\frac{\partial u_i}{\partial x_k} = \varepsilon_{jki} r_{ik} \quad (\text{because } \varepsilon_{jki} n_{ik} = 0 \text{ and } r_{ik} = -\frac{1}{2}\varepsilon_{ikj}\omega_j)$$

$$\frac{\partial \omega_j}{\partial x_j} = \varepsilon_{jki}\frac{\partial u_j}{\partial x_j \partial x_k} = 0 \qquad (\text{the vorticity } \omega_j \text{ is solenoidal}).$$

Equations (17) may be written as

$$\frac{\partial u_i}{\partial t} = \varepsilon_{ilk} u_l(\omega_k + 2T_k) - \frac{\partial}{\partial x_i}\left(\frac{1}{2}u_l u_l + G\right) - \frac{1}{\rho}\frac{\partial p}{\partial x_i} + \frac{1}{\rho}\frac{\partial f_{il}}{\partial x_l} \tag{40}$$

It is known that [13]

$$\varepsilon_{jqi}\varepsilon_{ilk} = \delta_{jl}\delta_{qk} - \delta_{jk}\delta_{ql};\ \varepsilon_{jqi}\frac{\partial}{\partial x_q}\left(\frac{\partial}{\partial x_q}\text{scalar}\right) = 0;\ \omega_k r_{lk} = -\frac{1}{2}\varepsilon_{lkm}(\omega_k\omega_m)$$

by permutation of indices k and m. And permuting again, the sign changes, thus the preceding term is equal to zero, and we have [13]

$$(\omega_k + 2T_k)\frac{\partial u_l}{\partial x_k} = (\omega_k + 2T_k)(n_{lk} + r_{lk}) = (\omega_k + 2T_k)n_{lk} + 2T_k r_{lk}$$

The equations of momentum in terms of vorticity then are obtained by operating $\varepsilon_{jqi}\frac{\partial}{\partial x_q}$ on the terms of equation (40), times ρ, giving

$$\rho\frac{d\omega_j}{dt} = \rho(\omega_k + 2T_k)n_{jk} - \rho(\omega_j + 2T_j)\frac{\partial u_k}{\partial x_k} - \varepsilon_{jki}T_k\rho\omega_i$$

$$+\varepsilon_{jqi}\frac{1}{\rho}\frac{\partial\rho}{\partial x_q}\frac{\partial p}{\partial x_i}-\varepsilon_{jqi}\frac{1}{\rho}\frac{\partial\rho}{\partial x_q}\frac{\partial f_{ik}}{\partial x_k}+\varepsilon_{jqi}\frac{\partial}{\partial x_q}\left(\frac{\partial f_{ik}}{\partial x_k}\right)=\mathcal{D}(l\omega_j) \tag{41}$$

It can be seen that for a viscous fluid

$$\varepsilon_{jqi}\frac{\partial}{\partial x_q}\left(\frac{\partial f_{ik}}{\partial x_k}\right)=2\varepsilon_{jqi}\frac{\partial}{\partial x_q}\left(n_{ik}\frac{\partial\mu}{\partial x_q}\right)+2\varepsilon_{jqi}\frac{\partial\mu}{\partial x_q}\frac{\partial n_{ik}}{\partial x_k}+\mu\frac{\partial^2\omega_j}{\partial x_k\partial x_k} \tag{42}$$

The vorticity components ω_j are separated into a mean vorticity Ω_j and a fluctuation of vorticity ω'_j; the rate of strain and the rotation tensor are divided into mean and fluctuating parts, all based on the separation of velocity ([6], [7])

$$\frac{\partial\tilde{u}_i}{\partial x_k}=N_{ik}+R_{ik};\quad N_{ik}=\frac{1}{2}\left(\frac{\partial\tilde{u}_i}{\partial x_k}+\frac{\partial\tilde{u}_k}{\partial x_i}\right);\quad R_{ik}=\frac{1}{2}\left(\frac{\partial\tilde{u}_i}{\partial x_k}-\frac{\partial\tilde{u}_k}{\partial x_i}\right)$$

$$\omega_j=\Omega_j+\omega'_j;\quad \Omega_j=\varepsilon_{jki}\frac{\partial\tilde{u}_i}{\partial x_k}=\varepsilon_{jki}R_{ik};\omega'_j=\varepsilon_{jki}\frac{\partial u'_i}{\partial x_k}\Rightarrow\frac{\partial\Omega_j}{\partial x_j}=\frac{\partial\omega'_j}{\partial x_j}=0 \tag{43}$$

thus the mean and fluctuating vorticities are solenoidal, but $\overline{\rho\omega'_i}\neq 0$.

We obtain the statistical equations of vorticity by averaging the equations of vorticity (41), ([6],[7])

$$\bar{\rho}\frac{D\Omega_j}{Dt}+\frac{D}{Dt}\overline{\rho\,\omega'_j}+\overline{\rho\,\omega'_j}\frac{\partial\tilde{u}_k}{\partial x_k}+\frac{\partial}{\partial x_k}\overline{\rho\,u'_k\,\omega'_j}=\bar{\rho}\left(\Omega_k+2T_k\right)N_{jk}$$

$$+\overline{\rho\omega'_k n'_{jk}}-\bar{\rho}\left(\Omega_j+2T_j\right)\frac{\partial\tilde{u}_k}{\partial x_k}-\overline{\rho\,\omega'_j\frac{\partial u'_k}{\partial x_k}}+\left(\Omega_k+2T_k\right)\overline{\rho n'_{jk}}$$

$$-\left(\Omega_j+2T_j\right)\overline{\rho\frac{\partial u'_k}{\partial x_k}}+\overline{\rho\,\omega'_k}N_{jk}-\overline{\rho\,\omega'_j}\frac{\partial\tilde{u}_k}{\partial x_k}+\varepsilon_{jqi}\overline{\frac{1}{\rho}\frac{\partial\rho}{\partial x_q}\frac{\partial p}{\partial x_i}}$$

$$-\varepsilon_{jki}T_k\left(\bar{\rho}\,\Omega_i+\overline{\rho\,\omega'_i}\right)-\varepsilon_{jqi}\overline{\frac{1}{\rho}\frac{\partial\rho}{\partial x_q}\frac{\partial f_{ik}}{\partial x_k}}+\varepsilon_{jqi}\frac{\partial}{\partial x_q}\left(\frac{\partial\overline{f_{ik}}}{\partial x_k}\right) \tag{44}$$

The last term on the left hand side of the equation is the turbulent diffusion of the vorticity. The two first terms on the right hand side correspond to the amplification or reduction and to a rotation of the mean vorticity by stretching or squeezing and rotation of the vortex by effects of the mean and of the turbulent strain rates. The two following terms represent the effects of the volume changes in the mean and the turbulent motions. The 9th term represents the mean effect of interaction between the gradient of density and the gradient of pressure, generating vorticity when these gradients are not colinear [11]. The last term corresponds to the mean diffusion of vorticity by viscosity, through (42).

Multiplying the terms of the equation of vorticity (41) by ω'_j and adding we obtain, after averaging, the statistical equations of the mean square of vorticity turbulent fluctuations multiplied by the density ([6], [7])

$$
\begin{aligned}
&\frac{D}{Dt}\overline{\rho\,\omega'_j\omega'_j}+\overline{\rho\,\omega'_j\omega'_j}\,\frac{\partial \tilde{u}_k}{\partial x_k}+\frac{\partial}{\partial x_k}\overline{\rho\, u'_k\omega'_j\omega'_j}+2\,\overline{\rho\, u'_k\omega'_j}\,\frac{\partial \Omega_j}{\partial x_k}+2\,\overline{\rho\,\omega'_j}\,\frac{D\Omega_j}{Dt}\\
&=2\,\overline{\rho\,\omega'_j\omega'_k}\,N_{jk}+2\,\overline{\rho\,\omega'_j\omega'_k n'_{jk}}-2\,\overline{\rho\,\omega'_j\omega'_j}\,\frac{\partial \tilde{u}_k}{\partial x_k}-2\,\overline{\rho\,\omega'_j\omega'_j\,\frac{\partial u'_k}{\partial x_k}}\\
&+2\left(\Omega_k+2T_k\right)\overline{\rho\,\omega'_j n'_{jk}}-2\left(\Omega_j+2T_j\right)\overline{\rho\,\omega'_j\,\frac{\partial u'_k}{\partial x_k}}\\
&+2\,\overline{\rho\,\omega'_j}\left(\Omega_k+2T_k\right)N_{jk}-2\,\overline{\rho\,\omega'_j}\left(\Omega_j+2T_j\right)\frac{\partial \tilde{u}_k}{\partial x_k}-2\varepsilon_{jki}T_k\left(\overline{\rho\,\omega'_j\omega'_i}-\overline{\rho\,\omega'_j}\,\Omega_i\right)\\
&+2\varepsilon_{jqi}\,\overline{\frac{\omega'_j}{\rho}\frac{\partial \rho}{\partial x_q}\frac{\partial p}{\partial x_i}}-2\varepsilon_{jqi}\,\overline{\frac{\omega'_j}{\rho}\frac{\partial \rho}{\partial x_q}\frac{\partial f_{ik}}{\partial x_k}}+2\varepsilon_{jqi}\,\overline{\omega'_j\frac{\partial}{\partial x_q}\left(\frac{\partial f_{ik}}{\partial x_k}\right)}
\end{aligned}
\qquad (45)
$$

For a viscous fluid, the last term reads

$$
\begin{aligned}
&2\varepsilon_{jqi}\,\overline{\omega'_j\frac{\partial}{\partial x_q}\left(\frac{\partial f_{ik}}{\partial x_k}\right)}=4\varepsilon_{jqi}\,\overline{\omega'_j\frac{\partial}{\partial x_q}\left(n_{ik}\frac{\partial \mu}{\partial x_k}\right)}+4\varepsilon_{jqi}\,\overline{\omega'_j\frac{\partial \mu}{\partial x_q}\frac{\partial n_{ik}}{\partial x_k}}\\
&2\,\overline{\mu\omega'_j}\,\frac{\partial^2 \Omega_j}{\partial x_k\,\partial x_k}+\mu\frac{\partial^2\,\overline{\omega'_j\,\omega'_j}}{\partial x_k\,\partial x_k}-2\mu\,\overline{\frac{\partial \omega'_j}{\partial x_k}\frac{\partial \omega'_j}{\partial x_k}}
\end{aligned}
\qquad (46)
$$

in which the last term is a dissipation, by the effects of viscosity.

When there is no summation on the index j, (45) represents the three equations for the square of each component of the turbulent vorticities ω'_j, times the density. When this summation is made, (45) represents the addition of these square values times the density. Since the enstrophy is the square of the vorticity, (45) is related to the enstrophy of the turbulent fluctuations times the density, but not exactly in mean value because

$$\overline{\rho\,\omega_j\,\omega_j} \equiv \bar{\rho}\,\Omega_j\,\Omega_j + 2\Omega_j\,\overline{\rho\,\omega'_j} + \overline{\rho\,\omega'_j\,\omega'_j}$$

Multiplying the terms of the equations of vorticity (41) by u'_i, and those of the equations of motion (18) by ω'_j, adding and averaging, we obtain the statistical equations of the mean turbulent fluxes of vorticity ([6], [7])

$$\frac{D}{Dt}\overline{\rho\,u'_i\,\omega'_j} + \overline{\rho\,u'_i\,\omega'_j}\,\frac{\partial \tilde{u}_k}{\partial x_k} + \overline{\rho\,u'_k\omega'_j}\,\frac{\partial \tilde{u}_i}{\partial x_k} + \overline{\rho\,u'_k u'_i}\,\frac{\partial \Omega_j}{\partial x_k}$$
$$+\frac{\partial}{\partial x_k}\left(\overline{\rho\,u'_k\,u'_i\,\omega'_j} - \overline{\omega'_j\,f'_{ik}} + \overline{\omega'_j\,p'}\,\delta_{ik}\right) + \overline{\rho\,\omega'_j}\,\frac{D\tilde{u}_i}{Dt}$$
$$= \overline{\rho\,\omega'_j}\,g_i - 2\varepsilon_{ikl}\,T_k\,\overline{\rho\,u'_l\,\omega'_j} + \overline{p'\frac{\partial \omega'_j}{\partial x_i}} - \overline{f'_{ik}\frac{\partial \omega'_j}{\partial x_k}} - \overline{\omega'_j}\,\frac{\partial \bar{p}}{\partial x_i}$$
$$+\overline{\rho\,u'_i\left(\omega_k + 2T_k\right)n_{jk}} - \overline{\rho\,u'_i\left(\omega_j + 2T_j\right)\frac{\partial u_k}{\partial x_k}} + \varepsilon_{jql}\,\overline{\frac{u'_i}{\rho}\,\frac{\partial \rho}{\partial x_q}\,\frac{\partial p}{\partial x_l}}$$
$$-\varepsilon_{jkl}\,T_k\,\overline{\rho\,u'_i\,\omega'_l} - \varepsilon_{jql}\,\overline{\frac{u'_i}{\rho}\,\frac{\partial \rho}{\partial x_q}\,\frac{\partial f_{lk}}{\partial x_k}} + \varepsilon_{jql}\,\overline{u'_i\,\frac{\partial}{\partial x_q}\left(\frac{\partial f_{lk}}{\partial x_k}\right)} \tag{47}$$

Since the helicity per unit volume is the scalar product of vorticity-velocity-density, when indices i and j are equal, the equation represent the helicity of turbulent fluctuations. When the fluid is viscous, the last term reads

$$\varepsilon_{jql}\,\overline{u'_i\,\frac{\partial}{\partial x_q}\left(\frac{\partial f_{lk}}{\partial x_k}\right)} = 2\varepsilon_{jql}\,\overline{u'_i\,\frac{\partial}{\partial x_q}\left(n_{lk}\,\frac{\partial \mu}{\partial x_k}\right)} + 2\varepsilon_{jql}\,\overline{u'_i\,\frac{\partial \mu}{\partial x_q}\,\frac{\partial n_{lk}}{\partial x_k}}$$
$$+\overline{\mu\,u'_i}\,\frac{\partial^2 \Omega_j}{\partial x_k\,\partial x_k} - \mu\,\frac{\partial}{\partial x_k}\overline{\left(\omega'_j\,\frac{\partial u'_i}{\partial x_k}\right)} + \mu\,\frac{\partial^2 \overline{\left(u'_i\,\omega'_j\right)}}{\partial x_k\,\partial x_k} - \mu\,\overline{\frac{\partial u'_i}{\partial x_k}\,\frac{\partial \omega'_j}{\partial x_k}} \tag{48}$$

The densities of the mean turbulent fluxes of vorticity are related to the gradients of the Reynolds stresses

$$\varepsilon_{ijk}\overline{\rho u'_j \omega'_k} = -\frac{\partial}{\partial x_j}\overline{\rho u'_i u'_j} + \overline{\rho \frac{\partial}{\partial x_i}\left(\frac{1}{2}u'_j u'_j\right)} + \overline{u'_i \frac{\partial}{\partial x_j}\rho u'_j} \qquad (49)$$

In the case of isovolumetric, stationary motions and for some relatively simple flows, the two last terms are negligible and the mean vorticity turbulent fluxes are equal to the gradients of Reynolds stresses [13].

4. Conclusion

When the density is constant, the "B averages" reduce to the "Reynolds averages" as well as the statistical equations. In the general case, the density is variable and the present statistical equations are applicable to: compressible flows up to supersonic speeds and in gravity fields, and flows with heat and mass transfer. An extension has been made to turbulent reacting flows with combustion by P.A Libby, F.A. Williams [11] and R.W. Bilger [1] using the "B average" method which leads to additional terms in the present statistical equations.

References

[1] Bilger R.W., 1980. Turbulent Flows with nonpremixed reactants. In : Turbulent Reacting Flows. Topics in Appl. Physics. Vol. 44, pp. 65-III. Springer-Verlag.

[2] Blackadar A.K., 1950. The transformation of energy by the large scale stress in the atmosphere. New York Univ. Meteor. Papers I, 4.

[3] Favre A.J.A., 1958. Equations statistiques des gaz turbulents. C.R. Academie des Sciences, T.246 : p.2576; p.2723; p.2839; p.3216.

[4] Favre A.J.A, 1965. Equations des gaz turbulents compressibles: Journal de Mécanique. Vol. 4, n°3 ; n° 4.

[5] Favre A.J.A., 1969. Statistical equations of turbulent gases. Prob. of Hydrodynamics and Continuum Mechanics. Soc. for Ind. and Appl. Mathematics. Philadelphia, pp. 232, 266.

[6] Favre A.J.A., 1975. Equations statistiques des fluides turbulents compressibles. Proc. Fifth Canad. Congr. Appl. Mechanics. Fredericton N.B. Canada; pp. G3-34.

[7] Favre A.J.A. 1976. Equations fondamentales des fluides à masse volumique variable en écoulements turbulents. In: La Turbulence en Mécanique des Fluides, A. Favre, L.S.G. Kovasznay, R. Dumas, J. Gaviglio, M. Coantic. Gauthiers-Villars, Paris.

[8] Favre A.J.A., 1983. Turbulence: space-time statistical properties and behaviour in supersonic flows. The Physics of Fluids. Vol. 26, N° 10.

[9] Favre A.J.A., 1988. De la Causalité à la Finalité à propos de la Turbulence., with Henri and Jean Guitton, A. Lichnerowicz, E. Wolff. Maloine, Paris.

[10] Hesselberg Th., 1926. Beits. Phys. Freien. Atmosph. 12, pp. 141-160.

[11] Libby P.A., Williams F.A., 1980. Turbulent reacting Flows. Top. in Applied Physics. Vol. 44. pp. I-43. Springer-Verlag.

[12] Van Mieghem J., 1943. Thermodynamique atmosphérique, I. Théorie. Inst. Roy. Météor. Belgique, Miscellanées, Fasc. XII.

[13] Tennekes H., Lumley J.L., 1972. A First Course in Turbulence. The MIT Press. Cambridge, Mass.

A DENSITY-WEIGHTED CONDITIONAL CLOSURE MODEL FOR DIFFUSION FLAMES

Johannes Janicka

Fachgebiet Energie- und Kraftwerkstechnik
Technische Hochschule Darmstadt
Petersenstrasse 30
6100 Darmstadt, Germany

ABSTRACT

A second-order closure model for the prediction of diffusion flames is presented. The model combines density-weighted and conditional averaging. The closure assumptions for the intermittency contributions are described. Finally comparisons between predictions and measurements for cold flows and diffusion flames are shown.

1. Introduction

Models of turbulent flames consist of a group of submodels. A turbulence model providing information about turbulent time and length scales, a combustion model describing the details of chemical reaction and a coupling model, taking into consideration the interaction between chemical reactions and turbulent fluctuations.

Research activities in the recent years were focussed mainly on the combustion model. To improve the prediction quality of the entire model a substantial advance in the development of the turbulence model is necessary. Up to now, mainly two equation models and standard Reynolds-stress models have been applied for the calculation of velocity and scalar moments as well as turbulent time and length scales.

The development of turbulence models with a higher level of sophistication such as multiple length scale models or conditional models including intermittency is still in its initial stage.

Intermittency and super layer effects are important in flames, mainly for two reasons. Firstly, the chemical reactions take place in the super layer (turbulent/nonturbulent) up to distances of x/D = 20 - 50 for nearly all fuels. Secondly, increase of viscosity causes an increase in the super-layer thickness, since its extension is of the order of the Kolmogorov microscale.

This work intends to present a density-weighted conditional model for the prediction of diffusion flames. The closure concept for the conditional variables [3] is based on an earlier closure for intermittency factor, mean velocities, Reynolds-stress components and dissipation rate [8].

The equations for conditional moments contain explicitly terms accounting for the relative motion of turbulent and non-turbulent zones and the transport of various quantities through them. Some of those terms are non-closed on the second-order level and their closure will be discussed. The resulting conditional Reynolds-stress model will be evaluated by comparison of calculated properties of thin shear flows with experimental results.

2. Conditional equations

The definition of conditional moments requires the notion of a non-negative discriminator variable $\phi(x_i, t)$ such that $\phi \geq \epsilon > 0$ corresponds to turbulent flow at the point (x_i, t) considered and $\phi < \epsilon$ to non-turbulent. If randomness and nonzero vorticity are accepted as properties defining the turbulent state of the flow, the scalar ϕ is given by

$$\phi(x_i, t) = \omega'_\alpha \omega'_\alpha \tag{1}$$

where ω'_α is the vorticity fluctuation. The threshold ϵ for the discrimination must be larger than zero for finite Reynolds numbers, because of the elliptic nature of the differential equations governing the motion. Then the indicator function $I(x_i, t)$can be defined

$$I(x_i, t) = \begin{cases} 0 & \text{for } \phi(x_i, t) < \epsilon \\ 1 & \text{for } \phi(x_i, t) \geq \epsilon. \end{cases} \tag{2}$$

Due to temperature increase and change in number of moles, density variations up to a factor of 10 take place in flames.

Despite some basic objections with respect to pressure-velocity correlations and third order moments [2,7] Favre-averaging is applied to deal with the density variation problem. In view of the smaller number of non-closed terms density-weighted averaging [5] is more recommendable.

To deduce transport equations, different kinds of averaging procedures have to be applied. Conventional averaged turbulent-zone moments are defined by

$$\overline{\widehat{\phi}}, \qquad \phi = \overline{\widehat{\phi}} + \phi^1 \tag{3}$$

whereas non-turbulent zone moments are given by

$$\widehat{\overline{\phi}}, \qquad \phi = \widehat{\overline{\phi}} + \phi^0. \tag{4}$$

Turbulent and non-turbulent Favre-averaged moments can be expressed by

$$\widetilde{\overline{\widehat{\phi}}} = \frac{\overline{\rho I \phi}}{\overline{\rho}\widetilde{\gamma}}, \qquad \phi = \widetilde{\overline{\widehat{\phi}}} + \phi^{11} \tag{5}$$

$$\widetilde{\widehat{\overline{\phi}}} = \frac{\overline{\rho(1-I)\phi}}{\overline{\rho}(1-\widetilde{\gamma})}, \qquad \phi = \widetilde{\widehat{\overline{\phi}}} + \phi^{00}. \tag{6}$$

The transport equation for the density-weighted intermittency factor following the proposals of [1] and [8] may be written as

$$\frac{\partial\widetilde{\gamma}}{\partial t} + \overline{u}_k\frac{\partial\widetilde{\gamma}}{\partial x_k} = -\frac{\partial}{\partial x_k}\overbrace{(\widetilde{\gamma}(1-\widetilde{\gamma})(\widetilde{\overline{\widehat{u_k}}} - \widetilde{\widehat{\overline{u_k}}}))}^{D_{\gamma k}} + S_\gamma. \tag{7}$$

All density-weighted conditional moment equations can be cast in a form analogous to the equations for the unconditional moments plus a collection of interface terms. A complete closure model requires the equations for the turbulent zone and non-turbulent zone mean velocity

$$\begin{aligned}
\overline{\rho}\frac{\partial\widetilde{\overline{\widehat{u}}}_i}{\partial t} + \overline{\rho}\widetilde{u}_k\frac{\partial\widetilde{\overline{\widehat{u}}}_i}{\partial x_k} + \frac{\partial}{\partial x_k}(\overline{\widetilde{\widehat{\overline{\rho}u_k^{11}u_i^{11}}}}) &= -\frac{\partial\overline{\widehat{p}}}{\partial x_i}\frac{\overline{\rho}}{\overline{\overline{\widehat{\rho}}}} + g_i\overline{\rho} \\
&\quad + \frac{\partial}{\partial x_k}(\overline{\rho}\nu\frac{\partial\widetilde{\overline{\widehat{u}}}_i}{\partial x_k}) + \widetilde{\overline{\widehat{F_i}}} \\
\widetilde{\overline{\widehat{F_i}}} &= -\overline{\rho}(1-\widetilde{\gamma})(\widetilde{\overline{\widehat{u_k}}} - \widetilde{\widehat{\overline{u_k}}})\frac{\partial\widetilde{\overline{\widehat{u}}}_i}{\partial x_k} - \overline{\widetilde{\widehat{\overline{\rho}u_i^{11}u_k^{11}}}}\frac{\partial ln\widetilde{\gamma}}{\partial x_k} + \widetilde{\overline{\widehat{FR_i}}} \\
\widetilde{\overline{\widehat{FR_i}}} &= \frac{1}{\widetilde{\overline{\gamma}}}(\overline{u_i^{11}\rho V\delta(s) + p^1 n_i\delta(s)} - \overline{\rho\nu\frac{\partial u_i^{11}}{\partial x_k}n_k\delta(s)} \\
&\quad - \frac{\partial}{\partial x_k}\overline{(\nu\rho u_i^{11}n_k\delta(s)} + \nu\overline{\frac{\partial u_i^{11}}{\partial x_k}I\frac{\partial\rho}{\partial x_k}} - \nu\frac{\partial\widetilde{\overline{\widehat{u}}}_i}{\partial x_k}\widetilde{\gamma}\frac{\partial\overline{\rho}}{\partial x_k})
\end{aligned} \tag{8}$$

$$
\begin{aligned}
\bar{\rho}\frac{\partial \overline{\overline{\tilde{u}}}_i}{\partial t} + \overline{\rho}\tilde{u}_k\frac{\partial \overline{\overline{\tilde{u}}}_i}{\partial x_k} + \frac{\partial}{\partial x_k}(\overline{\widetilde{\widetilde{\rho u_k^{00} u_i^{00}}}}) &= -\frac{\partial \overline{\overline{p}}}{\partial x_i}\frac{\overline{\rho}}{\overline{\overline{\widehat{\rho}}}} + g_i\overline{\rho} \\
&\quad + \frac{\partial}{\partial x_k}(\overline{\rho}\nu\frac{\partial \overline{\overline{\tilde{u}}}_i}{\partial x_k}) + \overline{\overline{\widetilde{F_i}}} \\
\overline{\overline{\widetilde{F_i}}} &= \overline{\rho}\tilde{\gamma}(\overline{\overline{\widetilde{u_k}}} - \overline{\overline{\widehat{u_k}}})\frac{\partial \overline{\overline{\tilde{u}}}_i}{\partial x_k} - \overline{\widetilde{\widetilde{\rho u_i^{00} u_k^{00}}}}\frac{\partial \ln\tilde{\gamma}}{\partial x_k} + \overline{\widetilde{\widetilde{FR_i}}} \\
\overline{\widetilde{\widetilde{FR_i}}} &= \frac{1}{(1-\tilde{\gamma})}(-\overline{u_i^{00}\rho V\delta(s)} + \overline{p^0 n_i\delta(s)} - \overline{\rho\nu\frac{\partial u_i^{00}}{\partial x_k}n_k\delta(s)} \\
&\quad - \frac{\partial}{\partial x_k}\overline{(\nu\rho u_i^{00} n_k\delta(s)} + \nu\overline{\frac{\partial u_i^{00}}{\partial x_k}I\frac{\partial \rho}{\partial x_k}} - \nu\frac{\partial \overline{\overline{\tilde{u}}}_i}{\partial x_k}\tilde{\gamma}\frac{\partial \overline{\rho}}{\partial x_k}). \qquad (9)
\end{aligned}
$$

V is the relative velocity of the turbulent/non-turbulent interface. The intermittency source $V\delta(s)$ is the rate of entrainment of non-turbulent fluid per unit volume [1].

The turbulent-zone stress tensor satisfies the equation

$$
\begin{aligned}
\bar{\rho}\frac{\partial \overline{\widetilde{\widehat{u_i^{11}u_j^{11}}}}}{\partial t} + \overline{\rho}\tilde{u}_k\frac{\partial}{\partial x_k}(\overline{\widetilde{\widehat{u_i^{11}u_j^{11}}}}) &= \frac{\partial}{\partial x_k}[\nu\overline{\rho}\frac{\partial}{\partial x_k}(\overline{\widetilde{\widehat{u_i^{11}u_j^{11}}}} \\
&\quad -\overline{\rho}\overline{\widetilde{\widehat{u_k^{11}u_j^{11}u_i^{11}}}} - \frac{\overline{\rho}}{\overline{\overline{\widehat{\rho}}}}(\delta_{ik}\overline{\widehat{p^1 u_j^{11}}} + \delta_{jk}\overline{\widehat{p^1 u_i^{11}}})] - \overline{\rho}(\overline{\widetilde{\widehat{u_i^{11}u_k^{11}}}}\frac{\partial \overline{\overline{\tilde{u}}}_j}{\partial x_k} \\
&\quad + \overline{\widetilde{\widehat{u_j^{11}u_k^{11}}}}\frac{\partial \overline{\overline{\tilde{u}}}_i}{\partial x_k}) + \frac{\overline{\rho}}{\overline{\overline{\widehat{\rho}}}}\overline{\widehat{p^1(\frac{\partial u_i^{11}}{\partial x_j} + \frac{\partial u_j^{11}}{\partial x_i})}} - 2\nu\overline{\rho}\overline{\widetilde{\widehat{\frac{\partial u_i^{11}}{\partial x_k}\frac{\partial u_j^{11}}{\partial x_k}}}} \\
&\quad - \frac{1}{\overline{\overline{\tilde{\gamma}}}}(\overline{u_i^{11}I}\frac{\partial \overline{\overline{p}}}{\partial x_j} + \overline{u_j^{11}I}\frac{\partial \overline{\overline{p}}}{\partial x_i}) + \overline{\overline{\widetilde{F_{ij}}}} \\
\overline{\overline{\widetilde{F_{ij}}}} &= -\overline{\rho}(1-\tilde{\gamma})(\overline{\overline{\widetilde{u_k}}} - \overline{\overline{\widehat{u_k}}})\frac{\partial}{\partial x_k}(\overline{\widetilde{\widehat{u_i^{11}u_j^{11}}}}) + [2\nu\overline{\rho}\frac{\partial}{\partial x_k}\overline{\widetilde{\widehat{u_i^{11}u_j^{11}}}} \\
&\quad -\overline{\rho}\overline{\widetilde{\widehat{u_k^{11}u_j^{11}u_i^{11}}}} - \frac{\overline{\rho}}{\overline{\overline{\widehat{\rho}}}}(\delta_{ik}\overline{\widehat{p^1 u_j^{11}}} + \delta_{jk}\overline{\widehat{p^1 u_i^{11}}})]\frac{\partial \ln\tilde{\gamma}}{\partial x_k} \\
&\quad + \frac{1}{\overline{\overline{\tilde{\gamma}}}}[\overline{\widehat{u_i^{11}u_j^{11}}(-s_\gamma\rho} + \overline{\rho}\nu\frac{\partial^2\gamma}{\partial x_k^2})] + \overline{\widetilde{\widehat{FR_{ij}}}} \\
\overline{\widetilde{\widehat{FR_{ij}}}} &= \frac{1}{\overline{\overline{\tilde{\gamma}}}}[\overline{(u_i^{11}u_j^{11}\rho\delta(s)V)} + \overline{p^1\delta(s)(n_i u_j^{11} + n_j u_i^{11})}
\end{aligned}
$$

$$
\begin{aligned}
&-\frac{\partial}{\partial x_k}\overline{(\rho\nu u_i^{11}u_j^{11}n_k\delta(s))}-\frac{\partial}{\partial x_k}\overline{(u_i^{11}u_j^{11})\rho\nu n_k\delta(s))}\\
&+\overline{u_i^{11}I\nu\frac{\partial\rho}{\partial x_k}}\frac{\partial\widetilde{\widetilde{\widehat{u}}}_j}{\partial x_k}+\overline{u_j^{11}I\nu\frac{\partial\rho}{\partial x_k}}\frac{\partial\widetilde{\widetilde{\widehat{u}}}_i}{\partial x_k}+\frac{\partial}{\partial x_k}(\nu\widetilde{\widehat{u_i^{11}u_j^{11}}}\widetilde{\gamma}\frac{\partial\overline{\rho}}{\partial x_k}\\
&\qquad\qquad\qquad\qquad\qquad\qquad\qquad\qquad-\overline{\nu u_i^{11}u_j^{11}I\frac{\partial\rho}{\partial x_k}})\\
&\qquad\qquad\qquad\qquad\qquad\qquad\qquad\qquad+\nu\widetilde{\widehat{u_i^{11}u_j^{11}}}\frac{\partial\widetilde{\gamma}}{\partial x_k}\frac{\partial\overline{\rho}}{\partial x_k}].
\end{aligned}\tag{10}
$$

Non-turbulent density-weighted stress components can be connected with the turbulent stress tensor by algebraic relations

$$
\widetilde{\widehat{\overline{u_i^{00}u_j^{00}}}} = (c_{\gamma 8}\delta_{ij}+c_{\gamma 9}(1-\delta_{ij}))\widetilde{\gamma}\widetilde{\widehat{u_i^{11}u_j^{11}}}. \tag{11}
$$

The modelfunctions $c_{\gamma 8}$ and $c_{\gamma 10}$ can be taken from Table I.

The dissipation rate for the turbulent-zone using standard closure methods for the conventional terms [16,14] is governed by

$$
\begin{aligned}
\overline{\rho}\widetilde{\widehat{u}}_i\frac{\partial\widetilde{\widetilde{\epsilon}}}{\partial x_i} &= \frac{\partial}{\partial x_i}(\overline{\rho}c_{t\epsilon}\frac{\widetilde{\widetilde{k}}}{\widetilde{\widetilde{\epsilon}}}\widetilde{\widehat{u_i^{11}u_i^{11}}}\frac{\partial\widetilde{\widetilde{\epsilon}}}{\partial x_i})+\overline{\rho}[-\psi_1\frac{\widetilde{\widetilde{\epsilon}}}{\widetilde{\widetilde{k}}}\widetilde{\widehat{u_i^{11}u_j^{11}}}\frac{\partial\widetilde{\widetilde{\widehat{u}}}_i}{\partial x_j}\\
&-\psi_0\frac{\widetilde{\widetilde{\epsilon}}^2}{\widetilde{\widetilde{k}}}+\psi_2 g_i\frac{\widetilde{\widehat{\overline{\rho^1 u_i^1}}}}{(\widetilde{\overline{\rho}})^2}\rho_\infty\frac{\widetilde{\widetilde{\epsilon}}}{\widetilde{\widetilde{k}}}+c_{\gamma 1}\widetilde{\widehat{u_j^{11}u_j^{11}}}\frac{\widetilde{\widetilde{k}}}{\widetilde{\widetilde{\epsilon}}}(1-\widetilde{\gamma})\frac{\partial ln\widetilde{\gamma}}{\partial x_j}\frac{\partial\widetilde{\widetilde{\epsilon}}}{\partial x_j}\\
&+(2\nu\frac{\partial\widetilde{\widetilde{\epsilon}}}{\partial x_j}-\widetilde{\widehat{\epsilon^{11}u_j^{11}}})\frac{\partial ln\widetilde{\gamma}}{\partial x_j}+\frac{1}{\widetilde{\widetilde{\gamma}}}[\widetilde{\widetilde{\epsilon}}(-\overline{S_\gamma}+\nu\frac{\partial^2\widetilde{\gamma}}{\partial x_j{}^2})]+\widetilde{\widehat{\overline{FR}}}_\epsilon.
\end{aligned}\tag{12}
$$

3. Closure Model

The model for the standard non-closed terms in the stress equations are taken from [9] and [10].

The source term contributions of the turbulent and non-turbulent velocity are coupled by the intermittency source

$$
\widetilde{\widehat{\overline{FR}}}_i(1-\gamma) = -\widetilde{\widehat{\overline{FR}}}_i\gamma - S_\gamma(\widetilde{\widetilde{\widehat{u}}}_i-\widetilde{\widetilde{\widehat{u}}}_i). \tag{13}
$$

The amount of momentum transferred depends on the turbulent time scale and the crossing frequency of the interface which is given by S_γ. Then the following expression emerges

$$\overline{\widetilde{\widehat{FR_i}}} = -(\widehat{\widetilde{\tilde{u}}}_i - \widetilde{\widetilde{u}}_i)[(1-\gamma)^3 c_{\gamma 5}\frac{\widehat{\widetilde{\tilde{\epsilon}}}}{\widehat{\widetilde{\tilde{k}}}} + \frac{S_\gamma}{\gamma}. \tag{14}$$

The intermittency source and divergence terms, following the proposal of [1], are given by

$$D_{\gamma j} = c_{\gamma 1}(1-\tilde{\gamma})\frac{\overline{\widetilde{\widehat{u_k^{11}u_k^{11}}}}}{\widehat{\widetilde{\tilde{\epsilon}}}}\overline{\widetilde{\widehat{u_k^{11}u_k^{11}}}}\frac{\partial\tilde{\gamma}}{\partial x_j} \tag{15}$$

$$S_\gamma = \tilde{\gamma}(1-\tilde{\gamma})[-c_{\gamma 2}\frac{\widehat{\widetilde{\tilde{b}}}_{ij}}{\overline{\widetilde{\widehat{u_k^{11}u_k^{11}}}}}\frac{\partial\widehat{\widetilde{\tilde{u}}}_j}{\partial x_i} - c_{\gamma 3}\frac{\widehat{\widetilde{\tilde{\epsilon}}}}{\overline{\widetilde{\widehat{u_k^{11}u_k^{11}}}}}] + c_{\gamma 4}\frac{[\overline{\widetilde{\widehat{u_k^{11}u_k^{11}}}}]^2}{\widehat{\widetilde{\tilde{\epsilon}}}}\frac{\partial\gamma}{\partial x_k}\frac{\partial\gamma}{\partial x_k}. \tag{16}$$

The modelfunctions are listed in Table I.

Forming non-dimensional groups the intermittency source of the stress equations can be expressed by

$$\overline{\widetilde{\widehat{FR_{ij}}}} = c_{\gamma 6}(1-\tilde{\gamma})\widetilde{\tilde{\epsilon}}\frac{\overline{\widetilde{\widehat{u_i^{11}u_j^{11}}}}}{\widehat{\widetilde{\tilde{k}}}}. \tag{17}$$

The dissipation rate in the non-turbulent zone is negligible, thus the propagation of the interface provides a gain of dissipation rate. The model for $\overline{\widetilde{\widehat{FR_\epsilon}}}$ is then

$$\overline{\widetilde{\widehat{FR_\epsilon}}} = c_{\gamma 7}(1-\tilde{\gamma})\frac{\widehat{\widetilde{\tilde{\epsilon}}}^2}{\widehat{\widetilde{\tilde{k}}}}. \tag{18}$$

4. Combustion Model

The high Damköhler number normally present in diffusion flames suggests the assumption of an infinitely fast one-step reaction. Using the Shvab-Zeldowich formalism, linear combination of the element mass fractions for the chemical components results in the well-known mixture fraction equation [17].

To derive the statistical moments from the mixture fraction, its pdf has to be known. In this paper the β -function is used to evaluate the mixture fraction pdf from the conditional Favre-averaged mean value $\widehat{\widetilde{\tilde{f}}}$ and variance $\overline{\widetilde{\widehat{f^{11}f^{11}}}}$.

5. Cold Flow Results

Prediction models for turbulent combustion systems have to allow a precise description of different types of turbulent shear layers.

This universality is of special significance with respect to applications in complex three-dimensional flow systems combining all types of shear layers.

The prediction quality of the proposed model for round and plane jets is shown in an earlier paper [8]. In this study calculation results for boundary layes and wakes are presented.

The intermittency factor for the boundary Layer (Fig. 1) and the wake (Fig. 3) agrees well with the experiments [13,4]. With respect to the wake measurements it has to be taken into account that the intermittency factor was determined by a temperature threshold whereas the boundary layer plots are based on vorticity measurements. Due to the low Reynolds-number of the experiments, superlayer effects are not negligible. The temperature threshold method considers parts of this superlayer whereas the model predicts only the turbulent and non-turbulent-zone without super-layer effects. Figs. 2 and 4 show the relative magnitude and form of the conditional stress component in the main stream direction. The measurements compare well with the predictions.

6. Flame Results

In Fig. 5 the radial distribution of density, mixture fraction and intermittency factor in a H_2-air diffusion flame is plotted. The density profiles show the advantage of conditional averaging for diffusion flame prediction. The mean density changes by an order of magnitude but the density variation in turbulent-respectively non-turbulent-zone remains small. By applying conditional averaging the discussion with respect to the appropriate averaging method loses some of its importance.

Figs. 6 and 7 show an example for the application of the proposed model with respect to the prediction of buoyancy controlled CH_4-air diffusion flames. The results for mean velocity and stress component in the main stream direction are in fair agreement with LDA-measurements. To compare with these measurements the density-weighted velocities are converted to conventional averaged velocities [10].

7. Conclusions

A closure model with conditional density-weighted statistics has been developed for the prediction of diffusion flames. As demonstrated by prediction/measurement comparisons, the proposed model allows for the calculation of conditional moments and intermittency factor over a wide range of different types of turbulent shear layers. Applying conditional averaging for diffusion flame prediction, density variations are considerably smaller in the turbulent-respectively non-turbulent-zone compared to variations of the mean density. A more elaborate formulation of the model requires detailed measurements of conditional moments in flames.

References

[1] Byggstoyl, S. and Kollmann, W.; 1981: Closure Model for Intermittent Turbulent Flows, Int. J. Heat and Mass Transfer, 24, pp. 1811 - 1822

[2] Chen, J. Y.; Gouldin, F. C. and Lumley, J. L.; 1987: Second-order Modeling of a Turbulent Nonpremixed H2-air Jet Flame with Intermittency and Conditional Averaging, Combust. Sci. and Tech., 53, pp. 235 - 257

[3] Dopazo, C.; 1977: On Conditioned Averages for Intermittent Turbulent Flows, J. Fluid Mech., 81, pp. 433 - 438

[4] Fabris, G.; 1979: Conditional Sampling Study of the Turbulent Wake of a Cylinder, Part 1, J. Fluid Mech., 94, pp. 673 - 709

[5] Favre, A.; 1965: The Equations of Compressible Turbulent Gases, USAF Contract AF61 (052)-772, AD 622097

[6] Hedley, T. B. and Keffer, J. F.; 1974: A Turbulent/Non-turbulent Decisions in an Intermittent Flow, J. Fluid Mech., 64, pp. 625 - 644

[7] Janicka, J. and Lumley, J. L.; 1980: Second-order Modeling in Non-constant Density Flows, Cornell University, Sibley School, Mech. and Aerospace Engineering Report FDA-81-01

[8] Janicka, J. and Kollmann W.; 1985: Reynolds Stress Closure for Conditional Variables, in "Turbulent Shear Flow 4" edited by L. J. S. Bradbury et al. Springer Verlag, pp. 73 - 86

[9] Janicka, J.; 1988: A Reynolds Stress Model for the Prediction of Diffusion Flames, 21st Symposium on Combustion, pp. 1409 - 1418

[10] Janicka, J.; 1988: On Modelfunctions of Reynolds Stress Models, Physics of Fluids, 31, pp. 49 - 55

[11] Knoche, K. F.; Stojanoff, C. G.; Altgeld, H.; Janicka, J. and El-Dib, A. W.; 1979: Geschwindigkeits- und Konzentrationsmessungen in turbulenten Wasserstoffdiffusionsflammen, "9. Deutscher Flammentag", VDI-Bericht Nr. 346, pp. 93 - 102

[12] Knoche, K. F.; Janicka, J. and Oebels, R.; 1981: Experimentelle und theoretische und Untersuchungen an turbulenten Methan-Luft-Diffusionsflammen, VDI-Bericht Nr. 423

[13] Kovasznay, L. S. G.; Kibens, V. and Blackwelder, R. F.; 1970: Largescale Motion in the Intermittent Region of a Turbulent Boundary Layer, J. Fluid Mech., 41, pp. 283 - 325

[14] Launder, B. E.; Reece, G. J. and Rodi, W.; 1975: Progress in the Development of a Reynolds Stress Turbulence Closure, J. Fluid Mech., 68, pp. 537 - 566

[15] La Rue, J. C. and Libby, P. A. 1974: Temperature Fluctuations in the Plane Turbulent Wake, Phys. Fluids, 17.

[16] Lumley, J. L.; 1978: Computational Modeling of Turbulent Flows, Advances in Applied Mechanics, 18, pp. 123 - 176

[17] Lockwood, F. C. and Naguib, A. S.; 1975: The Prediction of the Fluctuations in the Properties of Free Round-jet Turbulent Diffusion Flames, Comb. and Flame, 24, pp. 109 - 124

$c_{\gamma 1} = 0.225$	$c_{\gamma 4} = 0.2$	$c_{\gamma 7} = 2.0$
$c_{\gamma 2} = 0.0135$	$c_{\gamma 5} = 7.5$	$c_{\gamma 8} = 0.5$
$c_{\gamma 3} = 4.25$	$c_{\gamma 6} = 1.0$	$c_{\gamma 9} = 0.1$

Table I: Model constants

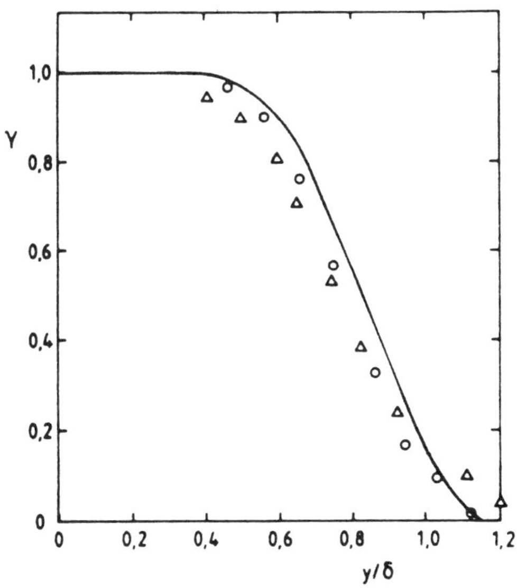

Fig. 1.: Intermittency factor in boundary layers.
Measurements: o *Kovasznay et al. 1970 [13],* △ *Hedley and Keffer 1974 [6].*
Predictions: —.

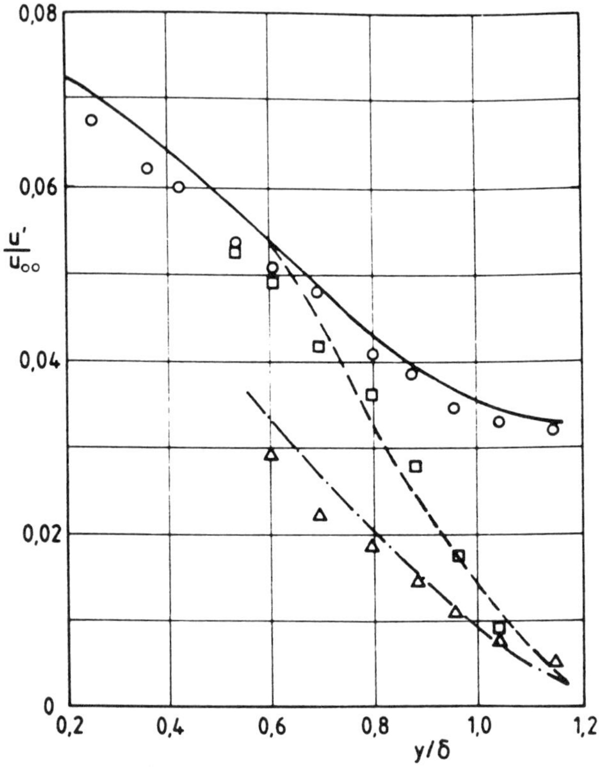

Fig. 2.: Normalized fluctuations in boundary layers
Measurements: Kovasznay et al. 1970 [13]) ○ u^{11}/u_∞; □ u'/u_∞; △ u^{00}/u_∞. *Predictions:* — .

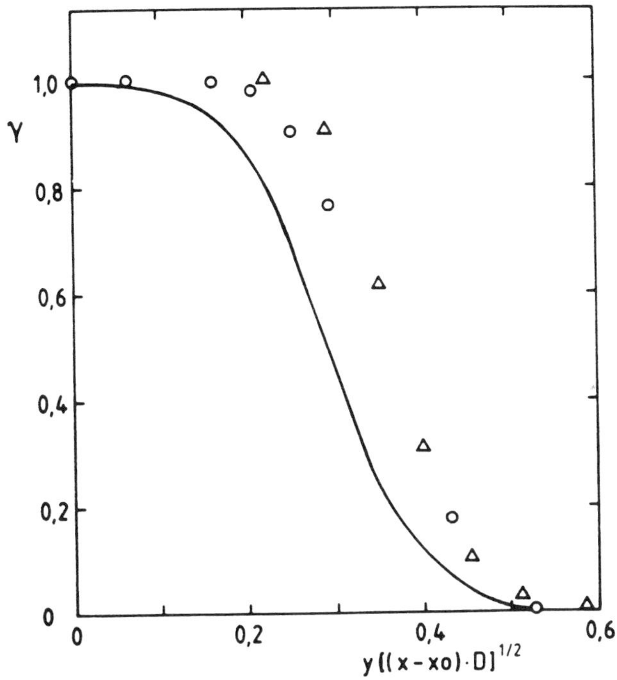

Fig. 3.: Intermittency factor in wakes
Measurements: ○ La Rue and Libby 1974 [15], △ Fabris 1979 [4].
Predictions: — .

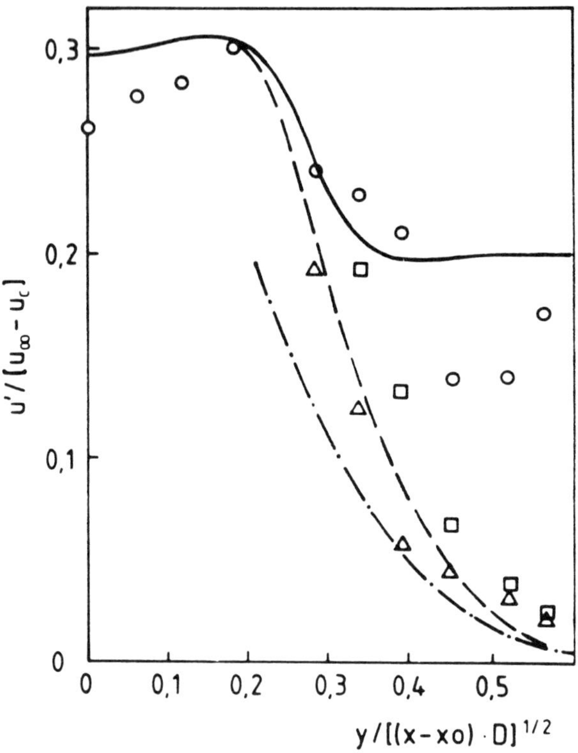

Fig. 4.: Normalized fluctuations in wakes
Measurements: Fabris 1979 [4] ○ $u^{11}/(u_\infty - u_c)$; □ $u'/(u_\infty - u_c)$; △ $u^{00}/(u_\infty - u_c)$.
Predictions: — .

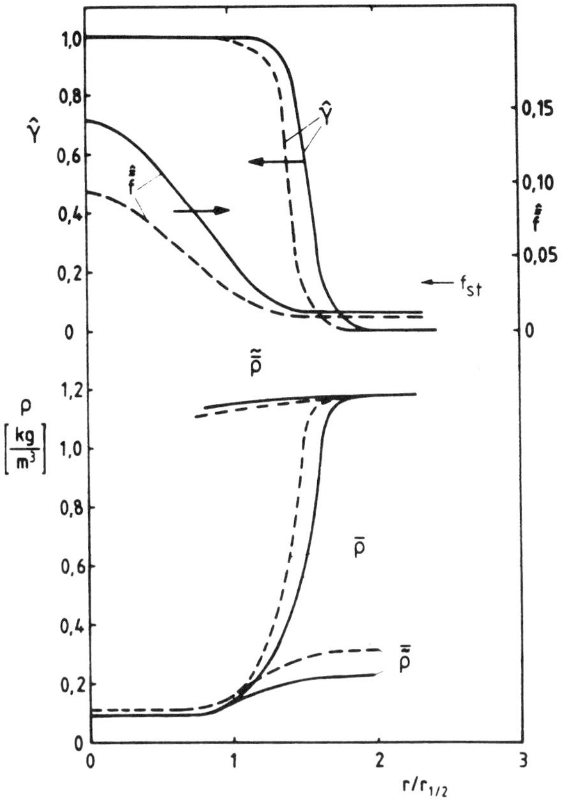

Fig. 5.: Radial distribution of densities
($\overline{\overline{\rho}}$: turbulent zone, $\widehat{\overline{\rho}}$: non-turbulent zone, $\overline{\rho}$: mean), mixture fraction $\tilde{\tilde{\hat{f}}}$ and intermittency factor $\tilde{\gamma}$ in a H_2-air diffusion flame — $x/D = 30$, - - $x/D = 50$.

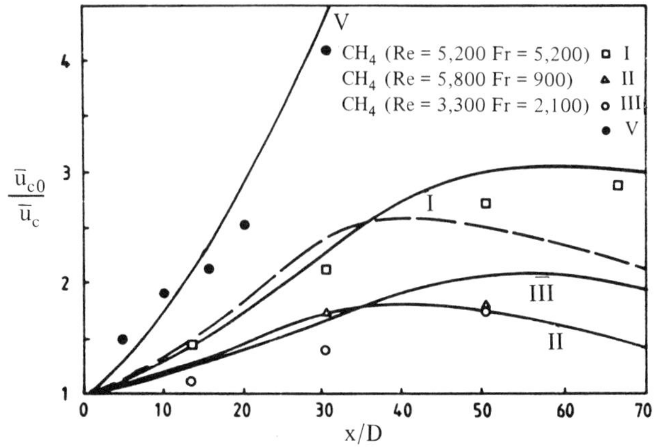

Fig. 6.: Mean velocity distribution on the axis of CH_4-air diffusion flames
Measurements: Knoche et al. 1981 [11]□, △, ○, ●
Predictions: I, II, III, IV

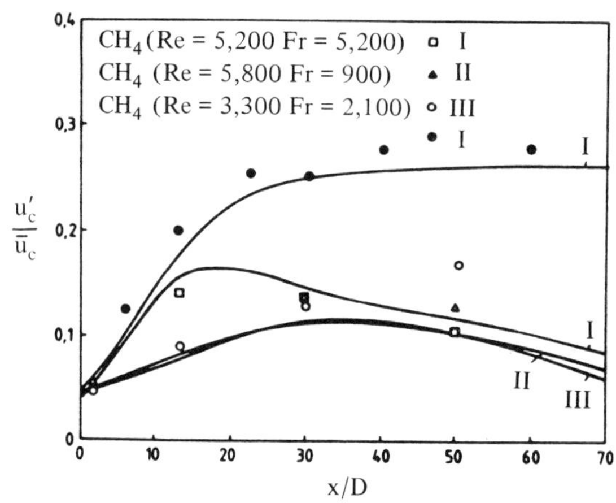

Fig. 7.: Normalized fluctuations on the axis of CH_4-air diffusion flames
Measurements: Knoche et al. 1981 [11] □, △, ○, ●
Predictions: I, II, III, IV

INTERACTION BETWEEN CHEMICAL REACTION AND TURBULENCE IN SUPERSONIC NONPREMIXED H_2-AIR COMBUSTION[1]

R. Villasenor, J.-Y. Chen[2], *and R. W. Pitz*

Vanderbilt University
Nashville, TN 37235

ABSTRACT

A numerical model for supersonic turbulent nonpremixed hydrogen-air jet flames is developed for exploring the interaction between chemical reaction and turbulence. The numerical code employs an implicit finite-volume, lower-upper successive overrelaxation (LU-SSOR) scheme for solving density-averaged Navier-Stokes equations with chemical reactions in axisymmetric flows. The effects of turbulence on chemical reaction are included by approximations appropriate in the limit of weak turbulence. Comparison of numerical results with and without the influence of turbulence on the chemical kinetics reveals that even under the condition of weak turbulence, the effect of turbulence on chemistry can be significant.

1. Introduction

With the recent renewed interest in hypersonic vehicles, there is substantial interest in developing efficient numerical codes to model supersonic turbulent flames for H_2-Air combustion. In the past few years, a number of numerical schemes capable of solving the complete 2-*d* Navier-Stokes equations coupled with chemical reaction has surfaced for both confined and unconfined compressible flows (Shuen and Yoon, 1988; Eklund *et al.*, 1989; Tsai and Hsieh, 1990). The focus of these codes has been on the numerical efficiency rather than

---

[1] This Work was supported by the U.S. DOE, Office of Basic Energy Sciences, Division of Chemical Sciences, and by NSF grant CBT/8657131

[2] Sandia National Laboratories, Livermore, California

on the development of physical models for turbulent combustion. The importance of turbulence on chemical reactions has been largely ignored by making the assumption that the level of turbulence is small due to the weak mixing occurring in supersonic flows. This assumption totally neglects the important role of turbulent mixing, which can have significant influence on the chemical reaction. Recent experimental data (Chen *et al.*, 1990) for temperature and species concentrations in a supersonic combustor indicate that the species and temperature fluctuations can be as high as 40% and 20%, respectively. Ignorance of these fluctuation effects on the mean chemical reaction rates could lead to an erroneous flow description that might yield an incorrect prediction of minor species, and consequently incorrect ignition point.

To adequately simulate the behavior of turbulent flames, the effects of turbulence on the chemical reaction must be included. Recently, a numerical study (Villasenor *et al.*, 1990) has been carried out for a perfectly expanded jet flow without shock waves so that the flow can be treated as a parabolic flow. The main effort in this study was focused on the development of a modified k-ϵ model to account for the compressibility effect on the mixing rate and on the potential departure from equilibrium chemistry. A partial equilibrium chemical model was used so that the combustion process can be described only by two scalars, a mixing scalar and a reaction progress variable. Interactions between turbulence and chemical kinetics are modeled by an assumed joint probability density function (*pdf*) between the mixing scalar and the reaction scalar. Noticeable difference in the predicted minor species have been observed when the nonequilibrium chemistry is considered as well as when the effects of turbulence are included.

The present effort differs and complements the above-mentioned study in two areas. First, the flow is not assumed parabolic, and thus possible shock wave interactions with combustion are included. Second, Chen and Kollmann (1990a) demonstrated that the partial equilibrium assumption breaks down at low temperatures; therefore, simplified chemical model is unable to predict correct ignition location. In the present study, a detailed chemical kinetic scheme is incorporated so that the ignition process can be predicted.

In order to explore the potential effects of turbulence on chemical reactions in supersonic flows, we develop approximations that account for the influence of temperature and species fluctuations in the limit of weak turbulence. Although the fluctuations of species concentrations and temperature are intimately related to one another, modeling the cross correlations are extremely difficult. Hence, as a first step to understanding the effects of fluctuations on chemistry, the temperature fluctuations and the species fluctuations will be decoupled, and the individual effects on the flow structure will be studied.

The paper is organized as follows. First, the governing equations are presented in their nondimensional forms so that the characteristic parameters, such as Mach number, Reynolds number, etc., can be easily identified. Second, the algebraic turbulent mixing length model is briefly described. Third, the approximations that account for the effects of temperature and species fluctuations on the chemical processes are derived. Finally, numerical results are presented and compared with the experimental data (Evans *et al.*, 1978) for a supersonic axisymmetric co-flowing jet, and the implications in future model development are offered.

2. Governing Equations

For two-dimensional flows, the density-averaged Navier-Stokes and the species transport equations can be cast into nondimensional form as

$$\frac{\partial \mathbf{Q}}{\partial t} + \frac{\partial}{\partial x}(\mathbf{F} - \mathbf{F}_v) + \frac{\partial}{\partial y}(\mathbf{G} - \mathbf{G}_v) = \mathbf{S}. \tag{1}$$

The elements of the vectors in Eq. (1) are defined by

$$\mathbf{Q} = y^m \begin{pmatrix} \bar{\rho} \\ \bar{\rho}\tilde{u} \\ \bar{\rho}\tilde{v} \\ \bar{\rho}\tilde{e}_t \\ \vdots \\ \bar{\rho}\tilde{Y}_i \end{pmatrix}, \quad \mathbf{F} = y^m \begin{pmatrix} \bar{\rho}\tilde{u} \\ \bar{\rho}\tilde{u}^2 + \bar{P} \\ \bar{\rho}\tilde{u}\tilde{v} \\ \tilde{u}[\bar{\rho}\tilde{e}_t + \bar{P}] \\ \vdots \\ \bar{\rho}\tilde{u}\tilde{Y}_i \end{pmatrix},$$

$$\mathbf{G} = y^m \begin{pmatrix} \bar{\rho}\tilde{v} \\ \bar{\rho}\tilde{u}\tilde{v} \\ \bar{\rho}\tilde{v}^2 + \bar{P} \\ \tilde{v}[\bar{\rho}\tilde{e}_t + \bar{P}] \\ \vdots \\ \bar{\rho}\tilde{v}\tilde{Y}_i \end{pmatrix}, \quad \mathbf{F}_v = y^m \begin{pmatrix} 0 \\ Re_\infty^{-1}(\bar{\tau}_{xx})_T \\ Re_\infty^{-1}(\bar{\tau}_{yx})_T \\ \Lambda_1 \\ \vdots \\ -\bar{\rho}^* \left(\frac{\tilde{D}_n}{l a_\infty}\right)_T \frac{\partial \tilde{Y}_i}{\partial x} \end{pmatrix},$$

$$\mathbf{G_v} = y^m \begin{pmatrix} 0 \\ Re_\infty^{-1}(\bar{\tau}_{yx})_T \\ Re_\infty^{-1}(\bar{\tau}_{yy})_T \\ \Lambda_2 \\ \vdots \\ -\bar{\rho}\left(\frac{\tilde{D}_n}{l a_\infty}\right)_T \frac{\partial \tilde{Y}_i}{\partial y} \end{pmatrix}, \mathbf{S} = \begin{pmatrix} 0 \\ 0 \\ \delta_{1m}[\bar{P} + Re_\infty^{-1}(\bar{\tau}_{\theta\theta})_T] \\ 0 \\ \vdots \\ y^m \frac{l}{\rho_\infty a_\infty} \overline{\dot{w}_n} \end{pmatrix}, \tag{2}$$

where

$$\Lambda_1 = Re_\infty^{-1}\left[\tilde{u}(\bar{\tau}_{xx})_T + \tilde{v}(\bar{\tau}_{yx})_T + \frac{\tilde{\lambda}_T \frac{\partial \tilde{T}}{\partial x}}{\widetilde{Pr}_\infty(\tilde{\gamma}_\infty - 1)}\right] + \bar{\rho}\sum_{i=1}^{N} \tilde{h}(\tilde{D}_i)_T \frac{\partial \tilde{Y}_i}{\partial x}$$

and

$$\Lambda_2 = Re_\infty^{-1}\left[\tilde{u}(\bar{\tau}_{yx})_T + \tilde{v}(\bar{\tau}_{yy})_T + \frac{\tilde{\lambda}_T \frac{\partial \tilde{T}}{\partial y}}{\widetilde{Pr}_\infty(\tilde{\gamma}_\infty - 1)}\right] + \bar{\rho}\sum_{i=1}^{N} \tilde{h}(\tilde{D}_i)_T \frac{\partial \tilde{Y}_i}{\partial y}.$$

The stress tensor components are given by

$$\begin{aligned} \bar{\tau}_{xx} &= \tilde{\mu}_T\left\{2\frac{\partial \tilde{u}}{\partial x} - \frac{2}{3}\left[\frac{1}{y^m}\frac{\partial}{\partial y}(y^m \tilde{v}) + \frac{\partial \tilde{u}}{\partial x}\right]\right\}, \\ \bar{\tau}_{yy} &= \tilde{\mu}_T\left\{2\frac{\partial \tilde{v}}{\partial y} - \frac{2}{3}\left[\frac{1}{y^m}\frac{\partial}{\partial y}(y^m \tilde{v}) + \frac{\partial \tilde{u}}{\partial x}\right]\right\}, \\ \bar{\tau}_{\theta\theta} &= \tilde{\mu}_T\left\{2\frac{\tilde{v}}{y} - \frac{2}{3}\left[\frac{1}{y^m}\frac{\partial}{\partial y}(y^m \tilde{v}) + \frac{\partial \tilde{u}}{\partial x}\right]\right\}, \\ \bar{\tau}_{yx} &= \tilde{\mu}_T\left\{\frac{\partial \tilde{u}}{\partial y} + \frac{\partial \tilde{v}}{\partial x}\right\}. \end{aligned} \tag{3}$$

For axisymmetric flows $m = 1$, whereas the value of 0 is taken for planar flows. The subscript T denotes both laminar and turbulent components. In the above equations, the outer jet stream properties are used for normalization.

3. Turbulence Closure

The turbulence closure is provided by the Eggers (1971) algebraic mixing length model which is defined as $\mu_t = c_e \rho u_{cl} z$, where c_e is a model constant (=0.0075 in this work), u_{cl} is the streamwise velocity on the jet centerline, and z is the width of the mixing layer. This width is defined as the radial distance between the two velocities $u_{0.95}$ and $u_{0.50}$ given by $u_{0.95} = u_\infty + 0.95(u_{cl} - u_\infty)$ and $u_{0.50} = u_\infty + 0.50(u_{cl} - u_\infty)$, where u_∞ is the outer jet velocity.

4. Influence of Turbulence on Combustion

To describe the chemistry processes, a thoroughly studied H_2-Air kinetic mechanism (Hitch *et al.*, 1986) for use in supersonic flows is chosen. The mechanism contains eighteen steps and eight species (H_2, O_2, OH, H_2O, H, O, HO_2, and H_2O_2) with N_2 coexisting as an inert gas. To illustrate the effects of turbulent fluctuations on the mean chemical reaction rates, let's consider the following bimolecular reaction in forward direction only $A + B \to C + D$ with the production rate for species A defined as

$$\dot{w}_A = -\rho^2 k_f \frac{Y_A Y_B}{M_A M_B}, \qquad k_f = c_f T^\alpha e^{-\frac{T_a}{T}}, \tag{4}$$

where c_f denotes the pre-exponential factor and $T_a \equiv E_a/R$ with E_a being the activation energy. The mean source term can be expanded as

$$\overline{\dot{w}_A} = \frac{-1}{M_A M_B} \Big\{ \bar{\rho} \tilde{Y}_A \tilde{Y}_B (\widetilde{\rho k_f}) + \overline{\rho Y_A'' Y_B''} (\widetilde{\rho k_f}) +$$

$$\overline{\rho (\rho k_f)'' Y_B''} \tilde{Y}_A + \overline{\rho (\rho k_f)'' Y_A''} \tilde{Y}_B + \overline{\rho (\rho k_f)'' Y_A'' Y_B''} \Big\}, \tag{5}$$

which demonstrates the complexity of the influence of turbulence on the mean chemical reaction. In particular, the cross correlations of temperature and mass fraction are difficult to model; therefore, in this work, we will consider fluctuations in concentration and temperature separately. Although this simplification may be unrealistic, useful insight can be gained by observing the isolated effects of temperature and concentration fluctuations on the structure of the flow.

First, a segregation parameter can be used to describe the concentration fluctuations. For simplicity, the term $(\rho k_f)''$ are assumed to be negligible and Eq. (5) is reduced to

$$\overline{\dot{w}_A} \approx \frac{-\bar{\rho}^2 \bar{k}_f}{M_A M_B} \tilde{Y}_A \tilde{Y}_B \Big[1 - \alpha_{AB}\Big], \tag{6}$$

where α_{AB} is the segregation parameter defined as $\alpha_{AB} = -\widetilde{Y_A'' Y_B''}/\tilde{Y}_A \tilde{Y}_B$, and $\bar{k}_f = k_f(\tilde{T})\theta_f^*$, with θ_f^* to be defined later. It is clear that the segregation parameter represents the influence of the species fluctuations on the mean chemical reaction rate. Likewise, θ_f^* represents the effects of temperature fluctuations on the mean chemical reaction rate. When the source term is evaluated with mean flow properties (i.e., laminar chemistry model), $\theta_f^*=1$ and $\alpha_{ij} = 0$.

Recently, the nature of segregation parameters has been explored by Chen and Kollmann (1990b) for hydrogen-air combustion of turbulent subsonic flames. Their analysis showed that the segregation parameters can be positive or negative in a multi-species combusting flow, and more importantly, the signs of segregation parameters can be correlated with the thermochemical state of the species. For example, the H atom has its peak concentration around the stoichiometric point and so does the OH radical; therefore, the segregation parameter $\alpha_{OH,H}$ should be negative as OH and H are positively correlated. To explore the effects of species fluctuations on the reacting flow, the magnitude of the segregation parameters is specified and the equilibrium thermochemical relation is used to determine the signs for the segregation parameters.

Next, we consider the effect of temperature fluctuations on the reaction rates. Since the reaction rate constant is exclusively a function of temperature, one can express the mean rate constant as $\bar{k}_f = \int k_f P(T) dT$ with $P(T)$ being the *pdf* of temperature. To study the effects of temperature fluctuations in the limit of weak turbulence, we assume that the *pdf*, $P(T)$, can be approximated by a Gaussian distribution with the standard deviation σ_T. Strictly speaking, this pdf is not suitable for temperature which is a bounded scalar. However, restricting to problems with small turbulence fluctuation levels (i.e., 5% or less temperature variation) this *pdf* may suffice to represent the temperature fluctuations. Now θ_f^* is defined as the ratio between $\bar{k}_f$ and $k_f(\bar{T})$

$$\theta_f^* \equiv \frac{1}{\sqrt{2\pi}\sigma_T} \int_{-\infty}^{\infty} \frac{T^\alpha}{\bar{T}} exp\left[\frac{T_a}{\bar{T}} - \frac{T_a}{T} - \frac{(T-\bar{T})^2}{2\sigma_T^2}\right] dT, \tag{7}$$

which gives us a quantitative measure of the influence of temperature fluctuations on the mean chemical reaction rate. A value of 0.05 was chosen for $\sigma_T/\bar{T}$ in the present study.

5. Results and Discussions

The current numerical code is a modified version of the LU-SSOR scheme written by Shuen and Yoon (1988) for supersonic premixed flames in Cartesian coordinates. For reacting supersonic axisymmetric jets, the flow conditions in the experiments of Evans *et al.* (1978) are used to evaluate the performance of the present model in axisymmetric flows. To be consistent with the experimental data, the outside diameter of the H_2 injector is used to normalize the radial profiles and the corresponding contour plots. Three numerical simulations are performed. In Case 1, calculations are carried out with laminar chemistry model, i.e., without the influence of any turbulent fluctuations. In Case 2, the magnitude of all segregation parameters is set to 30% but without temperature fluctuations. In Case 3, a 5% level of temperature fluctuations is considered with no concentration fluctuations.

To assess the model's capability in predicting the turbulent mixing, we examine a mixing scalar, *the mixture fraction*, f, which is defined as the normalized mass fraction of an atomic element originating from one of the input streams, usually the fuel stream. With the assumption of equal diffusivity, f is independent of the progress of chemical reaction and it indicates how well the two stream has been mixed. A comparison of the predicted and measured mixture fraction distributions are plotted at two locations in Fig. 1 for Case 3. As can be seen from Fig. 1a, the predictions are in satisfactory agreement with the experimental data at x/d=8.26. However, significant disagreement between the predictions and the experimental data are noticed at x/d=15.5 as presented in Fig. 1b. Also presented are the corresponding mixture fraction profiles obtained by Villasenor *et al.* (1990) with a parabolic code using a modified $k - \epsilon - g$ turbulence model to account for the compressibility effect on the mixing rate. Two solutions, parabolic with compressibility effects (*P.C.E.*) and parabolic with no compressibility effects (*P.N.C.E.*) are included for comparison. It is clear that both the P.N.C.E. model and the present model overpredict the turbulent mixing rate resulting in lower mixture fraction values compared to the data. This suggests that in order to predict the correct spreading rates the compressibility effects need to be included in the calculation.

To appreciate the effects of turbulence on combustion, the corresponding minor species at the axial location x/d=8.26 are presented in Figs. 2a, 2b and 2c for Case 1, 2 and 3 respectively. It is worth noting that by including fluctuations (either temperature or species), the OH concentration is enhanced relative to the O atom or H atom concentrations, but 5% temperature fluctuations lead to larger OH. This can be explained by the fact that the most temperature-dependent step $H + O_2 \rightleftharpoons OH + O$ is greatly enhanced in the forward direction by the temperature fluctuation.

An interesting feature of Case 2 is that the minor species profiles are lower than those obtained with laminar chemistry. This can be best understood in a bimolecular reacting system where the segregation parameter is always positive. From Eq. (6), the mean consumption rate reaches its maximum value only when the segregation paramcter approaches zero. Consequently, any unmixed fluids tend to decrease the mean chemical reaction rate. However, in multicomponent reacting systems, the segregation parameters can be positive, negative, or zero. It is interesting to notice that despite the complexity of our present chemical reaction mechanism, the overall contribution of these correlations on the species fluctuation terms tends to reduce the reaction rates, a similar trend that is well-understood in a simple bimolecular system.

The diffusion of oxygen into the hydrogen jet is another issue of interest. Isopleths for mass fraction of O_2 for Cases 1, 2, and 3 are shown in Figs. 3a, 3b, and 3c, respectively. Before chemical reaction can occur both fuel and oxidizer have to thoroughly mix to the molecular level. The mixing region is established in the vicinity of the injector lip thickness (from r/d=0.33 to 0.5) where the flow is slightly compressed. Case 2 presents the largest penetration of oxygen, but the opposite trend occurs when the reaction rates are modified by the temperature fluctuations. From these results, it is inferred that ignition is enhanced by temperature fluctuations, and inhibited by concentration fluctuations. This finding is consistent with the observation obtained from minor species profiles as reaction rates of two-body reactions (chain-branching steps) are intensified by temperature fluctuations, whereas the three-body reactions (radical termination steps) are less temperature sensitive.

6. Conclusions

The effects of species and temperature fluctuations on chemical processes have been independently explored for a supersonic turbulent reacting jet under weak turbulence. First, comparisons of predictions and experimental data (Evans *et al.*, 1978) show that the Eggers (1971) algebraic turbulence model gives a poor prediction of the mixing process in supersonic jet flames, and future improvements are needed. Second, comparison of numerical results with and without the influence of turbulence on the chemical kinetics reveals that even under the condition of weak turbulence, the effect of turbulence on chemistry can be significant. For example, the OH concentrations are increased relative to the O atom and the H atom when either temperature or species fluctuations are included. Due to the thick nozzle lip, the flame ignites near the injector lip in all three cases. Although the ignition position appears unchanged, flame ignition is promoted when temperature fluctuations are considered, based on both computed radical concentrations and reduced penetration of oxidizer into the fuel jet.

Acknowledgments

A note of appreciation is expressed to Drs. Peter Tsai and Beverly Duncan at Sverdrup Technology, Inc. for their permission and assistance of using their LU-SSOR computer codes, and to Dr. Mohammad Farshchi of Nielsen Engineering & Research, Inc. for his enlightened discussions. Also the valuable comments of Professor A. M. Mellor at Vanderbilt University are gratefully acknowledged.

References

[1] Chen, J.-Y., and Kollmann, W., "Chemical Models for Pdf Modeling of Hydrogen-Air Nonpremixed Turbulent Flames," *Combustion and Flame*, Vol. 79, pp. 75-99, 1990a.

[2] Chen, J.-Y., and Kollmann, W.,"Segregation Parameters and Pair-Exchange Mixing Models for Turbulent Nonpremixed Flames,"presented at the 23rd Symposium (International) on Combustion, Orleans, France, July 1990b.

[3] Chen, T. S., Wehrmeyer, J. A., Pitz, R. W., Jarret, O., and Northam, G. B., "Simultaneous Temperature and Multi-Species Measurements in a Supersonic Reacting Flow by a KrF Excimer Laser," Eastern States Meeting of the Combustion Institute, Orlando, Florida, Dec. 3-5, 1990.

[4] Eggers, J. M., "Turbulent Mixing of Coaxial Compressible Hydrogen-Air Jets," *NASA* TN D-6487, Sept., 1971.

[5] Eklund, D. R., Drummond, J. P., and Hassan, H. A., "Numerical Modeling of Turbulent Supersonic Reacting Coaxial Jets," 27th Aerospace Sciences Meeting, *AIAA* Paper No. 89-0660, Jan. 1989.

[6] Evans, J. S., Schexnayder, C. J., and Beach, H. L., "Application of a Two-Dimensional Parabolic Computer Program to

Prediction of Turbulent Reacting Flows," NASA TP-1169, pp. 1-56, 1978.

[7] Hitch, B. D., Laxter, W. R., Senser, D. W., and Sojka, P. E., "On the Selection of H_2-Air Kinetic Mechanisms for Use in Supersonic Combustor Modeling," presented at the Spring 1986 Technical Meeting of the Central States Section of the Combustion Institute, Cleveland, OH, May 1986.

[8] Shuen, J. S., and Yoon, S., "Numerical Study of Chemically Reacting Flows Using an LU Scheme," 26th Aerospace Sciences Meeting, *AIAA* Paper No. 88-0436, Jan. 1988.

[9] Tsai, Y. L., and Hsieh, K. C., "Comparative Study of Computational Efficiency of Two LU Schemes for Non-Equilibrium Reacting Flows," 28th Aerospace Sciences Meeting, *AIAA* Paper No. 90-0396, Jan. 1990.

[10] Villasenor, R., Chen, J. Y., and Pitz, R. W., "Modeling Ideally Expanded Supersonic Turbulent Jet Flows with Non-premixed H_2-Air Combustion," 28th Aerospace Sciences Meeting, *AIAA* Paper No. 90-0640, Jan. 1990.

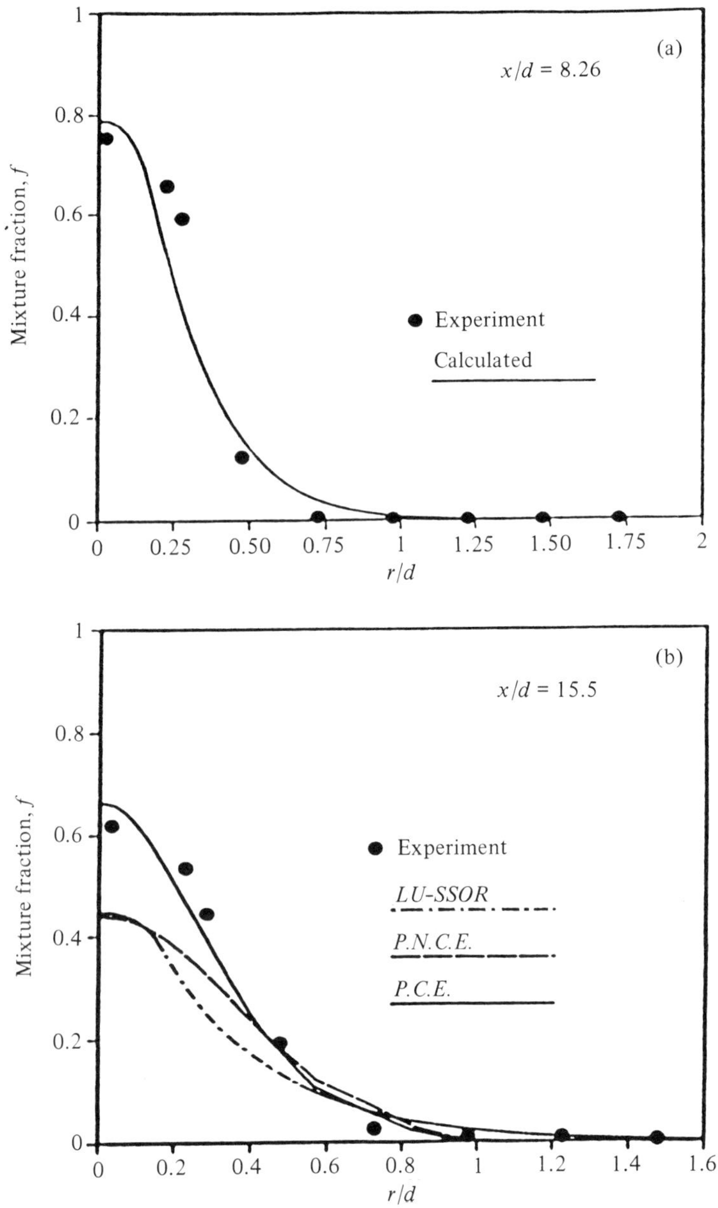

Fig. 1. Radial profiles of mixture fraction for Case 3 (5% temperature fluctuation). (a) x/d=8.26; (b) x/d=15.5 also the corresponding mixture fraction profiles obtained by Villasenor *et al.* (1990) with a parabolic code showing the compressibility effect.

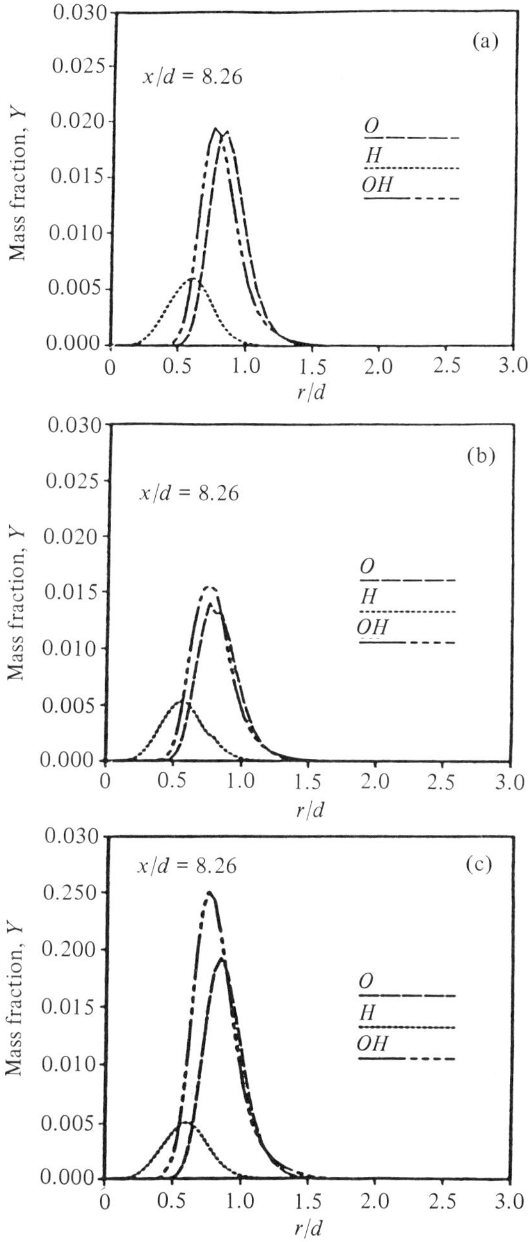

Fig. 2. Minor species concentration radial distribution at x/d=8.26. (a) Case 1: laminar chemistry; (b) Case 2: with 30% species fluctuation; (c) Case 3: with 5% temperature fluctuation.

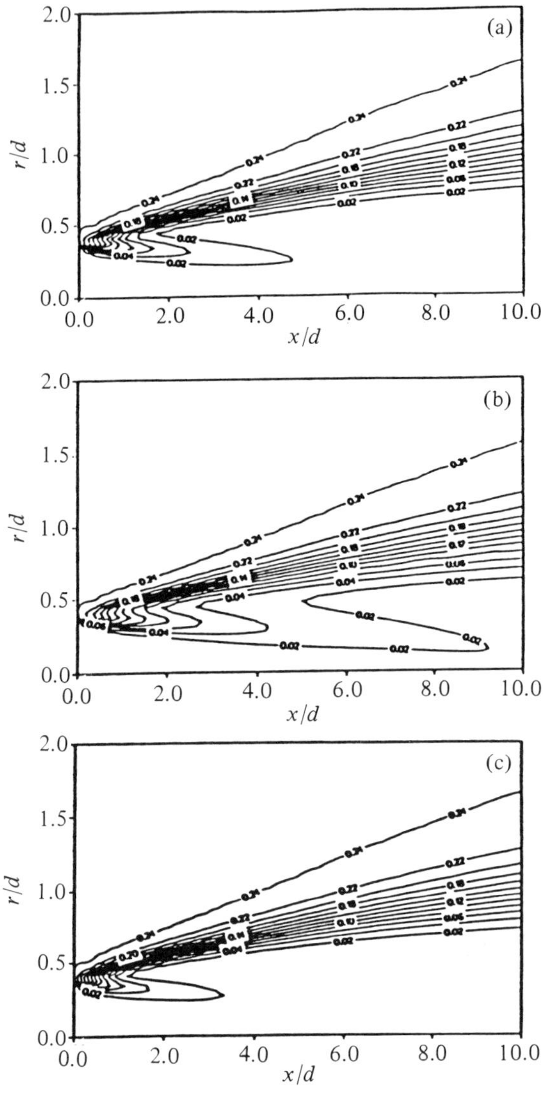

Fig. 3. Contour Plot of O_2 mass fraction showing the penetration of oxygen into the hydrogen jet. (a) Case 1: laminar chemistry; (b) Case 2: with 30% species fluctuation; (c) Case 3: with 5% temperature fluctuation.

COMPRESSIBLE TURBULENCE AND SHOCK WAVES

S. K. Lele, S. Lee and P. Moin

Stanford University
Stanford, California 94305

ABSTRACT

If the velocity scale associated with the turbulent fluctuations is a substantial fraction of the mean speed of sound, compressibility effects are expected to arise. This direct effect of compressibility, which is also the most well documented, appears to be responsible for the reduced spreading rate of turbulent mixing layers under compressible conditions, and the slower decay of hypersonic wakes. A different circumstance where the compressibility effects can become significant occurs when an otherwise quasi-incompressible turbulent flow is subjected to a rapidly changing environment. For example a turbulent flow passing through a shock wave or a steep expansion wave may suddenly enter in a state of disequilibrium. The sudden change may also generate significant fluctuations of acoustic and entropy modes. Thus the recovery of the turbulence from this disequilibrium may also exhibit compressibility effects. Other phenomena where the global behavior of the flow is strongly affected by compressibility include shock induced separations of a turbulent flow, phenomena involving the Ranque-Hilsh effect and other effects requiring chemical or thermodynamic nonequilibrium. The problem of shock wave boundary layer interaction in a hypersonic free-stream involves all of the above processes.

In the following we discuss two circumstances leading to significant compressibility effects. Detailed discussion of these has been published elsewhere [1, 2] and we restrict our discussion to a summary of these results. The reader is referred to the original references for a more complete discussion.

Let us consider the circumstance where the turbulent velocity fluctuations are a significant fraction of the mean speed of sound. Is it possible for a turbulent flow, in a natural course of its evolution, to generate and maintain velocity fluctuations comparable to (an appropriately defined) mean speed of sound? It would appear that turbulent compressible mixing layers behave this way. The evidence from the experiments [3, 4] suggests that as the convective

Mach number (M_c) is increased the peak turbulence intensities normalized by the velocity difference across the mixing layer decrease, but even with this decrease the Mach number of the fluctuations does increase (at least for $M_c < 1$). It may be noted that in most quasi-equilibrium flows a large fraction of the turbulence production is balanced by the local dissipation and it is the small imbalance which controls the streamwise evolution of the flow. Thus even a small shift in the production-dissipation balance due to effects of compressibility can cause a significant alteration of the streamwise growth rate. It has been suggested that compressibility alters the production mechanisms due to the strong Mach number dependence of hydrodynamic instabilities [5, 6]. The alternative to this is the proposal that it is the dissipation processes which are altered by compressibility effects [7, 8]. In Zeman's proposal [7] the increased dissipation is attributed to intermittent occurrence of eddy-shocklets in the turbulent flow. Sarkar et al. [8] argue for the increased dissipation on the basis of a small fluctuation Mach number expansion of the governing equations. The scaling of this additional dissipation, i.e. the dilatational dissipation, with the fluctuation Mach number is very different in these two proposals. It is important then to address the question of whether eddy-shocklets do arise in a naturally evolving turbulent flow and if they do arise what is their statistical contribution to the overall dilatation dissipation.

We have addressed this question by conducting direct numerical simulations of decaying isotropic compressible turbulence. The details of the simulations may be found in [1]. The list of parameters which affect the subsequent evolution is large even for this idealized flow [9, 10] and to some extent this is due to the absence of any turbulence production mechanism. The simulations are confined to those cases where the initial velocity fluctuations are incompressible and have Gaussian statistics, and the pressure and density are uniform. The fluctuation Mach number (M_t) and the Taylor microscale Reynolds number (R_λ) are the two parameters that are systematically varied in the simulations. While the absence of any density or pressure fluctuations in the initial state is quite unnatural, the flow recovers from this initial state during a short acoustic transient. The acoustic transient establishes the equipartition of dilatational kinetic energy with the compressible potential energy, expected for linear acoustic fluctuations. The acoustic transient is followed by the establishment of nonlinear energy transfer and the velocity derivative skewness achieves values around -0.5. After this time we regard the evolution of the turbulence to be 'natural'. During this natural decay we observe that the dilatation field acquires negative skewness and flatness values far in excess of the Gaussian value of 3. The vorticity field continues to be near-Gaussian during this evolution. A negative skewness of the dilatation suggests preferential steepening of compressions and a large flatness indicates the increased intermittency of regions with significant (positive or negative) dilatation. For simulations where the initial fluctuation Mach number is 0.5 or higher the dilatation flatness attains values as high as 10–15. An examination of the instantaneous flow fields from these runs shows the presence of eddy-shocklets. The simulations also show that the probability of occurrence of the shocklets increases as the fluctuation Mach number and/or Reynolds number are increased. The relative contribution of the extreme negative dilatation regions to the overall dilatational dissipation also increases with these two parameters. For the case with the initial $R_\lambda = 50.$ and $M_t = 0.61$ up to 20–25% of the dilatational dissipation arises from the negative dilatation regions beyond three standard deviations of dilatation (which occupy only about 1% of the flow domain). This suggests that eddy-shocklets are dynamically significant in the local energy balance and should be included in a parametrization of subgrid-scale processes. Their relevance to the evolution of the flow depends on the significance of the overall dilatational dissipation in the energy balance.

We now consider the second mechanism through which compressibility may influence the evolution of turbulence. Consider the passage of quasi-incompressible turbulence through a normal shock wave. Fluid elements are compressed in the direction normal to the shock wave as they flow through it. This leads to an increase of mean square vorticity downstream of the shock wave. An increase is also expected for the turbulent kinetic energy. However, since the deformation process is anisotropic, a turbulent field which may be isotropic upstream of the shock becomes anisotropic just downstream. Since the passage time of turbulent eddies through the shock is typically short compared to the time scale of the turbulent eddies it becomes possible to approximate the process of shock-turbulence interaction as a linear process. This linear analysis can be carried out either as a kinematic deformation of an incompressible turbulent field (see [11]) or in terms of the detailed interaction of individual vorticity, entropy and acoustic waves with a shock wave. This latter approach is elucidated by Ribner [12], Moore [13] and has been been pursued by others subsequently. We have conducted direct numerical simulations of shock-turbulence interaction. Detailed discussion of the simulation results and comparisons with the linear analyses may be found in [2]. The simulations in [2] considered the passage of two-dimensional turbulence through a normal shock wave. Our recent work has extended the simulations to the three-dimensional case. The simulation results have been compared with the predictions of the linear analyses. The comparisons show that the vorticity amplification through the shock is quite accurately predicted by the linear theory. The evolution of turbulent kinetic energy, however, shows qualitative differences. The simulations show that in the region downstream of the shock the turbulent kinetic energy does not decay monotonically, but rather undergoes a rapid increase, in a region well outside the intermittent zone containing the unsteady shock front, which is followed by a gradual decay. This re-adjustment of turbulence is a compressible phenomenon, as borne out by the fact that the principal cause of it is the pressure-work term in the turbulent kinetic energy equation. The simulations also show that the turbulence length scales and in particular the Taylor microscale decrease as the turbulence passes through the shock. This behavior is completely consistent with the linear analyses. The length scale increase reported by recent experiments is quite anomalous, and remains unexplained at present.

Acknowledgements

This work was performed under a grant from AFOSR; the computational resources were provided by NASA-Ames.

References

[1] Lee, S., Lele, S. K., and Moin, P., 1991, *Eddy-shocklets in decaying compressible turbulence*, to appear in Phys. Fluids A.

[2] Lee, S., Lele, S. K., and Moin, P., 1991, *Dircct simulation and analysis of shock-turbulence interaction,* AIAA paper, AIAA-91-0523, Reno.

[3] Ikawa, H. and Kubota, T., 1975, AIAA J., 13, p. 566.

[4] Elliot, G. and Samimy, M., 1990, Phys. Fluids A, 2, p. 1231.

[5] Papamoschou, D. and Roshko, A., 1988, J. Fluid Mech., 197, p. 453.

[6] Morkovin, M. V., 1990, this volume.

[7] Zeman, O., 1990, Phys. Fluids A, 2, p. 178.

[8] Sarkar, S., Erlebacher, G., Hussaini, M. Y., and Kreiss, H. O., 1989, ICASE Report 89-79.

[9] Erlebacher, G., Hussaini, M. Y., Kreiss, H. O., and Sarkar, S., 1990, ICASE Report 90-15.

[10] Blaisdell, G., 1990, Ph. D. Thesis, Mechanical Engineering, Stanford University.

[11] Lee, M. J., 1989, Phys. Fluids A, 1, p. 1541.

[12] Ribner, H. S., 1953, NACA TN-2864, also see NACA TN-3255, 1954.

[13] Moore, F. K., 1953, NACA TN-2879.

OCCURRENCE OF CAUSTICS FOR HIGH FREQUENCY ACOUSTIC WAVES PROPAGATING THROUGH TURBULENT FIELDS*

Ph. Blanc-Benon, D. Juvé , G. Comte-Bellot

Laboratoire de Mécanique des Fluides et d'Acoustique, C.N.R.S.
Ecole Centrale de Lyon, BP 163, 69131 Ecully Cedex, France

ABSTRACT

Theoretical investigations of wave propagation in random media usually rely on a statistical approach. The interaction between the wave and the medium is expressed in terms of a random refractive index related to the fluctuations of the medium, which has been introduced in the Helmholtz equation. The equations which govern the second and higher moments of the transmitted sound pressure field are then deduced. Using a closure hypothesis for the index fluctuation, solutions may be obtained either by some analytical developments or by numerical integration. It is only recently that studies have been reported that make use of computer generated fields to simulate wave propagation in random media. However they are still limited to scalar random fields. Moreover, in acoustics, the presence of velocity components introduces additional effects of wave convection which cannot be described a priori by such a simplified approach.

In this paper we describe a different technique, which does take into account fluctuations in velocity fields and avoids purely statistical theory and not well founded hypotheses. Assuming that the turbulent field is frozen during the transit time of the acoustic wave, the medium can be modelled by a sequence of independent realizations of a random field. Each realization of the field is generated by a superposition of a finite number of discrete random Fourier modes. The amplitude of these modes is chosen in order to obtain a pre-defined energy spectum. Isotropic fields with scalar or vectorial fluctuations can be obtained with roughly the same techniques. Then we consider the deterministic propagation of acoustic waves (in the geometrical approximation) through individual realizations of the simulated turbulent field. We integrate the equations of geometrical acoustics to describe the trajectories of rays as well as the evolution of the cross section of an elementary ray tube. The points where this section vanishes define the position of caustics. We illustrate the focusing and the defocusing effects of the inhomogeneous media by plotting the distorted rays. We accumulate statistics over an ensemble of realizations to estimate the probability density function (p.d.f.) of occurrence of the first caustic at a given distance from the source. Using initially plane waves propagating in 2D Gaussian correlated temperature or velocity random fields, we interpret the formation of caustics and the differences between scalar and vectorial inhomogeneous media.

In the scalar case, our simulations confirm the theoretical analysis of Kulkarny and White. The probability density of distance along a ray to the nearest caustic is given by a universal curve with a single distance scale parameter. For each p.d.f. we can notice that up to a certain length along the ray there is no caustic. Then we observe a sharp peak at a fixed distance and an exponential decay at larger distances. For a Gaussian correlation function $R_\mu(r) = \mu^2 exp(-r^2/L^2)$, the theoretical scaling factor $\sigma_T^{2/3}$ suggested by Kulkarny and White is:

$$\sigma_T^{2/3} = 12^{1/3}\,\pi^{1/6}\,\mu^{2/3}\;/L$$

where μ is the rms value of the index of refraction.

In the vectorial case, we extend these results using a new scaling parameter based on the component of the velocity field $v_1'(\vec{x})$ in the direction of propagation. According to the Helmholtz equation, this component gives the most important contribution to the fluctuation of the index of refraction. With a Gaussian correlation function $R_{11}(r) = v'^2 \;\; exp(-r^2/L^2)$ we get after some computation:

$$\sigma_V^{2/3} = 60^{1/3}\,\pi^{1/6}\,\mu^{2/3}\;/L$$

with $\mu = v'/c_0$, v' is the rms value of the velocity and c_0 the sound speed. Using this scaling parameter, excellent agreement with numerical experiments is obtained.

Extensions of this numerical simulation to consider more realistic turbulent fields involving several characteristic length scales and three dimensional aspects are in progress. In addition to the geometrical acoustics point of view, the parabolic approximation can be used to study the influence of the frequency on the pressure field.

---

*The full text of this paper is to appear in *Theoretical & Computational Fluid Dynamics.*

DIRECT SIMULATION OF COMPRESSIBLE TURBULENCE IN A SHEAR FLOW *

S. Sarkar, G. Erlebacher, and M.Y. Hussaini [1]

Institute for Computer Applications in Science and Engineering
NASA Langley Research Center
Hampton, VA 23665

ABSTRACT

The purpose of this study is to investigate compressibility effects on the turbulence in homogeneous shear flow. Physical experiments have not been and perhaps cannot be performed for homogeneous shear flows at flow speeds which are sufficiently high to introduce compressible effects on the turbulence. However, direct numerical simulation (DNS) of this problem could provide meaningful data, especially since DNS for the incompressible problem ([1], [2]) has been successful in giving realistic flow fields akin to those obtained in physical experiments. The present simulations of the homogeneous shear problem augment and extend our previous results ([3], [4]) for compressible isotropic turbulence.

The compressible Navier-Stokes equations are written in a frame of reference moving with the mean flow $\overline{u}$. We use a Fourier collocation method for the spatial discretization of the governing equations. A third order, low storage Runge-Kutta scheme is used for advancing the solution in time. Up to 128^3 points are used for the simulations. The simulated turbulent fields have Taylor microscale Reynolds numbers Re_λ up to 35 and turbulent Mach numbers M_t up to 0.6. The results are summarized below.

We find that an increase in initial compressibility level, either due to increased turbulent Mach number or increased dilatational fraction of the velocity field *decreases* the growth of turbulent kinetic energy (and all its components) in the case of homogeneous shear flow. This is reminiscent of the reduction in turbulent velocity intensities observed in experiments on supersonic free shear

[1] This research was supported by the National Aeronautics and Space Administration under NASA Contract No. NAS1-18605 while the authors were in residence at the Institute for Computer Applications in Science and Engineering (ICASE), NASA Langley Research Center, Hampton, VA 23665.

layers. In order to explain the phenomenon of reduced growth rate of kinetic energy, we consider the equation governing the kinetic energy of turbulence in homogeneous shear which is

$$\frac{d}{dt}(\overline{\rho}K) = -S\overline{\rho}\widetilde{u_1''u_2''} - \overline{\rho}\epsilon_s - \overline{\rho}\epsilon_c + \overline{p'd'} \tag{1}$$

It is pertinent to study the last two terms in Eq. (1), the compressible dissipation ϵ_c and the pressure-dilatation $\overline{p'd'}$, since they represent the explicit influence of the non-solenoidal nature of the fluctuating velocity field in the kinetic energy budget. The compressible dissipation ϵ_c is always positive (as expected) and its magnitude increases with the turbulent Mach number. The pressure-dilatation $\overline{p'd'}$ is highly oscillatory. Though $\overline{p'd'}$ can be both positive or negative, it tends to be predominantly negative. These DNS results along with the kinetic energy equation (1) show that both ϵ_c and $\overline{p'd'}$ have a dissipative effect on the turbulent kinetic energy contributing to its reduced growth. The predominantly negative values of $\overline{p'd'}$ in homogeneous shear is in contrast to the predominantly positive values reported [4] for the case of decaying isotropic turbulence. We are able to explain the different signs of the pressure-dilatation by theoretical consideration of the equations for the pressure variance and density variance.

Two results of the analysis of [4] which were validated previously in isotropic turbulence are now checked in homogeneous shear flow. These results are, first, the compressible dissipation obeys the scaling $\epsilon_c/\epsilon_s = \alpha M_t^2$, and second, there is a tendency toward equipartition between the kinetic and potential energies of the fluctuating compressible mode. Both of these results have now been substantiated in the case of homogeneous shear.

We have investigated pdf's and higher-order moments of the velocity derivatives. The skewness of the dilatation reaches values of about -0.5, signifying that local compressions are more likely than local expansions in the flow. The dilatational field has larger flatness than the vorticity field, indicating that it is more intermittent than the vorticity. In our simulations, the maximum magnitude of dilatation in the compression regions attains values which are about an order of magnitude larger than the rms dilatation. Such high ratios had been earlier observed in isotropic turbulence [5], and these high values had been associated with the dynamical significance of eddy shocklets. However, in order to assess the importance of these high dilatation regions it is necessary not only to identify the magnitude of the maximum dilatation but also to calculate the contribution of these regions to the rms dilatation. We find that, the compressive regions with a magnitude of dilatation larger than 3 times the rms dilatation contribute less than 5% to the variance of dilatation. This suggests that shocklet dissipation is not the dominant contributory mechanism to the compressible dissipation for $M_t < 0.5$.

Acknowledgements

On the occasion of John Lumley's 60th birthday, SS is pleased to thank John for teaching him turbulence and motivating him to pursue research in the field of turbulence. The authors wish to thank Tom Gatski and Charles Speziale for their comments on a preliminary draft of this manuscript, and Tom Zang for helpful discussions.

References

[1] Rogallo, R.S. (1981). Numerical Experiments in Homogeneous Turbulence. *NASA TM 81315.*

[2] Rogers, M.M., and Moin, P. (1987). The Structure of the Vorticity Field in Homogeneous Turbulent Flows. *J. Fluid Mech.*, **176**, 33.

[3] Erlebacher, G., Hussaini, M.Y., Kreiss, H.O., and Sarkar, S. (1990) The Analysis and Simulation of Compressible Turbulence. *Theoret. Comput. Fluid Dynamics*, **2**, 73.

[4] Sarkar, S., Erlebacher, G., Hussaini, M.Y., and Kreiss, H.O. (1989) The Analysis and Modeling of Dilatational Terms in Compressible Turbulence. *ICASE Report No. 89-79*, *J. Fluid Mech* (in press).

[5] Lee, S., Lele, S.K., and Moin, P. (1990) Eddy-shocklets in Decaying Compressible Turbulence. *CTR Manuscript 117.*

---

*The full text of this paper is to appear in *Theoretical & Computational Fluid Dynamics.*

THE RAPID EXPANSION OF A TURBULENT BOUNDARY LAYER IN A SUPERSONIC FLOW*

Douglas R. Smith and Alexander J. Smits[1]

Princeton University
Princeton, New Jersey 08544

ABSTRACT

Rapid distortion approximations (RDA) may be used to simplify the Reynolds Stress equations under the satisfaction of rapid distortion criteria as suggested by Dussauge and Gaviglio (1987). The evolution of Reynolds stresses along streamlines in a boundary layer may then be calculated. This calculation was made for the distortion of a supersonic flow through a 20° expansion (Figure 1) and the results were compared with experimental observation.

The rapid distortion approximations neglect diffusive and dissipative terms while retaining the production and pressure terms. The retained terms are modelled as functions of the Reynolds stress tensor and gradients of the mean flow. Models for the pressure-strain term were based on the work by Lumley (1978) and Shih and Lumley (1985).

The experimental measurements revealed a boundary layer downstream of the distortion which was dramatically different from its upstream counterpart. The mean velocity profile did not possess a logarithmic region. The velocity fluctuations decreased while the mass-flux fluctuations remained initially unchanged, and the strong Reynolds analogy showed that the change in the mass-flux fluctuations seems to follow the change in the density fluctuations.

Within the limits of the rapid distortion approximations, the Reynolds stress equations made surprisingly accurate predictions of the Reynolds stresses in a compressible boundary layer (Figure 2). The performance of three pressure-strain models was evaluated, and it was found that the calculation was mostly insensitive to the model chosen. The influence of dilatation alone was also investigated, and the results indicate that compressibility has a strong influence on the Reynolds stress.

---

[1]This work was supported by the U. S. Air Force under AFOSR contract 89-0420. Monitored by Dr. James McMichael.

References

[1] Dussauge, J. P. and Gaviglio, J., 1987, *The rapid expansion of a supersonic turbulent flow: role of bulk dilatation*, J. Fluid Mech., 174, pp. 81-112.

[2] Lumley, J. L., 1978, *Computational modeling of turbulent flows*, Adv. Appl. Mech., 18, pp. 124-176.

[3] Shih, T. H. and Lumley, J. L., 1985, *Modeling of pressure correlation terms in Reynolds stress and scalar flux equations*, Rep. FDA-85-3, Sibley School of Mech. and Aero. Eng., Cornell University.

---

*The full text of this paper is to appear in *Theoretical & Computational Fluid Dynamics.*

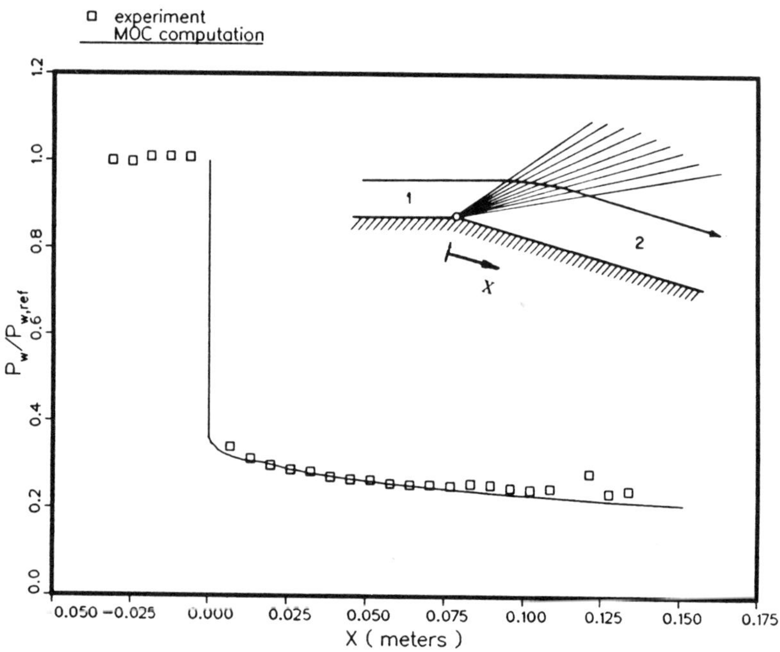

$P_{w,ref}$ = 22790.0 N/m²

Figure 1 **Comparison between experiment wall pressure and computed wall pressure.**

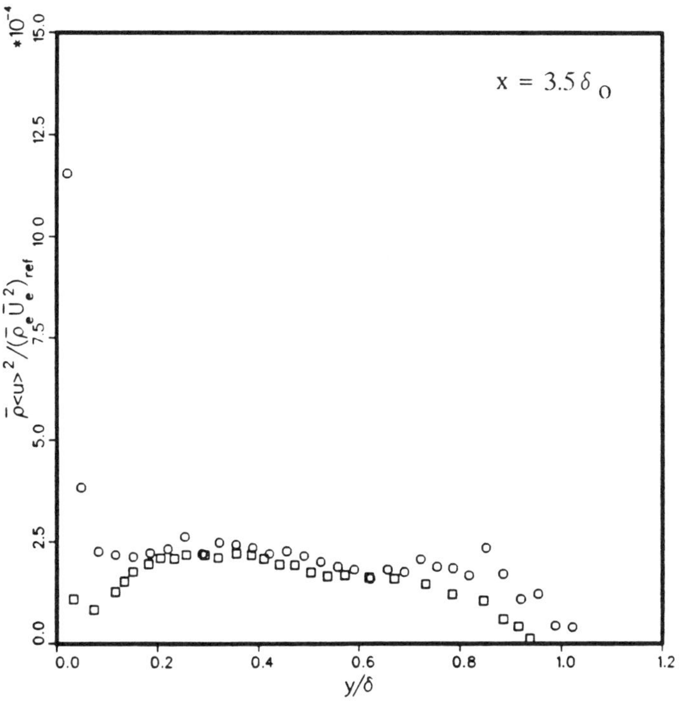

Figure 2 **Comparison between the experimental, o, and computed, □, longitudinal Reynolds stress, Shih and Lumley (1985) model.**

PART 4

TURBULENT SHEAR FLOWS

STRUCTURAL GENESIS IN WALL-BOUNDED TURBULENT FLOWS

Sidney Leibovich[1, 2]

Cornell University
Ithaca, New York 14853

ABSTRACT

Coherent structures in turbulent flows are usually attributed to patterns of concentrated vorticity. Reynolds averaging is not well suited to the eduction of vortical structures. On the other hand, the averaging that defines the "Generalized Lagrangian Mean" (or GLM) of Andrews and McIntyre captures the mean vortex dynamics in precisely the same way the instantaneous equations of motion describe the instantaneous vorticity dynamics. This suggests advantages might be expected to accrue from the use of this alternative averaging scheme. The possibilities of using the GLM formalism in turbulent flows is discussed, together with other thoughts on coherent structures, which suggest that we may think of the latter as associated with heteroclinic cycles arising from generalized resonances.

1. Introduction

A developing recognition of the prevalence of coherent structures in turbulent flows has resulted in practical progress, mostly by suggesting methods of controlling flows by manipulation of their characteristic structures. This has developed despite a notable lack of agreement on what is meant by a "coherent structure", and scant

---

[1] This work was supported by the US Air Force Office of Scientific Research under grants AFOSR-89-0226, monitored by Dr. J. McMichael, and AFOSR-89-0346, monitored by Dr. L. Sakell.

[2] This paper is dedicated to John Lumley on the occasion of his 60th birthday, in appreciation for his friendship and comradeship .

progress in incorporating these structural elements into theoretical representations of turbulence (the notable exception to my mind is the admirable blend of modelling and analysis in Aubry et al., 1988). Until there is an effective way to include the effects of coherent structures in a predictive scheme with a firm connection to the physics, this branch of turbulence research will remain essentially an empirical body of knowledge.

Despite the lack of an agreed upon definition of coherent structures, one thing seems clear, whatever these things are, you know one when you (literally) see one. That is, these are things that form patterns, and if the flow is visualized, these patterns are recognizable and occur repetitively. Patterns, arrangements of (here, usually visualized) elements, are most recognizable when they possess a high degree of spatial symmetry. From this perspective, there is an association of coherent structures with the appearance of localized spatial symmetry in flows. This is one theme of this paper, which attempts to deal with this by presenting a primitive coherent structure that can exist (at so far as the Navier-Stokes equations are concerned) in rotating pipe flows. The key element is strong symmetry that is not permanent but repetitive, and it arises in our example from a heteroclinic connection of approximations to three unstable, exact, solutions of the Navier-Stokes equations.

The second theme of the paper concerns practical ways of describing coherent structures in predictive, yet physically plausible, mathematical models. To be "practical", we suppose that the model should describe only statistical aspects of the flow *and* its structural elements. Success of the Craik-Leibovich (CL) theory of Langmuir circulations (see Leibovich, 1983, for a review), which predicts structures similar to the streaks in the near-wall region of turbulent boundary layers, suggests that extensions of a theory of this kind may be useful in describing structures in turbulent flows. The generalization of the CL theory is known (Leibovich, 1980) to be derivable from the generalized Lagrangian-mean (GLM) theory of Andrews and McIntyre (1978).

The GLM theory describes mean motions and is therefore conceptually equivalent to Reynolds averaging. On the other hand, the GLM describes Lagrangian aspects of the motion in an Eulerian setting, and is able to capture structural aspects of the Langmuir circulation flows arising as an instability of a "structureless flow". It seems likely that the CL theory is effective at this because the mean structures which arise are clearly due to vorticity rotation and stretching, and the GLM theory describes *mean* vorticity kinematics in the same way as the instantaneous vorticity kinematics are described. This may be contrasted with Reynolds averaging, which masks the vorticity kinematics revealed by the instantaneous equations of motion.

Like the Reynolds averaged equations, the GLM equations are not closed, and closure requires modelling of terms analogous to the Reynolds stresses. In the GLM, these take the form of a vector, the so-called "pseudomomentum", arising from unresolved fluctuations, as well as modelling of correlations of these fluctuations with fluctuating *viscous* forces. Phillips (1989, 1990,1991) has discussed certain specific models which he has proposed might be appropriate for a GLM treatment of turbulent channel flow. In this paper, possible strategies for the use of this formalism to educe coherent structures, as well as some of the difficulties which may be encountered, are discussed from a basic point of view.

2. The Generalized Lagrangian-Mean

Atmospheric motions are strongly affected by wave activity, and the waves generate a rectified effect, leading to mean atmospheric motions. The search for a rational way to separate "waves" from "mean flow", and to define wave–mean flow interaction, was a long one culminating in the generalized Lagrangian mean theory of Andrews and McIntyre (1978) (which I shall abbreviate as A & M).

The latter paper gives both the history of the search and a comprehensive exposition of the GLM theory. Here the structure of this theory is reviewed in the context of a constant density fluid and is contrasted to Eulerian–mean, or Reynolds, averaging.

Before beginning, it should be emphasized that the GLM is simply an alternative averaging scheme, and does not provide, as a matter of principle, any more information than Reynolds averaging. In particular, assumptions must be made to close the GLM equations, and this usage of the formulation differs in spirit and strategy from that contemplated by the developers of the theory. The only rationale for the consideration of the GLM as an alternative formulation for (mean) turbulent flows is the prospect that it may reflect vorticity dynamics in a more natural way than does Reynolds averaging, and therefore capture structural features of a turbulent flow more economically. By contrast, Reynolds averaging seems unsuitable for such a purpose. To be of practical use, however, it does seem that the GLM has some prospect of utility only if modelling can be done to close the problem.

2.1 Basic formulation

Assume the Eulerian velocity field $\mathbf{u}(\mathbf{x}, t)$ is known. Then we may construct the trajectory,

$$\begin{aligned} \mathbf{x} &= \mathbf{X}(\mathbf{a}, t), \quad \text{where} \\ \frac{d\mathbf{X}}{\mathbf{d}\mathbf{t}} &= \mathbf{u}(\mathbf{X}, t). \end{aligned}$$

of a particle which, at some reference time such as $t = 0$ is located at position $\mathbf{x} = \mathbf{a}$. Of course, the choice made here for the Lagrangian label, $\mathbf{a}$, is a natural one.

Suppose now we choose to shift to an alternative Lagrangian label in the following way. Consider an observer moving with a different "reference" flow (at the moment, quite arbitrarily), described by the velocity field $\hat{\mathbf{u}}$, then, a particle with Lagrangian coordinate $\mathbf{a}$ follows the trajectory

$$\begin{aligned} \hat{\mathbf{x}} &= \hat{\mathbf{X}}(\mathbf{a}, t), \quad \text{where} \\ \frac{d\hat{\mathbf{X}}}{dt} &= \hat{\mathbf{u}}(\hat{\mathbf{X}}, t). \end{aligned}$$

Both trajectories begin at the same place, $\mathbf{a}$, but one follows the path of a fluid particle, and the second follows the arbitrarily prescribed reference path.

Now we *choose* to describe the position of the fluid particle from the perspective of the observer following the "hatted" trajectory, and write

$$\mathbf{x} = \mathbf{X}(\mathbf{a}, t) = \hat{\mathbf{X}}(\mathbf{a}, t) + \xi(\mathbf{a}, t) \tag{1}$$

so ξ is the particle vector displacement measured from the observer's current position. Since $\hat{\mathbf{x}} = \hat{\mathbf{X}}(\mathbf{a}, t)$, we can invert to get $\mathbf{a} = \hat{\mathbf{X}}^{-1}(\hat{\mathbf{x}}, t)$, so expression (1) may be reexpressed in terms of $\hat{\mathbf{x}}$ instead of $\mathbf{a}$. This replacement of the particle label then describes the actual particle position in the form

$$\mathbf{x} = \mathbf{X}(\hat{\mathbf{x}}, t) = \hat{\mathbf{x}} + \xi(\hat{\mathbf{x}}, t) \tag{2}$$

which produces the mapping from the observer's position to the particle's position

$$\hat{\mathbf{x}} \mapsto \hat{\mathbf{x}} + \xi(\hat{\mathbf{x}}, t).$$

The meaning of $\xi(\hat{\mathbf{x}}, t)$ is therefore the following: at any position $\hat{\mathbf{x}}$ and time t in the fluid, $\xi(\hat{\mathbf{x}}, t)$ is the position of the fluid particle, relative to $\hat{\mathbf{x}}$, which at $t = 0$ was at the position found by tracing the reference flow trajectory backwards through a time interval t from $\hat{\mathbf{x}}$.

The total derivative of (2) yields

$$\frac{\partial \xi}{\partial t} + \hat{\mathbf{u}} \cdot \nabla_{\hat{\mathbf{x}}} \xi = \mathbf{u}(\mathbf{X}, t) - \hat{\mathbf{u}}(\hat{\mathbf{X}}, t) \quad (3)$$

$$= \mathbf{u}(\hat{\mathbf{x}} + \xi(\hat{\mathbf{x}}, t), t) - \hat{\mathbf{u}}(\hat{\mathbf{x}}, t). \quad (4)$$

This gives a *field* description of particle positions, since the $\hat{\mathbf{x}}$ coordinates are just Eulerian in character. If we define

$$\frac{\hat{D}()}{\hat{D}t} \equiv \frac{\partial()}{\partial t} + \hat{\mathbf{u}} \cdot \nabla_{\hat{\mathbf{x}}}(), \quad (5)$$

and, for any field $\phi(\mathbf{x})$

$$\phi^\xi \equiv \phi(\hat{\mathbf{x}} + \xi(\hat{\mathbf{x}}, t), t), \quad (6)$$

then

$$\frac{\hat{D}}{\hat{D}t}\phi^\xi = (\frac{\mathrm{D}\phi}{\mathrm{D}t})^\xi. \quad (7)$$

Here

$$\frac{\mathrm{D}()}{\mathrm{D}t} \equiv \frac{\partial()}{\partial t} + \mathbf{u} \cdot \nabla_{\hat{\mathbf{x}}}()$$

is the usual material derivative as it is expressed in an Eulerian framework. The left–hand side is the rate of change following the reference path of a particle quantity, and the right–hand side is the material rate of change of the fluid particle with displacement ξ.

Up to this point, the reference trajectory is entirely arbitrary. Now invoke any averaging operation that commutes with time or space differentiation (like the customary ones of time, space, or ensemble averages). Let the average be denoted by an overbar, and *require*

$$\bar{\xi} = 0. \quad (8)$$

This defines a particular reference trajectory, one for which the average displacement from this reference path of any quantity evaluated at the true Lagrangian position of a fluid particle is zero. This would seem to be a natural definition of "mean Lagrangian" motion, and is the one adopted by A & M. Accordingly, for any quantity, f, related to the fluid motion, set

$$\bar{f}^L \equiv \overline{f^\xi},$$

which defines the (generalized) Lagrangian mean, or "GLM". In particular, in view of (4) and (8) the Lagrangian mean velocity field is $\overline{\mathbf{u}^\xi} = \bar{\mathbf{u}}^L = \hat{\mathbf{u}}$, and

$$\frac{\hat{D}}{\hat{D}t} = \frac{\partial()}{\partial t} + \bar{\mathbf{u}}^L \cdot \nabla_{\hat{\mathbf{x}}} \equiv \frac{\bar{\mathrm{D}}^L}{\mathrm{D}t},$$

which I shall call the "pseudomaterial derivative". With this definition (4) becomes

$$\frac{\bar{\mathrm{D}}^L}{\mathrm{D}t}\xi = \mathbf{u}(\hat{\mathbf{x}} + \xi(\hat{\mathbf{x}}, t), t) - \bar{\mathbf{u}}^L(\hat{\mathbf{x}}, t). \tag{9}$$

The equations governing the generalized Lagrangian mean velocity field are found (see A & M) by premultiplying the instantaneous momentum equations by the tensor $\nabla_{\hat{\mathbf{x}}}\mathbf{X}$ and averaging, and by calculating the pseudomaterial derivative of the Jacobian, J, of the tranformation $\hat{\mathbf{x}} \mapsto \mathbf{X}(\hat{\mathbf{x}}, t) = \hat{\mathbf{x}} + \xi(\hat{\mathbf{x}}, t)$, so

$$J = \det(\nabla_{\hat{\mathbf{x}}}\mathbf{X}) = \det(\mathbf{I} + \nabla_{\hat{\mathbf{x}}}\xi) \tag{10}$$

(where $\mathbf{I}$ is the identity matrix) followed by enforcement of mass conservation (we will assume here that the fluid is incompressible, but the general theory of A & M is not restricted by this assumption). The first gives GLM momentum equations, and the second expresses GLM mass conservation. These steps yield the GLM momentum equations (and from here on, we drop the hat on $\mathbf{x}$),

$$\frac{\bar{\mathrm{D}}^L(\bar{\mathbf{u}}^L - \mathsf{p})}{\mathrm{D}t} + \nabla\bar{\mathbf{u}}^L \cdot (\bar{\mathbf{u}}^L - \mathsf{p}) = -\nabla\pi + \mathcal{F}, \tag{11}$$

where

$$\mathsf{p} \equiv -\overline{(\nabla\xi \cdot \frac{\bar{\mathrm{D}}^L\xi}{\mathrm{D}t})}, \tag{12}$$

is called the "pseudomomentum",

$$\mathcal{F} \equiv \nu\overline{\nabla\mathbf{X} \cdot \nabla^2\mathbf{u}^\xi} \tag{13}$$

is the contribution of the viscous force, and

$$\bar{\pi} \equiv \frac{\bar{p}^L}{\rho} + \frac{1}{2}\overline{|\mathbf{u}^\xi|^2},$$

is a modified pressure head. The equation for mass conservation is

$$\frac{\bar{\mathrm{D}}^L \log J}{\mathrm{D}t} + \nabla \cdot \bar{\mathbf{u}}^L = 0. \tag{14}$$

A&M show that $\overline{J} = J$, so J, although formed from instantaneous variables, is automatically a mean quantity.

It is also important to relate $\bar{\mathbf{u}}^L$ to the usual Eulerian mean velocity vector, and this is done through the "Stokes drift" velocity, $\bar{\mathbf{u}}^S$, defined to be

$$\bar{\mathbf{u}}^S \equiv \bar{\mathbf{u}}^L - \bar{\mathbf{u}}. \tag{15}$$

This definition reduces, for small perterbations of a fluid at rest, to the classical definition of mass drift given by Stokes.

Equations (11) and (14), together with the definitions for $\mathcal{F}$, p, and J, govern the generalized Lagrangian mean variables for a constant density fluid, in the form given by A&M. Notice, first, that the generalized Lagrangian–mean velocity, $\bar{\mathbf{u}}^L$, is not solenoidal, so the GLM motion is rather like the flow of a compressible fluid with "density" J determined by the fluctuating field. Second, the equations clearly are not closed. To form the quantities $\mathcal{F}$, p, and J, one needs to know the behavior of fluctuations about the mean, which are given by ξ. The role of the Reynolds stress tensor is played by the single vector, p, which is a simplification of sorts; but one immediately pays for simplification of the inertial correlations by lack of closure of the mean viscous force, and by the introduction of the term J. A naive count suggests that closure requires that seven terms be supplied in the GLM, versus only six components of the Reynolds stress tensor when Reynolds averaging is used. In fact, because of the implicit way the fluctuations enter this theory, this undercounts the number of expressions requiring modelling.

Now return to (11), and let

$$\mathbf{v} \equiv \bar{\mathbf{u}}^L - \mathsf{p}, \tag{16}$$
$$\Pi \equiv \bar{\pi} + \mathsf{p} \cdot \mathbf{v}, \tag{17}$$

then (11), expressed in terms of $\mathbf{v}$, becomes

$$\frac{\partial \mathbf{v}}{\partial t} + \mathbf{v} \cdot \nabla \mathbf{v} = -\frac{1}{\rho} \nabla \Pi + \mathsf{p} \times \omega + \mathcal{F} \tag{18}$$

where

$$\omega = \nabla \times \mathbf{v} \tag{19}$$

is like a mean vorticity, which I will call the "pseudovorticity". The GLM equations in the form (19) might be called "Langmuir circulation form", because this precisely the same as the equations derived for the Craik-Leibovich theory of Langmuir circulations (Leibovich, 1977), with p replaced by $\bar{\mathbf{u}}^S$, and $\mathbf{v}$ replaced by the Eulerian–mean velocity, $\bar{\mathbf{u}}$. For fluctuations which are purely irrotational, as in Langmuir circulations (see Leibovich, 1980), the pseudomomentum, p, and the Stokes drift velocity vector, $\mathbf{u}^S$, are the same. In such cases $\mathbf{v} = \bar{\mathbf{u}}$ is just the mean Eulerian velocity vector, so ω is exactly the Eulerian mean vorticity vector. When working with the Langmuir circulatation form of the GLM equations, the continuity equation is conveniently expressed in terms of $\mathbf{v}$ as

$$\nabla \cdot \mathbf{v} = \Sigma \tag{20}$$

$$\Sigma = -\left(\frac{1}{J}\right)\left(\frac{\partial J}{\partial t} + \mathbf{v} \cdot \nabla J + \nabla \cdot (\mathsf{p}J)\right). \tag{21}$$

There is an alternative form for Σ, obtained from (15) and the fact that $\bar{\mathbf{u}}$ is solenoidal. This is

$$\Sigma = \nabla \cdot (\bar{\mathbf{u}}^S - \mathsf{p}), \tag{22}$$

which shows that there is no need to deal with both $\bar{\mathbf{u}}^S$ and with J, and since we have must have information about the former, as previously explained, this alternative representation of Σ in (21) is preferable.

There also is another form of (11), rather similar to (19) but written in terms of $\bar{\mathbf{u}}^L$ rather than $\mathbf{v}$. This is

$$\frac{\bar{\mathrm{D}}^L \bar{\mathbf{u}}^L}{\mathrm{D}t} = -\frac{1}{\rho}\nabla\hat{\Pi} + \bar{\mathbf{u}}^L \times \mathrm{curl}\mathsf{p} + \frac{\partial \mathsf{p}}{\partial t} + \mathcal{F} \tag{23}$$

where

$$\hat{\Pi} \equiv \bar{\pi} + \frac{1}{2}\overline{|(\bar{\mathbf{u}}^L)^2|} - \mathsf{p} \cdot \bar{\mathbf{u}}^L, \tag{24}$$

but this seems less useful than (19).

2.2 Pseudovorticity kinematics

The Eulerian–mean vorticity defined by Reynolds averaging has no simple conservation properties even when viscosity is ignored. In contrast, the pseudovorticity defined in the previous section satisfies (as shown by A & M) a Kelvin-type of theorem, and in the absence of viscosity, pseudovorticity lines move with the Lagrangian–mean velocity. This is easily seen from (18) with F ignored: its curl yields the equation for pseudovorticity,

$$\frac{\partial \omega}{\partial t} - \mathrm{curl}(\bar{\mathbf{u}}^L \times \omega) = 0, \tag{25}$$

and the corresponding Kelvin theorem,

$$\frac{\bar{\mathrm{D}}^L}{\mathrm{D}t} \oint_L \mathbf{v} \cdot d\mathbf{x}. \tag{26}$$

These simple mean vortex kinematical properties, so akin to the instantaneous kinematics, suggest that the GLM averaging process, preserving Lagrangian features as it does, is more "natural" in the discussion of mean vorticity dynamics, and as a consequence likely to be better suited to the eduction of coherent structures arising from vortex line deformations.

2.3 Strategies for application to turbulent flows

Provided the Jacobian, J, of the transformation between true Lagrangian and the reference (GLM) positions remains nonzero everywhere during the course of the evolution of the fluid system, the extraction of the generalized Lagrangian mean, as it has been described, is well–determined and is in principle exact. Its prospects for utility, however, depend on the development of a method of modelling that permits one to close the equations for the Lagrangian mean and thereby to predict its evolution in the absence of complete knowledge of the instantaneous motion.

Development of models for the terms requiring closure is more difficult than it is for Reynolds averaging because

1. Lagrangian statistics are not directly measurable in the laboratory, and so the correlations needed for GLM closure are not available.

2. For the same reason, the intuition which has developed from the laboratory and which has been a principle source of guidance for the development of models for the Reynolds closures, cannot be appealed to here with confidence.

3. There are more terms to model. At a minimum, one must model $\mathsf{p}, \mathcal{F}$, and either J or the Stokes drift, $\bar{\mathbf{u}}^S$. In fact, it is

the latter that is indispensible, for without the Stokes drift, it is not possible to return to the Eulerian mean velocity components; and if this is not possible, checks against experimental data are precluded.

4. The GLM theory is highly implicit. As a consequence, it seems unlikely that higher order models can be constructed. If a first-order GLM model proves useful, then the GLM theory will have a role to play, otherwise this is unlikely.

We have elected to try two approaches to the closure problem. First, we developed the forms of the correlations that would be appropriate if the Lagrangian displacements remain small. This is a procedure in the spirit of the developers of the GLM, who were primarily interested in fluctuations caused by waves of fairly small amplitude. Phillips (1989, 1990) proceeded along these lines while working with us, and has continued on this path since.

One faces difficulties which must be overcome following this approach. John Lumley has addressed these, and I expect a full description to be given in a subsequent publication. In section 3, I shall briefly discuss how they might be used in structural eduction.

There is a second approach that we are trying, with an entirely different set of difficulties. Clearly, if one knows the Eulerian velocity field at all times, then all the information needed to construct the GLM is known. So another procedure is to take a numerical simulation of a flow containing fluctuations, and, treating this as though it were exact, to extract the "exact" GLM statistics. Here one is trying to develop intuition about quantities accessible to numerical but not to physical experiment. Scott Jones is working with us towards this end. This is work in progress, and I expect that we shall be able to give an account in due course.

3. Modelling Structural Genesis of Turbulent Channel Flow

We conceive that "expected" structures arise in problems like channel flows which have two homogeneous directions by bifurcation of a GLM "structureless equilibrium" which has the same symmetry. The bifurcation can break the symmetry, producing "structures". This is discussed, in general terms and incompletely, in this section for channel flow.

3.1 Structureless channel flow

To test this concept, and whether the GLM formulation is useful in structural eduction, we must first close the GLM equations by the introduction of models for the unresolved fluctuations (the displacement field ξ) and the correlations which depend on them. This is necessary both for structureless flows, as well as for flows with structure. To see what is required by way of modelling, we first set up the "exact" problem for structureless channel flow as expressed in the GLM framework.

Consider a channel flow with coordinate x_1 in the streamwise direction, x_2 in the spanwise direction, and x_3 measured normal to the lower wall. A structureless flow depends only on x_3, with $\mathrm{v} = (v_1, 0, v_3)$, so the pseudovorticity $\omega = (0, \partial v_1/\partial x_3, 0)$. Furthermore, the streamwise gradient of the Lagrangian mean pressure is the same as the streamwise Eulerian mean pressure, and is a constant which may be expressed as $-\rho u_\tau^2/h$, where u_τ is the usual friction velocity, and h is the channel half-height.

Mass conservation (21) requires

$$v_3 = \bar{u}_3^S - \mathsf{p}_3 \tag{27}$$

and so, in contrast to the purely Eulerian case, there is in general a non–zero mean velocity normal to the wall (in the channel interior only, of course — this component vanishes *at* the wall itself). The existence of this component implies, again in contrast to fully-developed (or structureless) Eulerian–mean flow in a channel, there is a contribution to the streamwise analog to the mean acceleration. Taking this into account, the exact streamwise component of (19) reduces to

$$\bar{u}_3^S v_1{}' = \frac{u_\tau^2}{h} + \mathcal{F}_1. \tag{28}$$

Here we have set

$$()' \equiv \frac{d}{dx_3}.$$

Thus, except very near the wall where $\mathcal{F}$ cannot be ignored, $\bar{u}_3^S v_1{}'$ is a constant. Of course, this does not immediately tell us very much, since $\bar{u}_3^S$ is a functional of $v_1{}'$, which must be supplied by a closure hypothesis, to which we now turn.

3.2 Closure

The Eulerian–mean flow $(U(x_3), 0, 0)$ is known in channel flow. The Eulerian velocity fluctuations are typically small compared to the Eulerian–mean, but it is not clear that ξ is small compared to the large scales, and may even scale with the channel half–width. Similarly, it is not clear that $|\nabla\xi|$, which is a measure of the "slope" of GLM displacements, or alternatively, a measure of the total strain associated with fluctuations, is a small number. Nevertheless, if we assume that displacements and displacement gradients are small, then we can make partially analytic estimates, and so we shall do so despite the lack of a rational (or even an intuitive!) basis for the assumption. In the end, we will ignore the restriction of small perturbations, and use the functional *form* of the models derived this way without requiring small displacements.

Given this starting point, $\bar{\mathbf{u}}^L = \bar{\mathbf{u}} + O(\epsilon^2) = \mathbf{v} + O(\epsilon^2)$ where ϵ is a small parameter that measures ξ, and so knowledge of $\bar{\mathbf{u}}$ allows approximation of $\bar{\mathbf{u}}^L$. Under these assumptions, equation (9) of section 2 can be linearized and integrated to yield

$$\xi = \int_{t'}^{t} \{\mathbf{u}'(\mathbf{x}' + \mathbf{r}(s)) + (t-s)\mathbf{u}'(\mathbf{x}' + \mathbf{r}(s)) \cdot \nabla\bar{\mathbf{u}}\}ds, \qquad (29)$$

where $\mathbf{u}'$ is the fluctuating Eulerian–mean velocity,

$$\mathbf{r}(t) \equiv (t - t')\mathbf{U},$$

and ξ has been initialized to vanish at $t = t', \mathbf{x} = \mathbf{x}'$. The terms requiring closure, such as p, can now be formally expressed in terms of more familiar Eulerian–mean correlations. Furthermore, assuming the displacements are small compared to the radius of curvature of the mean profiles, we may neglect the variation of the mean velocity over the range of integration in (29), and, rewrite that equation in the form

$$\xi = \int_{t'}^{t} \mathbf{u}'(\mathbf{x}' + \mathbf{r}(s))ds + \{\int_{t'}^{t} (t-s)\mathbf{u}'(\mathbf{x}' + \mathbf{r}(s))ds\} \cdot \nabla\mathbf{v}, \qquad (30)$$

where we have replaced $\bar{\mathbf{u}}$ by $\mathbf{v}$ where it explicitly appears in (29).

With these replacements, we may evaluate the various terms requiring closure in terms of Eulerian correlations. For example, given the assumptions leading to (30), the form taken by the pseudomomentum is

$$\begin{aligned} -\mathsf{p}_i &= \mathsf{P}_i^{(0)} + \mathsf{P}_{ijk}^{(1)} v_{j,k} + \mathsf{P}_{ikl}^{(2)} v_{j,k} v_{j,l} \\ &+ \mathsf{P}_{ikl}^{(3)} v_{k,j} v_{j,l} + \mathsf{P}_{ikl}^{(4)} v_{j,m} v_{j,l} v_{m,k} \\ &+ \mathsf{P}_{jk}^{(5)} v_{j,ki} + \mathsf{P}_{jk}^{(6)} v_{l,k} v_{l,ji} + \mathsf{P}_{jk}^{(7)} v_{l,m} v_{m,k} v_{l,ji} \end{aligned} \tag{31}$$

In an entirely similar way, the Stokes drift is

$$\bar{\mathrm{u}}_i^S = \mathcal{S}_i^{(0)} + \mathcal{S}_{ijk}^{(1)} v_{j,k} + v_{i,jk}\{\mathcal{S}_{jk}^{(2)} + \mathcal{S}_{jl}^{(3)} v_{k,l} + \mathcal{S}_{lm}^{(4)} v_{k,l} v_{k,m}\}. \tag{32}$$

Here each $\mathsf{P}^{(\mathrm{n})}$ and $\mathcal{S}^{(n)}$ is a two–point correlation tensor between fluctuating Eulerian velocity components and their derivatives at nonzero time delay. Each is a function of the time delay, $t - t'$, the x_1 separation, $U(x_3)(t - t')$, and the x_3 position. Favre and his coworkers have measured some of the kinds of space-time correlations that appear in (32). The speed $U(x_3)$ is not much different from the speed at "optimal delay", as defined by Favre et al. (1957), and Favre (1965).

We now drop (at the suggestion of John Lumley) the second derivatives of $\mathbf{v}$ in (32), on the grounds that the development of this formula assumed negligible variation of velocity gradients in the solution for displacements. This leads to the simplified expression

$$-\mathsf{p}_i = \mathsf{P}_i^{(0)} + \mathsf{P}_{ijk}^{(1)} v_{j,k} + \mathsf{P}_{ikl}^{(2)} v_{j,k} v_{j,l} + \mathsf{P}_{ikl}^{(3)} v_{k,j} v_{j,l}. \tag{33}$$

Lumley (private communication) has constructed models for the tensors appearing in (33) in terms of single point correlations, and the local integral time scale. The considerations entering into these models require lengthy explanations, which Lumley will explain elsewhere. For now, we note the relevant results as they apply to structureless flows only. In this case, the only nonnegligible term p is p_3. Furthermore, $\mathbf{v} = (v_1, 0, v_3)$, where $v_1 = O(U(x_3))$, $v_3 \ll v_1$, so the last term in (33) is negligible, and

$$-\mathsf{p}_3 \approx \mathsf{P}_3^{(0)} + \mathsf{P}_{313}^{(1)} v_{1,3} + \mathsf{P}_{333}^{(2)} v_{1,3}^2.$$

Figure 1 shows the behavior of the coefficients appearing in p_3 for structureless flows, according to Lumley's models. In a similar way, we can model J. This is not required for closure, but there is some interest in having an estimate of this scalar function, because the

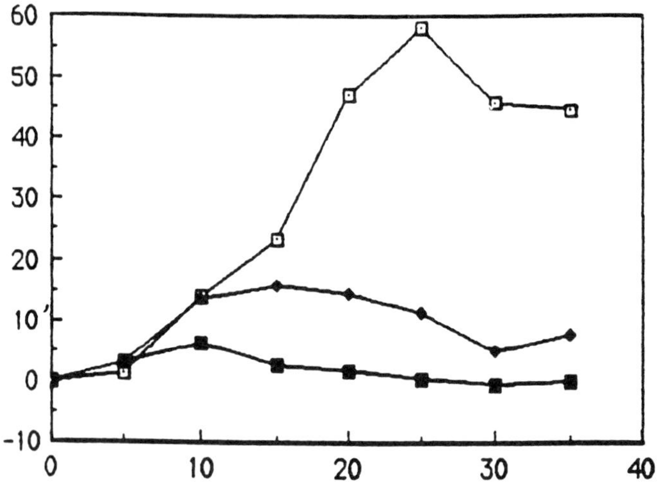

Figure 1: Coeffients appearing in Lumley's models for p_3, restricted to structureless flows. Filled squares, $\mathsf{P}_3^{(0)}$; diamonds, $\mathsf{P}_{313}^{(1)}$; open squares, $\mathsf{P}_{333}^{(2)}$. Abscissa plotted in wall units.

GLM formalism breaks down if J vanishes. Vanishing of a modelled form for J does not imply that a singularity vitiating the GLM formulation has in fact occurred, but it should make one give the matter further thought. With no approximation, from (10), we have

$$J = 1 + \overline{\frac{1}{2}(\nabla \cdot \xi)^2 - \frac{1}{2}\nabla\xi : \nabla\xi + \det \xi_{i,j}}$$

and the small disturbance assumption can be shown to imply that $\nabla \cdot \xi = 0$, and since $\det \xi_{i,j} = O(|\xi|^3)$, J is approximately given by

$$J = 1 - \frac{1}{2}\overline{\nabla\xi : \nabla\xi}. \tag{34}$$

3.3 Structural eduction

Now that we have a form for the model of $\bar{\mathbf{u}}^S$, we can write an equation for $v_1{}'$ from (28). Note that $v_{3,3}$ is $O(\epsilon^2)$ according to (21), and each of the coefficient tensors $\mathcal{S}^{(n)}$ is itself of $O(\epsilon^2)$, so for structureless flow, $\bar{u}_3^S \approx \mathcal{S}_3^{(0)} + \mathcal{S}_{313}^{(1)} v_{1,3}$. If we set

$$\phi = v_1{}'$$

then

$$\mathcal{S}_3^{(0)}\phi + \mathcal{S}_{313}^{(1)}\phi^2 = \frac{u_\tau^2}{h} + \mathcal{F}_1. \tag{35}$$

If $\mathcal{F}_1$ were modelled by an eddy viscosity, ν_t, then the equation above becomes a Ricatti equation for ϕ,

$$\nu_t \phi' = \mathcal{S}_3^{(0)}\phi + \mathcal{S}_{313}^{(1)}\phi^2 - \frac{u_\tau^2}{h}. \tag{36}$$

Assuming the structureless flow has been found, "structures" are sought as bifurcations of the structureless flow that involve variations in x_2, and possibly x_1, and velocity perturbations with a component in the 2-direction. The solution of (36), and the search for other, structured, solutions of the GLM equations must follow the introduction of specific models for the coefficients $\mathcal{S}^{(n)}$, and will not be pursued here.

4. Coherent Structures

The thesis advanced here is that coherent structures are recognizable because the arrangements of visible or measurable elements is *locally* regular. This local regularity reflects an approximate symmetry. A very simple example, not related to turbulence, is a wave group which might be represented by the nonlinear Schrödinger equation. Here one sees a pattern consisting of nearly regularly spaced wave crests confined to a local region of space defined by the wave envelope.

4.1 Patterns as Walks on Nets

Think now about a turbulent flow, and to fix ideas, think of a direct numerical simulation of turbulent channel flow, of the kind carried out by Kim, Moin, and Moser (1987). Such a flow has two homogeneous directions, and periodicity with a specified period in each direction is assumed. The quadratic nonlinearity of the Navier-Stokes equations implies that wavenumbers forming triads are linked together. Thus, if the spatial periods in the two homogeneous directions x_1 and x_2 are λ_1 and λ_2, respectively, then the *exact* solution will have all Fourier components $(2\pi l x_1)/\lambda_1$, $(2\pi m x_2)/\lambda_2$, $\pm(l,m) = 1, 2, \ldots$. Each Fourier component, say m_k, is linked to all (x_2) Fourier components, but only through other pairs that satisfy the triad relation $m_k = \pm m_p \pm m_q$. Triad coupling is the fundamental linkage, as is well known, and we may then think of the connections between different Fourier modes as taking place on a "net" made up of elementary triads, with each Fourier component pair (i.e., both l and m) as a node in this infinite net. Of course, in any numerical simulation, a truncation occurs which limits the number of nodes to a finite number.

Now each node in the net is endowed with a special symmetry, since it corresponds to a particular spatially periodic motion. Furthermore, although it does not seem to be widely recognized, each of these nodes corresponds to an *exact* solution of the Navier-Stokes equations. (This was shown for the slightly more complex problem of helical solutions of the Navier-Stokes equations by Mahalov, Titi, and Leibovich, 1990. The demonstration given in that paper carries over in an even simpler way to the present discussion.) The time–dependence of each of these nodes is in general very complicated, and the existing evidence (Mahalov and Leibovich, 1990) shows that the solutions may be chaotic in time. Furthermore, for any given Reynolds number, most of these exact solutions are unstable when other modes are included, and for high enough Reynolds number it is likely that *all* are unstable. Nevertheless, each of these individual modes is an invariant subspace of the Navier-Stokes equations, and it is known in dynamical systems theory that it is important to understand unstable equilibria and unstable subspaces, and that these frequently play a key role as organizing centers for the full dynamical system.

An example is given below, which has features that are similar to other low-dimensional dynamical systems. It contains three unstable fixed points, each of which is a node in a particularly simple net, since it is made up of a single triad. The three unstable subspaces are linked by a heteroclinic orbit. Given initial data that lies in one of the invariant subspace, the system stays in this subspace (by definition), and executes a motion no more complicated than a periodic modulation of a primary oscillation, so the signal at a fixed spatial point would have at most two independent frequencies. If the motion is started on the heteroclinic orbit, then one of the three subspaces will be approached asymptotically. If the system is started with any other initial data, then it is attracted to the heteroclinic cycle, and visits *each* of the three subspaces, hovering around each one for an increasingly long time, then moving to the next, and so forth.

What is seen when the system is near a particular subspace is a motion with a spatial pattern and temporal signature peculiar to this subspace. This pattern then breaks up as the system moves away from the vicinity of this node, and a confused transient state (so far as spatial and temporal behavior) takes over for a long time, until the system approaches the next node. It pauses in the vicinity of this second node for a long time, acquiring its characteristic spatial pattern and temporal signature, and so forth. It is natural to consider the pattern impressed on the system when it is near an unstable node to be a "coherent structure". It is worth noting that, in contrast to solutions assuming periodicity, the triads here are *resonant*, and the nodes do not have commensurate wavelengths.

This example then makes patterns appear for irregular periods of time, and to have irregular lifetimes. On the other hand, the patterns are not localized in space, and we should expect coherent structures to appear irregularly in both time and space. The simple model can be generalized, however, to include spatial modulation, and *then* we would expect spatial localization, rather like the solitary structures now familiar in nonlinear dispersive wave propagation problems. This generalization of the example is being attempted by Mahalov and myself, and we hope to report on it in due course.

This very simple example suggests a conceptual framework for coherent structures. For example, in a DNS of turbulent channel flow, most if not all of the nodes are excited, and the computer may be pictured as a centipede walking on a net, with millions of legs, each one of which steps on one node. But not with equal weight, and if for some reason, a few nodes bear most of the weight, then the patterns connected with these nodes dominate for a period of time. If the number of nodes bearing significant weight are sufficiently small, a coherent structure will be seen. The expected instability of nodes ensures that none of these patterns is permanent, and the system will shift to other structures in a process that is repetitive, but irregularly so. In our simple example, the dominant modes rise from the background because they are in temporal (as well as spatial) resonance. Precise temporal resonance is precluded, as a rule, in a DNS because of the periodic boundary conditions. On the other hand, it may be that certain modes are selected because they are near resonance, being only slightly detuned, or there may some hitherto unknown fully nonlinear generalization of what is conventionally understood as "resonance" which makes the selections.

4.2 A Primitive Example

I will describe, very briefly, a flow that might be thought of as a primitive "coherent structure". It occurs in rotating pipe flow (or "RPF"), and leads to pattern hopping in which three well defined patterns replace each other repetitively, and the sequence, if perturbed by an infinitesimal amount of noise, will repeat intermittently, at random times, with each pattern being recognizable for an irregular duration.

RPF flows, consisting of the superposition of ordinary Hagen Poiseuille flow ("HPF") in an infinitely long circular pipe, together with a rigid rotation about the pipe axis of symmetry, are exact solutions of the Navier-Stokes . The linear stability of this flow has been thoroughly explored (Pedley, 1968, 1969; Mackrodt, 1976; Cotton and Salwen; 1981), who showed that although solid body rotation in the absence of other motion is absolutely stable, and HPF is now generally believed to be linearly stable for all Reynolds numbers, the combination of the two is linearly unstable. Although certainly different in important respects from concentrated vortices, this fully diffused motion is a particularly convenient starting point to begin to develop a feeling for the possibilities of the nonlinear evolution and the equilibration of motion in unstable vortices.

For a variety of reasons, RPF has been the subject of a number of recent efforts (Akylas and Demburger, 1984; Topolsky and Akylas, 1988; Mahalov et al., 1990; Landman, 1990; Mahalov and Leibovich, 1990a; Mahalov and Leibovich, 1990b; Yang and Leibovich; 1990) which have addressed the nonlinear dynamics. The system depends on two dimensionless control parameters, R, a Reynolds number based on pipe radius and the peak axial velocity of the undisturbed flow, and Ω, a Reynolds number based on the pipe radius and the linear speed of the rotating pipe wall.

We may summarize the general findings by reference to figure 2, which gives the marginal stability boundary ("MSB") in the parameter space (Ω, R), and is due originally to Mackrodt (1976). The papers of Akylas and his coauthors concern, in part, the computation of the Landau constant at several locations on the marginal stability boundary of figure 6, and does the paper by Yang and Leibovich. The Landau constant results in the first two papers unreliable, as pointed out by Mahalov and Leibovich (1988), although the conclusion of Topolsky and Akylas that the conditions tested lead to supercritical bifurcation is correct. Yang and Leibovich show that this conclusion holds for a very wide range of conditions and most likely

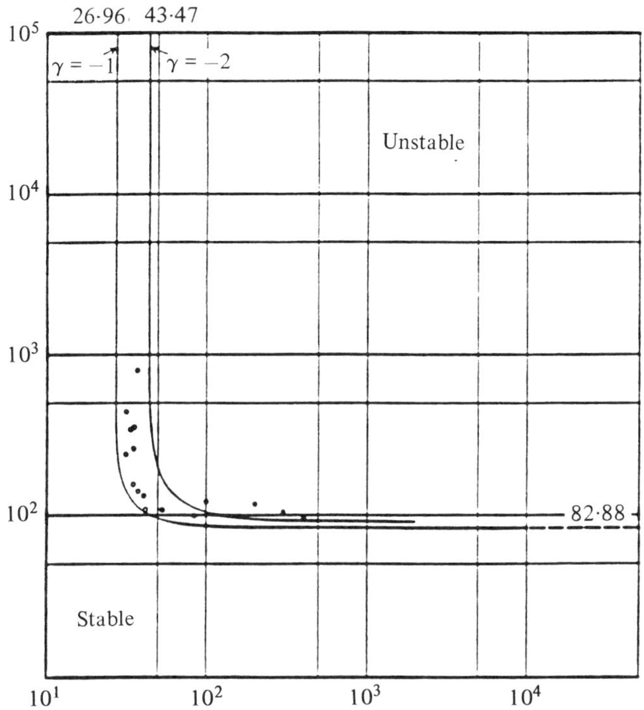

Figure 2: Stability boundary for rotating pipe flow, after Mackrodt (1976). Ordinate, R, abscissa, Ω.

along the entire marginal stability boundary. The latter authors also derive the coefficients of the Ginsburg-Landau equations, and discuss the stability of its Stokes wave solution all along the marginal stability boundary. Topolsky and Akylas (1988) and Landman (1990) numerically integrate the Navier-Stokes equaitions to determine the full nonlinear behavior of solutions bifurcating at a small number of points on the MSB. Their solutions are for fixed axial pressure gradient, and are restricted to helical disturbances. These are of the form

$$Q(r, \theta, z, t) = q_1(r, t)e^{i(\alpha z + m\theta)} \tag{37}$$

where Q represents any perturbation quantity expressed in cylindrical (r, θ, z) coordinates, and these comprise two-parameter (α, m) invariant subspaces of exact solutions of the Navier-Stokes equations (Mahalov et al., 1990). A given problem depends on the four parameters (R, Ω, α, m), and symmetry dictates that to a solution for any given set there correspond solutions, with the same time history, with the signs of any two of the parameters changed.

With the (slight) exception of the work by Yang and Leibovich, who allow for spatial modulation, the work mentioned so far concerns motion in helical subspaces. These depend on only two space coordinate, radius and helical position, and so they are not truly three-dimensional. Mahalov and Leibovich (M & L, 1990b), in contrast, consider the resonant interaction of a wave triad, and the solution is fully three-dimensional. Because the work may be applicable to other kinds of flows (such as plane Poiseuille flow), I will discuss this problem briefly.

Pedley (1969) considered thhe linearized problem in the limit $\alpha \to 0$, $\Omega \to \infty$ with $\alpha\Omega$ and R fixed. He showed that, provided $\alpha m < 0$, the flow may be unstable, and in such cases the neutral modes are described by

$$4(\alpha\Omega)^2 + 4\alpha\Omega m R + c(m) = 0, \tag{38}$$

the $c(m)$ being constants that he calculates from his theory. (The first four are: $c(1) = (4.36)^6$, $c(2) = (5.67)^6$, $c(3) = (6.91)^6$, $c(4) = (8.12)^6$.) Cotton and Salwen (1981) find (38) to be an exceptionally accurate formula for a much wider range of α and Ω than is suggested by the formal conditions for validity of the asymptotic theory. Cotton and Salwen also show that the frequency of neutral modes on the high R portion of the upper neutral branch (where Pedley's results should be valid), as determined numerically, are well fitted by the formula

$$\omega = -m\Omega - \alpha R. \tag{39}$$

It is easy to show (Mahalov and Leibovich (1990b) that equations (38) and (39) allow one to locate approximate conditions where three neutral curves intersect with wavenumbers and frequencies satisfying the conditions for three-wave resonance,

$$\alpha_1 = \alpha_2 + \alpha_3; n_1 = n_2 + n_3; \omega_1 = \omega_2 + \omega_3, \tag{40}$$

and in fact an infinite number of resonant conditions satisfying (38) and (39) and (40) may be found.

If $A_j(t) = a_j(t)\exp(i\omega_j t), j = 1, 2, 3$, are the complex amplitudes of three resonantly interacting modes near a point of resonance at which the frequencies are ω_i , then the interacting modes are described by the equations

$$\dot{a}_1 \quad = \quad \mu_1 a_1 + B_1 a_2 a_3 + a_1 \sum_{j=1}^{3} C_{1j}|a_j|^2 \tag{41}$$

$$\dot{a}_2 = \mu_2 a_1 + B_2 a_1 a*_3 + a_2 \sum_{j=1}^{3} C_{2j} |a_j|^2 \qquad (42)$$

$$\dot{a}_3 = \mu_3 a_1 + B_3 a_1 a*_2 + a_3 \sum_{j=1}^{3} C_{3j} |a_j|^2. \qquad (43)$$

The asterisk means complex conjuate, and the μ_i represent complex growth or damping rates which are nonzero at points away from resonance. These values, as well as the complex coupling coeffients B_i and C_{ij} have been computed for several resonant points for RPF. The equations are standard, except the cubic terms are usually ignored. Mahalov and I take the view that it is best to retain them, reasoning that the cubic system has invariant subspaces (one a_i nonzero, the others vanishing, representing Hopf bifurcations for the ith mode by itself, and being examples of the invariant helical subspaces studied by Mahalov et al., 1990) and the dynamical system is therefore degenerate if they are omitted.

A complete characterization of the possible dynamics represented by equations (43) is not known, but it is clear that there are many possibilities. Two that Mahalov and I have found are: (i) stationary solutions for the a_i's, which in general implies either three (incommensurate) frequency mixed mode states, or periodic motion for the a_i, which leads to four incommensurate frequencies for physical quantities; and (ii), attraction to a heteroclinic orbit connecting the three fixed points in the invariant subspaces already mentioned. We have found the former at $R = 91$, at which a $(n_1, n_2, n_3) = (2, 1, 1)$ resonance occurs. At $R = 111$, where a $(3, 2, 1)$ resonance occurs, provides an example of the second kind of possibility. Figure 3 shows the attraction to a heteroclinic cycle for this case, as found by numerical integration of (43). Here the projection is onto the modulus of a_1, a_2, and a_3, where a_1 is "out of the paper", and a_3 is "upwards". The heteroclinic orbit is from the subspaces $\{a_2 = a_3 = 0 \mapsto \{a_1 = a_3 = 0\} \mapsto \{a_2 = a_1 = 0\}$. Figure 4 shows a segment from the time traces of the real parts of a_1, a_2, and a_3, respectively. Recalling that each subspace being visited has a distinct symmetry, the "full" system would display a high degree of spatial symmetry while in the vicinity of a given invariant subspace (and the duration of visitation will increase with time as the heteroclinic cycle is approached more and more closely), followed by a confused transitional state in which all three spatial forms are in play, followed by a long visit to a second distinct subspace and, so forth. This is

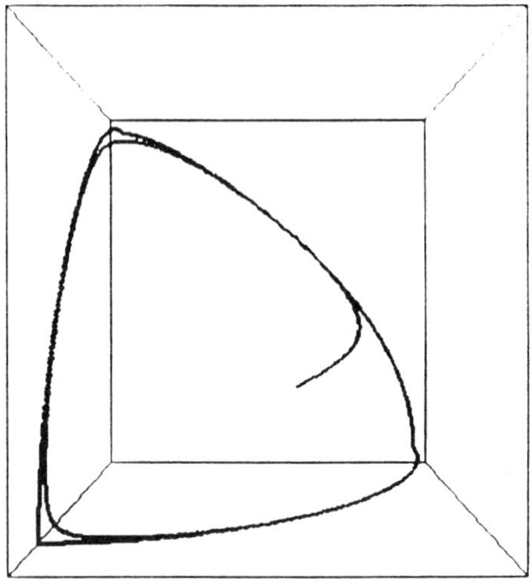

Figure 3: Heteroclinic cycle connecting three invariant RPF helical subspaces at a (3,2,1) resonance.

suggestive of the episodic appearances of coherent structures in both transitional and fully developed turbulent flows.

Acknowledgements

I am indebted to Alex Mahalov, who has provided figures 3 and 4, and on whose work the final section of the paper is based. I am also indebted to the person we honor in this volume, John Lumley, for his contributions to section 3.2.

References

[1] Akylas, T.R. and Demburger, J.P. 1984 The effect of rigid rotation on the finite-amplitude stability of pipe flow at high Reynolds number, *J. Fluid Mech.*, **148**,193-205.

[2] Andrews, D.G. and McIntryre, M.E. 1978 An exact theory of nonlinear waves on a Lagrangian-mean flow. *J. Fluid Mech.*, **89**, 609-646.

[3] Aubry, N., Holmes, P., Lumley, J.L. and Stone, E. 1988 The dynamics of coherent structures in the wall region of a turbulent boundary layer, *J. Fluid Mech.*, **192**, 115-173.

[4] Cotton, F.W. and Salwen, H. 1981 Linear stability of rotating Hagen-Poiseuille flow, *J. Fluid Mech.*, **108**, 101-125.

[5] Craik, A.D.D. 1982 The generalized Lagrangian– mean equations and hydrodynamic stability, *J. Fluid Mech.*, **125**, 27-35.

[6] Craik, A.D.D. 1982 Wave-induced longidtudinal– vortex instability in shear flows *J. Fluid Mech.*, **125**, 37-52.

[7] Favre, A.J., Gaviglio, J., and Dumas, R. 1957 Space-time double correlations and spectra in a turbulent boundary layer, *J. Fluid Mech.*, **2**, 313-342.

[8] Favre, A.J.1965 Review on space-time correlations in turbulent fluids, *J. Applied Mech.*, **32**, 241-257.

[9] Kim, J., Moin, P., and Moser, R.D. 1987 Turbulence statistics in fully developed channel flow at low Reynolds number, *J. Fluid Mech.*, **177**, 133-166.

[10] Landman, M.J. 1990 Time dependent helical waves in rotating pipe flow, *J. Fluid Mech.*,**221**, 289-310.

[11] Leibovich, S. 1977 Convective instability of stably stratified water in the ocean, *J. Fluid Mech.*, **82**, 561-585.

[12] Leibovich, S. 1980 On wave-current interaction theories of Langmuir circulations, *J. Fluid Mech.*, **99**, 715-724.

[13] Leibovich, S.1983 The form and dynamics of Langmuir circulations, *Ann. Rev. Fluid Mech.*, **15**, 391-427.

[14] Mackrodt, P-A. 1976 Stability of Hagen-Poiseuille flow with superimposed rigid rotation, *J. Fluid Mech.*, **73**, 153-164.

[15] Mahalov, A. and Leibovich, S. 1988 Amplitude expansion for viscous rotating pipe flow near a degenerate bifurcation point, *Bull. Am. Phys. Soc.*, **33**, 2247.

[16] Mahalov, A., Titi, E.S., and Leibovich, S. 1990 Invariant helical subspaces of the Navier-Stokes equations, *Archive for Rational Mechanics and Analysis*. In press.

[17] Mahalov, A., and Leibovich, S. 1990a Weakly nonlinear analysis of rotating Hagen-Poiseuille flow. *European J. Mech. B*. In press.

[18] Mahalov, A., and Leibovich, S. 1990b Resonant interactions in rotating pipe flow. Submitted for publication.

[19] Pedley, T.J. 1968 On the instability of rapidly rotating shear flows to non-axisymmetric disturbances, *J. Fluid Mech.*, **31**, 603-607.

[20] Pedley, T.J. 1969 On the instability of viscous flow in a rapidly rotating pipe, *J. Fluid Mech.*, **35**, 97-115.

[21] Phillips, W.R.C. 1989 The generalized Lagrangian–mean equation: turbulent channel flow and streamwise vortices. In *Near wall turbulence: Zoran Zaric Memorial Seminar 1988*, S. Kline (ed.), Hemisphere, 736.

[22] Phillips, W.R.C. 1990 Coherent structures and the generalized Lagrangian–mean equation, *Appl. Mech. Rev.*, **43**, S227. (See also, this volume.)

[23] Phillips, W.R.C. 1991 This volume.

[24] Topolsky, N. and Akylas, T.R. 1988 Nonlinear spiral waves in rotating pipe flow, *J. Fluid Mech.*, **190**, 39-54.

[25] Yang, Z. and Leibovich, S. 1990 Nonlinear dynamics near athe stability margin in rotating pipe flow. In review.

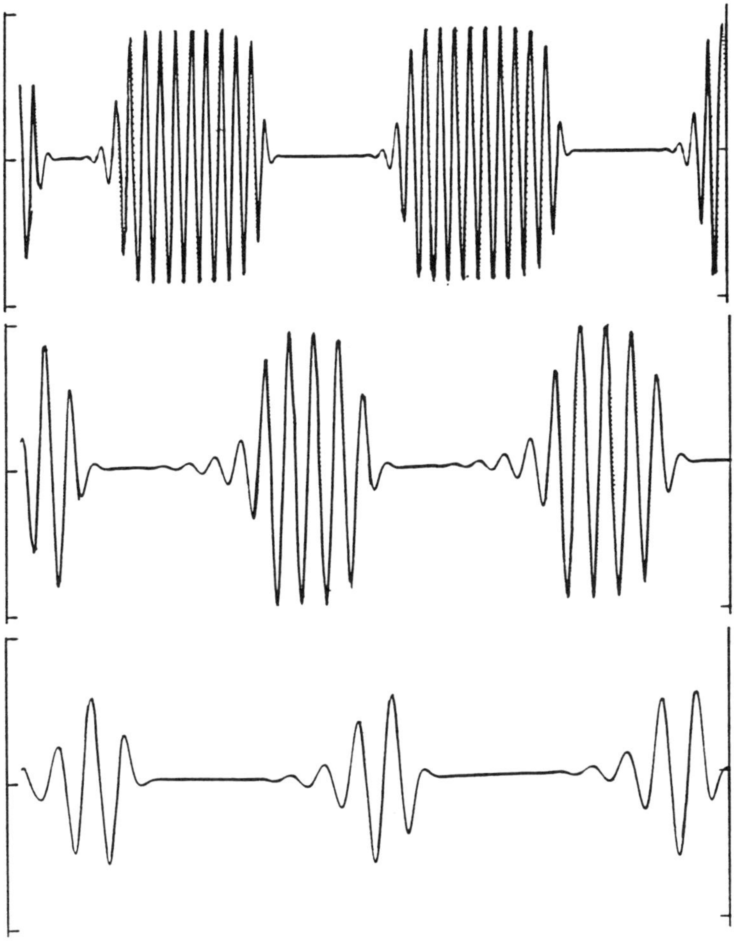

Figure 4: Time history of the three resonantly related amplitudes (real part only) for (3,2,1) the resonance in RPF shown in the preceding figure. Top trace, real part of $a_1(t)$, middle, $a_2(t)$; bottom, $a_3(t)$.

SOME PRELIMINARY EXPERIMENTS CONCERNING THERMAL DISPERSION IN A JET

Z. Warhaft [1, 2]

Cornell University
Ithaca, NY 14853

ABSTRACT

Two fine heated rings of different diameters are placed concentrically in an air jet, providing an analogue of a line source pair experiment in grid turbulence (Warhaft 1984). The mixing of the two sources is determined by inference. The evaluation of the cross correlation coefficient, ρ, and the segregation coefficient, α, is found to be strongly dependent on where the rings are located relative to the jet origin. A particularly novel finding occurs when the rings are placed well inside the potential core (0.5 D from the jet). Here ρ is initially close to unity, then becomes negative and finally grows to unity by about $x/d \sim 10$. The effect of the rings on the velocity field is also studied and it is shown that they suppress the turbulence intensity when placed close to the jet but have little effect on the turbulence when placed further downstream. The results, which are preliminary, are contrasted with more traditional methods of introducing scalars into jets, such as warming the jet relative to the ambient air.

1. Introduction

An attractive aspect of the study of turbulent mixing of passive scalars is that the scalar can be introduced into the flow at varying scales relative to that of the velocity field. The subsequent evolution of the scalar field under the various input conditions may then shed new insight into the nature of the velocity field itself. This provides impetus for the study of scalar mixing over and above the purely pragmatic, which is concerned with pollution, chemical reactions and the like. Of course any basic insight into scalar mixing should be of help in practical problems.

Yet varying the way a scalar is introduced into a particular flow has received relatively little attention; much of the previous work having introduced the scalar evenly throughout the flow. For example, mixing in turbulent jets has been traditionally studied by either warming the whole jet relative to the surroundings (e.g. Corrsin and Uberoi 1950) or by having a jet of pure species A mixing with the surroundings of pure species B (eg. Dowling and Dimotakis 1990 and references therein).

---

[1] Dedicated to John Lumley on the occasion of his sixtieth birthday.

[2] Supported by the U.S. Department of Energy (Basic Energy Sciences, DE-FG02-88ER13929

Here we give preliminary results of the mixing of passive temperature inhomogeneities in a jet. The thermal field is caused by means of placing thin heated wire rings, axisymmetrically, downstream from the jet origin (fig. 1). The diameter of the rings and their placing from the jet is varied so that the effect of initial conditions on the subsequent evolution of the scalar mixing can be studied. The placing of a thin wire ring in a jet is analogous to placing a line source in homogeneous turbulence, a subject that has received much attention (Warhaft 1984 and references cited therein). We can anticipate that the thermal field will be more complex than that of a line source since as the flow evolves the thermal field from one side of the flow will begin to mix with that from the other side. This does not occur with a line source in grid turbulence. So far as I can determine there has been no previous study of a heated ring source in a jet.

Although a single ring in a jet is deserving of thorough study we add a further element to our experiment. In a previous investigation grid turbulence (Warhaft 1984) we showed that when two parallel line sources were placed in the flow their thermal fields over-lapped (fig. 2). We were able to deduce, by a method of inference (Warhaft 1981), the cross correlation between the two scalar fields. Thus this experiment was equivalent to introducing two different passive scalars (with the same Schmidt numbers) into a flow and determining their mixing, a subject of great importance particularly to the atmosphere where mixing (and reaction) usually involves species in trace amounts. The jet analogy to our grid experiment is to place two rings in the flow (fig. 1). As in the previous studies (Warhaft 1981, 1984) the cross correlation, ρ, between the thermal fields coming from each ring is determined by operating ring 1 and ring 2 separately and then operating them together. Under the assumptions of statistical stationarity, and that the scalar fields are passive, we determine the cross correlation from the relation

$$\rho = \frac{\overline{\theta_b^2} - \overline{\theta_1^2} - \overline{\theta_2^2}}{2(\overline{\theta_1^2})^{1/2}(\overline{\theta_2^2})^{1/2}} \qquad (1)$$

where $\overline{\theta_b^2} = \overline{(\theta_1+\theta_2)^2}$ is the variance with both rings operating and $\overline{\theta_1^2}$ and $\overline{\theta_2^2}$ are the variances of each ring operating separately. In the grid turbulence experiment the line sources were fine enough, and the turbulence was sufficiently developed such that the wires did not have any effect on the velocity field. Here we will show that there is a significant effect on the velocity field when the rings are placed in the potential core of

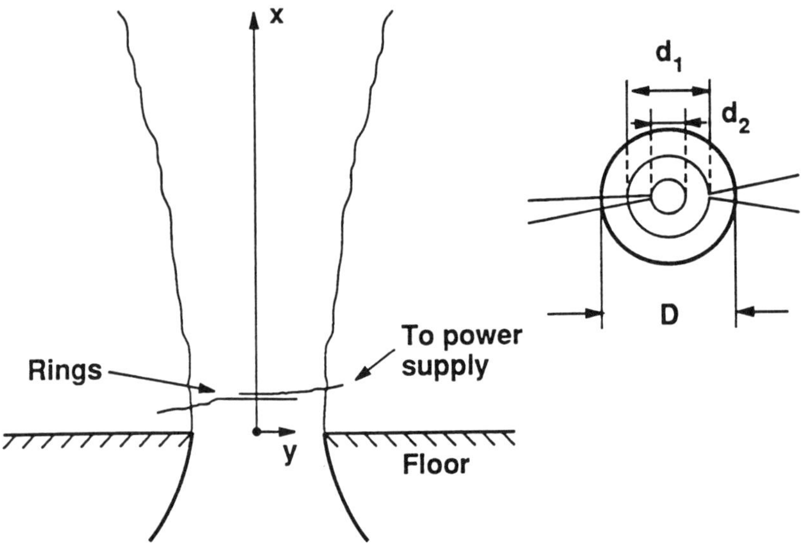

Figure 1. Side and plan view of the jet of diameter D showing the fine wire rings (of diameters d_1 and d_2). The rings were placed concentrically above the jet, both in the same plane. They were suspended in the flow by means of their leads which were held by a clamp, outside the jet.

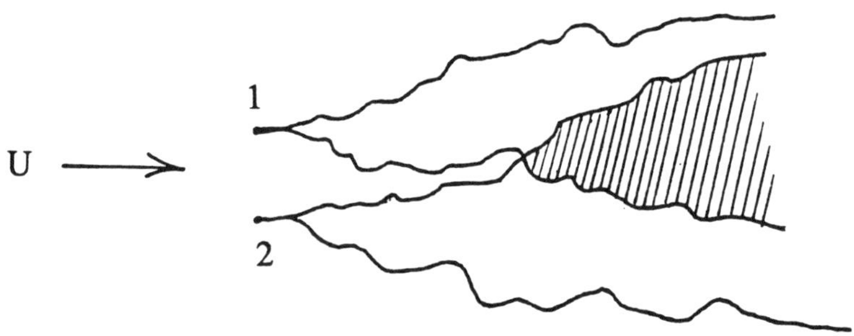

Figure 2. Instantaneous thermal fields for two thermal line sources in a turbulent flow showing the region of overlap or interference (hatched region). This situation was studied by Warhaft 1984. Here we present the jet analogue using two fine wire rings.

the jet. However the inference method for determining ρ is still valid so long as both rings are in the flow at all times and that heating the ring itself does not affect the velocity field (ie. the thermal field is passive).

Another important parameter in turbulent mixing and chemical reaction is the segregation parameter $\alpha = \overline{ab} / \overline{A}\,\overline{B}$ where $\overline{A}$ and $\overline{B}$ are the mean values of the reacting species A and B and $\overline{ab}$ is the covariance of their fluctuations. This parameter is very important in determining the depletion rate, R, of reactants (eg. Komori et.al. 1990), defined as

$$R = k\,AB[1 + \alpha]$$

where k is the reaction rate constant. In our experiment α will be determined as $\overline{\theta_1\theta_2}/T_1T_2$ where T_1 and T_2 are the mean values of the thermal field (relative to ambient) and $\overline{\theta_1\theta_2}$ is the covariance of their fluctuations.

The origin of the experiment owes much to the recent theoretical analysis of Kerstein (1990) who has considered and predicted the way two species would mix in a jet when introduced in a manner similar to that described here.

The results are of a preliminary nature, only single point measurements have been carried out and only the longitudinal velocity, and temperature, have been measured. No spectral or p.d.f. analysis will be reported here, although some preliminary work has been done in this direction. Moreover, we have not yet completed a systematic variation of the input conditions. Nevertheless the initial results appear to be rather novel and I feel they should be of interest to the turbulence mixing community.

2. Apparatus

The air jet was originally constructed by N. R. Panchapakesan (see Panchapakesan and Lumley 1990) who did extensive far field measurements for both an air and helium jet. The jet was mounted vertically and its exit was flush to the floor of the measuring chamber (fig. 1). Various jet sizes were used. In the present work 1.5 cm and 3.0 cm diameter jets were employed and the velocity profile at the nozzle exit was close to top hat. Compressed air was used to supply the jet and this ensured a low turbulence level (of less than 0.1%) at the nozzle exit. For the 1.5 cm jet the air speed was 7.1 m/s yielding a Reynolds number $Re_j = U_jD/\nu$ of 6,600 where U_j is the jet velocity at the exit, D is the jet diameter and ν is the kinematic viscosity of the air. For the 3 cm jet U_j was 7.2 m/s and thus Re_j = 13,500.

The temperature rings were made from 0.127 mm nichrome wire which was fashioned into a circle. The conductors leading to the rings were the same nichrome wire but were coated with copper to decrease their electrical resistance so that there was no heating above the leads. The diameter of the leads was approximately 0.16 mm. The Reynolds number, Re_W, based on the diameter of the wire varied, depending on where they were placed in the flow but their maximum value was 54 when a ring was in the potential core of the 3.0 cm jet. Transverse profiles across the rings normal to the leads produced essentially the same results as across the rings, along the leads ie., in spite of the existence of the leads, the flow and the thermal field was essentially axisymmetric.

In this preliminary study we describe two experiments. For the 3 cm jet a 1.9 cm diameter and a 1.0 cm diameter ring was placed axisymmetrically at 1.5 cm, ie., at $x_0/D = 0.5$ from the jet exit (fig. 1) (here x_0 is the distance of the ring's from the jet). For the 1.5 cm diameter jet a 1.9 cm and 1.25 cm diameter ring was placed at 3D from the jet exit. In both experiments the rings were held in the horizontal plane by their leads which were attached to a clamp and stand, outside the jet. Slight vibration of the rings was observed when they were placed at $x/D = 3$, due to the fluid turbulence but no vibration was observed when they were at $x/D = 0.5$. For the measurements done with the 1.9 cm and 1.25 cm diameter ring pair (1.5 cm jet) the dc power in the rings was 5.5 and 7 watts respectively. For the 1.9 cm and 1.0 cm ring pair (3 cm jet) the power was 10.4 and 4.7 watts respectively. Variation of the power did not affect the normalized results. Longitudinal velocity (mean and fluctuation) measurements as well as mean and fluctuating temperature measurements were made using standard hot wire techniques (e.g. Warhaft 1984).

3. Results

It will be shown below that mixing in a jet occurs very rapidly (compared to grid turbulence), principally because of the presence of the shear (Komori et.al. 1990). We define two dilute scalars in a third medium (here the two temperature fields in air) as being completely mixed when $\rho = 1$. Once the correlation coefficient has reached 1 the subsequent evolution of the scalar fluctuations should be the same as heated jet or a jet of pure species A mixing with species B. For this reason we confine our study to the region where $\rho < 1$ and this inevitably leads to a study of the near field of the jet.

a. The velocity field

We begin with the experiment where two rings of 1.9 cm and 1.0 cm diameter were placed 0.5 D above a 3.0 cm jet. Thus both rings were in the potential core. The development of the centerline rms longitudinal velocity fluctuations $(\overline{u^2})^{1/2}$ normalized by the centerline mean velocity, U_c, is show in fig. 3. Although close to the jet exit the velocity field is very sensitive to initial conditions, the general form of the profiles is consistent with other measurements in the near field (eg. Crow and Champagne 1971). The most striking observation is, however, that the presence of the rings severely **diminishes** the intensity of the velocity fluctuations. Each ring placed separately in the flow caused the fluctuation intensity to diminish, with the smaller ring having the more pronounced effect. When the rings were placed further downstream (at distances greater than x/d ~ 3) the velocity fluctuation intensity was relatively unaffected by their presence. It appears that the rings act to suppress the instability in the potential core of the jet. The suppression mechanism may be similar to the recent experiment of Strykowski and Sreenivasan (1990) who showed that vortex shedding behind a cylinder can be suppressed or diminished by judicious placement of a second, smaller, cylinder downstream from the first. There also appear to be differences to the Strykowski and Sreenavasan work. The Re of our rings are quite high, above 50, where as in the cylinder experiment it was low. Yet we found that when the velocity of the jet was reduced, the same effect occurred, ie, it was relatively insensitive to Reynolds number. When the rings were in place the velocity spectra showed a general diminution at low wave numbers (in their energy containing range) but there was no evidence of any sharp peaks. Some spectra showed a relative enhancement at high wave numbers when the rings were in place. The results also appear to be insensitive to whether the rings were heated or not. Strykowski and Sreenivasan found a marked change in their results when the control cylinder was heated.

Figure 4 shows the centerline mean velocity normalized by U_j. There is evidence that the rings are delaying the transition point at which the centerline velocity begins to diminish (at the end of the potential core) and this is consistent with the lower turbulence intensity (fig. 3).

Figure 5 shows the transverse velocity intensity profiles for the rings at 0.5 D (the case of figs. 3 and 4). The traverse is at x/D= 3.5, ie. at 3D above the rings. The arrows show the location of the rings. The reduction of the intensity occurs across the central core of the flow. This should be contrasted to the case in which the same rings were placed at $x_o/D = 3$ above the jet (fig. 6). Again the measurements were made at x/D = 3 from the rings. The rings do not cause any significant change in the velocity intensity. These and other measurements suggest that the suppression of the velocity fluctuations only occurs when the rings are placed close to the jet exit, ie. well inside the potential core. As stated in the introduction, a detailed analysis of the phenomenon is now in progress.

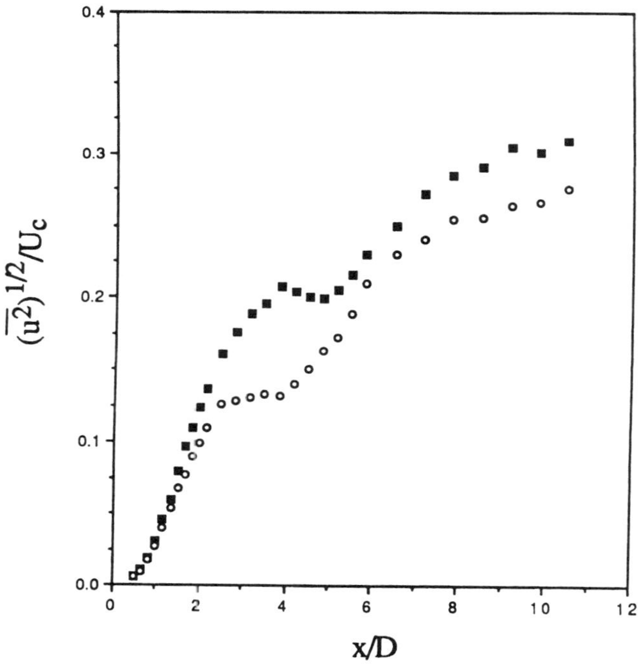

Figure 3. The centerline turbulence intensity profiles for the 3 cm diameter jet, with $U_j = 7.0$ m/s. Open circles are with the fine rings in the flow at $x_o/D = 0.5$ and closed squares are without the rings in the flow. The rings are 1.9 cm 1.0 cm diameter. See text for Reynolds numbers. There is no current in the rings.

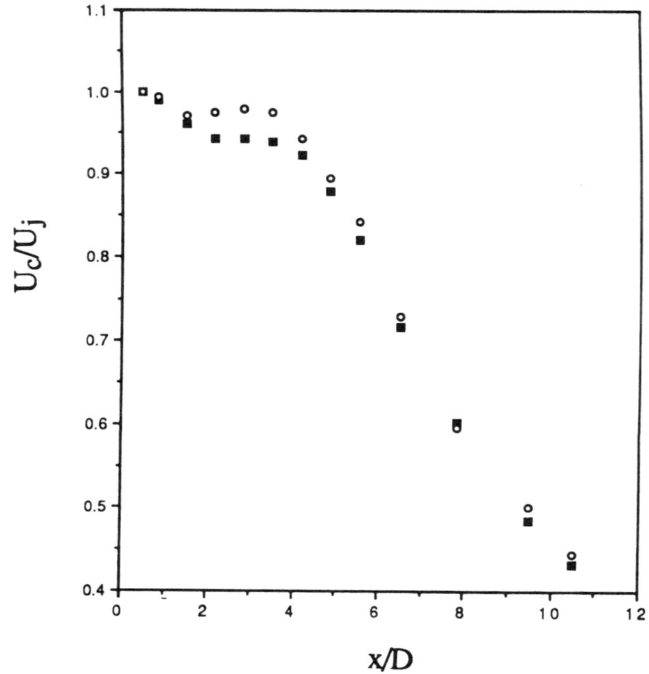

Figure 4. The centerline mean velocity normalized by the jet exit velocity for the same data as figure 3. Same symbols on figure 3.

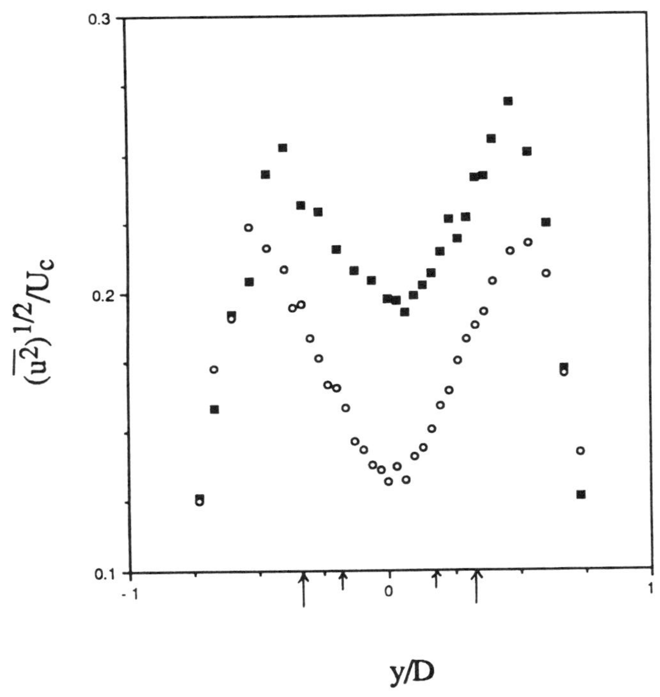

Figure 5. Transverse turbulence rms(u) profile normalized by the centerline velocity, at x/D = 3.5. Symbols and conditions are the same as for figure 3. The arrows indicate the locations of the rings.

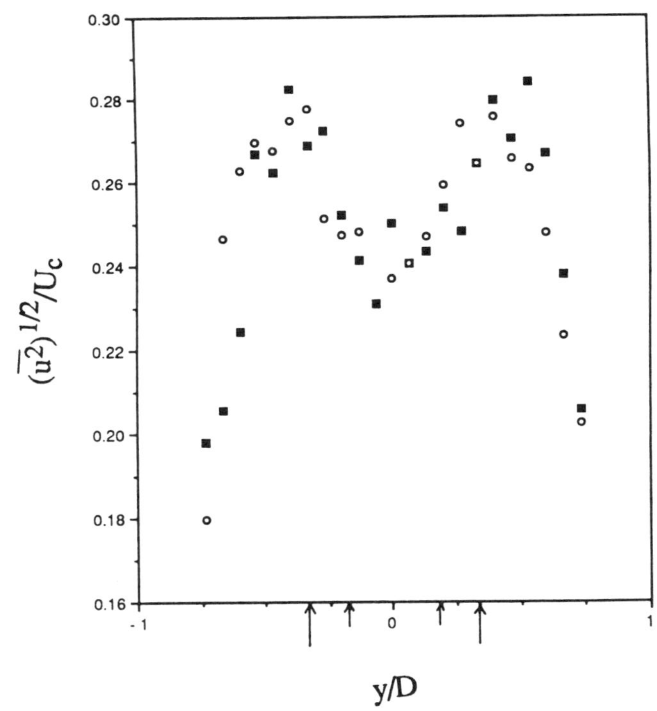

Figure 6. Transverse turbulence rms(u) profile normalized by the centerline velocity, at x/D = 6.0. The rings are at $x_o/D = 3.0$, otherwise the conditions are the same as for figures 3-5.

b. The temperature field

We consider first the 1.9 cm and 1.25 cm rings placed 3.0 D from the 1.5 cm jet. Figure 7 shows the ratio of the rms to mean temperature, $(\overline{\theta^2})^{1/2}/T$, measured along the jet axis and also directly above each of the rings (ie. at 0.95 cm and 0.5 cm from the jet centerline). Here T is the mean temperature of the flow with respect to ambient. Each ring is operated separately. Along the centerline, close to the rings, $(\overline{\theta^2})^{1/2}/T$ is initially high relative to its value directly above the rings, as would be expected. However by about x/D = 5, ie. only two jet diameters downstream from the rings, the ratio is approximately the same for the measurements taken directly above the rings or along the centerline of the jet. By about 12 x/D, $(\overline{\theta^2})^{1/2}/T$ has reached its asymptotic value of approximately 0.2, a value consistent with other passive scalar mixing experiments of heated jets or of jets of species A mixing with species B (Dowling and Dimotakis 1990). Beyond this x/D we would expect that the thermal field from the heated rings (be they operating separately or together) to be the same as the previous heated or concentration jet experiments.

Figure 8 shows the cross correlation, ρ, (squares) and the segregation parameter, α, (circles) along the centerline of the jet for the same configuration as fig. 7. Note the ρ, inferred by operating each ring separately and then together (equation 1), reaches its asymptotic value of 1 by about x/D = 12, which is consistent with the $(\overline{\theta^2})^{1/2}/T$ profiles ie. beyond this point the thermal fields from the two rings are completely mixed. α starts at a very high value and decreases to zero by about x/D = 10. Positive values of α (indicating enhanced depletion rates of two chemically reacting species) do not appear to have been previously measured (Bilger et.al. 1985).

For the rings placed at 0.5D from the jet (fig. 9) ρ and α both asymptote to the same values (of 1 and 0 respectively) as for the rings placed higher in the flow (fig. 8). (Note: the small ring was smaller (1.0 cm as opposed to 1.25 cm) and the nozzle was twice the diameter, than for the measurements of figs. 7 and 8 but this difference will not affect the salient qualitative findings. Other measurements (not shown here) have confirmed that moving the same pair of rings vertically in the same jet will produce the same qualitative differences to be described here.) Notice also that the asymptotic values of ρ and α are achieved at about the same x/D from the rings whether they are placed at x/D= 0.5 or x/D= 3 ie. ρ and α

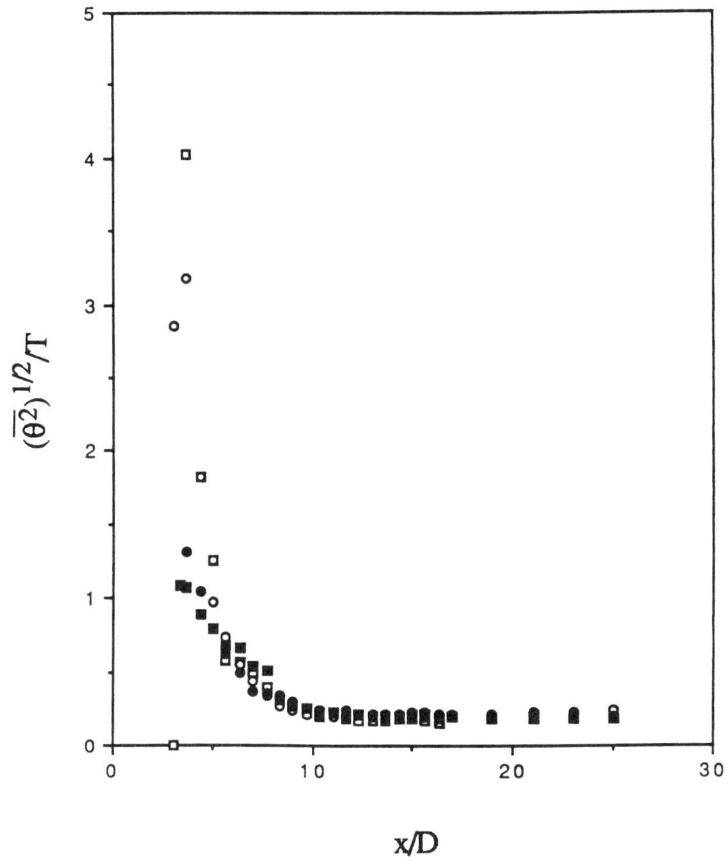

Figure 7. The ratio of the rms to mean temperature along the centerline, and above each heated ring (ie. at y/D = +0.417 and y/D = +0.633). Jet diameter is 1.5 cm and the 1.25 cm and 1.9 cm diameter rings are 3D above the jet. Open squares, centerline for 1.9 cm diameter ring, open circles centerline for 1.25 cm diameter ring. Filled squares, y/D = +.633 (ie. above the 1.9 cm diameter ring); filled circles, y/D = +.417 (ie. above the 1.25 cm diameter ring). The rings are operated separately. See text for Reynolds numbers.

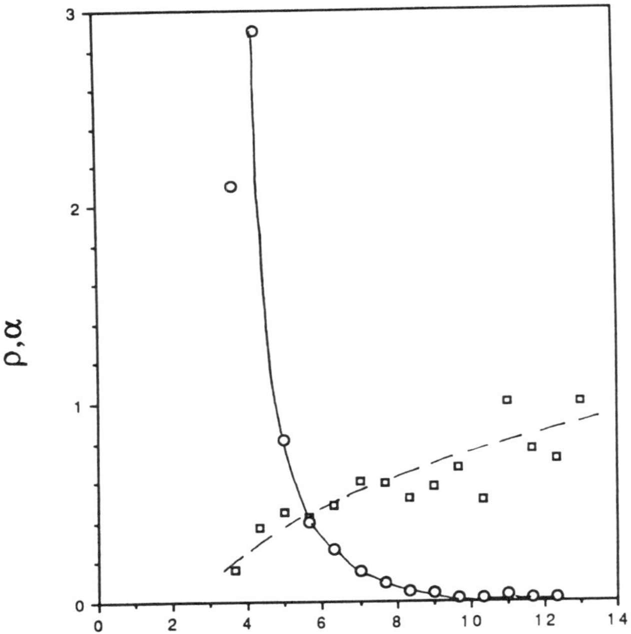

Figure 8. The cross-correlation coefficient, ρ, (squares) and the segregation coefficient, α, (circles) for the 1.25 cm. and 1.9 cm diameter rings placed at 3D above the 1.5 cm jet, ie. the same configuration as figure 7.

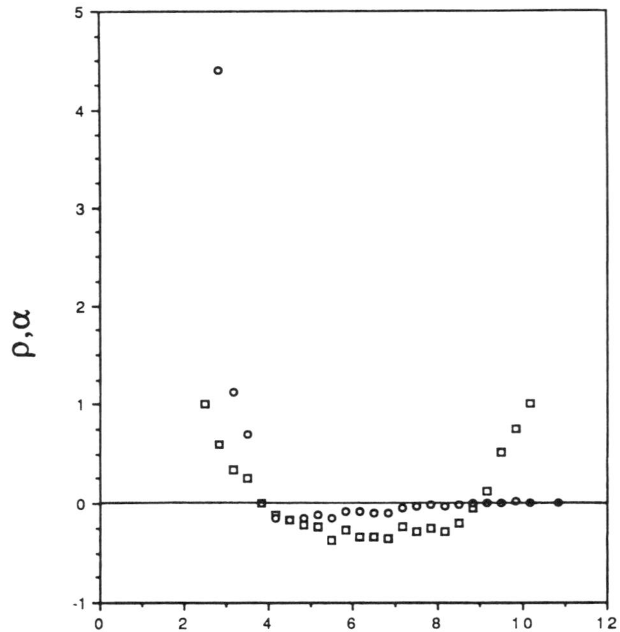

Figure 9. ρ (squares) and α (circles) for the 1.0 cm and 1.9 cm diameter rings placed 0.5D above the 3 cm jet ie. the same configuration as figures 3-5.

reach the values of 1 and 0 about 10 x/D from the rings in both cases (figs. 8 and 9).

It is the region close to the rings where the most significant differences occur. While for the case where the rings are at $x_o/D = 3$, ρ is always positive and is always growing (fig. 8), for the rings at $x_o/D = 0.5$, ρ is initially 1, then diminishes to a negative value of approximately -0.5 before it ascends to 1 (fig. 9). A negative initial value of ρ, growing to a positive value has previously been observed (Warhaft 1984) and physically it suggests that initially the two signals are flapping, roughly in synchronism, past the sensor (negative correlation) and then this correlation grows until the two signals are completely mixed (ρ=1). However, here, (fig. 9) we observe that close to the thermal sources the signals appear to be unmixing, starting from a value where they are perfectly correlated. Here the signal was very intermittent with only occasional "bursts" of temperature signal. Apparently when large disturbances occur here they are so violent that the temperature wake from each ring is swept past the sensor at the same moment. This has significant practical implications since it provides perfect mixing very early in the flow. Notice that α begins with a large positive value but instead of decreasing to 0 from positive values (as for fig. 8) it becomes negative and finally asymptotes to zero from below.

Finally we show some transverse profiles for the configuration of fig. 9, ie. for the 1.0 cm and 1.9 cm diameter rings at $x_o/D = 0.5$ from the 3 cm jet. Figure 10 shows the mean temperature profiles 3D from the rings (ie. 3.5 D from the jet). The open squares and closed triangles are for the 1.0cm and 1.9cm rings respectively. As would be expected this close to the source they are double peaked, the maximum occurring directly above the rings. Notice there is some asymmetry but this has little effect on the normalized covariance quantities, ρ and α shown in fig. 11, which have remarkable symmetry. Note the very rapid variation of ρ and α in the transverse direction. The minimum of ρ, between the two rings, is consistent with the previous findings concerning two line sources in grid turbulence (Warhaft 1984, fig. 13). The approach to unity at large transverse distances from the line sources was also observed in the grid turbulence studies. Notice that α and ρ have approximately the same values across the flow indicating that $(\overline{\theta_1^2})^{1/2}(\overline{\theta_2^2})^{1/2} = T_1T_2$ for this downstream location. This does not occur further downstream (fig. 9). Finally fig. 12 shows that as the thermal field evolves, the strong peak in the center disappears and centerline value of ρ first decreases (fig. 12a) and then finally increases (fig. 12b), as was shown in fig. 9.

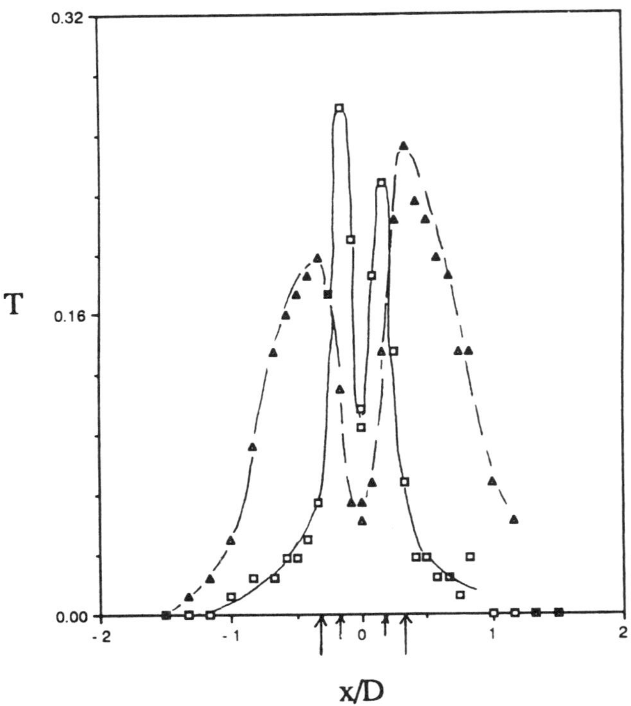

Figure 10. Transverse profile of the mean temperature T at x/D = 3.5 from the jet. The 1.0 cm and 1.9 cm rings are 0.5D above the 3 cm jet. (Same configuration as fig. 9). The arrows indicate the locations of the rings. The open squares and closed triangles are for the 1.0cm and 1.9cm diameter rings respectively.

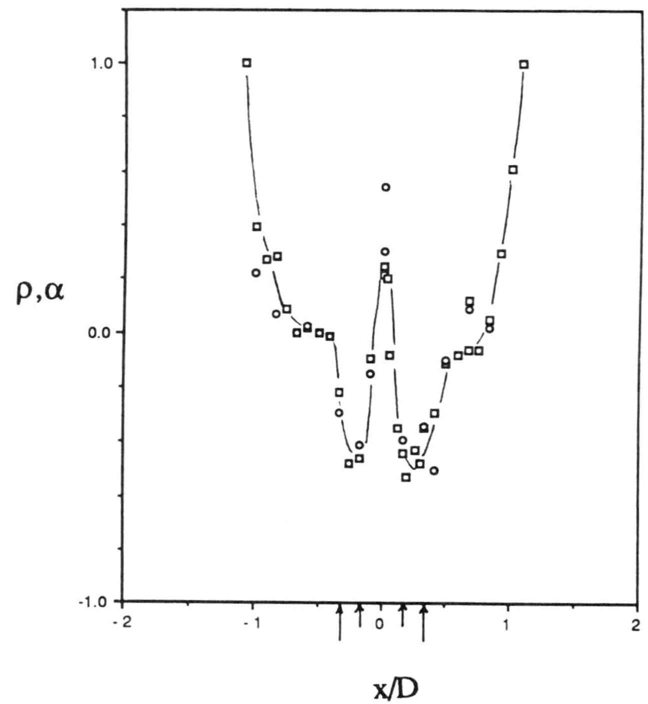

Figure 11. Transverse profiles of ρ (squares) and α (circles) at x/D = 3.5 for the same conditions as figure 10.

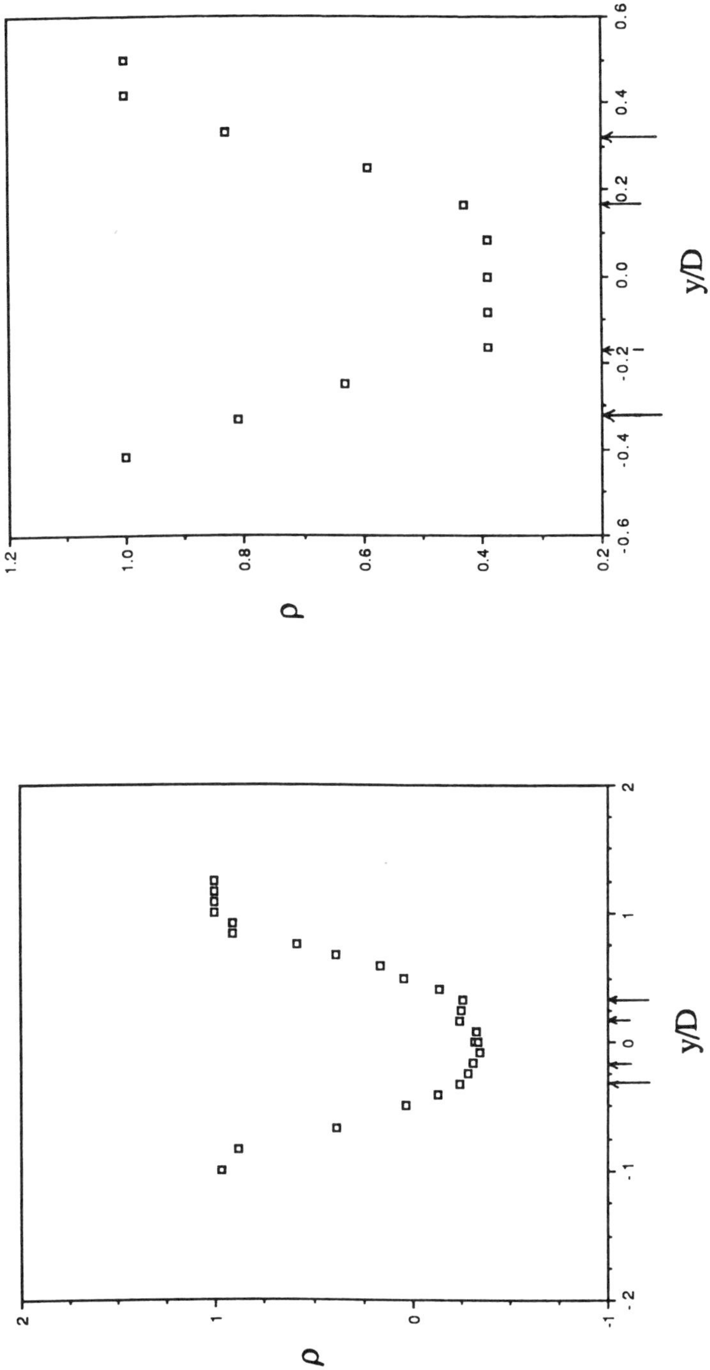

Figure 12. Transverse profiles of ρ at x/D = 6.5(a) and x/D = 9.5(b) for the same conditions as figure 10.

4. Concluding Remarks

Our preliminary study has indicated a rich variety of effects that can occur when a scalar is fed into a jet using thin circular line sources. Further experiment and analysis will attempt to elucidate the mechanism whereby the rings cause a diminution of the turbulence intensity of the jet. The suppression mechanism may be similar to that observed by Strykowski and Sreenivasan (1990) for a small cylinder in the wake of another cylinder. Our work should also shed light on the use of flame stabilizers, which are frequently used in combustion. The study of the cross correlation between the sources shows that in the near field, it is very sensitive to the placing of the rings in the flow but by about 10 jet diameters from the rings the sensitivity to initial conditions is lost, ie. $\rho \rightarrow 1$, $\alpha \rightarrow 0$ and $(\overline{\theta^2})^{1/2}/T \rightarrow 0.2$ (the asymptotic value found for heated jets). These limits are reached much more quickly than the theoretical predictions of Kerstein (1990) although his results are qualitatively similar to ours. Our findings also show that ρ and α can vary dramatically both across the flow and along it. In particular we have shown that α may have large positive values and that ρ may decrease from a value of 1 to negative values will then asymptote to 1 again as the flow evolves.

Finally, I wish to warmly thank John Lumley for his encouragement and support for my experiments on scalar mixing, a subject that has occupied me since I did my first experiment with him on the subject in the mid seventies (Warhaft and Lumley 1978). The work described here has direct origins in that paper.

The experiment described here was done with the enthusiastic and most able help from Erin O'Neill and Prateep Siamwalla, both in their junior year at Cornell. I am most grateful for their assistance. I also thank Allan Kerstein, N. R. Panchapakesan, Stephen Pope and Charles Williamson for discussions and, Ed Jordan for his technical assistance.

References

[1] Corrsin, S. and Uberoi, M.S., 1950, *Further experiments on the flow and heat transfer in a heated turbulent air jet*, NACA Rep. 998.

[2] Crow, S.C. and Champagne, F.H., 1971, *Orderly structure in jet turbulence*, J. Fluid Mech. **48**, 547.

[3] Dowling, D.R. and Dimotakis, P.E., 1990, *Similarity of the concentration field of gas-phase turbulent jets*, J. Fluid Mech. **218**, 109.

[4] Komori, S., Hunt, J.C.R., Kunzaki, T. and Murakami, Y., 1990, *The effects of turbulent mixing on the correlation between the*

species concentration fluctuations in non-premixed reacting flows, (submitted to J. Fluid Mech.).

[5] Kerstein, A.R., 1990, *Linear eddy modelling of turbulent transport, Part 3. Mixing and differential molecular diffusion in round jets*, J. Fluid Mech. **216**, 411.

[6] Panchapakesan, N.R. and Lumley, J.L., 1990, *Turbulence measurements in axisymmetric jets of air and helium. Part 1. Air jet*, (submitted to J. Fluid Mech.).

[7] Strykowski, P.J. and Sreenivasan, K.R., 1990, *On the formation and suppression of vortex 'shedding' at low Reynolds numbers*, J. Fluid Mech. **218**, 71.

[8] Warhaft, Z., 1984, *The interference of thermal fields from line sources in grid turbulence*, J. Fluid Mech. **144**, 363.

[9] Warhaft, Z., 1981, *The use of dual heat injection to infer scalar covariance decay in grid turbulence*, J. Fluid Mech., **104**, 93.

[10] Warhaft, Z. and Lumley, J.L., 1978, *An experimental study of the decay of temperature fluctuations in grid-generated turbulence*, J. Fluid Mech., **88**, 659.

THE VORTEX-LATTICE STRUCTURE OF TURBULENT-SHEAR FLOWS

M. Lesieur, P. Comte and X. Normand

Institut de Mécanique de Grenoble[1],[2]
BP 53 X - 38041 Grenoble-Cedex, France.

ABSTRACT

We perform numerical experiments on *natural* and *controlled* transition in two types of shear flows: a periodic incompressible mixing layer, and a supersonic (Mach 5) boundary layer developing on an insulated flat plate.

In the mixing-layer case, a direct-numerical simulation is developed, starting with a hyperbolic-tangent velocity profile. When the flow is forced initially by a small 3D isotropic residual turbulence (natural transition), it develops a vortex-lattice structure of the Kelvin-Helmholtz billows: this subharmonic secondary instability, corresponding to the helical-pairing instability studied by Pierrehumbert and Widnall (J. Fluid Mech., **114**, pp 59-82, 1982), is in our calculation more amplified than the two-dimensional pairing. This is at variance with the predictions of the secondary instability theory, and yields a highly three-dimensional structure of the flow. A quasi two-dimensional perturbation (controlled transition), on the contrary, leads to longitudinal hairpin vortices strained between quasi two-dimensional billows, as found in the laboratory experiments of Bernal and Roshko (J. Fluid Mech., **170**, pp 499-525, 1986).

The compressible turbulent boundary layer at Mach 5 is numerically out of reach if one restricts to direct-numerical simulations. We have developed a large-eddy simulation using eddy-coefficients based on the second-order structure function of the velocity. This allows us to have a controlled-transition experiment, where the flow is forced upstream by a set of two-dimensional waves (second mode), to which a small 3D

---

[1] Unité Mixte de recherche CNRS, Institut National Polytechnique de Grenoble, Université Joseph Fourier.

[2] This work was supported by D.R.E.T. under contract 88/150, by CNES/Avions Marcel Dassault (HERMES European Space Programme), and by GDR-CNRS *Mécanique des Fluides Numérique*. Part of the calculations were done on a grant of the Centre de Calcul Vectoriel pour la Recherche.

white-noise perturbation is superposed. It develops large-scale Λ-shaped structures elongated in the flow direction, and staggered in the spanwise direction. These vortices scale on the boundary-layer thickness, and thus grow downstream, undergoing some sort of merging. It seems that compressibility is here the catalyst which triggers the appearance of the subharmonic secondary mode, even in a controlled transition experiment.

It is therefore possible that **within** turbulent shear flows (free or wall-bounded) a universal vortex structure on a lattice, as envisaged by Chorin (Comm. Pure and Appl. Maths, **39**, pp S47-S65, 1986) for isotropic turbulence, **exists**.

1. Three-dimensionalization of a temporal mixing layer

The mixing layer between two parallel flows of different velocity is well understood numerically[1] in two dimensions, either in the temporal (periodic in the flow direction) or spatially-growing case (see Comte, 1989, and Lesieur, 1990) for a review. These 2D calculations show in particular how the *coherent structures* form by roll up of the initial vortex sheet after the development of the linear instability, and undergo subsequent pairings. It was shown also numerically by Lesieur et al (1988) that the coherent vortices are *unpredictable* (or *chaotic*): if one considers two flows consisting in two rows of Kelvin-Helmholtz vortices slightly decorrelated, the difference between the two fields will grow exponentially. As shown by Lesieur (1990, p 358), this is due to a *subharmonic decorrelation*: if the two rows are submitted to subharmonic perturbations which are out of phase, the pairings will occur in each flow between different vortices, which, as a result, will induce a strong decorrelation between the two fields. Furthermore, the coherent vortices are (still in 2D), responsible for an enstrophy cascade towards the small scales, with a kinetic energy spectrum close to k^{-3}. It is thus natural to talk of *two-dimensional turbulence* for the 2D mixing layer, since it possesses the two major ingredients one requires of turbulence, that is, *unpredictability* (or *chaos*) and broad-band spatial spectrum. Another important point, demonstrated by this example, is that the coherent structures participate in turbulence: thus, their existence is not in contradiction with a statistical approach, as stressed by Lumley (1989).

Unfortunately, the reality is not so nicely two-dimensional[2]. Labo-

[1] That is, with the aid of direct-numerical simulations of the Navier-Stokes equations.

[2] *Nicely*, from the numerical viewpoint, since 2D calculations can be

ratory experiments show the development of 3D instabilities[3], under the form for instance of thin ***hairpin*** vortex filaments strained longitudinally between the Kelvin-Helmholtz vortices (Breidenthal, 1981, Bernal and Roshko, 1986, Lasheras and Choi, 1988). Also, the kinetic energy spectra measured in these experiments follow a Kolmogorov $k^{-5/3}$ law for scales smaller than the vorticity thickness. This cannot be an inverse-energy cascade of 2D turbulence (which would develop in larger scales), and therefore can only correspond to a 3D kinetic energy cascade due to the development of 3D instabilities.

In order to understand at least some of these phenomena, we have carried out direct-numerical simulations of the 3D Navier-Stokes equations (with uniform density) applied to a temporal mixing layer in a cubic box. Periodicity is assumed in the x and z directions, with free-slip boundary conditions on the lateral boundaries. We focus on the transition stage, and start initially with a hyperbolic-tangent velocity profile $U \tanh 2y/\delta_i$, to which are superposed two random isotropic perturbations of wide spectrum, peaking at the fundamental mode $k_a = 2\pi/7\delta_i$: the first perturbation is two-dimensional (in the x, y plane), of kinetic energy $\epsilon_{2D} U^2$. The second perturbation is three-dimensional, of kinetic energy $\epsilon_{3D} U^2$. The initial Reynolds number is $R_e = U\delta_i/\nu = 100$. The numerical methods are of the classical pseudo-spectral type developed by Orszag, and we solve in parallel a passive-scalar equation (with a Schmidt number of 1), taking the initial passive scalar distribution identical to the basic velocity: the passive scalar will play the role of a numerical dye, allowing us to see the two flows mix. The resolution of the computation corresponds to 128^3 grid points in physical space.

1.1 Natural transition

If $\epsilon_{3D} = 0$, the problem is two-dimensional, and corresponds to the study done in Lesieur et al. (1988): 2D Kelvin-Helmholtz billows form at about 15 δ_i/U, and pair at about 30 δ_i/U. Now, we show calculations with $\epsilon_{2D} = 0$ and $\epsilon_{3D} = 10^{-4}$, that is, with a purely three-dimensional perturbation: this corresponds to a *natural transition* numerical experiment. The following figures are taken at $t = 30\ \delta_i/U$, characteristic of the 2D pairing. Figures 1a, 1b, and 1c show a top view of respectively the vortex-field topology, the vortex lines and the passive scalar at the interface: the Kelvin-Helmholtz billows are severely distorted (in phase opposition) in the spanwise direction, and pair therefore only in the central region of the span. There are also vortex filaments which escape to the attraction of the big billows in the region of pairing, detach, and

---

performed now on a workstation.

[3] This is fortunate for the super-computer developers.

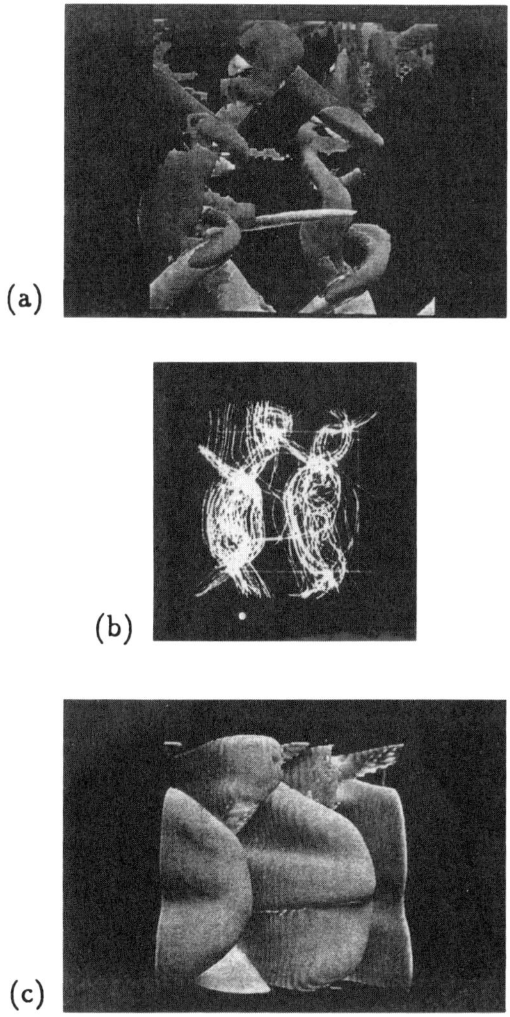

Figure 1: top view of the mixing layer in the natural transition numerical calculation: a) vortex field, b) vortex lines, c) numerical dye at the interface.

are stretched away by the mean flow, giving rise to thin longitudinal vortices.

This highly three-dimensional Λ-shaped structure of the Kelvin-Helmholtz billows may be interpreted as resulting from the development of a secondary subharmonic instability: assume that the Kelvin-Helmholtz billows oscillate in vertical y, z planes and out of phase[4], as

[4] which corresponds to a spanwise sine modulation of the longitudinal subharmonic perturbation

indicated on Figure 2. Thus, the vorticity of the peaks P will be (to the lowest order, see last section) convected along the upper flow, while that of the valleys V will be convected by the lower flow. This will lead to pairings as indicated on Figure 2b, resulting into a vortex-lattice structure. This instability, named ***helical-pairing instability***, had been proposed by Pierrehumbert and Widnall (1982). However, the 3D secondary instability theory predicts that it is less amplified than the two-dimensional pairing. It is thus surprising to see this instability developing from a small random perturbation, and ousting the two-dimensional pairing. This calculation is the first numerical evidence of the helical-pairing instability, in numerical simulations starting from a random perturbation. The helical-pairing instability had already been displayed by Meiburg (1985) in a simulation using vortex methods. Recent 3D simulations of a spatially-growing mixing layer (Fouillet, 1990, by finite-difference methods, and Fiedler, 1990, by vortex methods) show the same vortex patterns, with branchings of the Kelvin-Helmholtz billows. These branchings have been observed in the laboratory by Browand and Troutt (1980).

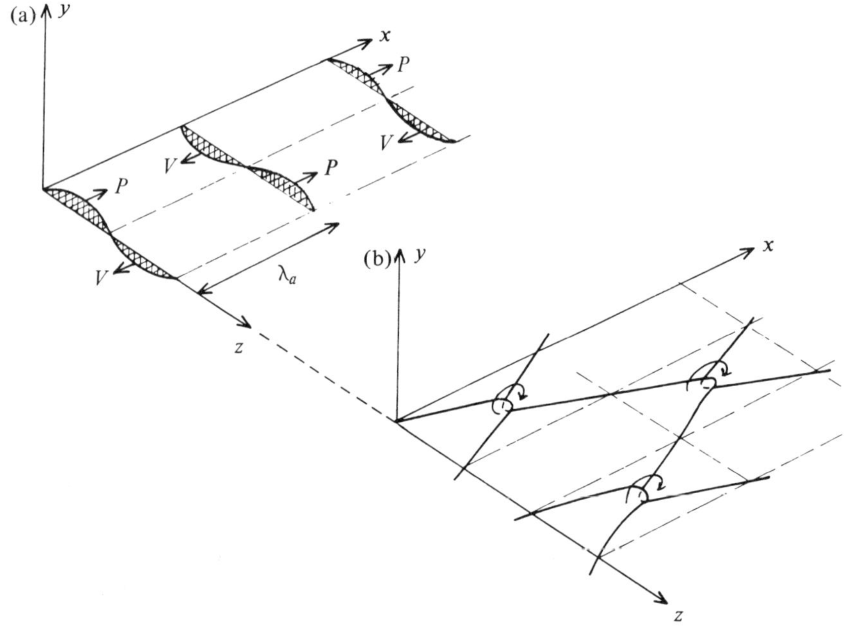

Figure 2: schematic representation of the helical-pairing instability, leading to a vortex-lattice structure.

1.2 Controlled transition

Afterwards, we have redone the calculation with a quasi two-dimensional perturbation, taking $\epsilon_{2D} = 10^{-4}$ and $\epsilon_{3D} = 10^{-5}$. In this case, the 4 initial Kelvin-Helmholtz billows undergo a quasi two-dimensional pairing, and the two remaining billows oscillate slightly in phase. Meanwhile, hairpin vortex filaments are stretched between the primary vortices. This is reminiscent of the *translative instability*, also discussed by Pierrehumbert and Widnall (1982). This instability may be explained as in Figure 2, but with vortex billows oscillating in phase. However, the thin longitudinal vortex filaments observed in the calculations and in the laboratory experiments do not seem to originate directly from the translative instability, but are simply stretched between the billows.

2. Supersonic boundary layer on a flat plate

The second study which will be presented here deals with the transition to turbulence in a supersonic boundary layer developing above an insulated flat plate. The Mach number is 5, which corresponds to high supersonic conditions, such as may be found in fast planes which are presently being developed. The numerical methods use a finite-differences scheme of high order proposed by Gottlieb and Turkel (1976), with a stretching of coordinates in the transverse direction, and permeable boundary conditions upstream and downstream (see Normand and Lesieur, 1990, for details). We have taken a very long computational domain: $400 \times 40 \times 20$ points respectively in the streamwise, transverse and spanwise directions. The Reynolds number, based on the upstream displacement thickness, is chosen equal to 10 000, in order to allow for the growth of the instabilities in the domain. At such a Reynolds number, no direct-numerical simulation is possible, and it is necessary to have recourse to a subgrid-scale model.

2.1 Structure-function based subgrid-scale model

The model we have used is based on the concepts of *spectral eddy-viscosity* and *spectral eddy-diffusivity* proposed by Kraichnan (1976, see also Lesieur, 1990, for a review). It had been mainly used for homogeneous turbulence, and corresponded then to a constant eddy-viscosity in physical space. A generalization of this concept, taking into account the intermittency and inhomogeneity of turbulence, has been proposed by Métais and Lesieur (1990): the turbulence is supposed to be locally, at a position $\vec{x}$, isotropic at scales smaller than the grid mesh Δx. A local kinetic energy spectrum $E_{\vec{x}}(k_c)$ (with $k_c = \pi/\Delta x$) is then calculated in terms of the local 2nd order velocity structure function

$< \|\vec{u}(\vec{x},t) - \vec{u}(\vec{x}+\vec{r},t)\|^2 >_{\|\vec{r}\|=\Delta x}$ of the large-scale velocity field (where $< . >$ is a proper spatial average upon points $\vec{x}+\vec{r}$ a distance Δx from $\vec{x}$). Then the isotropic EDQNM expression for the eddy-viscosity is used:

$$\nu_t(\vec{x}) = 0.4 \sqrt{\frac{E_{\vec{x}}(k_c)}{k_c}} \quad , \tag{1}$$

so we may construct an eddy-viscosity which is variable with the local turbulent activity. An analogous eddy-conductivity is built taking a turbulent Prandtl number equal to 0.6, as given by the isotropic incompressible studies (see Lesieur, 1990). Afterwards, these eddy-coefficients are substituted with the viscous coefficients into the full compressible Navier-Stokes equations. Notice that eq. (1) is a simplified form of the complete EDQNM spectral eddy-viscosity, without any cusp in the neighbourhood of the cutoff k_c. The coefficient $A = 0.4$ is adjusted as proposed by Leslie and Quarini (1979), in order to have an eddy-viscous flux $2\nu_t \int_0^{k_c} p^2 E(p)dp$ equal to the kinetic energy flux ϵ in the Kolmogorov cascade ($E(k) = C_K \epsilon^{2/3} k^{-5/3}$) identical to the kinetic energy flux ϵ. This gives $A = (2/3)C_K^{-3/2}$, where C_K is the Kolmogorov constant taken here equal to 1.4. Notice that the same adjustment is made in the determination of an analogous eddy-viscosity in the frame of the so-called *Renormalization-Group techniques* (see Yakhot and Orszag, 1986). Notice also that the use of these incompressible eddy-coefficients for compressible calculations may be only justified by assuming that the subgrid-scales are negligibly affected by compressibility. Notice finally that our model gives, in the incompressible case, a better agreement with laboratory experiments than the Smagorinsky model, for two incompressible flows: behind a backward-facing step (Silveira Neto et al., 1990), and channel flow (Comte et al., 1990).

2.2 Controlled transition

In three dimensions, it was not possible to run the code at high Mach number in conditions of *natural transition*, that is, with a laminar velocity profile upstream forced by a small white-noise perturbations. However, it was possible to use the code for *controlled transition*, the upstream flow consisting of a set of two-dimensional waves forced by the white noise. The two-dimensional waves (second mode) were obtained with the same code in a 2D calculation. Figure 3a shows a horizontal section of the longitudinal velocity at the top of the layer: it indicates clearly how the two-dimensional waves upstream evolve into longitudinal streaks. The vortex structure corresponding to the latter is shown on Figure 3b: it is clear that Λ-shaped vortices staggered in the spanwise direction have formed, which corresponds to a secondary subharmonic instability of the second mode. Their size increases as they travel downstream, and they might undergo some sort of merging, as we have observed (with the same code) in a 3D periodic boundary layer at Mach 0.5. They are submitted to self-induction and are lifted from the wall, as side views of the vortices indicate. It seems here that the effect of compressibility, as compared to the incompressible boundary layer, is to elongate the vortices in the flow direction, and to select the staggered pattern: indeed, our above-quoted low Mach number computations show an aligned pattern of Λ-vortices in the controlled-transition simulations (Normand, 1990).

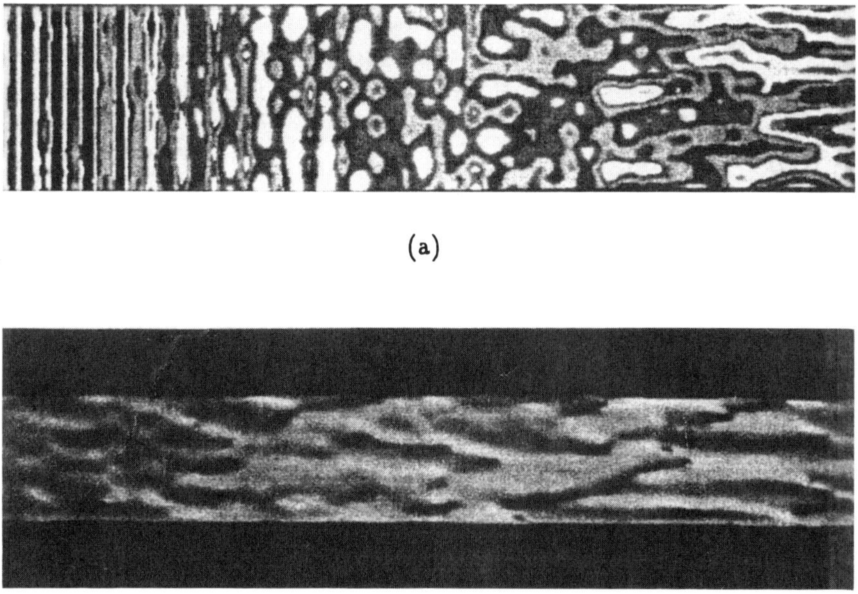

(a)

(b)

Figure 3: boundary-layer at Mach 5; horizontal cross-section of the longitudinal velocity on the top of the layer (a), top view of the vortex structure in the transitional region (b).

3. Conclusion and discussion

We have shown, using three-dimensional direct-numerical simulations, that a periodic mixing layer, developing from a hyperbolic-tangent velocity profile, could be subject to a violent three-dimensional instability leading to a vortex-lattice structure if the initial perturbation was random and three-dimensionally isotropic. This behaviour may be explained in terms of a secondary subharmonic instability, and corresponds to the helical-pairing instability, discovered by Pierrehumbert and Widnall (1982): our calculations show that, in a ***natural transition*** numerical experiment, this instability is more amplified than the 2D pairing result which is at variance with the secondary-instability theory predictions. The existence of this vortex-lattice structure suports a model proposed by Chorin (1986) for 3D isotropic turbulence. On the contrary, calculations with a quasi two-dimensional random perturbation (*controlled transition*) show a 2D pairing of the large billows, while thinner longitudinal hairpin vortices are stretched in the direction of the basic shear.

In the high supersonic (Mach 5) boundary layer developing on an insulated flat plate, numerical experiments of *controlled transition* using large-eddy simulation techniques show the emergence of the staggered vortex structure, whose development must be favoured by compressibility.

It seems therefore that the turbulence structure in shear flows (free or wall-bounded) may depend heavily upon the nature of the residual incoming turbulence. It is possible that the staggered structure of large hairpin vortices (vortex lattice) could be found in most of the turbulent flows in *natural conditions*. The questions which arise are for instance: to which extent does this structure contribute to the build up of the Kolmogorov cascade at high Reynolds number? Does the appearance of this structure correspond to the transition to small-scale turbulence? Are the turbulent fluxes affected by this structure? What kind of initial conditions may disrupt it and possibly modify the stresses? And finally: what kind of new simple models may be derived from the increased knowledge the computer gave us on the intimate nature of turbulence?

AKNOWLEDGEMENTS

One of the authors (M. Lesieur) is pleased to note numerous discussions with John Lumley during these last few years, on statistical theories, stratified turbulence, coherent structures, boundary-layers, chaos, and the dissipativity of turbulence (the so-called: $\epsilon = u^3/l$ discussion).

REFERENCES

Bernal, L.P. and Roshko, A., 1986, "Streamwise vortex structure in plane mixing layers", J. Fluid Mech., **170**, pp 499-525.

Breidenthal, R., 1981, "Structure in turbulent mixing layers and wakes using a chemical reaction", J. Fluid Mech., **109**, pp 1-24.

Browand, F.K. and Troutt, T.R., 1980, "A note on spanwise structure in the two-dimensional mixing layer" J. Fluid Mech., **93**, pp 325-336.

Browand, F.K. and Ho, C.M., 1983, "The mixing layer: an example of quasi two-dimensional turbulence", in *Two-dimensional turbulence*, J. Mec. Theor. Appl., Suppl. , R. Moreau ed., pp 99-120.

Chorin, A.J., 1986, "Turbulence and vortex stretching on a lattice", Comm. Pure and Appl. Maths, **XXXIX**, pp S47-S65.

Comte, P., 1989, *Etude par simulation numérique de la transition à la turbulence en écoulement cisaillé libre*, Thèse de l' Institut National Polytechnique de Grenoble.

Comte, P., Lee, S.S., Cabot, W. and Moin, P., 1990, "A subgrid-scale model based upon the second-order velocity structure function", Proc. Summer program 1990, Center for Turbulence Research, Stanford, California.

Fiedler, H.,1990, private communication.

Fouillet, Y.,1990, private communication.

Gottlieb, D. and Turkel, E., 1976, "Dissipative two-four methods for time-dependent problems", Math. Comp., **30**, pp 703-723.

Kraichnan, R.H., 1976, "Eddy viscosity in two and three dimensions", J. Atmos. Sci., **33** , pp 1521-1536.

Lasheras, J.C. and Choi, H., 1988, "Three-dimensional instability of a plane free shear layer: an experimental study of the formation and evolution of streamwise vortices, J. Fluid Mech., **189**, pp 53-86.

Lesieur, M., 1990, *Turbulence in Fluids* (second revised edition), Kluwer Publishers, Dordrecht.

Lesieur, M., Staquet, C., Le Roy, P. and Comte, P., 1988, "The mixing layer and its coherence examined from the point of view of two-dimensional turbulence, J. Fluid Mech., **192**, pp 511-534.

Leslie, D.C. and Quarini, G.L., 1979, "The application of turbulence theory to the formulation of subgrid modelling procedures", J. Fluid Mech., **91** , pp 65-91.

Lumley, J.L., 1989, in *Wither turbulence? Turbulence at the crossroads*, Springer Verlag Lecture Notes in Physics, **357**, pp 49-57, J. Lumley, ed.

Meiburg, E., 1985, PhD thesis, University of Karlsruhe.

Métais, O. and Lesieur, M., 1990, "Spectral large-eddy simulations of isotropic and stably-stratified turbulence", submitted to J. Fluid Mech.

Normand, X., 1990, *Transition à la turbulence dans les écoulements cisaillés libres ou pariétaux*, Thèse de l'Institut National Polytechnique de Grenoble.

Normand, X. and Lesieur, M., 1990, "Numerical experiments on transition in the compressible boundary layer over an insulated flat plate", Theor. Comp. Fluid Dynamics, in press.

Pierrehumbert, R.T. et Widnall, S.E., 1982, "The two and three-dimensional instabilities of the spatially periodic shear layer", J. Fluid Mech., **114**, pp 59-82.

Silveira Neto, A., Grand, D., Métais, O. and Lesieur, M., 1990, "Large-eddy simulation of the turbulent flow in the downstream region of a backward-facing step", submitted to Phys. Rev. Letters.

Yakhot, V. and Orszag, S., 1986, "Renormalization Group (RNG) methods for turbulence closure", J. Scientific Computing, **1**, pp 3-52.

SCALAR TRANSPORT IN A LONGITUDINAL VORTICITY SYSTEM IN BOUNDARY LAYERS

Joseph T. C. Liu[1]
Brown University
Providence, RI 02912

ABSTRACT

Scalar transport in a longitudinal vorticity system, exemplified by nonlinear spatially developing Gortler vortices, is discussed. Analogy with the longitudinal momentum problem for unity Prandtl and Schmidt numbers is explored. Mushroom-like flow visualization patterns and enhanced surface heat transfer rates are explainable under this analogy.

1. Introduction

Scalar transport in turbulent shear flows is of considerable interest in both engineering and geophysical disciplines, including environmental considerations. A student in turbulent shear flows would be introduced to the subject on scalar transport in Tennekes and Lumley (1972). Some advanced treatment of the subject from theoretical as well as computational modeling points of view are given in Lumley (1974, 1975, 1978).

One of the aims in coherent structure research (e.g. Liu 1988, 1989; Aubry, et al. 1988) is that the flow field associated with the large-scale structures (and hence, the transported scalars) could be "deterministically" described from dynamics prior to the usual Reynolds' averaging. The parametrization of such deterministically obtained large-scale structures in the Reynolds mean flow description in turn, would shed light on how to lessen the semiempiricism in modeling turbulent shear flows and the associated scalar transport. The possible removal of empiricism in turbulence modeling from the description of motions contributed by dominating eddies would

[1] This work was initiated through support from the Fluid, Particulate and Hydraulic Systems Program of the National Science Foundation to the Division of Engineering at Brown University.

be of enormous technological interest. This is in no small way attributable to the fact that direct numerical simulation of practical engineering problems is not yet achievable.

In the following, we give a modest account of the scalar transport in a prototype longitudinal vorticity system in a wall bounded flow.

2. Discussion of the Problem

Blackwelder (1983) pointed out the amazing similarities between the longitudinal vorticity system set up in the boundary layer on a slightly curved wall (Gortler vortices) and that in the wall region of the turbulent boundary layer. This led to quantitative experimentation on the transitional Gortler vortex behavior by Swearingen and Blackwelder (1987). A similar counter rotating vorticity system is also enduced by Guezennec, et al. (1987) in wall-bounded turbulent flows. Using an unsteady boundary layer analogy, Sabry and Liu (1988, 1989) showed how the momentum problem could be "modeled" with simple calculations. They started from initially amplified, but linear, Gortler vortices consistent with experiments and showed how the high shear layer was developed and migrated to the outer edges of the boundary layer. They also obtained development of mushroom-like structural features for the streamwise velocity component that were consistent with the quantitative measurements of Swearingen and Blackwelder (1987).

The mushroom-like structures were also obtained in flow visualization studies in air (e.g. Ito 1985) as well as in water (Peerhossaini and Wesfried 1988). The leading question is thus, how are the passive scalars transported in such a longitudinal vorticity system and what is the relation between the visualized mushroom-like scalar concentration structures and those of the iso-velocity contours of the streamwise velocity, measured by use of hot wires (Swearingen and Blackwelder 1987), which also appear in the form of mushroom like structures.

The curiosity does not stop there. We should also ask about the behavior of another similarly transported scalar: "heat". This may answer questions with regard to the heat transfer problem associated with streamwise vortices in a boundary layer (Pauley and Eaton 1988).

Our answers, of course, would be peculiar to the longitudinal vorticity system associated with the very active system which developed from initially amplified Gortler vortices. Nevertheless, similarities with the educed wall-bounded turbulent flow problem have already been mentioned.

3. Scalar Transport and Longitudinal Momentum Transport Analogy

Possible similarities between concentration and streamwise velocity structures were already given a clue by the flow visualization studies and hot wire measurements (although one needs to be careful about similar boundary and streamwise-initial conditions). We shall develop these similarities quantitatively.

For most practical situations, the Reynolds number is large and the boundary layer thickness δ is much smaller than the radius (R) of the slightly curved concave surface. Furthermore, the spatial development problem is describable by parabolized viscous effects (Floryan and Saric 1982; Hall 1983). Scale analysis also shows that the streamwise-pressure gradient is of higher order. In this case, the continuity and momentum equations, to order δ/R, for an incompressible fluid are of the form:

$$\nabla \cdot \mathbf{u} = 0,$$

$$\mathbf{u} \cdot \nabla u = \nu \nabla^2 u,$$

$$\mathbf{u} \cdot \nabla v + (u^2/R) = -(1/\rho)\ \partial p/\partial y + \nu \nabla^2 v, \qquad (1)$$

$$\mathbf{u} \cdot \nabla w = -(1/\rho)\ \partial p/\partial z + \nu \nabla^2 w,$$

where $\mathbf{u}$ is the vector velocity having components (u, v, w) in, respectively, the streamwise, normal to the wall and spanwise directions (x, y, z), $\nabla \cdot$ is the three dimensional divergence operator, the advective operator $\mathbf{u} \cdot \nabla$ is three dimensional, ∇^2 is the parabolized Laplacian in the yz cross sectional plane, ρ is the fluid density (taken as constant), p is the pressure and ν is the kinematic viscosity.

The reduction to "incompressible" form is, of course, obtained only with the full appreciation of "compressibility" effects. The incompressible form of the energy (static enthalpy) equation, obtained under small Mach number ($M^2 \to 0$) and small temperature loading ($\Delta T/T \to 0$) assumptions, is in the form

$$\mathbf{u} \cdot \nabla T = (\nu/Pr)\nabla^2 T, \tag{2}$$

where T is the static temperature and Pr is the Prandtl number, which is the ratio of ν to the thermal diffusivity κ. In the above form of the energy equation, the work done due to pressure gradients as well as the rate of viscous dissipation do not appear as they are of order $M^2 \to 0$. The transport properties in the energy equation (and momentum equations) are taken as uniform for a unifrom basic state because of the small temperature loading $\Delta T/T \to 0$ assumption.

In the absence of chemical reactions, the species concentration equation for an effective binary mixture is

$$\mathbf{u} \cdot \nabla K_i = (\nu/Sc)\ \nabla^2 K_i, \tag{3}$$

where K_i is the mass fraction of the ith species, Sc is the Schmidt number and is the ratio of ν to the binary diffusion coefficient D_{12}.

We can already notice the similarities between energy, concentration equations and the x-momentum equation. We remark here that the "thin boundary layer" approximation rendered curvature effects to be of higher order in the x-momentum equation; otherwise the curvature effects there would preclude the possibility of similarity with the energy and concentration equations.

We need to bring out the similarities more formally by defining the following dimensionless quantities:

$$U = \mathbf{u}/\ u_e\ \ (\mathbf{U}=\mathbf{u}/u_e),\ \ \theta=(T-T_w)/(T_e-T_w),$$
$$Z_i=(K_i-K_{iw})/(K_{ie}-K_{iw}) \tag{4}$$

where the subscript e denotes conditions at the outer edge of the boundary layer and subscript w that at the wall. Thus the x-momentum, energy and concentration equations become

$$\mathbf{U} \cdot \nabla U = \nu\nabla^2 U$$
$$\mathbf{U} \cdot \nabla \theta = (\nu/Pr)\nabla^2 \theta \tag{5}$$

$$U \cdot \nabla Z_i = (\nu/Sc)\nabla^2 Z_i,$$

the momentum problem, which is not fully stated here, is presumed "known".

The wall boundary conditions are $U=\theta=Z_i=0$, while those at the outer edge of the boundary layer are $U=\theta=Z_i=1$. We have not split the flow quantities into a Reynolds mean plus a perturbation part. The dependent variables here are thus still the total flow quantities (as was measured and visualized). The spanwise periodicity enters through the spanwise boundary conditions with periodicity set by the momentum problem. The spanwise wavelength would correspond, say, to the observed robust wavelength λ (or wavenumber $\alpha=(2\pi)/\lambda$). The streamwise initial conditions at x_o would be $U=U\ (x_o;\ y,\ z)$, $\theta=\theta(x_o;\ y,\ z)$, $Z_i=Z_i(x_o;\ y,\ z)$.

Thus, formally, if the boundary and initial conditions for U, θ and Z_i are similar, and if $Pr=Sc=1$, then

$$\theta(x, y, z) = Z_i(x, y, z) = U(x, y, z); \tag{6}$$

similarities also exist between the surface heat and mass transfer rates and the skin friction.

Although the smoke injection boundary and upstream initial conditions differ from those of the x-momentum, there is the tendency of iso-concentration profiles to develop into mushroom-like structures in air (e.g., Ito 1985), where Pr, Sc are close to unity. For a liquid system (e.g., Peerhossaini and Wesfried 1988), the iso-concentration profiles also has the tendency to form mushroom-like structures. Though modified, the iso-K_i structures are still similar to that of the iso-U structures even though the iso-concentration structures are Sc is rather large. In the nonlinear stages of the longitudinal vorticity system development, scalar advection by eddy motions compete and even overwhelm molecular diffusion. Even if similarity cannot prevail, there is a one-to-one correspondence between iso-concentration profiles and the iso-U velocity profiles. This can be established by computation of the nonlinear momentum problem and the uncoupled *linear* scalar transport equation subject to realistic initial and boundary conditions.

We might mention some progress made in this direction. The nonlinear, spatial evolution of Gortler vortices was computed recently by K. Lee*, (Lee and Liu 1990, Liu 1990)

*Graduate student, Division of Engineering, Brown University.

starting with a Gortler vortex upstream, corresponding to the observed wavelength and amplitude (Swearingen and Blackwelder 1987). It is necessary, as in the experiments, to follow the development of a single, robust wavelength. The linearized theory (Floryan and Saric 1982) is seen as the best "curve fit" for the experimentally measured x-component velocity, while giving the relationship among other velocity components. As such, the linear theory provides the initial shape distribution of a disturbance that satisfies the dynamical equations even in an approximate sense. In this situation, Lee found (Lee and Liu 1990) that the downstream nonlinear solution is not sensitive to the upstream x station where the initial conditions are imposed (provided a definite λ is followed locally, as in the experiments). The comparisons with experiments of computed structural details as well as skin friction development are amazingly good.

The scalar transport problem has also been computed by Lee with similar boundary and initial conditions (Liu and Lee 1990; Liu 1990). Numerical results for Pr=Sc=0.72 are close to those for Pr=Sc=1, as is expected. For Pr=Sc=7, as for a liquid, the iso-temperature or iso-concentration profiles are modified somewhat due to molecular effects, but the advective transport by the longitudinal eddies appear to dominate in this active system.

The potential of quantitative interpretation of the streamwise velocity through nonintrusive concentration measurements remains to be explored through solving the scalar transport problem with real "injection" initial and boundary conditions. The possible control of surface heat transfer rates through manipulation of advecting longitudinal eddies also remains to be studied, after preliminary encouraging potentials demonstrated by Liu and Sabry (1990).

Acknowledgement

We have come to this Symposium to honor a much respected colleague, who has inspired us through his accomplishments and his example, . . .

References

[1] Aubry, N., Holmes, P., Lumley, J.L., and Stone, E., 1988, *The dynamics of coherent structures in the wall region of a turbulent boundary layer*, J. Fluid Mech. **192**, 115.

[2] Blackwelder, R. F., 1983, *Analogies between transitional and turbulent boundary layers*, Phys. Fluids **26**, 2807.

[3] Floryan, J. M. and Saric, W. S. 1982, *Stability of Gortler vortices in boundary layers*, AIAA J. **20**, 316.

[4] Guezennec, Y. G., Piomelli, U., and Kim, J. 1987, *Conditionally averaged structures in wall-bounded turbulent flows*, Proc. Center Turb. Res. Summer Program, 263, NASA/A mes.

[5] Hall, P., 1983, *The linear development of Gortler vortices in growing boundary layers*, J. Fluid Mech. **130**, 141.

[6] Ito, A., 1985, *Breakdown of longitudinal vortices along a concave wall, J. Japan, Soc. Aero. Space Sci.* **33**, 166.

[7] Lee, K. and Liu, J. T. C. 1990, *Nonlinear evolution of Gortler vortices in a spatially growing boundary layer*, Bull. Am. Phys. Soc. 35, 2230 (abstract).

[8] Liu, J.T.C., 1988, *Contribution to the understanding of large-sale coherent structures in developing free turbulent shear flows.* Adv. Appl. Mech. **26**, 183-309.

[9] Liu, J.T.C., 1989, *Coherent structures in transitional and turbulent free shear flows.* Ann. Rev. Fluid Mech. **21**, 285-315.

[10] Liu, J.T.C., 1990, *Heat and mass transfer in nonlinear spatially developing Gortler vortices*, EUROMECH 261, Colloquium on Gortler Vortex Flows, 11, ISITEM, Nantes.

[11] Liu, J. T. C. and Lee, K., 1990, *Advected scalars in nonlinear, spatially developing Gortler vortices*, Bull. Am. Phys. Soc. 35, 2230 (abstract).

[12] Liu, J. T. C. and Sabry, A. S., 1990, *Concentration and heat transfer in nonlinear Gortler vortex flow and the analogy with longitudinal momentum transfer*, Proc. Royal Soc. Lond. A (in press).

[13] Lumley, J. L., 1974, *Computational modeling of turbulent transport*, Advances in Geophysics **18A**, 169, Academic, NY.

[14] Lumley, J. L., 1975, *Modeling turbulent flux of passive scalar quantities in inhomogeneous flow*, Phys. Fluids **18**, 619.

[15] Lumley, J. L., 1978, *Computational modeling of turbulent flows*, Advances in Applied Mechanics **18**, 124, Academic, NY.

[16] Pauley, W. R. and Eaton, J. K., 1988, *The fluid dynamics and heat transfer effects of streamwise vortices embedded in a turbulent boundary layer*, Stanford Univ, Thermal Sci. Div. Rep. MD-51.

[17] Peerhossaini, H. and Wesfried, J. E., 1988, *On the inner structure of streamwise Gortler rolls*, Int. J. Heat Fluid Flow **9**, 12.

[18] Sabry, A. S. and Liu, J. T. C., 1988, *Nonlinear development of Gortler vortices and the generation of high shear layers in the boundary layer*, Proc. Symp. in Honor of C. C. Lin, June 1987 (ed. D. J. Benney, F. H. Shu and C. Yuan), 175, World Scientific Singapore.

[19] Sabry, A. S. and Liu, J. T. C., 1989, *Longitudinal vorticity elements in boundary layers: their nonlinear development from initial Gortler vortices as a prototype problem*, Brown University, Division of Engineering Research Rep. FM89-11.

[20] Swearingen, J. D. and Blackwelder, R. F., 1987, *The growth and breakdown of streamwise vortices in the presence of the wall*, J. Fluid Mech. **182**, 255.

[21] Tennekes, H. and Lumley, J. L. 1972, A First Course in Turbulence, 223, MIT Press.

CONDITIONALLY SAMPLED VORTICITIES AND NEAR-WALL TURBULENCE STRUCTURE

Kwing-So Choi[1]

British Maritime Technology
Teddington, England

ABSTRACT

Vorticity maps of the turbulent boundary layer over the riblet surface during the near-wall burst, an event associated with a downwash of high-momentum fluid giving rise to a high wall-skin friction, were constructed using experimental data obtained in a wind tunnel at a constant adverse pressure gradient condition. They seem to support the conceptual model of near-wall turbulence activities proposed by Choi, depicting a downwash sequence close to the wall ($y^+ \simeq 15$) in an early part of the near-wall burst, followed by an up-wash sequence away from the wall ($y^+ \simeq 30$). The following downwash sequence was found to be shifted in spanwise direction by a half spacing of the near-wall structure, suggesting that the near-wall bursts take place in a staggered fashion.

1. Introduction

An extensive work on the near-wall turbulence structure has been conducted by many researchers in an effort to understand the drag reduction mechanisms by riblets in the turbulent boundary layers (Hooshmand et al. 1983; Gallagher & Thomas, 1984; Bacher & Smith, 1985; Bechert et al. 1985; Choi, 1985, 1986, 1987a, 1987b, 1989; Dinkelacker et al. 1987; Coustols & Cousteix, 1988; Pulles, 1988; Reidy & Anderson, 1988; Djenidi et al. 1988). It was however, Choi who pointed out the importance of the near-wall bursts in considering the turbulent drag reduction. He argued that a large part of the turbulent skin friction production is made during the events, which are

[1]This work has been supported by British Aerospace plc and the Department of Trade and Industry, U.K.

associated with a downwash of high-momentum fluid. This downwash was observed to take place between a pair of counter-rotating vortices (Choi, 1986). A conditional sampling technique was then used to capture the velocity field during the near-wall bursts (Choi, 1987b and 1989). It was found that the conditionally-sampled and ensemble averaged v-component velocity was negative (toward the wall surface) during the near-wall bursts and the corresponding u-component velocity was positive. These results confirmed the claim by Choi (1986) that the near-wall bursts are associated with a downwash of high-momentum fluid.

A further analysis of the flow field was carried out by Choi (1987b and 1989) who constructed the particle paths using experimental data within a near-wall region of the boundary layer. From this study a pair of counter-rotating vortex motion was detected during the near-wall bursts, further strengthening the argument for the model (see Fig. 1). In the Stage 1 of the model, a concentrated vortex filament in the turbulent boundary layer is deformed by the locally large u-velocity created by the downwash of high-momentum fluid between the pairs of counter-rotating longitudinal vortices. The deformed vortex filament then develops into hairpin loops by the self-induction mechanisms in the Stage 2. In the Stage 3, the hairpin loops develop further moving in the flow direction at approximately 45 degrees (Head & Bandyopadhyay, 1981) growing in size, and finally eject away from the wall surface. At this last stage, the neighbouring legs of hairpin loops form pairs of counter-rotating longitudinal vortices and the near-wall bursts of the next generation take place between the pairs. These pairs of vortices are also shown on the left (in the downstream) of the Stage 1, thus completing a cycle of the burst events. Note a difference in spanwise position of the near-wall bursts in the Stage 3, which is not in line with those in the Stage 1. In other words, the near-wall bursts take place in a staggered fashion, which is similar to some of the structure found in a transitional boundary layer (see Herbert, 1988). The development of the perturbed layer of vorticity into concentrated vortex filaments in the turbulent boundary layers and subsequent stretching

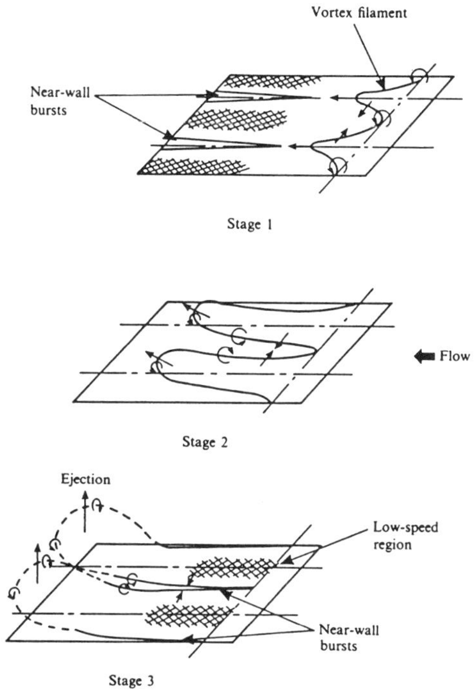

Figure 1 Conceptual model for the burst sequence

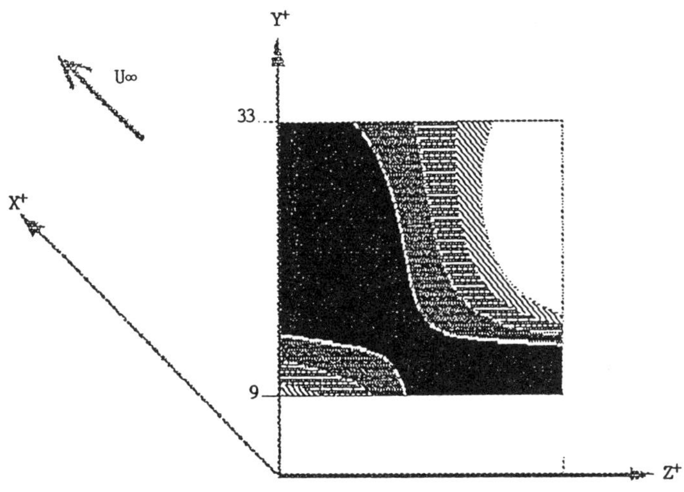

Figure 2 Coordinate system and the region of interest for the vorticity maps

into hairpin vortices are also found by Moin et al. (1986) in the Navier-Stokes calculation of a channel flow under a constant pressure gradient condition.

The mechanisms of turbulent drag reduction by riblets can be explained by using this conceptual model. When the riblets are placed over the wall surface in the turbulent boundary layer, the riblets would act like small longitudinal fences to prevent the spanwise movement of the legs of hairpin loops (see Stage 3, Figure 1) during the ejection. As a result, the legs of neighbouring hairpin loops cannot move closer to each other to produce a strong downwash of high-momentum fluid. This "premature" burst over the riblet surface, resulting in reduced intensity and shortened duration of the near-wall bursts (Choi, 1987b; Choi 1989), seems to be one of the mechanisms of the drag reduction due to riblets.

In these studies, we were able to examine a detailed structure of the near-wall turbulent boundary layer, especially the structure during the near-wall bursts, which seem to play an important role in the production of turbulent-skin friction. It is, however, felt that an additional study of vortex dynamics in the near-wall turbulent boundary layer is essential to make a better understanding of the turbulent structure associated with the near-wall bursts. For this reason, we have constructed the vorticity maps of the flow over the riblet surface during the near-wall bursts using the conditionally sampled and ensemble averaged velocity data. Only the streamwise component of the vorticity was considered here for the case of a constant adverse pressure gradient over a riblet surface.

2. Experimental Procedure

Experimental data were taken at the Royal Aerospace Establishment in Bedford, U.K. using the Boundary Layer Wind Tunnel. All of the three velocity signals u, v and w were conditionally sampled during the near-wall burst, and ensemble averaged typically for one hundred realisations at each data point. The detections of the near-wall burst were carried out using wall-skin friction signals from TSI 1471 surface film sensors mounted in the valleys of riblet surface. Velocities were measured using DISA miniature cross-wires, 55P63 and 55P64 with DISA 55M bridge units operating at constant temperature mode. All the experiments were conducted with a free-stream velocity of 3.0m/s under a constant adverse pressure gradient of $\beta = 3.1$, where the pressure gradient parameter β is defined as

$$\beta = \theta/\tau_\omega \cdot dp/dx \qquad (1)$$

The shape factor H of the velocity profile under this condition was 1.65 and the Reynolds number based on the momentum thickness was 4.8×10^3. Further details on the experimental facility, procedure and conditions can be found in Choi (1990).

3. Results and Discussions

Figure 2 shows the coordinate system and the region of interest where the vorticity maps are constructed. Each map covers vertical distance of 9 to 33 in y^+ and spanwise distance of 0 to 120 in z^+. Conditionally sampled and ensemble averaged w- and v-component velocities at 81 mesh points within this region (9 positions each for both vertical and spanwise directions) were firstly interpolated to produce 81 by 81 evaluation points using a third-order polynomial fitted through the velocity data by the least-square method. Then velocity gradients $\partial w/\partial y$ and $\partial v/\partial z$ were evaluated at these 6,561 points to obtain the streamwise vorticity $\xi = (\partial w/\partial y - \partial v/\partial z)$ for every four time steps of the velocity data. The longitudinal axis x^+ is aligned with the freestream velocity U, and the trigger sensor used for the conditional sampling is located at the origin. The burst signature of wall-skin friction detected by the sensor is shown in Figure 3 versus time t^+. The sequence numbers 1 to 77 affixed along the burst signature are for identifying the individual vorticity maps shown in Figure 4. The time interval between the successive maps is 5ms (t^+=3.0). The grey scales showing the sign and magnitude of the vorticity are given alongside the maps.

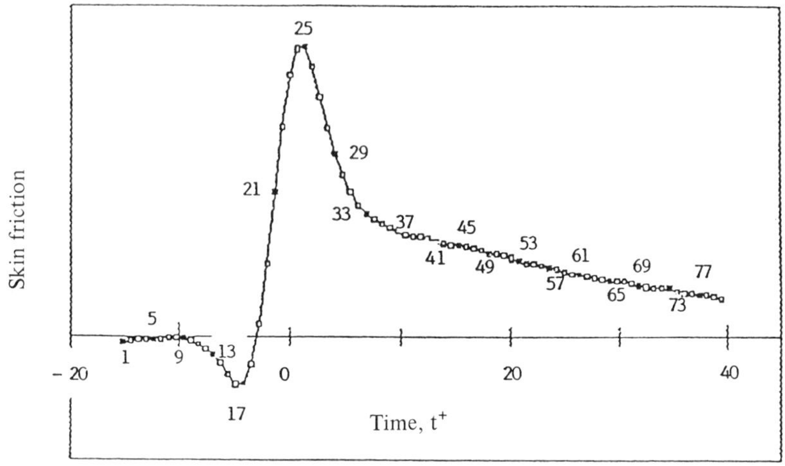

Figure 3 The burst signature of wall-skin friction

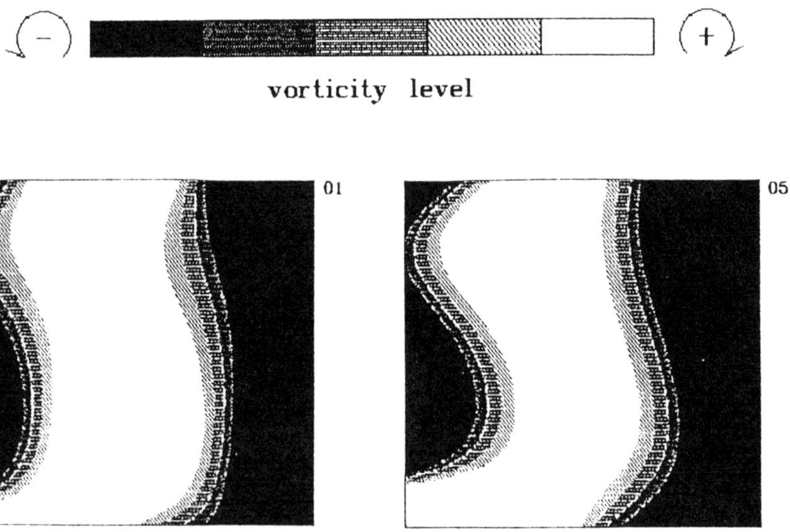

Figure 4 Maps of the streamwise vorticity during the near-wall burst. The time interval between the successive maps is 5ms (t^+=3.0). The sign and magnitude of the vorticity are indicated by the grey scale

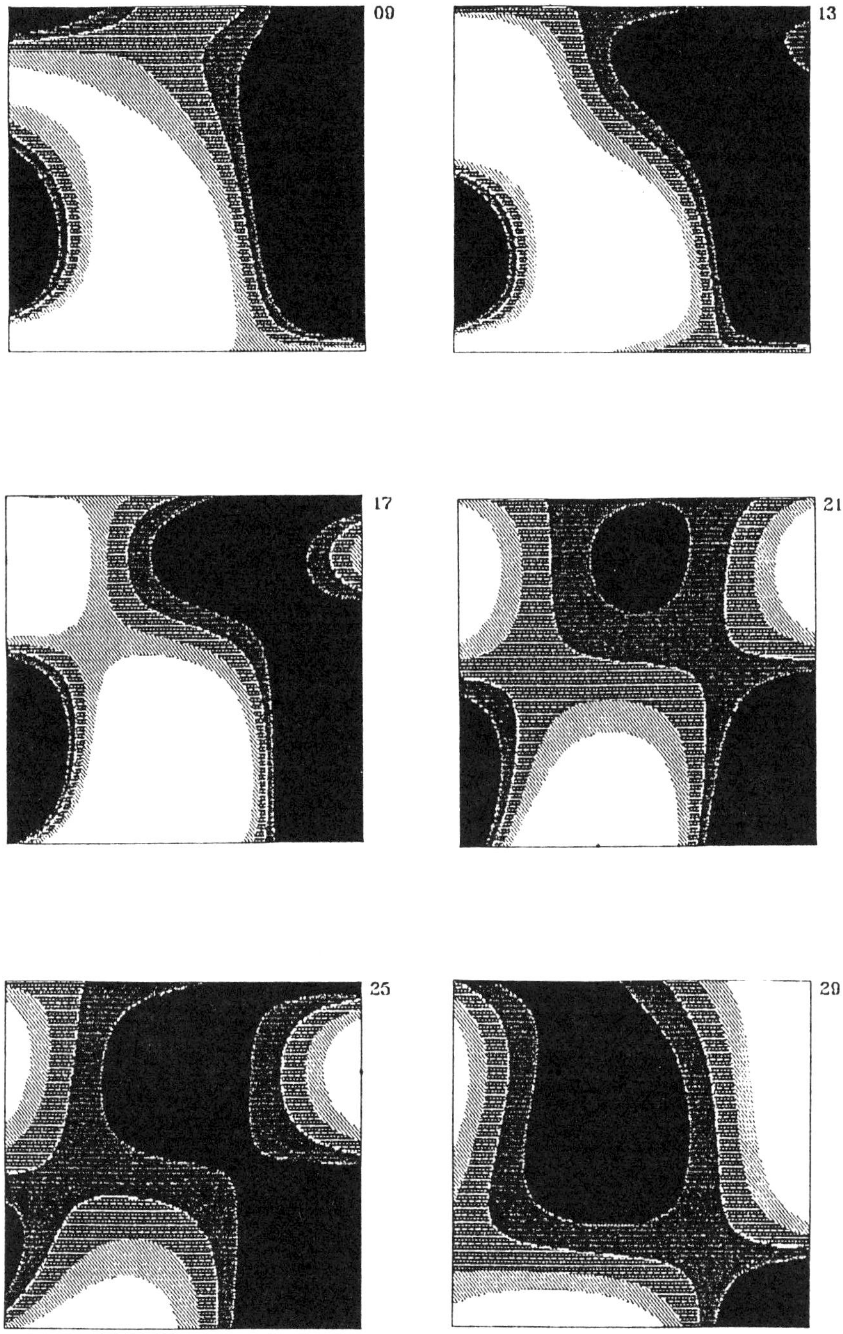

Figure 4 (continued)

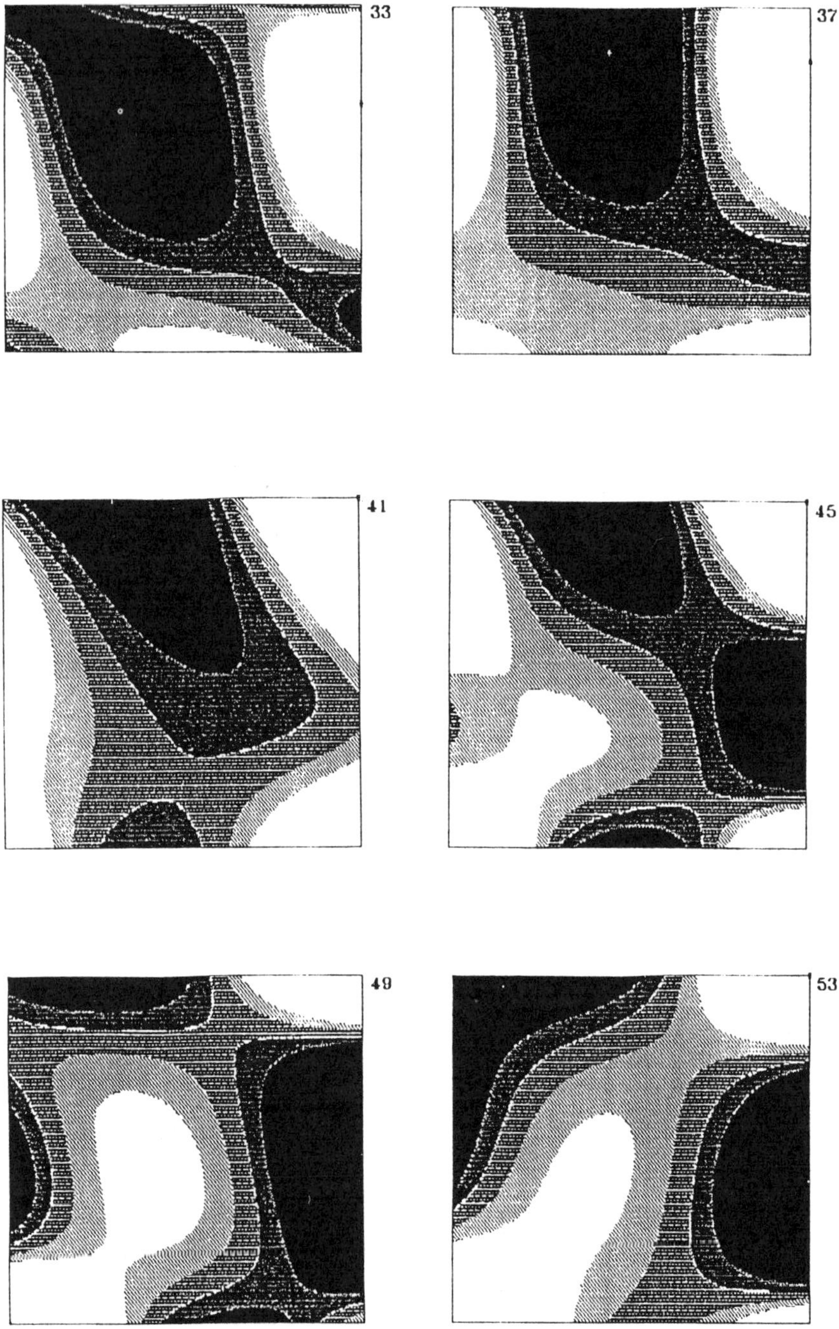

Figure 4 (continued)

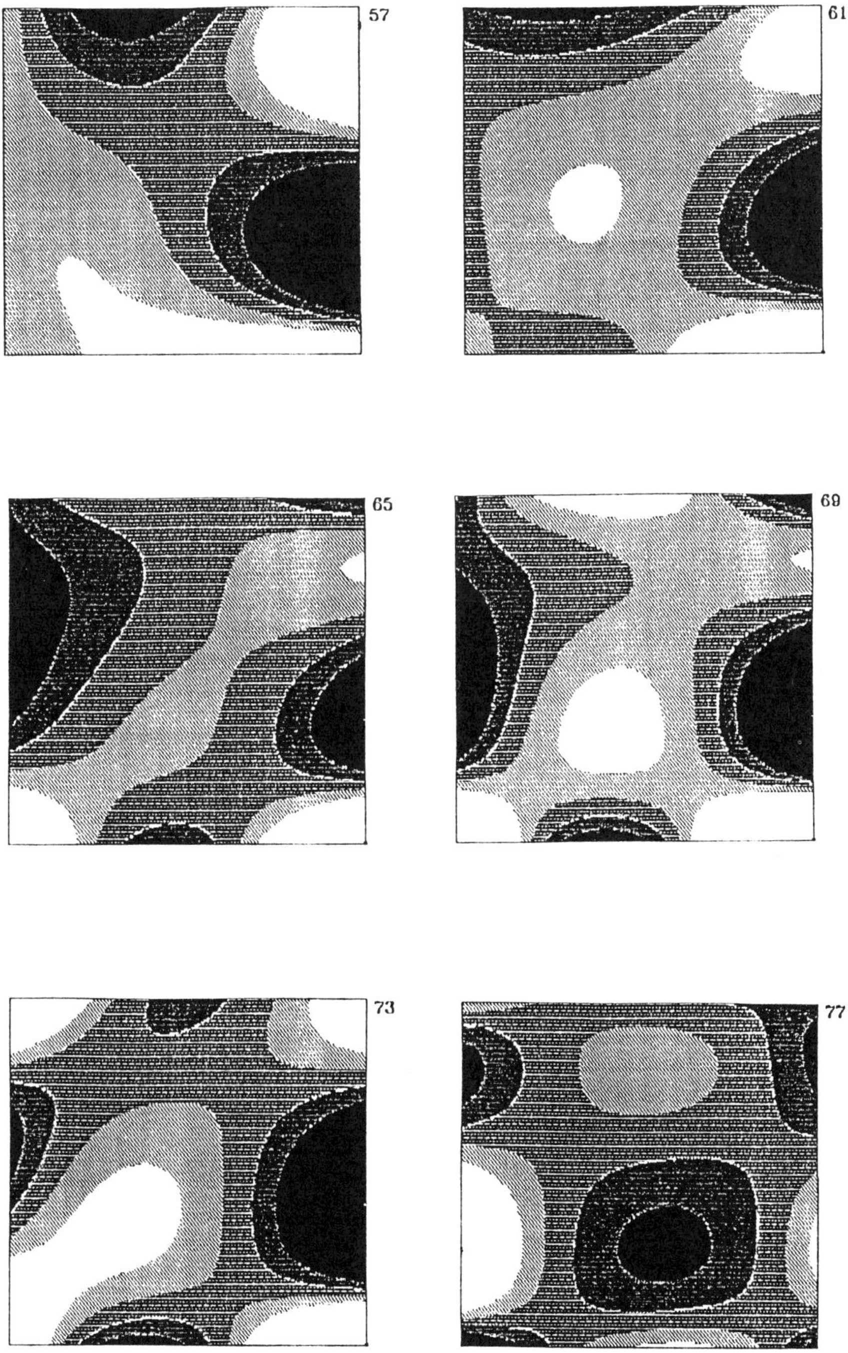

Figure 4 (continued)

In the sequence of vorticity maps from no. 1 to 25, there is a negative (counter-clockwise) vorticity near the y^+ axis, moving down along the y^+ axis toward the wall surface with time. The sign of the vorticity is such that the induction of the velocity at the centreline is negative (toward the wall surface). There is also a region of negative vorticity in the sequence from no. 1 to 25 of Figure 4 at around z^+=100, with a positive vorticity in between. It seems that the vorticity maps in this time sequence are consistent with the results of Choi (1987b and 1989) in that the near-wall bursts are associated with a downwash of high-momentum fluid. They also agree with that maximum negative v-component (vertical) velocity is observed at the sharpest (positive) slope point of the near-wall burst (the sequence no. 21 in Figure 3). A vortex motion associated with the downwash was also detected by Choi (1987b and 1989) in the trajectory of particles at a similar time sequence. The negative vorticity at z^+=100 and the positive vorticity at z^+=60 in Fig. 5 also produce a downwash of high-momentum fluid at around z^+=80. The spanwise distance between these two downwashes is similar to the mean spacing between the low-speed streaks observed by Kline et al.(1967) in the turbulent boundary layers under various pressure gradient conditions.

After the downwash phase of the near-wall bursts is over with the sequence no. 25 of Figure 3, the vorticity field is completely changed to produce an upwash of low-momentum fluid at the centreline (z^+=0) and at around z^+=80. This can be seen in the sequence between no. 29 and 45, in which a positive vorticity is shown close to the centreline and at z^+=100 (see Fig. 6). A negative vorticity is seen at around z^+=60, suggesting an existence of a simultaneous upwash at z^+=80 and downwash at z^+=40. This phase corresponds to the sequence of "velocity reversal" shown by Choi (1987b and 1989), where the u- and v-component velocity near the centreline change their signs from positive/negative to negative/positive toward the end of the near-wall bursts.

The upwash phase is then followed by a downwash phase in which the high-momentum fluid is brought down toward the wall (sequence no. 49 to 77). This time, the downwash takes place at around z^+=40, but not at the

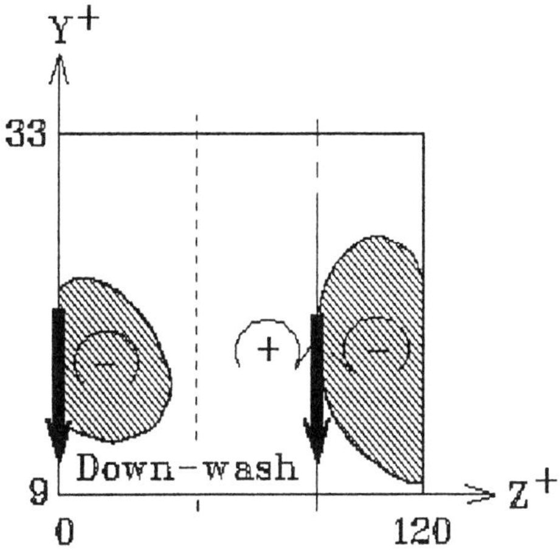

Figure 5 A vorticity map illustrating the downwash at $z^+=0$ and 80 in early stage of the near-wall burst

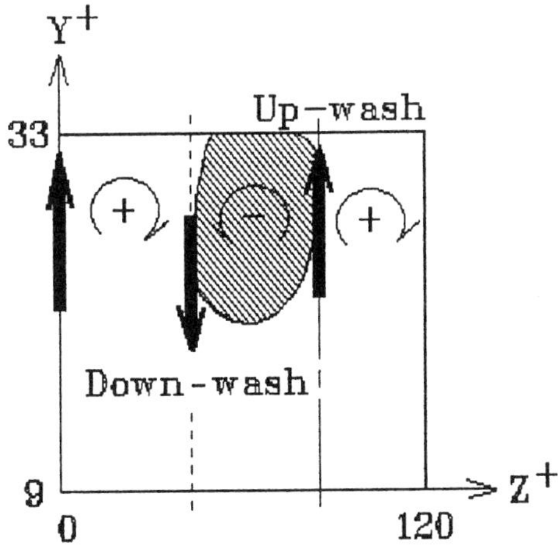

Figure 6 A vorticity map illustrating the upwash at $z^+=0$ and 80 in middle stage of the near-wall burst

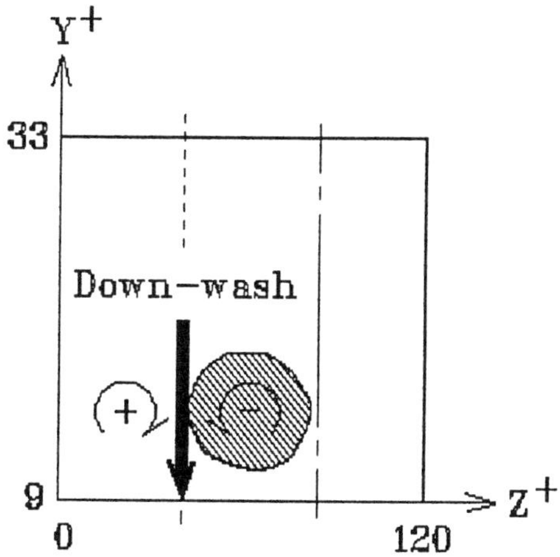

Figure 7 A vorticity map illustrating the downwash at $z^+=40$ in final stage of the near-wall burst

centreline (z^+=0) as in earlier sequence of the near-wall burst. The vorticities associated with this downwash (a negative vorticity at z^+=60 and a positive one at z^+=20) are found closer to the wall (see Figs. 5 and 7) than those in the upwash phase (see Fig. 6).

4. Concluding Remarks

The vorticity field in the near-wall region of the turbulent boundary layer revealed in the present investigation seems to support the conceptual model of the near-wall bursts proposed by Choi (1986). The downwash sequence (nos. 1-25) depicted in Fig. 5 corresponds to those during the near-wall bursts in the Stage 1 of the model (Fig. 1). The upwash sequence (nos. 29-45) which is associated with a positive vorticity near the centreline (see Fig. 6) can be identified with that in the hairpin loops shown in the Stage 2 of Fig. 1. The following downwash sequence (no. 49-77), can be seen between the stretched legs of neighbouring hairpin loops in the Stage 3. This time, the near-wall bursts take place at their upstream location between the original bursts. The vorticities associated with both of the downwashes are located close to the wall surface, since the near-wall bursts take place between the longitudinal vortices whose upstream ends are pinned down to the wall surface (see Stages 1 and 3 in Figure 1). On the other hand, the vorticities associated with the upwashes are located further away from the wall as they are a result of the developing hairpin loops as shown in the Stage 2.

Acknowledgements

The author would like to express his gratitude to Dr. R. Johnson of BMT for his help in the execution of experiments and data analysis. He would also like to extend his acknowledgements to Prof. M. Gaster of Cambridge University, Mr. M. Firmin and Mr. W. Sawyer of RAE, and Dr J.A.B. Wills and Dr. M.E. Davies of BMT for their encouragement and discussions.

References

[1] Bacher, E.V. & Smith, C.R., 1985, A combined visualisation-anemometry study of the turbulent drag reducing mechanisms of triangular micro-groove surface modifications. AIAA Paper 85-0548.

[2] Bechert, D.W. et al., 1985, On the drag reduction of the shark skin. AIAA Paper 85-0546.

[3] Choi, K.-S., 1985, Near-wall turbulence structure on a riblet wall. BMT Report. BMT Ltd., Teddington, Middlesex, U.K.

[4] Choi, K.-S., 1986, A new look at the near-wall turbulence structure. In Advances in Turbulence, ed. by Mathieu, J. and Comte-Bellot, G., Springer-Verlag.

[5] Choi, K.-S., 1987a, The wall pressure fluctuations of modified turbulent boundary layers with riblets. In Turbulence Management and Relaminarisation, ed. by Liepman, H.W. and Narasimha, R., Springer-Verlag.

[6] Choi, K.-S., 1987b, On physical mechanisms of turbulent drag reduction using riblets. In Transport Phenomena in Turbulent Flows, ed. by Hirata, M. and Kasagi, N., Hemisphere.

[7] Choi, K.-S., 1989, Near-wall structure of turbulent boundary layer with riblets. J. Fluid Mech., 208, 417.

[8] Choi, K.-S., 1990, Effects of longitudinal pressure gradients on turbulent drag reduction with riblets. In Turbulence Control by Passive Means, ed. by Coustols, E., Kluwer Academic Publications.

[9] Coustols, E. & Cousteix, J., 1988, Turbulent boundary layer manipulation in zero pressure gradient. 16th Congress of the International Council of the Aeronautical Sciences, Jerusalem.

[10] Dinkelacker, A. et al., 1987, On the possibility of drag reduction with the help of longitudinal ridges in the walls. In Turbulence Management and Relaminarisation, ed. by Liepman, H.W. and Narasimha, R., Springer-Verlag.

[11] Djenidi, L. et al., 1988, About the mechanism involved in a turbulent boundary layer over riblets. In Advances in Turbulence 2, ed. by Fernholz, H.-H. and Fiedler, H.E., Spring-Verlag.

[12] Gallagher, J.A. & Thomas, A.S.W., 1984, Turbulent boundary layer characteristics over streamwise grooves. AIAA Paper No. 84-2185.

[13] Head, M.R. & Bandyopadhyay, P., 1981, New aspects of turbulent boundary layer structure. J. Fluid Mech., 107, 297.

[14] Herbert, T., 1988, Secondary instability of boundary layers. Ann. Rev. Fluid Mech., 20, 487.

[15] Hooshmand, D. et al., 1983, An experimental study of changes in the structure of a turbulent boundary layer due to surface geometry changes. AIAA Paper 83-0230.

[16] Kline, S.J. et al., 1967, The structure of turbulent boundary layers. J. Fluid Mech., 30, 741.

[17] Moin, P., Leonard, A. & Kim, J., 1986, Evolution of a curved vortex filament into a vortex ring. Phys. Fluids, 29, 955.

[18] Pulles, C.J.A., 1988, Drag reduction of turbulent boundary layers by means of grooved surfaces. Ph.D. Thesis. Technical University of Eindhoven.

[19] Reidy, L.W. & Anderson, G.W., 1988, Drag reduction for external and internal boundary layers using riblets and polymers. AIAA Paper 88-0138.

MODELING THE EFFECT OF INITIAL AND FREE-STREAM CONDITIONS ON CIRCULAR WAKES

Jacques Lewalle[1]

Department of Mechanical and Aerospace Engineering
Syracuse University
Syracuse, NY 13244-1240

ABSTRACT

The cascade-transport model combines the isotropic turbulent cascade with transport and production terms based on a spectral distribution of the eddy viscosity. Thus, the turbulence energy spectrum can be modeled in shear flows. Containing both spatial and spectral information, the model could be a crude approximation to the wavelet-transformed equations of motion. The initial conditions for the wake are specified by the momentum thickness, the turbulence energy and Reynolds number, and the spectral exponent on the large-eddy side. Multiple nearly-self-preserving solutions are obtained, differing by the energy spectra, wake radii, centerline turbulence levels and turbulence energy budgets. One such solution is attracting for a range of initial conditions, while some memory of the initial spectra is observed for the others. The transport of weak, persistent freestream turbulence into a memoryless wake does not introduce a dependence on initial or ambient conditions. One concludes that the absence of memory is set in the near-wake, and corresponds to a stable solution.

1. Introduction

The cascade-transport (c-t) model was derived and calibrated in Lewalle & Tavlarides (1990). Based on eddy viscosity modeling, the c-t model provides a

---

[1] Computations were carried out on the Alliant FX/80 of the Northeast Parallel Architecture Center (NPAC) at Syracuse University.

spectral decomposition of the turbulence energy. It is therefore possible to discriminate between eddies of different sizes in the transport and production of turbulence in a shear flow.

The turbulence model equations are used under the usual assumptions of large Reynolds number of the large eddies, dominant gradients in the y-direction and no streamwise pressure gradient. In cylindrical coordinates, the Reynolds averaging of the momentum equations gives

$$U\frac{\partial U}{\partial x} + V\frac{\partial U}{\partial r} = \frac{1}{r}\frac{\partial}{\partial r}(rR_{xr}) + \frac{1}{r}\frac{\partial}{\partial r}\left(\nu r\frac{\partial U}{\partial r}\right), \qquad (1)$$

where the Reynolds stress R_{xr} is unknown. The modeling of the stress with an eddy-viscosity yields

$$U\frac{\partial U}{\partial x} + V\frac{\partial U}{\partial r} = \frac{1}{r}\frac{\partial}{\partial r}\left((\nu_T+\nu)\, r\frac{\partial U}{\partial r}\right), \qquad (2)$$

together with the continuity equation. The closure consists in obtaining a value of the eddy viscosity ν_T.

In the cascade-transport (c-t) model, the turbulent eddies are distributed among discrete sizes separated by a factor 2. The wave-number scale is used accordingly, with $k_i = 2\, k_{i-1}$. To each eddy size, a typical velocity u_i is attached, in such a way that the kinetic energy per unit mass κ is obtained by summation of the spectral energy densities $E(k_i)$

$$\kappa = \log(2)\sum_1^n u_i^2 = \log(2)\sum_1^n k_i\, E(k_i). \qquad (3)$$

A direct estimate of the viscous dissipation rate is given by

$$\varepsilon = \sum 2\log(2)\,\nu\, k_i^2\, u_i^2. \qquad (4)$$

Turbulence statistics can be calculated from the spectrum $E(k_i)$. For example, assuming isotropy, the r.m.s. velocity fluctuations are given by

$$u' = \left(\frac{2\,\kappa}{3}\right)^{1/2}, \qquad (5)$$

the integral scale of the turbulence by

$$L = \frac{3\,\pi}{8\,\kappa}\int_0^\infty \frac{E(k)}{k}\, dk \qquad (6)$$

and the turbulence Reynolds number by

$$Re = \frac{u' L}{\nu} . \tag{7}$$

The Heisenberg spectral eddy viscosity is applicable at scales smaller than that of the energy-containing eddies. Heisenberg's formula is modified so that it can be used also for large eddies departing from the $k^{-5/3}$ distribution and when the spectrum is out of equilibrium because of turbulence energy production P. Thus

$$\nu_{Ti} = c_H\, u_i / k_i \;/\; \rho \tag{8}$$

forms the core of the model, with

$$\rho = [\, \varepsilon^{2/3}\, k_i^{-5/3} / E(k_i)\,]^B\, [1+\beta P/\varepsilon] , \tag{9}$$

for which the calibrated constants are reproduced below. The sum of all contributions

$$\nu_T = \sum \nu_{Ti} \tag{10}$$

is applicable to the transport terms (Reynolds stress, energy transport) while the spectral eddy viscosity ν_{Ti} governs the distribution of the turbulence energy production into the various modes

$$\frac{\partial u_i^2}{\partial t} \sim \frac{1}{r}\frac{\partial}{\partial r}\left((\nu+\nu_T)\frac{\partial u_i^2}{\partial r}\right) + \frac{\nu_{Ti}}{\log(2)}\left(\frac{\partial U}{\partial r}\right)^2 - 2\nu k_i^2 u_i^2 . \tag{11}$$

The model is completed by the addition of the transfer term between eddies of neighboring sizes (Desnyanski & Novikov, 1974). In the form calibrated in Lewalle & Tavlarides (1990), the model kinetic energy equations are

$$U\frac{\partial u_i^2}{\partial x} + V\frac{\partial u_i^2}{\partial r} = \frac{1}{r}\frac{\partial}{\partial r}\left((\nu+\nu_T)\frac{\partial u_i^2}{\partial r}\right) + \frac{\nu_{Ti}}{\log(2)}\left(\frac{\partial U}{\partial r}\right)^2 - 2\nu k_i^2 u_i^2$$
$$+ 2\alpha k_i u_i \left[\left(u_{i-1}^2 - 2u_{i+1} u_i\right) - 2^{1/3} C\left(u_{i-1} u_i - 2u_{i+1}^2\right)\right], \tag{12}$$

showing, in order, the convection, transport, production, dissipation and transfer of spectral energy. The number of modes between the energy containing range and the dissipative range is estimated by $\log_2(Re^{3/4})$. Several modes must be allowed on the large eddy side of the spectrum, as these govern the power-law decay exponent of homogeneous turbulence (see the discussion of initial conditions, below). Typically, 10 to 15 equations (12) are solved simultaneously with the mean flow equation (2), the continuity equation, the expression of the spectral eddy viscosity (8) and the relations (3) and (4).

In Lewalle & Tavlarides (1990), the model was calibrated on the homogeneous shear grid turbulence data of Champagne, Harris & Corrsin (1970) and Harris, Graham & Corrsin (1977). The best fit is given by the values:

c_H=0.345 ; α=0.800 ; C=0.330 ; β=1.02 ; B=.75 .

However, the modeling of round jets (Lewalle 1990 b) shows that these parameters yield too much energy production into the large eddies in axisymmetric configurations. The value B = .88 used for the round jet is also preferable here, and all results are presented on that basis. Changes in the calibrated values of the model constants will be discussed in section 4.

2. Numerical procedure

The code is based on the same Crank-Nicolson algorithm with upwind transverse advection used in κ-ε solutions. The implementation of eq. (12) alternates the advection-transport-production terms and the cascade-dissipation terms. The semi-implicit scheme for the first group is solved in parallel for all modes, while the semi-implicit scheme for the second group is solved in parallel at all points in the profile. Both procedures result in tri-diagonal systems of linear equations. The implementation on the Alliant FX/80 parallel computer leads to a 10:1 speed-up factor compared with the VAX-8820 version. No vector optimization was attempted. The increments in the streamwise marching scheme are selected dynamically to maintain diagonal dominance in all systems of equations.

The initial conditions for the mean velocity U_0

consist of a freestream Reynolds number based on an obstacle diameter d of about 50,000. The top-hat velocity defect U_S is taken arbitrarily as 30% of freestream velocity for all calculations, and corresponds to a momentum thickness $\theta=.46$ (normalized by the radius d/2 of the obstacle) given by

$$\theta^2 = \int U/U_0 \ (1-U/U_0)\ 2r\ dr. \qquad (13)$$

For the turbulence, the specification of initial conditions amounts to the imposition of an energy spectrum, also with a top-hat profile. Because of the sensitive dependence of the model on ε, it is imperative that the initial turbulence spectrum be fully developed. This is accomplished by letting an arbitrary input decay for two turnover times (as estimated by L/u'), at which stage approximate self-preservation is reached. These nearly-self-preserving spectra are selected for their easy reproducibility. With the viscosity already imposed in the mean flow, the initial turbulence is completely specified by its r.m.s. velocity, its Reynolds number Re or integral scale L, and the spectral exponent s such that $E(k) \sim k^s$ on the large eddy side.

Because of the important role played by 's' in the c-t model calculations, it may be recalled (Leith, 1967; Lesieur, Montmory & Chollet, 1987; Lewalle, 1990 a) that 's' is related to the turbulence decay exponent n (as in $u^2 \sim x^{-n}$) by the relation

$$n = 2(s+1)/(s+3). \qquad (14)$$

Then, in grid turbulence, $Re \sim x^{1-n}$. This trend can be compared to the wake scaling relation $Re \sim x^{-1/3}$, pointing to the possible difference in flow response to initial conditions with $n>4/3$ (decaying faster than the wake turbulence) or $n<4/3$ associated with persistent initial conditions.

As the Heisenberg model requires some level of turbulence, the edge of the wake is identified by the point at which the turbulence Reynolds number falls below 20; eddy viscosity is assumed to vanish beyond that point. All solutions presented below are independent of this cut-off value.

3. The axisymmetric wake

The wake development is now calculated as a function of the initial conditions. Initial r.m.s. turbulence velocities are taken as 5, 15 and 45 % of U_0, and the Reynolds number at 300, 900 and 2000. The latter value of Re corresponds to an integral scale comparable to the wake diameter, and can lead to excessive initial spreading. Such solutions are discarded when the r.m.s. velocity u' becomes too much larger than the centerline velocity defect U_s. The various combinations of initial parameters are summarized in table 1. In each case, two values of the 'decay' exponent n were used: n = 1.1 and n = 1.6, on either side of the critical n = 4/3 . Accordingly, most figures will be given in pairs.

Case	U_0	U_s/U_0	u'/U_0	Re	L/d
A	1	.3	.05	100	.1
B	1	.3	.15	300	.1
C	1	.3	.45	900	.1
D	1	.3	.15	900	.3
E	1	.3	.15	2000	.67

Table 1. The initial conditions as coded in the figure captions

The ratio of the centerline r.m.s. velocity to the velocity defect u'/U_s is one simple diagnostic of self-similar development of the wake. Figs. 1 and 2 show the evolution of u'/U_s down to 2000 obstacle diameters. For n=1.6 , even unlikely initial conditions appear to relax to a unique asymptote. In contrast, no such asymptote is apparent in the case n=1.1 .

The same feature is illustrated by the similarity plotting of $r_{1/2}$ (the radius of half velocity defect), non-dimensionalized by the momentum thickness θ, against $(x/\theta)^{1/3}$ (Figs. 3 and 4). While the n = 1.6 solution appears to relax to a similarity radius around 0.33, a stable range of values from 0.32 to 0.40 is observed for the n= 1.1 case. Thus, two categories of solutions are distinguished by their dependence, or lack thereof, on the initial conditions.

Of all initial turbulence parameters, 'n' appears to dominate the wake development. The assumption that self-preservation is associated with independence of initial conditions is promptly dispelled by the similarity plotting of the 1/2 profile spectra (Fig. 5). A representative sample of all results shows the evolution of the spectra from x/d = 20 to 500, for n=1.6 case B (no memory), n=1.1 case B (no memory, same width) and n=1.1 case D. Remnants of the initial conditions for n=1.1 are visible in the very-large-eddy range. The difference in spectral energies in the third example is due to the scaling on U_S (see below), the wider wake and the conservation of the momentum thickness. Imperfect superposition in the inertial range is for x/d = 20; the dissipative range is discussed next.

A similarity form of eqs. (2) and (12) can be carried out along classical lines, and is briefly summarized. The advection axial velocity scales as U_O, its derivatives as U_S. The scale for radial coordinates is taken as the centerline integral scale L, the axial distance x is its own scale. Then, conservation of the momentum defect, large Re, continuity and eq. (2) combine to make the transverse advection negligible, and to have ν_T scale as $U_S L$. Substitution in the turbulence equations (12) and wavenumber scaling with L yields approximate similarity if the spectral velocities scale on U_S. Similarity of spectral transfer results in $L \sim x^{1/3}$. The only departure from strict similarity is due to the viscous dissipation term, which decreases as $Re \sim x^{-1/3}$. This explains the progression of the dissipative cut-off to larger scales, and is identical to the near-preservation of grid turbulence.

The existence of multiple self-preserving solutions was predicted by George (1988), with some dependence on initial conditions affecting the turbulence statistics but not the mean velocity profiles as shown indeed on Figs. 6 and 7. Therefore mean velocity does not provide a diagnostic of the persistence of initial conditions. The turbulence r.m.s. velocity profiles (Figs. 8 and 9) are similar in shape, with the centerline values reflecting the scaling on the momentum defect.

Beside the spectra, the turbulence energy budgets (Figs. 10, 11 and 12) may be the best qualitative indicators of memory effects. Differences in numerical values are again due to the U_S scaling. In the outer part of the wake, the separate balances between production and dissipation, and between streamwise advection and radial transport, are well known (e.g. Tennekes & Lumley, 1972). The persistent case (Fig. 12) shows larger centerline advection, and smaller centerline transport to match; and smaller peak production of turbulence, which again may be a reflection of the more persistent advection.

As for the case of jets, the κ-ε coefficient

$$c_D = \nu_T \ \varepsilon \ / \ \kappa^2 \tag{15}$$

peaks in the regions of small energy production, which may be an artifact of eq. (9). In contrast to the jets, c_D now bottoms out at values larger than the accepted .09 in mid-profile (Fig. 13 and 14), and shows a lack of similarity for both values of 'n'. It is not clear if discretization errors account for the scatter.

4. Discussion

The model is currently in the trial phase. The closure (eq. 12) lacks a clear connection to the Navier-Stokes equations (Lee, 1981), and the modifications to the Heisenberg eddy-viscosity (eq. 9) have no theoretical basis. Without the support of a Fourier transform, the c-t model makes no pretense of being a two-point closure. The proper mathematical description of a field in physical space *and* in wave-number space is not addressed here. One may speculate that a wavelet transform (see e.g. Holschneider, 1988; Farge and Rabreau, 1988) of the equations of motion could admit Eq. 12 as a simple model.

Within the framework of the model as presented in section 1, the invariance of the calibration constants should be discussed, in particular the value of B. The homogeneous shear flow calibration B = .750 leads to the same difficulties in the round wake as it does in the round jet (Lewalle, 1990 b), namely excessive diffusivity (and value of the calculated c_D), resulting in overestimates of the turbulence energy. One may conjecture that the value B = .88 is characteristic of axisymmetric flows, and that it corresponds to a real decrease of the turbulence-producing ability of large eddies.

The tampering with model constants has been presented in many instances as a lack of generality of the model, a shortcoming. We may ask if the lack of generality can be traced to specific physical aspects of the turbulence.

Glaring differences in the shape of the spectra can be observed (for the c-t model) between the grid turbulence calibration and the shear flow. Even with significant production, the homogeneous shear spectra (Lewalle & Tavlarides, 1990) are not as peaky as in the wake, and exhibit an overextended $k^{-5/3}$ range. By selecting one type of experiment for calibration purposes, the modeler selects, among other turbulence statistics, a particular energy spectrum, which is now found to affect the evolution of the flow. The unpleasant conclusion is that we should not expect calibrations to hold in all flows where spectra could play a role. This applies to the κ-ε model and second order model constants as well as to the c-t model.

5. Wake in a turbulent freestream

Finally, the energy spectrum can be manipulated after the wake has been shed, and the wake development affected as a result. The modification of the spectrum is achieved by allowing a contamination of the wake by some persistent turbulence from the freestream.

The wake is initiated as in section 3, with n=1.6 only. Left to its own devices, the wake would tend to the self-preserving solution portrayed above.

Weak, persistent (n=1.1) grid-like turbulence is seeded in the freestream, and allowed to decay and be transported into the wake. A wide range of relative initial conditions was tried, from which Figs. 15 and 16 are representative. In either case, the wake is saturated with the freestream conditions by $x/d = 20$. In Fig. 15, the freestream turbulence has a length scale much larger than the wake's, and barely decays in the time frame of the calculation. The freestream turbulence shows in the 1/2 profile spectrum at $x/d = 20$, but is separated from the energy-containing range by a valley. As the flow develops, the peak erodes and the valley fills up, leading to a spectral profile which could be associated with lack of self-preservation. A variant on the same idea is summarized by Fig. 16, with smaller freestream eddies, noticeable decay and no spectral gap.

Contrary to expectations, such manipulations do not prevent the convergence to the memoryless solution. Except for minor smearing of the u'/U_S profiles at the very edge of the wake, all statistics converge to the limit solution associated with all cases n=1.6 and isolated cases n=1.1 in section 3. The conclusion is that, for all initial conditions tried so far, their relaxation in the near wake is sufficient to resist the later influence of the freestream turbulence.

6. Conclusions

The cascade-transport model is a first-order one-point closure of the turbulence equations. The discrimination between eddy sizes is reminiscent of two-point closures, but is postulated at each point rather than being obtained by Fourier transform. A description of the turbulence spectrum at various points is therefore possible.

In comparison to second-order closures, a number of physically desirable (Lumley, 1978) properties are missing. Anisotropy is not included in the cascade equations, which could prevent the use of the c-t model in strongly buoyant flows and possibly create difficulties in wall-bounded flows. Fluctuating pressure is present only to the extent that spectral transfer is. Pressure transport and rapid distortion of the turbulence are not included in the model. The absence of return to isotropy is not an issue in the isotropic model. Other physical requirements are ignored, such as realizability. Most limitations of first-order closures are not overcome by the c-t model, but the modeling of the energy spectrum distinguishes the c-t model from other first-order closures.

The main contribution of this study is to demonstrate the role of the very-large-eddies (VLE's), and to derive from them a new understanding of wakes and their lack of universality. Eddies larger and weaker than the energy-containing scales have received very little attention, no doubt due to the difficulty of measuring the 3-D spectrum accurately in that range, and also because of a lack of insight in their dynamical role. Typically, the VLE's are poorly resolved in finite difference simulations, cut-off in spectral simulations, and ignored in the calibration of one-point closures. The c-t model fills a niche between these options and the two-point closures, providing scale discrimination at a smaller computational cost than the EDQNM and similar models.

The computational results show that the VLE's are a determining factor in the development of self-preserving solutions for the axisymmetric wake. This conclusion may be model-dependent, but is compatible with physical differences observed in experimental data (Bevilaqua & Lykoudis, 1978) and not previously attainable by first or second order closures (see e.g. Taulbee & Lumley, 1981, and the data reviewed there). A steep slope of the VLE spectrum is a sufficient condition for asymptotic independence of initial conditions in the far wake. On the basis of the Reynolds number considerations developed in section 2, the spectral exponent $s=3$ ($n=4/3$) probably marks the transition to the history-dependent wake.

Memory-prone wakes are associated with the VLE spectrum such that $s<3$, a necessary but not sufficient condition. For a range of initial r.m.s. velocities and length scales, the mean and r.m.s. velocity profiles are not distinguishable from the case $s>3$, even if differences in the spectra are noticeable. In the presence of freestream turbulence, the transport of a memory-prone spectrum into the wake does not overcome the initial tendency to self-preservation.

Thus, one may speculate that the lack of memory is an attractor for circular wakes, with a limited basin of attraction. The size of the basin of attraction depends on the VLE spectrum. Long term memory of the initial conditions is, in this view, dependent on s being small enough and on the initial solution being outside the assumed basin of attraction of the self-similar solution. Self-preserving solutions are apparently not nudged out of the attractor by weak continuous input from the freestream: the wake falls ouside the attractor either in its formative stages (near wake) or later by brutal manipulation of the spectrum further downstream, not attempted here.

Further numerical and laboratory experimentation is needed before a reliable map of the self-preservation attractor can be drawn.

Acknowledgments

Prof. E.A. Bogucz provided helpful background information and an experienced sounding board for the reliability of model constants. Prof. H. Higuchi was an open mine of facts, and made me wish I had applied the model to a few dozen different flows already. Their interest in this project is appreciated.

References

[1] Bevilaqua, P.M. and Lykoudis, P.S., 1978, *Turbulence memory in self-preserving wakes*, J. Fluid Mech. **89**, 589.

[2] Champagne, F.H , Harris, V.G. and Corrsin, S., 1970, *Experiments on nearly homogeneous turbulent shear flow*, J. Fluid Mech. **41**, 81.

[3] Desnyansky, V.N., and Novikov, E.A., 1974, *Simulation of cascade process in turbulent flows*, J. Appl. Math. Mech. **38**, 468.

[4] Farge, M. and Rabreau G., 1988, *Wavelet transform to detect and analyze coherent structures in two-dimensional turbulent flows,* C.R. Acad. Sci. II **307**, 1479.

[5] George, W.K., 1988, *The self-preservation of turbulent flows and its relation to coherent structures and initial conditions,* in Advances in Turbulence, Hemisphere Press.

[6] Harris, V.G., Graham, J.A.H. and Corrsin, S., 1977, *Further experiments in nearly homogeneous turbulent shear flow,* J. Fluid Mech. **81**, 657.

[7] Holschneider, M., 1988, *On the wavelet transformation of fractal objects,* J. Stat. Phys. 50, 963.

[8] Lee, J., 1981, *Dynamical cascade models for Kolmogorov's inertial flow,* J. Fluid. Mech. **101**, 349.

[9] Leith, C.E., 1967, *Diffusion approximation to inertial energy transfer in isotropic turbulence,* Phys. Fluids **10**, 1409.

[10] Lesieur, M., Montmory, C. and Chollet, J.-P., 1987, *The decay of kinetic energy and temperature variance in three-dimensional isotropic turbulence,* Phys. Fluids **30**, 1278.

[11] Lewalle, J., 1990 a, *Decay of velocity and temperature fluctuations in grid turbulence,* AIAA Journal **28**, 106-112.

[12] Lewalle J., 1990 b, *A multiple-length-scale model for turbulent shear flows, Part 2, Calculation of plane and circular jets,* submitted to AIAA Journ.

[13] Lewalle, J. and Tavlarides, L.L., 1990, *A multiple-length-scale model for turbulent shear flows, Part I, Model derivation and calibration,* submitted to AIAA Journ.

[14] Lumley, J.L., 1978, *Computational modeling of turbulent flow*, Adv. Applied Mech. 18, 124.

[15] Taulbee, D.B. and Lumley, J.L, 1981, *Prediction of the turbulent wake with second-order closure*, Sibley School of Mechanical and Aerospace Engineering Report # FDA-81-04, Cornell University.

[16] Tennekes, H., and Lumley, J.L., 1972, *A first course in turbulence*, M.I.T. Press.

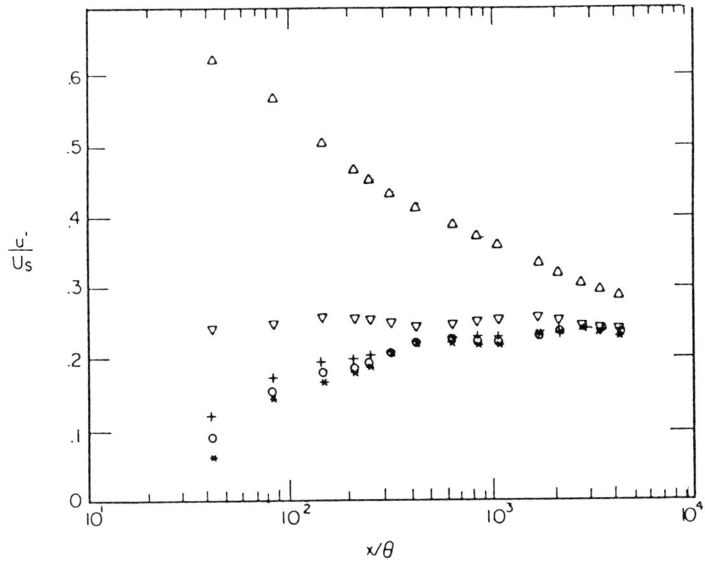

Figure 1. Asymptotic similarity for the centerline r.m.s. velocity fluctuations for n = 1.6.
* Case A; o Case B; + Case C; ∇ Case D; Δ Case E.
Last point corresponds to x/d = 2000.

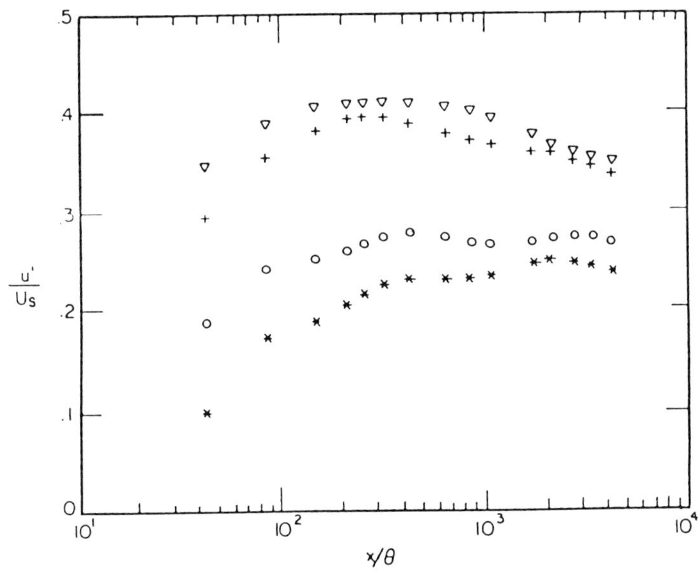

Figure 2. Long term memory for the centerline r.m.s. velocity fluctuations for n = 1.1.
Symbols as in Fig. 1. Case E not shown.

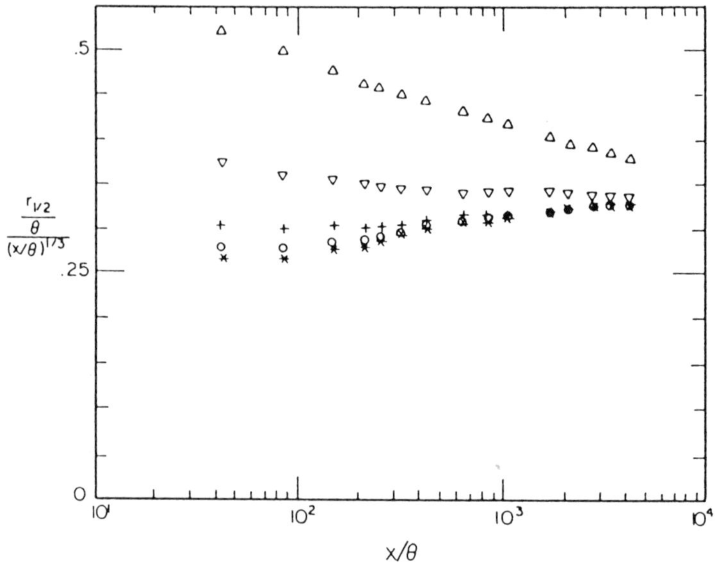

Figure 3. Similarity scaling for the wake width $[r_{1/2}/\theta]$ / $[x/\theta]^{1/3}$ for n=1.6.

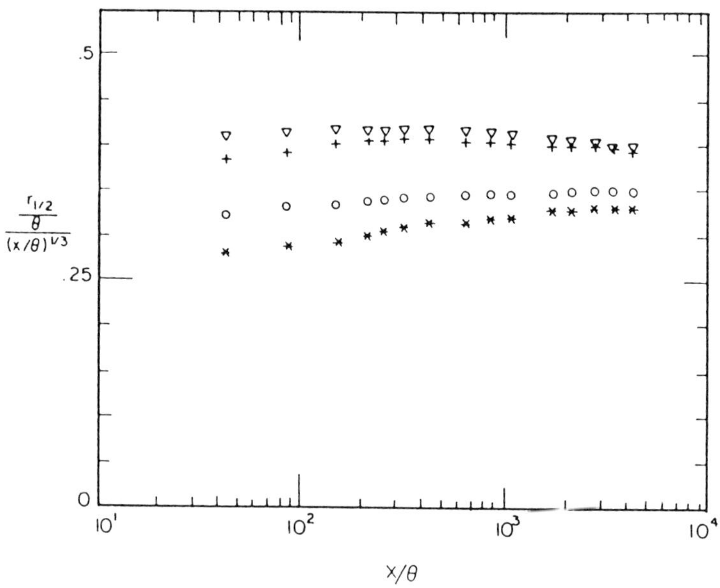

Figure 4. Similarity scaling for the wake width $[r_{1/2}/\theta]$ / $[x/\theta]^{1/3}$ for n=1.1.

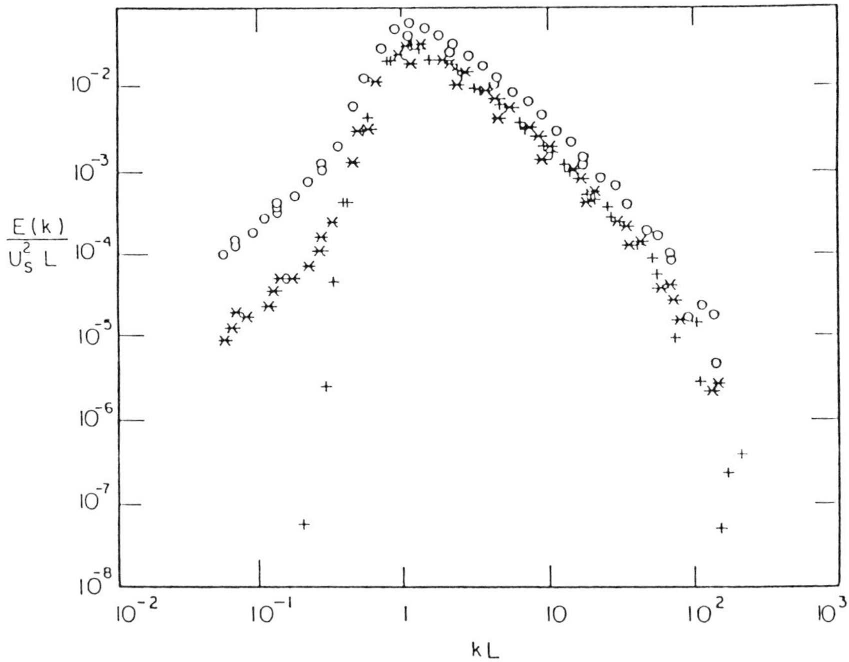

Figure 5. Similarity of the half-profile spectra at x/d=20, 100, 200, 500. + n=1.6 case B; ✱ n=1.1 case B: o n=1.1 case D. The remnants of initial conditions show on the left.

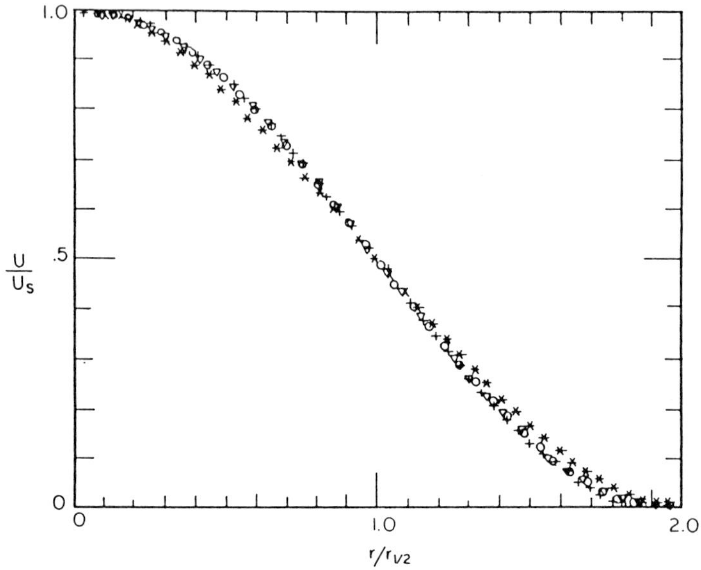

Figure 6. Similarity of the mean velocity profiles for n=1.6. * x/d=20; o x/d=100; + x/d=200; ∇ x/d=500.

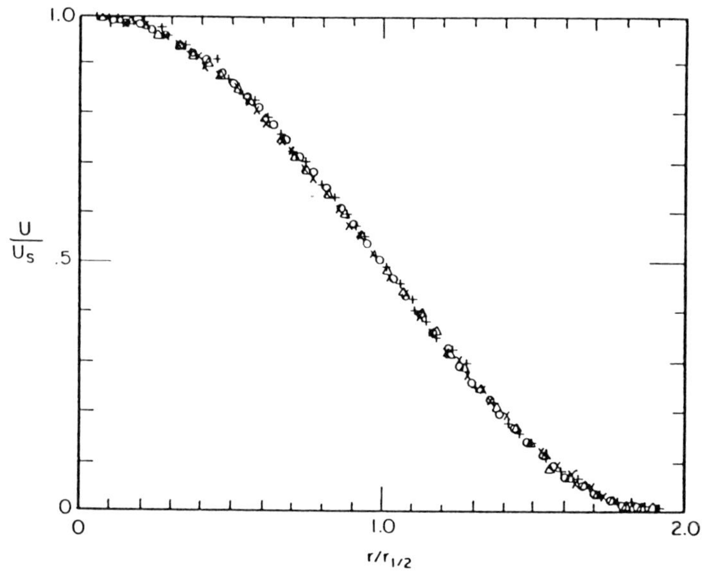

Figure 7. Similarity of the mean velocity profiles for n=1.1. Symbols as for Fig. 6.

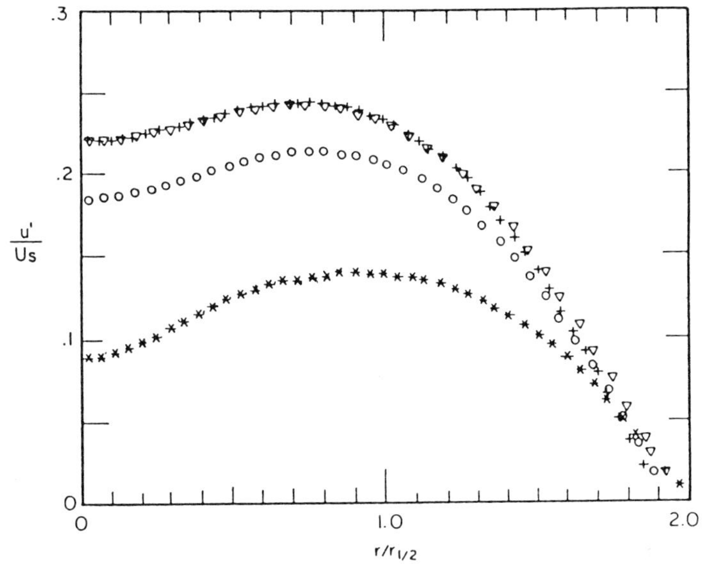

Figure 8. Asymptotic similarity for the r.m.s. turbulence fluctuations for n=1.6. Symbols as for Fig. 6.

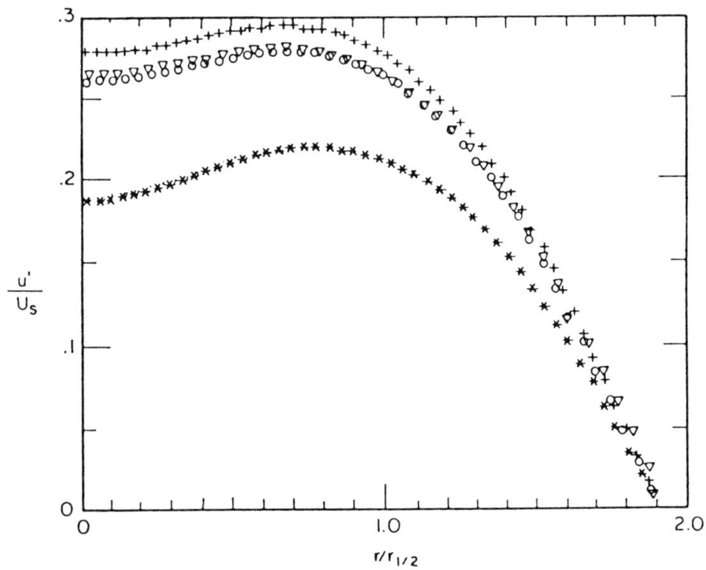

Figure 9. Lack of similarity for the r.m.s. turbulence fluctuations for n=1.1. Symbols as for Fig. 6.

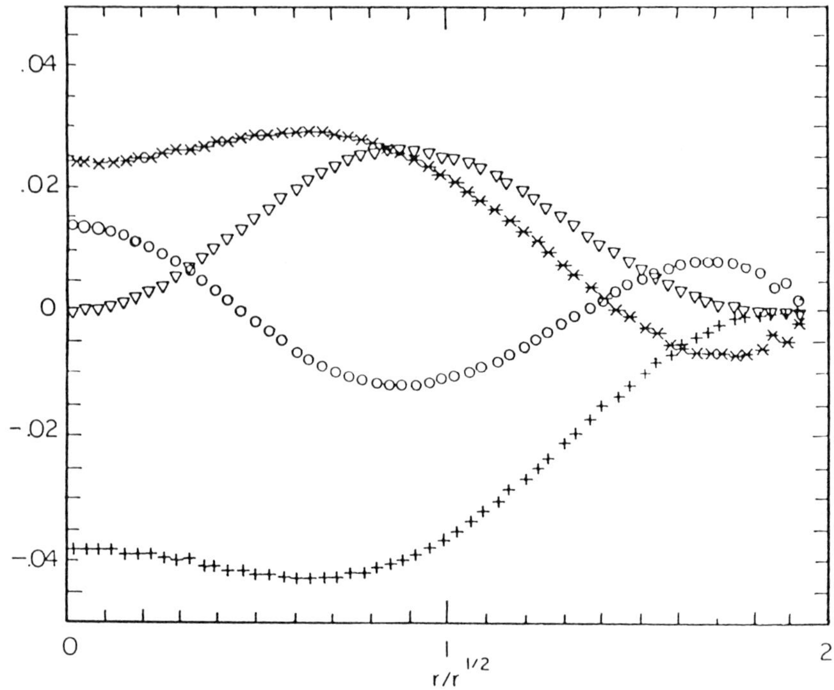

Figure 10. The turbulence energy budget at x/d=100 for n=1.6. case B, normalization with U_S^3/L.
* Streamwise advection; o radial transport;
+ dissipation; ∇ production.

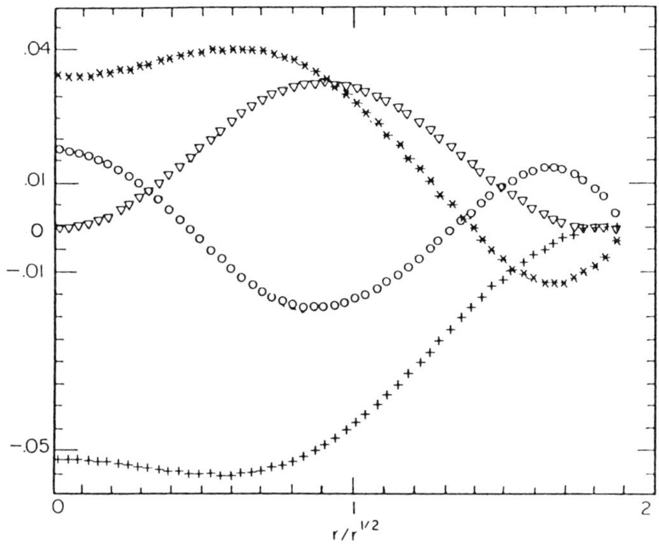

Figure 11. The turbulence energy budget at x/d=100 for n=1.1, case B. Symbols as for Fig. 10.

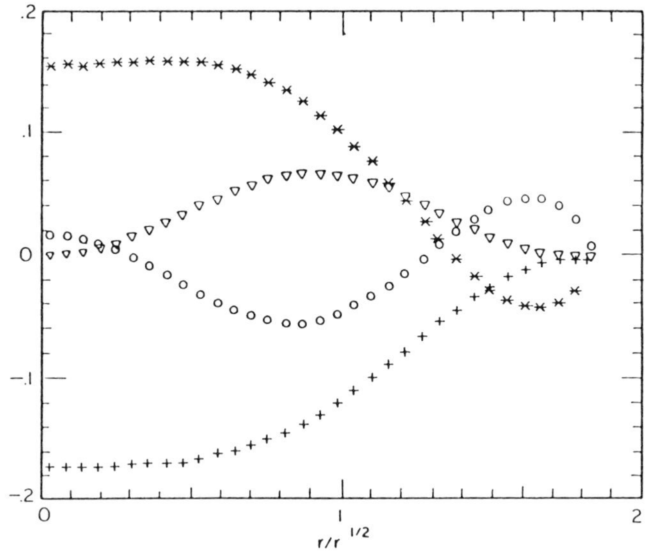

Figure 12. The turbulence energy budget at x/d=100 for n=1.1, case D. Symbols as for Fig. 10.

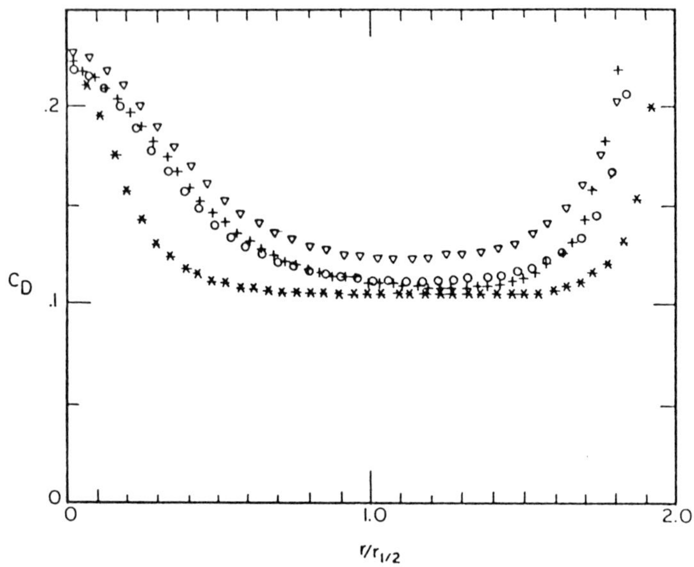

Figure 13. Profiles of $c_D = \nu_T \varepsilon / \kappa^2$ for n=1.6, case B. Symbols as for Fig. 6.

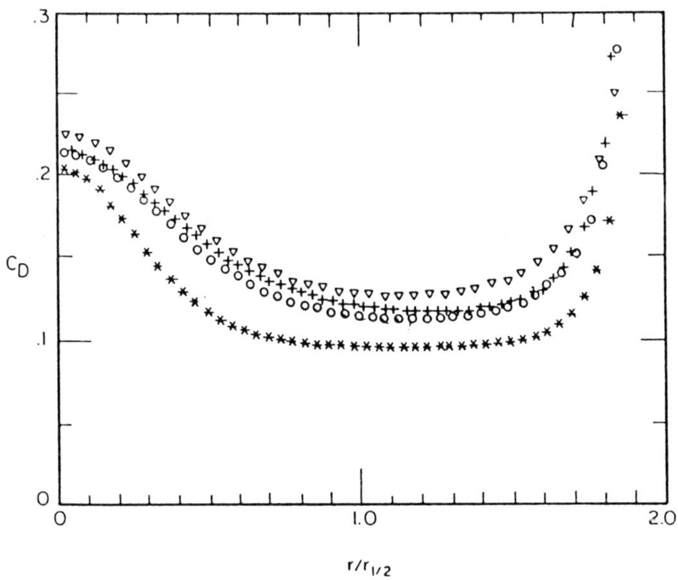

Figure 14. Profiles of $c_D = \nu_T \varepsilon / \kappa^2$ for n=1.1, case B. Symbols as for Fig. 6.

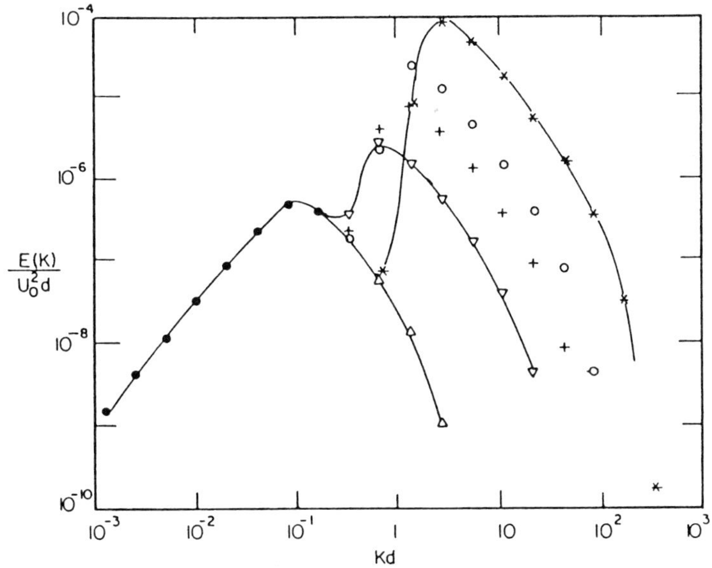

Figure 15. Effect of freestream turbulence on 1/2 profile spectra. * x/d = 20; o x/d = 100; + x/d = 200; ∇ x/d = 500; Δ freestream at x/d= 20 and 500; ● superposed data.

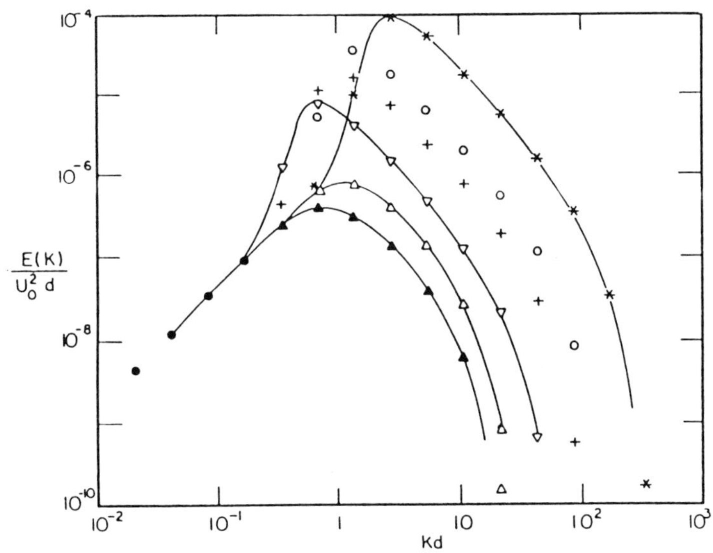

Figure 16. Effect of freestream turbulence on 1/2 profile spectra. * x/d = 20; o x/d = 100; + x/d = 200; ∇ x/d = 500; Δ freestream at x/d= 20; ▲ x/d = 500; ● superposed data.

ON THE AETIOLOGY OF SHEAR LAYER VORTICES*

W.R.C. Phillips [1]

Clarkson University, Potsdam NY 13699-5725

ABSTRACT

In an effort to isolate the mechanism by which streamwise structures form in turbulent wall layers, evolution equations were derived for the streamwise velocity and vorticity perturbations about a mean turbulent shear flow. 'Mean', of course, implies an average and the form of average is most important. One serious candidate is Reynolds-Hussain averaging, but because the instantaneous velocity is broken into mean, periodic and fluctuating components, difficulties arise in specifying terms which result from phase averages of fluctuating velocities and time averages of phase velocities. Such problems are less acute if the Generalized Lagrangian mean (GLM) equations of Andrews & McIntyre (1978) are employed and that is the approach taken. The GLM formulation - which is an exact mapping of the Navier Stokes equation - was constructed with wave/mean flow interactions in mind. In turbulent flow, however, the wave field is not explicit but exists as feedback from the the global or rectified effect of turbulence fluctuations throughout the layer. The net effect of this implicit wave field is captured formally in terms of space-time correlations at convection delay through two vector quantities, the Stokes drift and the pseudomomentum. Since these correlations have, for the most part, been measured, the resulting nonlinear evolution equations may be closed by appeal to experiment.

Fully developed turbulent channel flow, is considered, so that all mean quantities save the mean pressure are independent of the streamwise and, to leading order, spanwise directions. The intent is to study the instability of this flow to the wave field induced by the global effect of background turbulence. It is thus appropriate to postulate pre-existing time-independent Stokes drift and pseudomomentum fields but no Eulerian velocity perturbations, and pose the initial value problem corresponding to the application of a constant mean shear flow for all positive time. With appropriate boundary

[1]This work was supported in part by the US Office of Naval Research under SRO IV grant number N00014-85-K-0172 and in part by the National Science Foundation under grant CTS-9008477.

conditions, and assumed spanwise periodicity, the equations are then soluble numerically and depict the growth of what Phillips (1990) terms shear-layer vortices. But of particular interest is the mechanism by which these vortices form and their lateral spacing.

Of the mechanisms known to induce streamwise vortices in a mean shear flow, that which most closely captures the essence of the flow is the Craik-Leibovich mechanism (Leibovich 1977), although the details of the mechanism are here somewhat more complex than those which describe Langmuir circulations. The periodicity of shear layer vortices may be estimated on the assumption that it corresponds with the wavelength of the instantaneously most unstable linear disturbance. So if streaks are thought of as the footprints of these vortices, a formal way then exists to determine the spacing of streaks. Spacings well in accord with experiment are found and that not two vortices form per period, but four. Moreover an intense local shear layer is formed about the plane in which the convection velocity equals the mean Eulerian velocity, with the upper pair of vortices centered on this shear layer. Initially these vortices are the most intense but as the shear layer grows the lower pair gradually intensify. Eventually the lower pair dominate and at about $100t^+$ their intensity grows (explodes!) exponentially. The local shear layer is not obliterated but a distorted and spanwise periodic streamwise velocity-field emerges. Nonlinearities are now important and these act to limit the extent of the event and to stabilize the perturbation velocity field at a level somewhat below its peak.

References

[1] Andrews, D.G. and McIntyre, M.E. 1978 *An exact theory of non-linear waves on a Lagrangian-mean flow,* J. Fluid Mech.,**89**, 609.

[2] Leibovich, S, 1977 *Convective instability of stably stratified water in the ocean.* J. Fluid Mech.,**82**, 561.

[3] Phillips, W.R.C. 1990 *Coherent structures and the Generalized Lagrangian mean equation*, App. Mech. Rev, **43**, S227.

---

*The full text of this paper is to appear in *Theoretical & Computational Fluid Dynamics.*

PART 5

GENERAL TOPICS IN TURBULENCE

CHARACTERIZATION AND COMPRESSION OF TURBULENT SIGNALS AND IMAGES USING WAVELET-PACKETS

Lareef Zubair, K.R. Sreenivasan & M. Victor Wickerhauser

Yale University
New Haven, CT 06520

ABSTRACT

The newly introduced Wavelet-Packet transform (Coifman & Meyer 1989, Coifman et al. 1990) allows the decomposition of a signal as a function of the scale, the position and the frequency (or wavelength) optimally. Each Wavelet-Packet coefficient provides insight into the structure of the data locally and at the appropriate scale. We have applied this transform technique to one-dimensional data and two-dimensional images and report on its ability to characterize turbulence data with a few coefficients. We find that, overall, the Wavelet-Packet transform technique performs better than its competitors. That is, significant data compression ratios can be achieved without severely distorting the signal or the image.

1. Introduction

Turbulent motion has traditionally been decomposed in terms of Fourier modes, and one speaks of its frequency or wavenumber components. Our intuition about 'large' and 'small' scales of motion is substantially biased by Fourier description. There are several reasons why one should think of alternative descriptions. First, most turbulent flows (except for the hypothetical case of infinitely extended homogeneous turbulence) are finite in spatial extent, at least in one or two directions, so that expansions in terms of Fourier modes depend on mode-cancellation outside the flow domain – this being uneconomical in general. One therefore needs basis functions with compact support in real space. Secondly, a description of sharp gradients by Fourier modes is inefficient. Finally, turbulent flows (even the homogeneous ones) possess spatial structure of some sort, and the question of how to represent them efficiently is gaining increasing importance. The structure is stronger in some flows than in others and has better definition at some scales than at others, but it is clear that one should make a distinction between Fourier modes on the one hand and spatial structures on the other. Indeed, descriptive efforts in turbulence have always resorted to a variety of 'turbulent eddies' (see for example, Townsend 1956, 1976). Whatever the precise meaning of the term 'eddy', *it is clear that an eddy is not a Fourier mode*!

Lumley (Tennekes & Lumley 1972) recognized this distinction and wrote as follows (p. 259): 'The Fourier transform of a velocity field is a decomposition into waves of different wavelengths; each wave is associated with a single Fourier coefficient. An eddy, however, is associated with many Fourier coefficients and the phase relations among them. Fourier transforms are used because they are convenient (spectra can be measured easily); more sophisticated transforms are needed if one wants to decompose a velocity field into eddies instead of waves ...' In particular, figure 1 shows Lumley's schematic of an eddy in both real and Fourier spaces. It turns out that this eddy, which has the desired property of being spatially compact, is an example of what are now known as 'wavelets'. It is very much to Lumley's credit that he should have introduced wavelets to turbulence, albeit without using the name, at least as early as 1972! Yet, the broad recognition that wavelets possess useful mathematical properties that are appropriate for turbulence description is quite recent.

Formally, a wavelet is a spatially localized function which can be translated and rescaled while maintaining its shape (see, for example, Daubechies 1988). Wavelets in general are not local in Fourier space, and there is the so-called 'uncertainty relation' which tells us that spatial localization and wavenumber localization are complementary, and that there is a quantifiable trade off between them. Wavelet analysis provides a means for studying scaling and transient behavior of signals by using the dilates and translates of the wavelet as a basis. Decomposition of the signal into these basis functions corresponds to identifying the local scaling behavior of the signal. Different wavelets have been used to analyze different features of turbulent signals and images. Some examples are Argoul et al. (1989), Everson et al. (1989), Liandrat & Moret-Bailly (1990), Meneveau (1990).

While wavelets are important and their use in turbulence is bound to grow in several different directions, it is useful to recognize the many advantages possessed by Fourier representation (Strang 1989). At the least, our intuition of many periodic as well as non-periodic phenomena is based on Fourier decomposition. As an example in turbulence, the distribution of small scales might to some extent be described by modulated Fourier modes. It thus becomes apparent that new transforms, which can provide information about local frequency (or wavelength) in addition to that on scales, would be very valuable. Recently, such a transform, called the Wavelet-Packet transform, has been invented by Coifman & Meyer (1989). The transform provides local frequency and local scale information about data in one, two or three dimensions. It allows a signal to be segmented into dyadic intervals (1,2,4,8, ... data points), and the segmentation is arranged so that each 'near homogeneous' piece is decomposed into the basis that suits it best. A part of the attraction is that the transform also provides a choice of basis for different segments of data. Each of the functions in a basis can be localized (to different degrees) in the Fourier domain. The localization of the segment provides the position information. Although the algorithm allows uneven segmentation of the data and a choice of basis, the decomposition is nonredundant and complete. Further, fast algorithms requiring $O(N \log N)$ operations have been written (Beylkin et al. 1989).

The Wavelet-Packet coefficients can be arranged in the order of decreasing magnitude. If it happens that some of them are large and most others small, we can discard the small ones and achieve economical representation of the data. If the memory required to store the retained coefficients is small (relative to that required to store the original signal), then we have achieved efficient data compression. Data characterization and compression are related, but not identical. For example, for each Wavelet-Packet coefficient retained, its position, scale and frequency will also have to be stored. This 'overhead' makes data compression less efficient than data characterization. On the other hand, in turbulence as well as in many other applications, information on position, scale and frequency are very useful.

The Wavelet-Packet algorithm has already been implemented in the context of acoustic signals (Wickerhauser 1989). Here, we apply the Wavelet-Packet transform both for the characterization and for the compression of turbulent data. Because of space limitations, we describe Wavelet-Packets only briefly (section 2). Some comments on the index of performance are made in section 3. We briefly describe results for one-dimensional turbulent velocity signals (section 4), and for two-dimensional concentration fields of turbulent jets (section 5). A summary and conclusions are given in section 6. For greater details, reference must be made to Zubair et al. (1990) which we plan to publish shortly.

It should be mentioned that the spirit of image and data compression here is different from that of the Karhunen-Loéve transform procedure or the proper orthogonal decomposition – yet another concept introduced to turbulence by Lumley (1970). In the latter procedure, we are given a large collection of statistically similar objects, and the purpose is to pick out eigenfunctions that garner the most energy in the ensemble-average sense. Here, on the other hand, we are given a single image or a segment of data and asked to represent it best and most efficiently by means of various shapes and scales. The two questions are complementary to each other. (Note: A single image can be broken up into blocks and these blocks can be thought of as forming various members of the ensemble. The two techniques are thus related to each other.)

2. Wavelet-Packets

In general a collection of Wavelet-Packets consists of the translates, dilates and modulates of a mother Wavelet-Packet, w. The Wavelet-Packet w can be be dilated as $\sqrt{s}w(st)$, translated as $w(t-k)$ and modulated as $e^{\imath f t}w(t)$. In this section we describe the construction of Wavelet-Packets and some of its properties.

The starting point is a low pass filter sequence $h = h_j$, ($j = 1, 2, \ldots, N$, where N is the width of the filter). This filter h_j must possess some smoothness and finiteness conditions as described by Daubechies (1988). Define a complementary filter $g = g_j$ as

$$g_j = (-1)^j h_{1-j}.$$

We denote the 'convolution and subsampling' operation with h_j and g_j as H and G given by

$$H\zeta(i) = \sum_j h_{j-2i}\zeta(j)$$
$$G\zeta(i) = \sum_j g_{j-2i}\zeta(j). \tag{1}$$

Here $\zeta(j)$ represents the discrete data array. Following Coifman & Wickerhauser (1990) we use three indices for a Wavelet-Packet $w_{f,s,k}$, where the index f has a unique relationship with the frequency. The indices k and s describe the position and scale of the Wavelet-Packet. All the f values may be obtained using the following recursive scheme:

$$w_{2f,s+1,0}(t) = Hw_{f,s,0}$$
$$w_{2f+1,s+1,0}(t) = Gw_{f,s,0}. \tag{2}$$

The position of the Wavelet-Packet is obtained using $w_{f,s,k}(t) = w_{f,s,0}(t-k)$. The k index takes only the values possible at that scale. The resolution of the data limits the f index that can be obtained at any scale. The recursive construction scheme given by Eqs. 2 generates a hierarchy of Wavelet-Packets. This hierarchy is shown in figure 2. We show only the f, s indices in the figure.

The scale increases in dyadic steps $(1, 2, 4, \ldots)$ with the hierarchy. Each of the Wavelet-Packets generated by Eqs. 2 may be translated. This yields a large and redundant collection of basis functions. This collection is termed 'a library of Wavelet-Packets' by Coifman & Meyer (1989). It can be shown that the Wavelet-Packet library contains a multitude of orthonormal bases.

A measured signal is a sampled version of the original continuous variable. The sampling in data acquisition can be considered as the decomposition of the continuous variable into the Wavelet-Packet basis at the scale of the sampling interval. We set this basis as $w_{0,0,k}$

where k takes integer values up to the data length(L). The collection $w_{0,0,k}$ forms a complete orthonormal basis. This is the the first subset of the Wavelet-Packet library which we identify as a basis.

The property that the siblings are equivalent to their parent can be used to identify further subsets of the Wavelet-Packet library which form a basis. The siblings of $w_{0,0,k}$ are $w_{1,1,k}$ and $w_{0,1,k}$. Therefore the collection $w_{1,1,k}$ and $w_{0,1,k}$ where $k = 1, 2, \ldots, L/2$, form a second basis. (Since the siblings are at twice the scale of $w_{0,0,k}$, k in $w_{1,1,k}$ and $w_{0,1,k}$ takes only values up to $L/2$.) Similarly the collection of all Wavelet-Packets at any level in the hierarchy shown in figure 2, for admissible values of k, form a complete orthonormal basis. In the simplest case, if we start with filter coefficients $h_0 = \frac{1}{\sqrt{2}}$, $h_1 = \frac{1}{\sqrt{2}}$, then we generate the set of Walsh functions (e.g., Beauchamp 1984) as the Wavelet-Packets at the final level. Figure 3 shows the first eight Walsh functions. Figure 4 shows Wavelet-Packets of some regularity constructed with the Coiflet filter of range 30 (Zubair et al. 1990).

We are not restricted to using Wavelet-Packets of the same scale for all the basis functions. We may choose some Wavelet-Packets at some level in the hierarchy and the siblings of the rest of the Wavelet-Packets at other levels. Extending this reasoning we can make an orthonormal basis with Wavelet-Packets from several levels. All that has to be ensured is that every 'line of heredity' in the hierarchy is represented exactly once.

The optimal representation of the data within the library of Wavelet-Packet is obtained by using the so-called 'Best Basis Algorithm' (Coifman & Wickerhauser 1990). The choice poses two questions. How is the signal to be segmented? Which basis is to be used for each of these segments? These two questions have to be answered simultaneously. We need to evaluate all possible segmentation schemes and all basis functions for each of these segments to find the optimal representation.

This gives us an immense search space. The search is started at the smallest scale. The sampling in the data acquisition can be considered as the decomposition of the continuous signal into the Wavelet-Packet basis $w_{0,0,k}$ under translation k at the sampling interval. That is, we set the discrete signal as $x_{0,0,k}$. ($x_{f,s,k}$ denotes the coefficient of the corresponding Wavelet-Packet, $w_{f,s,k}$.)

We now compute the coefficients of the siblings ($x_{1,1,k}$ and

$x_{0,1,k}$). Since the siblings live on a scale twice as large as the parent so that the translation step is twice as large, we still have the same number of coefficients at both levels. We compute the coefficients of the data for all the Wavelet-Packets generated by translating the siblings.

Since the representation of the data as coefficients of the parent Wavelet-Packet or as coefficients of the siblings is equivalent, we could choose one over the other. We have used an 'entropy' criterion, which is essentially a 'least distance' criterion, to choose one over the other. The entropy is equivalent to the Shannon-Weaver measure of information of a sequence (Coifman & Wickerhauser 1990).

This criterion is the natural choice for data characterization algorithm. Other criteria such as counting the bits required to code different representations may be suited for data compression.

Keeping the minimum entropy criterion, we continue the search for all k values. Thereafter we compute the coefficients at the next level of the hierarchy. We compare the entropy of the new representation to the previous minimum for the segment of data under consideration and retain the representation having the lesser of the two entropies. This search is continued to the last level. The search algorithm ensures that the entire signal is covered by disjoint segments. Each segment customizes for itself the optimal basis from the many possibilties.

The signal is completely described by the coefficients in the optimal basis so determined. No information has been lost. To compress the data we rank the retained coefficients by their magnitude. The least significant ones are discarded. The image can be reconstructed by summing all the retained Wavelet-Packets weighted by their coefficients.

For a one-dimensional record of length N, the decomposition and search algorithm has a complexity of $O(rN \log_2 N)$, where r refers to the length of the filter. For images of size N^2 the algorithm has a complexity of $O(rN^2 \log_2 N^2)$.

3. Index of Performance

An index of performance used to quantify the compression achieved is the 'Coefficient Compression Ratio' (CCR). It is defined as the ratio of the total number of coefficients in the Wavelet-Packet decomposition to the number of coefficients in the retained basis. The compression ratio measures the ability of the transform to characterize data. Another index is the Bits-Per-Pixel (BPP) ratio. It is defined as the average number of bits needed to code each pixel. This index includes the overhead bits and is a true measure of the compression achieved.

To implement any algorithm an allowable distortion criterion has to be established. Both the information content and the subjective appearance of the image are degraded by the distortion. The information lost can be quantified by a measure such as the Signal-to-Noise-Ratio (SNR) or the Normed Mean-Square Error (NMSE) between the orignal and the reconstructed image. These are defined as :

$$NMSE = \frac{1}{p^2 NM} \sum_{m=0}^{M-1} \sum_{n=0}^{N-1} [f(i,j) - f'(i,j)]^2 \tag{3}$$

$$SNR = 10 \log_{10} \frac{1}{NMSE} \tag{4}$$

where f refers to the original image and f' to the reconstruction, and p refers to the peak value of the data. Definitions for one-dimensional data follow trivially.

4. Compression of One-Dimensional Data

The top trace in figure 5 shows the velocity measured in the atmospheric surface layer about 6.5 meters above the ground over a substantial stretch of a wheat field at the Connecticut Agricultural Experiment Station. The microscale Reynolds number is about 2000. Thus we expect a large scaling range. We have analyzed several data segments, and present results for one typical segment of 16384 points. The Wavelet-Packet transform was applied and the best basis functions for the data were determined as described in section 2. Reconstructions of the top trace were obtained by retaining some small number of coefficients. The second trace from the top was reconstructed using only about 5% of the coefficients. That is, the Coefficient Compression Ratio (CCR) was about 21. The reconstruction shows excellent fidelity to the original signal. The following traces are reconstructions with approximate CCR's of 41, 234 and 497 respectively. It is seen that the algorithm is able to capture local events efficiently and preserve local edges in the signal. The primary effect of increasing CCR appears to be the increasing absence of small scales; even for CCR's as high as 234 and 497, the large scale features are preserved. If the reconstruction were attempted by means of Fourier techniques, the sharp local variations would be smoothed out completely. Because of the Gibbs phenomenon, Fourier transform characterizes abrupt changes rather poorly.

Figure 6 is a plot of SNR as a function of CCR. As expected the quality of reconstructions degrades with CCR, but is still acceptable even at CCR's of order 500.

Similar reconstructions have been attempted for a variety of signals in laboratory flows, in particular those close to the wall in the boundary layer. The results are comparable in quality to those in figure 5. The effectiveness of such relatively high CCR's suggests the ubiquitous presence of strong structure in the signal. Notice that the energy of the reconstructed signals is a very large fraction of the original; even for CCR's of the order of a few hundreds, typically more than 90% of the energy is retained. Reconstructions with high CCR are by definition low-dimensional, and it is instructive to note that low-dimensional reconstructions contain most of the energy.

Computations performed on highly intermittent quantities such as the energy dissipation (strictly, $(\frac{du}{dt})^2$) yield poorer reconstructions. This is not surprising because a highly intermittent signal has poor spatial correlation (or structure). The Wavelet-Packet transform can only pick out the structure in signals, in the absence of which its performance will diminish.

Typically, decomposition and reconstruction of one dimensional records considered here take on the order of a few minutes on a VAX II/GPX.

5. Compression of Two-Dimensional Images

The extension of the Wavelet-Packet analysis to higher dimensions is quite straightforward and is described by Coifman & Meyer (1989). The two-dimensional basis functions are constructed as tensor products of one-dimensional Wavelet-Packets. For convenience of computation, we restrict attention to tensor products of Wavelet-Packets of the same scale. The rules for choosing different orthonormal basis functions are direct extensions of the one-dimensional case. As before, data compression is achieved by ranking the coefficients of the 'best basis' and discarding the least significant ones.

Figure 7 shows a two-dimensional map of the concentration field in a turbulent jet at a nozzle Reynolds number of about 4000. The gray level indicates the concentration of the fluorescing dye, and the image has been obtained by the laser-induced-fluorescence technique whose details can be found in Prasad & Sreenivasan (1989, 1990). The jet pictures obtained by Prasad and Sreenivasan are typically 1300 x 1030 pixels in extent, and each pixel is a twelve-bit word. To economize on the memory and computational requirements, we have analyzed various fractions of these images using Wavelet-Packets; for example, figure 7 is 1024 x 512 pixels in extent. We have also analyzed images at a lower Reynolds number of 2000; these latter images have a steady laminar region near the nozzle exit, transitional region somewhat downstream, and the turbulent regime far downstream. The analysis helps us to understand the downstream evolution of various scales in the concentration field. For brevity, we present results only for the higher Reynolds number jet. The figures are presented in gray scales. A representation via color is more instructive (and these were presented at the meeting), but exorbitant reproduction costs preclude this option.

Figure 8, 9 and 10 show reconstructions of figure 7 for various CCR's. The reconstructions are obtained by keeping smaller and smaller number of coefficients. The CCR values are 50, 100 and 160 respectively. It is clear that the reconstructed images retain many structural features tolerably well even at the highest CCR's.

The behavior of SNR as a function of CCR is similar to that for one-dimensional signals (figure 11). It may be helpful to include data on the the Signal-to-Noise Ratio as a function of BPP, or the average number of bits used to code each pixel (see section 3). This is given in figure 12.

We have by no means optimized the computer programs, and it is perhaps not as useful here to quote typical computation times as it is for the one-dimensional case. Yet the following number may give some indication. In the current version of our programs, decomposition and reconstruction of images of the size of figure 7 takes about eight minutes on a Cray Y-MP/832.

6. Summary and Conclusions

Turbulent signals and images possess some temporal and spatial structure. There is always some correlation between neighboring elements which, in some sense, implies that there is redundancy of information. This in principle allows data compression. We may also be willing to tolerate some specified degradation in data compression and reconstruction. Often, as in the case of turbulent jet images discussed in section 5, the degradation is not perceptible except for high CCR's. This is because the precision in data coding may surpass the discrimination of human eye. In any case, some degradation may indeed be tolerated depending on applications.

Several data compression techniques are now available. We have outlined one of them and briefly presented some results. The technique allows the definition of the structure in terms of its shape, size and frequency (or wavenumber). A comparison between the performance of the present technique and several others has been made. The exact comparison can be carried out at several levels, and depends on the precise parameters held fixed. For example, one can specify the maximum mean-square error permissible in the reconstruction, and ask questions about the speed of the method and the computer memory required. One can keep a fixed CCR and determine the SNR of reconstructions. Furthermore, when comparing with the Fourier representation, details of image segmentation will be crucial. For all these reasons, a proper comparison requires a lot of detail and will not be reported here; it can be found in Zubair et al. (1990). It suffices to say that, overall, the Wavelet-Packet transform performs somewhat better than its competitors. We believe that the greatest advantage of Wavelet-Packets lies in the explicit identification of spatial position, frequency as well as the scale of structures. This immediately suggests several possible applications in turbulence. These studies are currently under way.

Acknowledgements

We are thankful to Rafi Coifman for many valuable discussions and to DARPA(URI) and AFOSR for financial support. Computations on the Cray was done at the Pittsburgh Supercomputing Center.

References

[1] F. Argoul, A. Arneodo, G. Grasseau, Y. Gagne, E.J. Hopfinger & U. Frisch, 1989, *Wavelet analysis of turbulence reveals the multifractal nature of the Richardson cascade.* Nature, 338:51-53.

[2] K.G. Beauchamp, 1984, *Applications of Walsh and Related Functions.* Academic Press, Orlando.

[3] G. Beylkin, R.R. Coifman & V. Rokhlin, 1989, *Fast wavelet transforms and numerical algorithms.* Yale University, preprint.

[4] R.R. Coifman & M.V. Wickerhauser, 1990, *Best-adapted wavelet-packet bases.* Yale University, preprint.

[5] R.R. Coifman & Y. Meyer, 1989, *Orthonormal wavelet-packet basis.* Yale University, preprint.

[6] R.R. Coifman, Y. Meyer, S. Quake & M.V. Wickerhauser, 1990, *Signal processing and compression with wavelet-packets.* Yale University, preprint.

[7] I. Daubechies, 1988, *Orthonormal wavelets of compactly supported basis.* Comm. Pure Appl. Math., 41:909–996.

[8] R. Everson, L. Sirovich & K.R. Sreenivasan, 1990, *Wavelet analysis of the turbulent jet.* Physics Letters A, 145:314–322.

[9] J. Liandrat & F. Moret-Bailly, 1990, *The wavelet transform: some applications to fluid dynamics and turbulence.* Eur. J. Mech., B/Fluids, 9:1–19.

[10] C. Meneveau, 1990, *Dual spectra and mixed energy cascade of turbulence in the wavelet representation.* Johns Hopkins University, preprint.

[11] R.R. Prasad & K.R. Sreenivasan, 1989, *Determination of scalar interfaces in digital images of turbulent flows.* Experiments in Fluids, 7:259-264.

[12] R.R. Prasad & K.R. Sreenivasan, 1990, *The measurement and interpretation of fractal dimension of the scalar interface of in turbulent flows.* Phys. Fluids A., 2:792-807.

[13] G. Strang, 1989, *Wavelets and dilation equations: a brief introduction.* SIAM Rev., 31:614-627.

[14] H. Tennekes & J.L. Lumley, 1972, *A First Course in Turbulence.* M.I.T. Press, Cambridge, Massachusetts.

[15] A. A. Townsend, 1956, *The Structure of Turbulent Shear Flows.* Cambridge University Press. Second edition issued in 1976.

[16] M.V. Wickerhauser, 1989, *Acoustic signal compression with Walsh-type wavelet-packets.* Yale University, preprint.

[17] L. Zubair, K.R. Sreenivasan & M.V. Wickerhauser, 1990, *Compression of turbulence data and images using Wavelet-Packets.* Yale University, preprint.

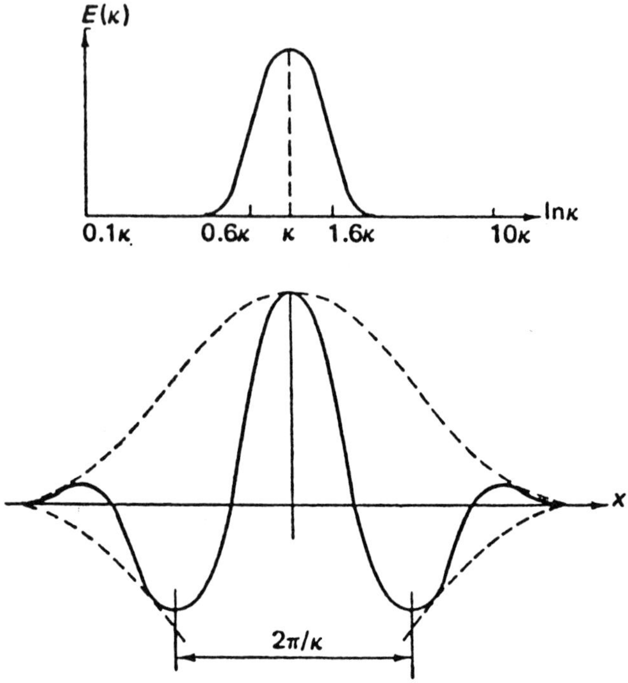

Figure 1: An eddy of wave number κ and wavelength $2\pi/\kappa$. From Tennekes & Lumley (1972).

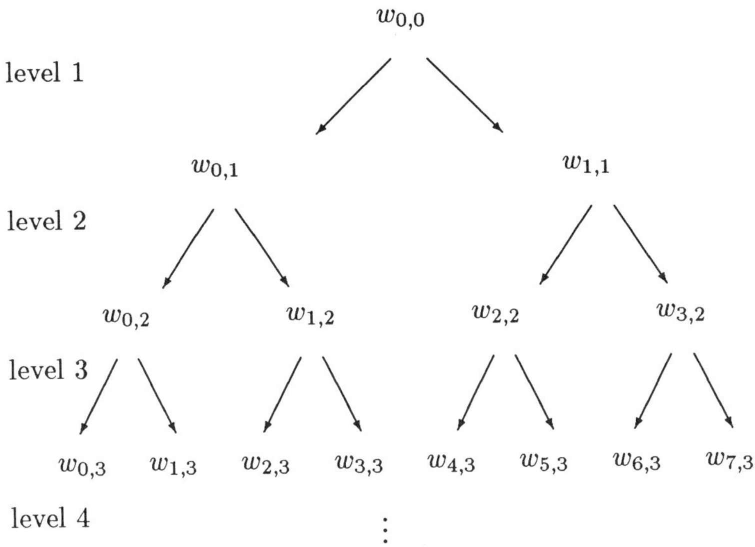

Figure 2: Hierarchy of Wavelet-Packets

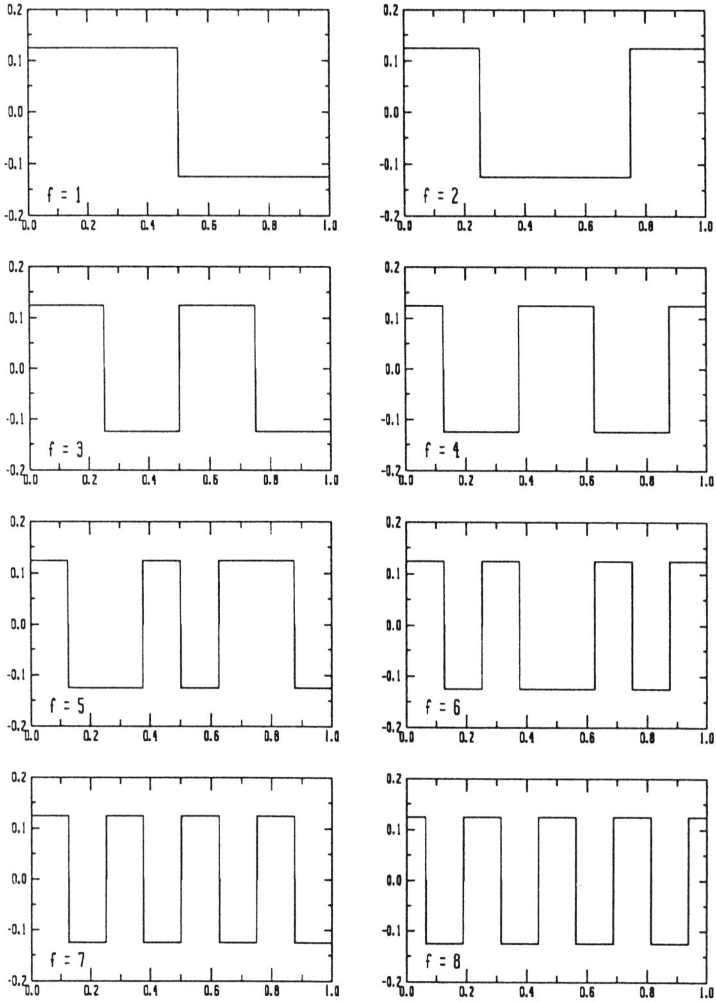

Figure 3: Walsh functions (see Beauchamp 1984).

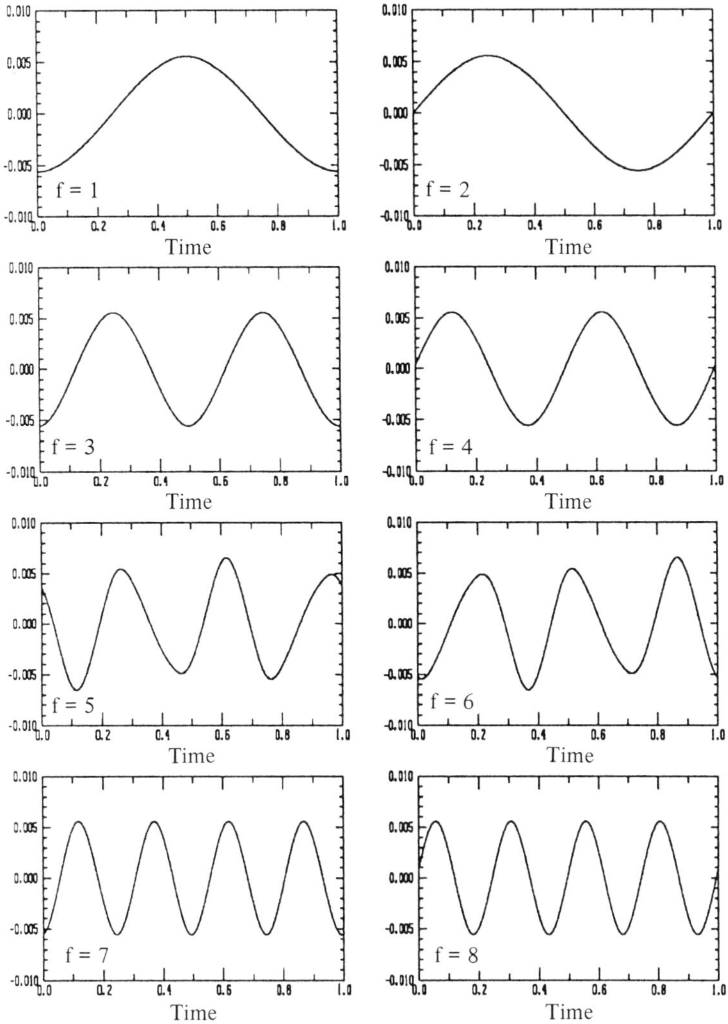

Figure 4: Coiflet functions (see Zubair et al. 1990).

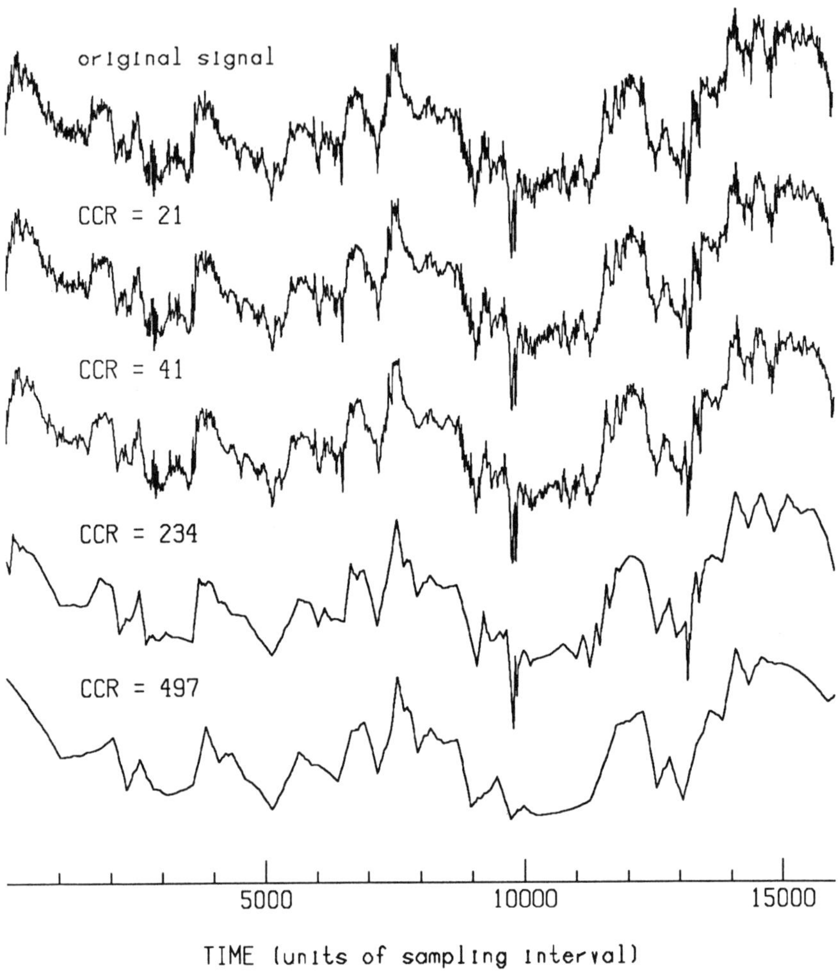

Figure 5: Compression and reconstruction of atmospheric turbulence velocity using Wavelet-Packet transforms.

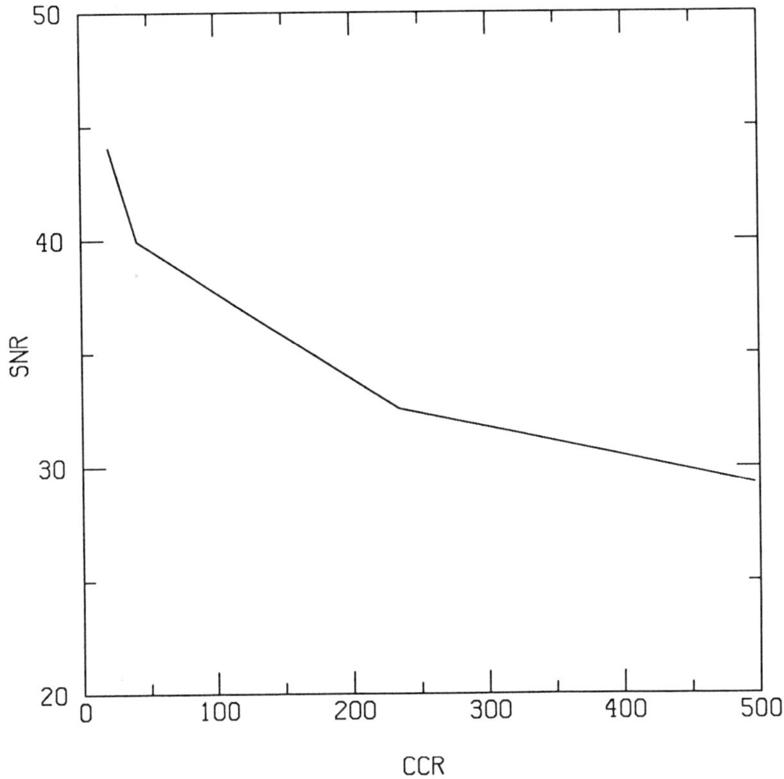

Figure 6: SNR as a function of CCR for one-dimensional velocity data

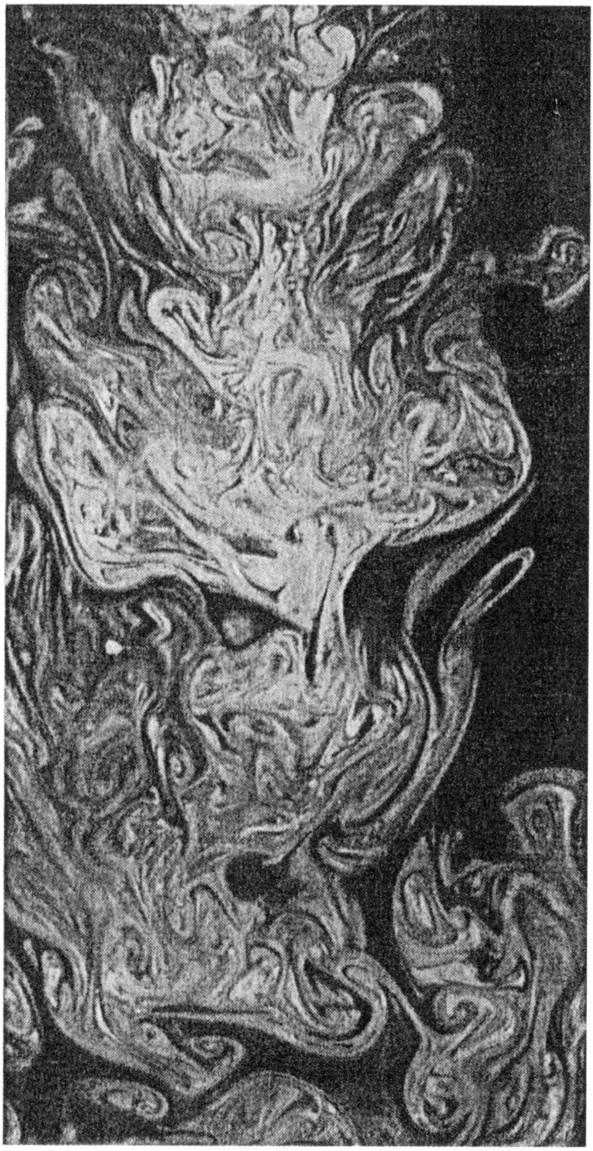

Figure 7: A two-dimensional map of the concentration field in an axisymmetric jet. A thin slice of the field has been obtained by laser-induced-fluorescence (see Prasad & Sreenivasan 1989, 1990).

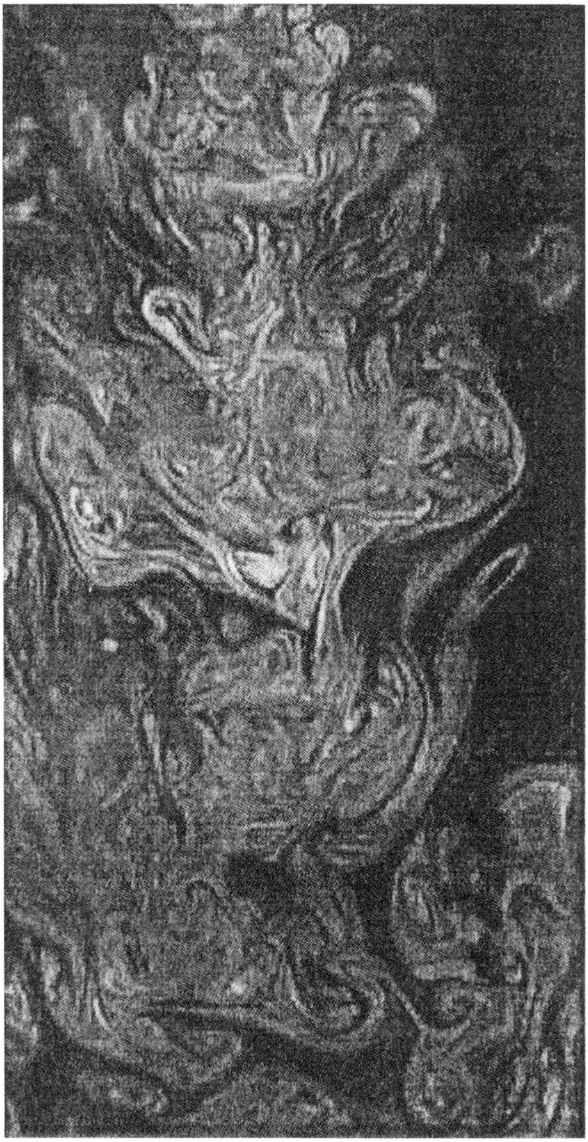

Figure 8: Wavelet-Packet transform reconstruction of figure 7 for a CCR = 50.

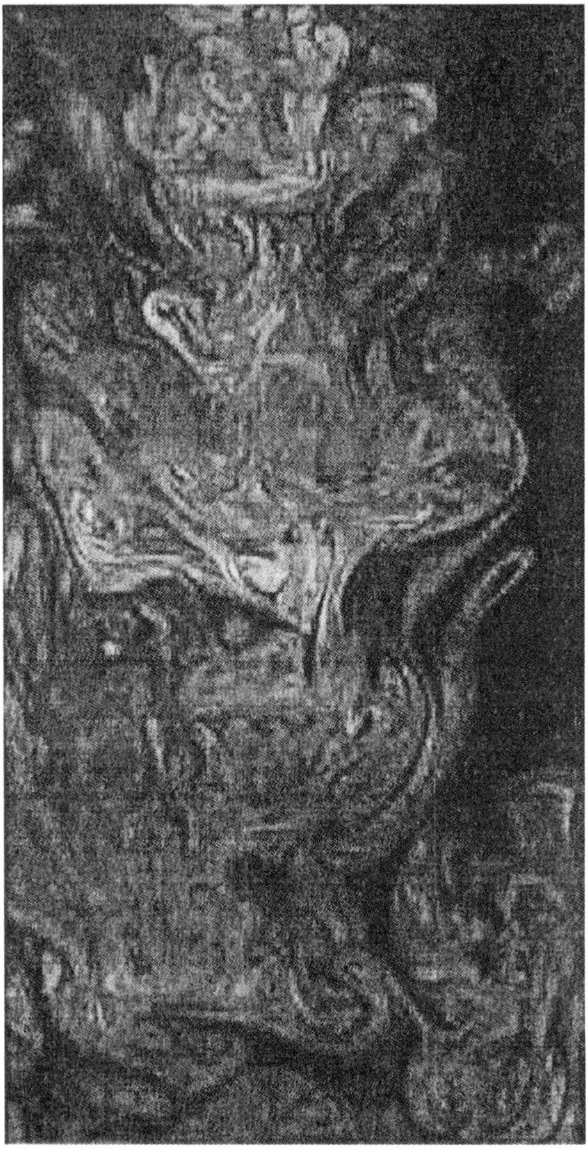

Figure 9: Wavelet-Packet transform reconstruction of figure 7 for a CCR = 100.

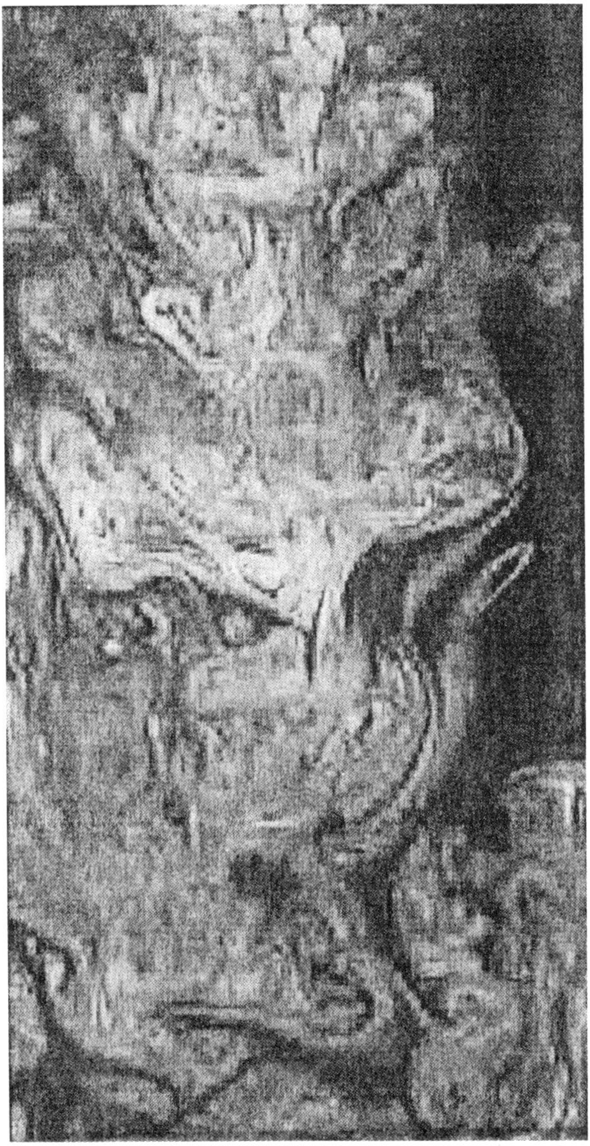

Figure 10: Wavelet-Packet transform reconstruction of figure 7 for a CCR = 160.

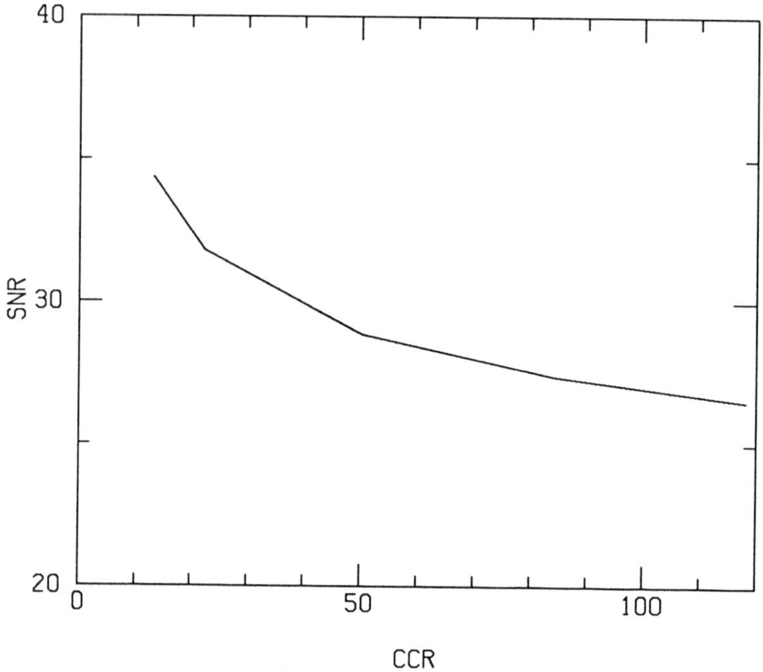

Figure 11: SNR as a function of CCR for a two-dimensional concentration field

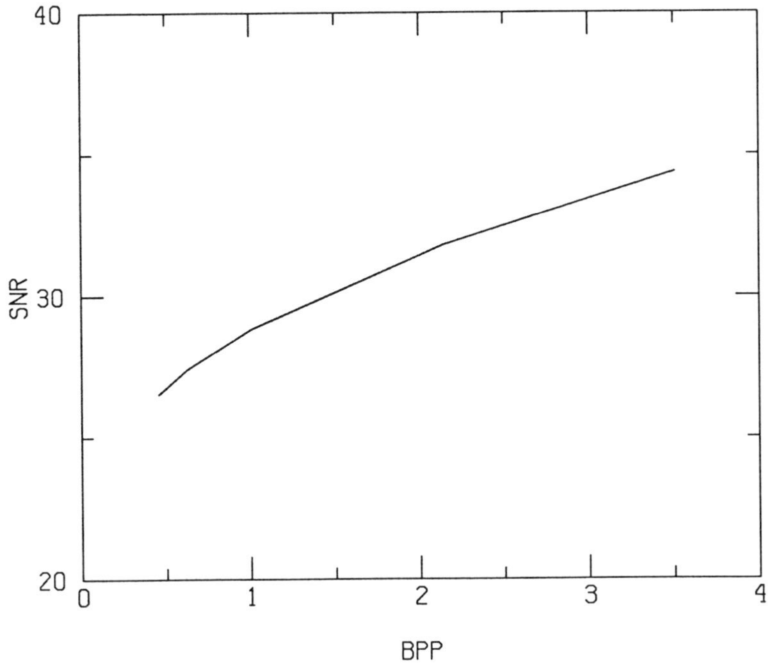

Figure 12: SNR as a function of BPP for a two-dimensional concentration field

SELF-PRESERVATION OF TEMPERATURE FLUCTUATIONS IN ISOTROPIC TURBULENCE

William K. George

University at Buffalo
State University of New York
Buffalo, NY 14260

ABSTRACT

The averaged spectral equations governing the decay of temperature fluctuations in an isotropic turbulent field are considered and found to admit to fully self-preserving solutions which retain a dependence on the initial conditions. The characteristic length scale for the decay is shown to be the scalar Taylor microscale, and the characteristic temperature is the square root of the scalar variance. In addition to recovering the well-known empirically established power law decay for the variance, the mechanical/temperature time scale ratio is shown to be constant. The invariants of scalar decay are discussed and a number of questions are raised which require further research.

1. Introduction

The decay of temperature fluctuations in isotropic turbulence has been a subject of considerable interest since the early 1950's. Corrsin (1951a,b) was the first to write equations describing the dynamics of the spectrum and correlation functions in isotropic flow. Monin and Yaglom (1975) provide a comprehensive review of attempts to scale and close these equations.

There have been numerous attempts over the years to establish the characteristics of the decay of temperature fluctuations introduced into grid-generated turbulence (Mills et al. 1958, Lanza and Schwarz 1966, Yeh and Van Atta 1973, Lin and Lin 1973, Warhaft and Lumley 1978). Warhaft and Lumley

(1978) showed that each attempt generated a decay rate and spectral shape uniquely determined by its initial conditions. In addition they showed that the mechanical/thermal time scale ratio defined by

$$r = \tau/\tau_\theta = (\overline{q^2}/\epsilon)/(\overline{\theta^2}/\epsilon_\theta) \tag{1}$$

was unlikely to achieve the expected equilibrium value of unity, but were unable to establish a physical reason as to why.

Recently George (1987a,b) has shown that the dynamics of isothermal decaying isotropic turbulence achieve a self-preserving state, and that the spectrum throughout decay can be scaled by a single length scale, the Taylor microscale defined by

$$\lambda^2 = 15\nu\ u^2/\epsilon \tag{2}$$

and a velocity scale defined from the kinetic energy by

$$u = \left[\frac{1}{3}\ \overline{q^2}\right]^{1/2} \tag{3}$$

A further consequence of the theory is that the turbulence energy undergoes a power law decay $u^2 \sim t^n$. Both n and the spectral shape are determined by the initial conditions.

It is the purpose of this paper to extend these self-preservation arguments to the scalar field. The result will be shown to again be that the spectrum (this time the scalar spectrum) scales with a single length scale, the thermal Taylor microscale λ_θ, and the scalar intensity, $\theta = (\overline{\theta^2})^{1/2}$. The consequences of self-preservation on other turbulence properties of interest will be explored. Finally, the results of the analysis will be shown to be reasonably consistent with experiment.

2. The Scalar Spectral Equation

The equation governing the evolution of a homogeneous, isotropic passive temperature field, $\theta(\underline{x},t)$, in a homogeneous, isotropic turbulence is given by (Monin and Yaglom 1975)

$$\frac{\partial}{\partial t} E_\theta = T_\theta - 2\ \alpha\ k^2 E_\theta \tag{4}$$

where α is the thermal diffusivity, $E_\theta = E_\theta(k,t)$ is the three-dimensional scalar spectral function (hereafter referred to as the scalar spectrum), and $T_\theta = T_\theta(k,t)$ is the scalar spectral energy transfer function. The integral of the scalar spectrum over all wavenumbers is one-half of the scalar variance, i.e.

$$\frac{1}{2}\theta^2 = \frac{1}{2}\overline{\theta^2} = \int_0^\infty E_\theta(k,t)dk \tag{5}$$

It can also be shown (Monin and Yaglom 1975) that the scalar dissipation rate, ϵ_θ, is given by

$$\epsilon_\theta = -\frac{d}{dt}\left[\frac{1}{2}\theta^2\right] = 2\alpha\int_0^\infty k^2E_\theta(k,t)dk \tag{6}$$

3. The Self-Preservation Analysis

We seek self-preserving solutions for which all of the terms in the scalar spectral equations remain in relative balance throughout the decay. We define scaling functions $E_{\theta s}(t)$, $T_{\theta s}(t)$ and $L_\theta(t)$ so that

$$\eta_\theta = kL_\theta(t) \tag{7}$$

$$E_\theta(k,t) = E_{\theta s}(t)f_\theta(\eta_\theta) \tag{8}$$

and

$$T_\theta(k,t) = T_{\theta s}(t)\,g_\theta(\eta_\theta) \tag{9}$$

These can be substituted directly into the scalar spectral equation to yield

$$\left[\dot{E}_{\theta s}\right]f_\theta + \left[\frac{E_{\theta s}\dot{L}_\theta}{L_\theta}\right]\eta_\theta f_\theta = \left[T_{s\theta}\right]g_\theta - \left[\frac{\alpha E_{\theta s}}{L_\theta^2}\right]2\eta_\theta^2 f_\theta \tag{10}$$

For self-preservation, all of the bracketed terms must have the <u>same</u> time-dependence. For convenience, the entire expression can be divided by the last bracketed term to yield

$$\left[\frac{\dot{E}_{s\theta}L_\theta^2}{\alpha E_{s\theta}}\right]f_\theta + \left[\frac{L_\theta\dot{L}_\theta}{\alpha}\right]\eta_\theta f'_\theta = \left[\frac{T_{s\theta}L_\theta^2}{\alpha E_{s\theta}}\right]g_\theta - [1]2\eta_\theta^2 f_\theta \tag{11}$$

Now since the last bracketed term is time-independent, all the others must also be time-independent for self-preservation. Therefore, the conditions for self-preservation are

$$\text{(i)} \qquad \frac{\dot{E}_{s\theta} L_\theta^2}{\alpha E_{s\theta}} = \text{constant} \tag{12}$$

$$\text{(ii)} \qquad \frac{\dot{L}_\theta L_\theta}{\alpha} = \text{constant} \tag{13}$$

$$\text{(iii)} \qquad \frac{T_{s\theta} L_\theta^2}{\alpha E_{s\theta}} = \text{constant} \tag{14}$$

Equation (13) can be integrated directly to yield

$$L_\theta^2 \propto \alpha \, (t-t_o) \tag{15a}$$

or

$$L_\theta^2 = 2A \, \alpha \, (t-t_o) \tag{15b}$$

where the origin in time, t=o, is suitably chosen to absorb the initial condition in the <u>mechanical energy equation</u> (see below), A is a coefficient determined by the initial conditions, and the factor of 2 is introduced for convenience. Thus the length scale increases with the square root of time measured from the origin of the scalar field, $t=t_o$.

Equation (12) can be integrated to yield

$$E_{s\theta} = \left[\frac{t-t_o}{t_1-t_o}\right]^q E_{s\theta}\Big|_{t-t_o=t_1-t_o} \tag{16}$$

where t_1 is an arbitrary reference time and q is an exponent determined by the initial conditions. Thus the scalar spectrum undergoes a power law decay.

The spectral decay constant can be related to the decay of the scalar variance by substituting the self-preserving forms into the integral of equation (5) to yield

$$\frac{1}{2}\,\theta^2 = \left[E_{s\theta} L_\theta^{-1}\right] \int_o^\infty f_\theta \, (\eta_\theta) d\eta_\theta \tag{17}$$

Thus,

$$E_{s\theta} \sim \theta^2 L_\theta \tag{18}$$

It follows immediately from equations (15)–(18) that

$$\theta^2 \propto (t-t_o)^m \tag{19}$$

where

$$m = q - \frac{1}{2} \tag{20}$$

Thus the scalar variance also undergoes a power law decay with an exponent which is determined by the initial conditions.

The length scale L_θ can be related to a physically meaningful length scale by using the dissipation integral relation. By substituting the self-preserving forms into equation (6) it follows that

$$\epsilon_\theta = \left[\alpha E_{s\theta} L_\theta^{-3}\right] 2 \int_0^\infty \eta_\theta^2 \, f_\theta(\eta_\theta) d\eta_\theta \tag{21}$$

or using equation (18),

$$\epsilon_\theta \propto \alpha\theta^2/L_\theta^2 \tag{22}$$

The scalar Taylor microscale, λ_θ, is defined from (Monin and Yaglom 1975)

$$\overline{\left[\frac{\partial\theta}{\partial x}\right]^2} = \frac{2\theta^2}{\lambda_\theta^2} \tag{23}$$

For an isotropic scalar field the dissipation can be shown to be

$$\epsilon_\theta = 3\alpha \left[\frac{\partial\theta}{\partial x}\right]^2 = 6\alpha \frac{\theta^2}{\lambda_\theta^2} \tag{24}$$

Comparison of equations (23) and (24) makes it immediately clear that

$$L_\theta \propto \lambda_\theta \tag{25}$$

so that the scalar Taylor microscale is proportional to the self-preservation length scale. Therefore without loss of generality we choose the constant of proportionality to be unity (and absorb a factor into f_θ), i.e.

$$L_\theta = \lambda_\theta \tag{26}$$

The constant of proportionality in equation (15b) can be shown to be uniquely related to the scalar variance decay exponent m. From equations (6) and (24),

$$\frac{d\frac{1}{2}\theta^2}{dt} = -\epsilon_\theta = -6\alpha\frac{\theta^2}{\lambda_\theta^2} \tag{27}$$

By substituting from equations (22) and (18b) for θ^2 and λ_θ it follows that

$$\lambda_\theta^2 = -\frac{12}{m}\alpha\,(t-t_o) \tag{28}$$

or

$$A = -6/m. \tag{29}$$

The scalar spectral transfer scaling function $T_{\theta s}$ can be evaluated using equations (14), (18) and (26) as

$$T_{s\theta} \propto \alpha\frac{\theta^2}{\lambda_\theta} \tag{30a}$$

or

$$T_{s\theta} = B_\theta\,\alpha\frac{\theta^2}{\lambda_\theta} \tag{30b}$$

where B_θ is the constant of proportionality.

Note that there is nothing in the self-preservation analysis which determines the value of the constants. Therefore these will be uniquely determined by the initial conditions. There will be a coupling, however, between the energy decay exponent m, the spectral transfer coefficient B_θ, and the shape of the energy spectrum since these are linked by the spectral energy equation. This coupling can be easily seen by substituting the self-preservation conditions (18), (28), and (30b) into equation (11) with the result that the scalar spectral equation becomes

$$m\left[12f_\theta - 2\eta_\theta^2 f_\theta + B_\theta g_\theta\right] + 6\eta_\theta f'_\theta + 6f_\theta = 0$$

In summary, the scalar spectral equation has been

found to admit to self-preserving solutions which have the following properties:

(i) The length scale is the Taylor microscale, λ_θ

(ii) The scalar variance undergoes a power law decay.

(iii) The Taylor microscale grows as the square root of time (measured from its own virtual origin).

(iv) The scalar spectrum and scalar spectral transfer function collapse at all wavenumbers when plotted as

$$\frac{E_\theta(k,t)}{\theta^2 \lambda_\theta} \quad \text{vs. } k\lambda_\theta$$

and

$$\frac{\lambda_\theta T_\theta(k,t)}{\alpha \theta^2} \quad \text{vs. } k\lambda_\theta$$

(v) The spectral shapes and constants are determined by the initial conditions.

4. The Time Scale and Taylor Microscale Ratios

It follows from equations (6), (19), and the power law decay for the energy that

$$r = m/n \tag{31}$$

where n is the energy decay exponent. Since n and m are determined by the initial conditions, so must be the time scale ratio, and <u>no universal value should be expected.</u>

A fact of immediate interest is that both the mechanical and scalar Taylor microscales grow as the square root of time, although with different virtual origins. Thus asymptotically,

$$\lambda_\theta/\lambda = \text{constant}, \tag{32}$$

the exact value being determined by the initial

conditions. The constant in equation (32) can be shown to be given uniquely by the time scale ratio r defined by equation (2). From equations (2), (23) and (3) it follows that

$$r = m/n = 3(\lambda^2/15\nu)/(\lambda_\theta^2/6\alpha)$$

$$= \frac{18}{15}\left[\frac{\alpha}{\nu}\right]\left[\frac{\lambda}{\lambda_\theta}\right]^2 \qquad (33)$$

so that

$$\left[\frac{\lambda}{\lambda_\theta}\right]^2 = \frac{5}{6}\left[\frac{m}{n}\right]\sigma = \frac{5}{6} r\sigma \qquad (34)$$

where σ is the Prandtl number.

5. The Spectral Transfer and Triple Correlations

As pointed out by George (1987b), a consequence of the spectral transfer scaling (or an equivalent analysis of the von Karman-Howarth equation) is that the triple correlation $\overline{u^2(x)\ u(x+\rho)}$ is given by

$$\overline{u^2(x)\ u(x+\rho)} = R_\lambda^{-1}\ u^3 k(\rho/\lambda) \qquad (35)$$

where $k(\rho/\lambda)$ is not the usual non-dimensional triple-correlation introduced by von Karman and Howarth because of the R_λ^{-1} in front of it. Thus the velocity skewness (and the derivative skewness as well) is inversely proportional to R_λ, i.e.

$$\frac{\overline{u^3}}{(\overline{u^2})^{3/2}} = \frac{\overline{u^3}}{u^3} \sim R_\lambda^{-1} \qquad (36)$$

A similar relation can be derived from equation (30b) for the velocity-scalar triple correlation (or equivalently, the scalar spectral transfer). Showing,

$$\overline{u(x)\ \theta(x)\ \theta(x+\rho)} = \frac{\alpha}{\lambda_\theta}\ \theta^2\ p(\rho/\lambda_\theta)$$

$$= \left[\frac{\alpha}{\nu}\right]\left[\frac{\nu}{u\lambda}\right]\left[\frac{\lambda}{\lambda_\theta}\right] u\theta^2\ p(\rho/\lambda_\theta) \qquad (37)$$

Thus the scalar-velocity triple correlation will depend inversely on the Reynolds number and the scalar to mechanical length scale ratio.

The scalar-velocity triple correlation can be expressed in terms of the time scale ratio r using equation (33) as

$$\overline{u\theta^2}/\sqrt{\overline{u^2}}\ \overline{\theta} = (5r/\sigma)^{1/2}\, R_\lambda^{-1} \tag{38}$$

Because $u \sim t^{n/2}$ and $\lambda \sim t^{1/2}$, $R_\lambda \sim t^{(n+1)/2}$; and it follows that

$$\frac{\overline{u\theta^2}}{u\theta^2} \sim t^{-(n+1)/2} \tag{39}$$

Thus the time variation of the scalar velocity triple correlation coefficient depends only on the exponent for the kinetic energy. Since $n < -1$ in all experiments to-date, the normalized triple correlation coefficient increases during decay, just as does that for the velocity alone (George 1987b).

6. Comparison with Experimental Data

The preceeding analysis has predicted with no assumption other than complete self-preservation, that the kinetic energy and mean square thermal variance should decay as some power of the time measured from appropriate virtual origins. Such power law decays have been long established experimentally in the absence of a supporting theory. For example, Figure (1) is reproduced from Wahrhaft and Lumley (1978), and illustrates both the power law decay of the thermal variance and the dependence on the initial conditions of the thermal fluctuations.

A second prediction of the theory is that the mechanical and thermal Taylor microscales should vary as the square root of time (or distance in a wind tunnel) when measured from their respective virtual origins. Since the Taylor microscale, however, is computed from the decay data (eg. using equation 30), and since a power law form for the thermal variance (or kinetic energy) implies directly a square root dependence, the Taylor microscale must behave in the appropriate manner and can therefore not provide independent confirmation. The same is true of relations like equations (32)-(34) relating the time scale ratio to the mechanical and thermal Taylor microscales, since these are satisfied identically once the power law behavior is established for the variances.

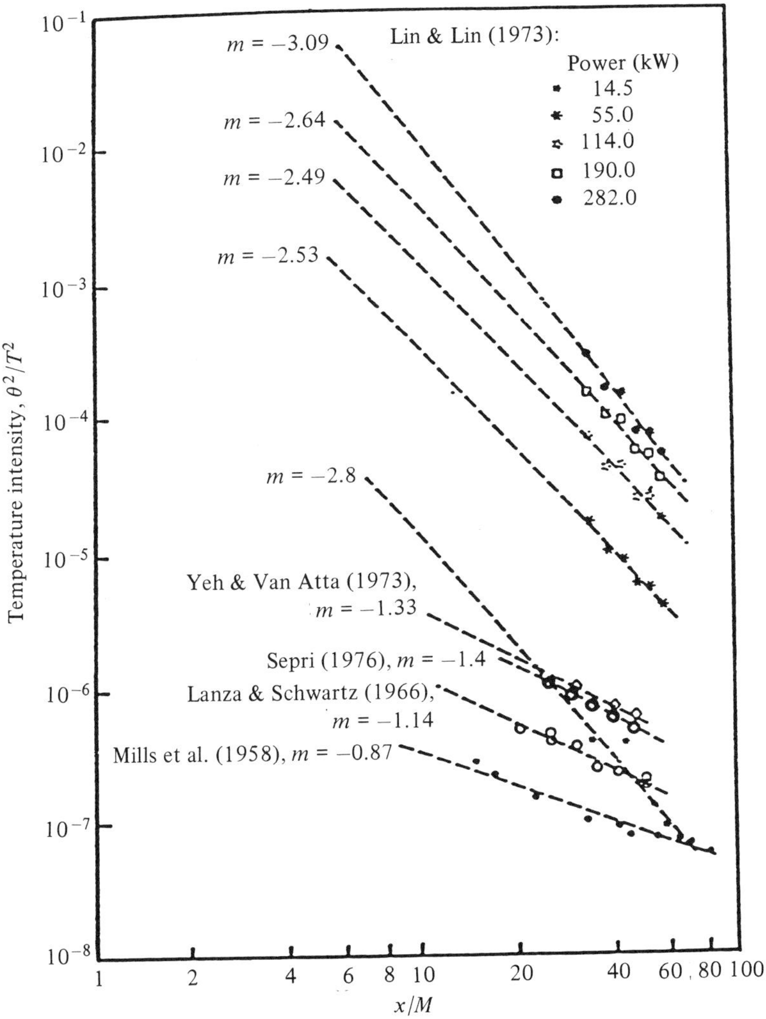

Figure 1. Decay of temperature fluctuations behind heated grid (from Warhaft and Lumley 1978).

The theory predicts that the turbulence can be characterized by a single length scale, the Taylor microscale. Thus, all integral scales must be proportional to it. While this would seem to provide a straightforward test of the theory, unfortunately the integral scale is one of the most difficult parameters to determine experimentally because of the large scales (or low wavenumbers) which determine it. A better experimental test of the proposed scaling laws (than the measured integral scales) is whether the velocity and temperature spectra for a single initial condition can be collapsed throughout the decay. Figures (2) and (3) show the one-dimensional spectral data for the heated grid experiment of Wahrhaft and Lumley (1978, Figures 4 and 10). normalized as $F^1_{11}/u^2\lambda$ versus $k\lambda$ and $F^1_{\theta\theta}$ versus $k\lambda_\theta$ which are the one-dimensional spectral counterparts of the proposed scaling laws. The collapse of the velocity spectral data is over the entire range of scales, including even the largest wavenumbers. This last fact is particularly satisfying in view of the problems cited above since the value of the spectra at the origin can be related directly to the integral scales. The collapse of the temperature spectra is less spectacular but is generally supportive of the theory.

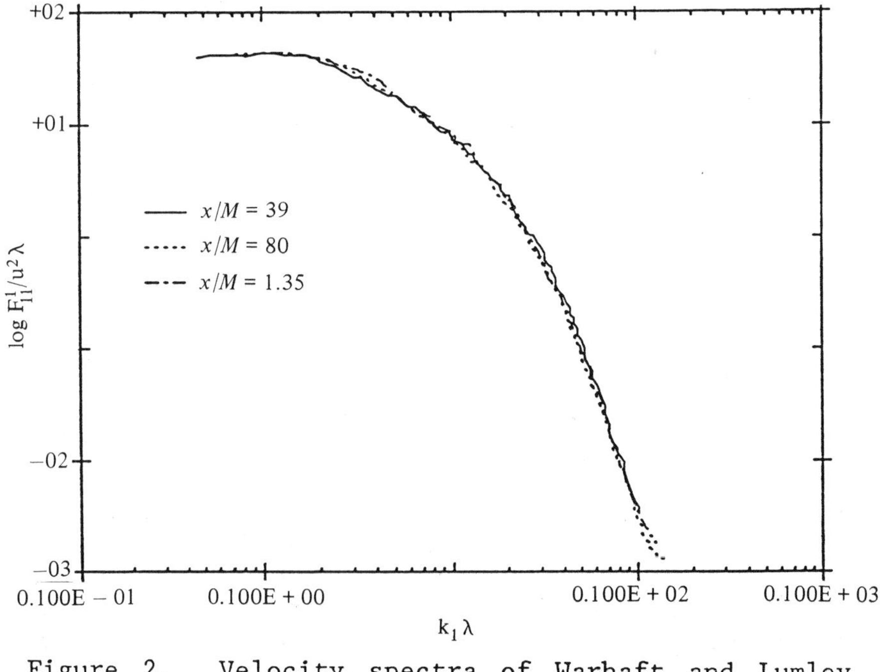

Figure 2. Velocity spectra of Warhaft and Lumley normalized in Taylor variables.

An additional feature of the proposed theory is that each method of generating turbulence will generate its own spectral and decay characteristics which will persist throughout the decay. The entire paper by Wahrhaft and Lumley (1978) is a documentation of this fact and thus this feature must be regarded as confirmed.

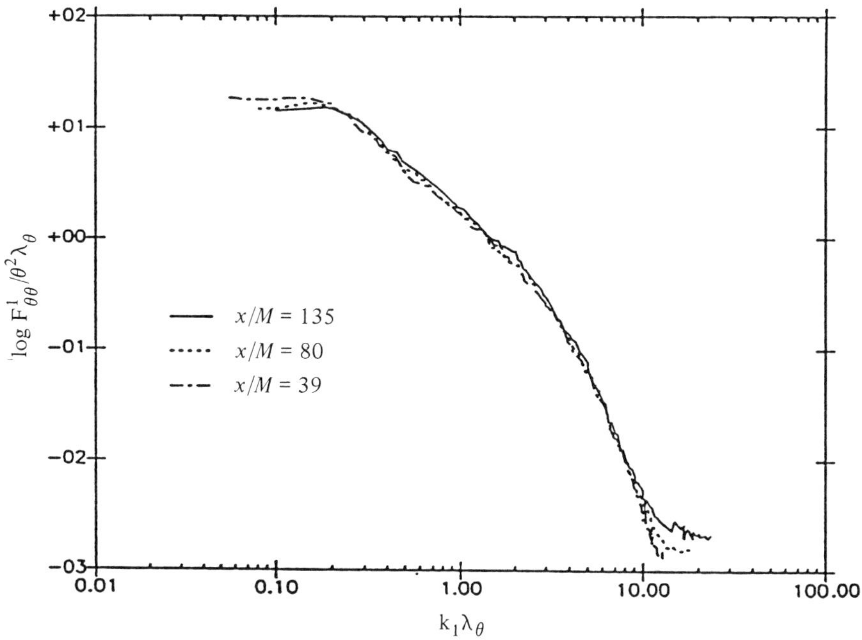

Figure 3. Temperature spectra of Warhaft and Lumley, normalized by Taylor variable.

Finally, the theory predicts that both the mechanical and thermal spectral transfers should scale inversely with the Reynolds number $u\lambda/\nu$. While the downstream development of these does not appear to have been studied, Mills et al. (1958) present measurements of the equivalent triple moments defined by equations (35) and (37), both of which should also show the inverse Reynolds number dependence. These correlations collapse well up to the peak value when plotted as $-p(r)\, R_\lambda$ and $-k(r)R_\lambda$ versus r/λ, but the collapse deteriorates considerably for larger separations, reflecting the same problem as for the double correlations in this experiment. The peak values, however, are in approximate agreement with the proposed scaling as shown in Table I which also includes the velocity-derivative skewness.

TABLE I

Experimental Verification of Predicted Scaling for the Triple Correlations and Velocity Derivative Skewness (from Mills et al. 1958)

x/M	17	32	45	69
$-Re_\lambda\ k_{max}$	1.48	1.63	–	1.41
$-\sigma Re_\lambda\ p_{max}$	0.95	1.09	–	1.07
$-Re_\lambda\ S_{\dot{u}}$	10.6	11.0	10.3	9.6

7. Summary and Conclusions

There is enough evidence to indicate that the predicted self-preserving behavior of the decay of temperature fluctuations behind a grid is reasonable. The theory predicts the oft-observed power law dependence for the decay of the temperature variation, the collapse of the spectra, and the observed dependence on the initial conditions. The evidence for the Reynolds number dependence of the non-linear spectral transfer terms is more speculative, and further experimental work would be helpful.

It is interesting that the theory presented here makes no assumptions regarding the relation of the scalar spectral transfer to the mechanical nature of the turbulence. Thus the existing arguments as to the nature of this dependence would appear to be unchanged, as well as the predictions from them (like the k^{-1} range, v. Monin and Yaglom, Vol. II).

On the other hand, it is easy to see that the existence of full self-preservation at all scales of motion negates the possibility of a universal equilibrium range, and with it the validity of the Kolmogorov theory for the university of the small scales (except possibly as the first term in an expansion about the infinite Reynolds number limit, see below). This has been discussed in detail by George (1987a,b) for the mechanical energy spectrum, and the arguments can easily be extended to the scalar spectrum.

A related question has to do with how the self-preservation arguments presented here are applicable in the limit of infinite Reynolds number. The full self-preservation presented here requires that the spectral transfer be forever dependent on Reynolds number. In effect, the spectral transfer adjusts itself so as to provide the required amount of energy dissipation ($\epsilon \sim \nu u^2/\ell^2$) to maintain similarity. This is the opposite of the conventional wisdom which requires that the decay rate be determined only by the energy-containing eddies ($\epsilon \sim u^3/\ell$). The Kolmogorov theory can be recovered in the limit of infinite Reynolds number only if $n \to -1$ in this limit (as speculated by George 1987a).

In summary, there is evidence that the proposed full self-preservation describes many of the observed features of the behavior of isotropic temperature fluctuations. The analysis has raised, however, a number of questions which require additional experimental or numerical simulation research.

References

[1] Corrsin, S. (1951a) On the spectrum of isotropic temperature fluctuations in an isotropic turbulence. J. Appl. Phys. 22, 4, 469-473.

[2] Corrsin, S. (1951b) The decay of isotropic temperature fluctuations in an isotropic turbulence. J. Aeronaut. Sci., 18, 6, 417-423.

[3] George, W.K. (1987a) A theory for the decay of homogeneous isotropic turbulence. Turbulence Research Laboratory Report 135, University at Buffalo, SUNY, Buffalo, NY (submitted to JFM).

[4] George, W.K. (1987b) The decay of homogeneous isotropic turbulence. Proc. of 2nd Int. Conf. on Transport Phenomena in Turbulent Flows, Tokyo.

[5] Lanza, J. and W. Schwarz (1966) unpublished.

[6] Lin, S.C. and S.C. Lin (1973) Study of strong temperature mixing in subsonic grid turbulence. Phys. Flds, 16, 1587-1598.

[7] Mills, R.R., A.L. Kistler, V. O'Brien, and S. Corrsin (1958) Turbulence and temperature fluctuations behind a heated grid. NACA Tech. Note 4288.

[8] Monin, A. S. and A.M. Yaglom (1975) Statistical Fluid Mechanics. Vol. II, MIT Press, Cambridge, MA.

[9] Wahrhaft, Z. and J. L. Lumley (1978) An experimental study of the decay of temperature fluctuations in grid-generated turbulence. JFM, 88, 4, 659-684.

[10] Yeh, T.T. and C.W. VanAtta (1973) Spectral transfer of scalar and velocity fields in heated-grid turbulence. JFM, 58, 233-261.

THE TURBULENCE CONTRIBUTION TO THE PRESSURE TERM IN THE SCALAR-FLUX EQUATIONS

T. Dakos and M.M. Gibson [1]

Imperial College of Science, Technology and Medicine
London SW7 2BX, England.

ABSTRACT

A theory of the pressure-scalar-gradient correlations is further developed. In an earlier paper the Navier-Stokes and scalar-transport equations were formally solved to obtain nonlinear expressions for the fluctuating parts of the pressure and the scalar. Four components of the pressure-scalar-gradient correlation were identified, and the term in the mean scalar gradient was modelled in terms of known variables. The problem of modelling the turbulence part is now addressed. For isotropic turbulence, a fourth-order cumulant discard approximation allows this contribution to be expressed in terms of single-point double products and spectra of the turbulent energy and scalar flux. Numerical calculations are made for assumed spectrum shapes. The results agree well with new experimental data.

1. Introduction

A major obstacle to the solution of the equations for the scalar fluxes $\langle u_i\vartheta\rangle$ is presented by the pressure correlations $\langle p\partial\vartheta/\partial x_i\rangle$. It is customary (Lumley 1978) to consider separately two components of this quantity, one of which involves gradients of the mean velocity while the other contains only turbulence quantities. In a previous paper (Dakos and Gibson 1987 ≡ DG1) we obtained an expression for $\langle p\partial\vartheta/\partial x_i\rangle$, following Weinstock's (1981, 1982) theoretical treatment of the corresponding terms in the Reynolds-stress equations. The method (for nearly homogeneous unidirectional flow with weak gradients) was to determine the fluctuating parts of p and ϑ by formal solution of the Navier-Stokes and scalar-transport equations. A new expression for $\langle p\partial\vartheta/\partial x_i\rangle$

[1] This work was supported by Grant No. GR/B83674 from the Science and Engineering Research Council.

was obtained by substituting p(**k**,t) and ϑ(**k**,t) in it's Fourier transform. This expression consisted of four terms, one of which contained the mean scalar gradient. A fourth-order cumulant discard approximation allowed this term to be expressed in terms of single-point covariances, and spectra of the turbulence energy and stress. It was then shown that to first order in anisotropy this general expression was also proportional to the anisotropy tensor $b_{ij} \equiv \langle u_i u_j \rangle / q^2 - \delta_{ij}/3$. The coefficient of proportionality was calculated for simple spectrum assumptions, just as Weinstock (1981, 1982) had evaluated the Rotta constant in the return-to-isotropy part of the pressure-rate-of-strain correlation. The method is not quite so readily applied to the other components of $\langle p\partial\vartheta/\partial x_i \rangle$, because the equations cannot be linearised by splitting the spectrum of $\langle u_i\vartheta \rangle$ into isotropic and non-isotropic parts. If, however, the analysis is restricted to isotropic turbulence, it is possible to obtain an expression for the turbulence part of $\langle p\partial\vartheta/\partial x_i \rangle$. This is the purpose of the present paper.

A familiar approximation for the term in question is an exact counterpart of Rotta's (1951) expression for the return of non-isotropic turbulence to isotropy, usually attributed to Monin (1965):

$$\left\langle \frac{p}{\rho} \frac{\partial \vartheta}{\partial x_i} \right\rangle_1 = - C_{\vartheta 1} \left(\frac{2\varepsilon}{q^2} \right) < u_i \vartheta > \tag{1}$$

where $C_{\vartheta 1}$ is a constant to be evaluated in the analysis. Many modellers have followed Launder (1976) in taking $C_{\vartheta 1} \approx 3$, which is the value required to produce plausible values of the scalar-flux ratio $\langle u_1\vartheta \rangle / \langle u_2\vartheta \rangle$ and the turbulent Prandtl number in two-dimensional shear flow. A value of about 3 is also implied by Shih and Lumley's (1986) expression based on the mechanical / thermal time-scale ratio when this quantity is taken as about 1.4, as in the homogeneous flow of Sirivat and Warhaft (1983). The experimental data are inconclusive on this point. Measurements reported by Sirivat and Warhaft (1983) indicate $C_{\vartheta 1}$ increasing from about 2 at low Reynolds number to approximately 3 in the higher Re experiments. The data of Budwig et al (1985) produce values as high as 4. But new measurements by Dakos (1990) are consistent with the theoretical estimate $C_{\vartheta 1} \approx 1.7$.

Only the main steps in the analysis can be presented here. For the finer details the reader is referred to the papers by Weinstock (for the calculation strategy) and to the thesis by Dakos (1990).

2. Analysis

2.1 Derivation of the Pressure Term

In DG1 we considered homogeneous turbulence in incompressible flow in the x_1 direction, with uniform gradients of mean velocity and the mean scalar in the x_2 direction. Starting from the Fourier transform of the Poisson equation for p we obtained

$$\frac{p(\mathbf{k},t)}{\rho} = N(\mathbf{k},t) + \left(\frac{2ik_1}{k^2}\right)u_2(\mathbf{k},t)\frac{dU_1}{dx_2} \tag{2}$$

where

$$N(\mathbf{k},t) = -\int d\mathbf{k}_1\left(\frac{k_ik_j}{k^2}\right)[u_i(\mathbf{k}_1)u_j(\mathbf{k}-\mathbf{k}_1)] \tag{3}$$

For the fluctuating part of a passive scalar in turbulence at high Reynolds numbers

$$\vartheta(\mathbf{k},t) = G_0(\mathbf{k},t)\vartheta(\mathbf{k},0) - \int_0^t dt_1 G_0(\mathbf{k},t-t_1)[(u_m\vartheta,_m)_{\mathbf{k},t_1} + u_2(\mathbf{k},t_1)T,_2] \tag{4}$$

where the Fourier transform of $u_m\vartheta_{,m}$ is evaluated at k and t_1, and

$$G_0(\mathbf{k},t-t_1) = \exp(-i\mathbf{k}.\mathbf{U}(t-t_1)) \tag{5}$$

Equation (4) is similar to the expression obtained by Weinstock (1981) for the fluctuating velocity component. The Fourier transform pairs for p and ϑ are then used to obtain

$$\left\langle p\frac{\partial\vartheta}{\partial x_i}\right\rangle = -\frac{(2\pi)^3}{V}\int ik_i < p(\mathbf{k},t)\vartheta^*(\mathbf{k},t) > d\mathbf{k} \tag{6}$$

where V is the volume of the system, which will cancel out later on, and $\vartheta^*(\mathbf{k},t)$ is the complex conjugate of the Fourier transform of ϑ. When $p(\mathbf{k},t)$ and $\vartheta^*(\mathbf{k},t)$ are substituted from (2) and (4), $\langle p\partial\vartheta/\partial x_i\rangle$ is found to consist of four components. If these are denoted by $\Phi_{i\vartheta j}$ so that

$$\Phi_{i\vartheta} \equiv \rho^{-1}\langle p\vartheta,_i\rangle = \Phi_{i\vartheta 1} + \Phi_{i\vartheta 2} + \Phi_{i\vartheta 3} + \Phi_{i\vartheta 4} \tag{7}$$

then

$$\Phi_{i\vartheta 1} = -\frac{(2\pi)^3}{V}\int ik_i < N(\mathbf{k},t)[G_0^*(\mathbf{k},t)\,\vartheta^*(\mathbf{k},0) - \int_0^t dt_1 G_0^*(\mathbf{k},t-t_1)(u_m\vartheta,_m)^*_{\mathbf{k},t_1}] > d\mathbf{k} \tag{8}$$

$$\Phi_{i\vartheta 2} = \frac{(2\pi)^3}{V}\int_0^t dt_1 \int ik_i < N(\mathbf{k},t)G_0^*(\mathbf{k},t-t_1)u_2^*(\mathbf{k},t_1) > T,_2 d\mathbf{k} \tag{9}$$

$$\Phi_{i\vartheta 3} = -\frac{(2\pi)^3}{V}\int ik_i < (2ik_1/k^2)u_2(\mathbf{k},t)[G_0^*(\mathbf{k},t)\vartheta^*(\mathbf{k},0) - \int_0^t dt_1 G_0^*(\mathbf{k},t-t_1)(u_m\vartheta,_m)^*_{\mathbf{k},t_1}] > U_{1',2} d\mathbf{k} \tag{10}$$

$$\Phi_{i\vartheta 4} = \frac{(2\pi)^3}{V}\int_0^t dt_1 \int ik_i < (2ik_1/k^2)u_2(\mathbf{k},t)u_2^*(\mathbf{k},t_1) G_0^*(\mathbf{k},t-t_1) > T,_2 U_{1},_2 d\mathbf{k} \tag{11}$$

This identification of the four contributions to $\langle p\vartheta,_i\rangle$ was a principal result in DG1. In that paper an approximate expression was found for the scalar-gradient term, $\Phi_{i\vartheta 2}$, in terms of measureable quantities. Now the turbulence component, or "slow" part, $\Phi_{i\vartheta 1}$, is to be treated.

2.2 $\phi_{i\vartheta 1}$ in Isotropic Turbulence

For time separations, t, much larger than the integral time scale of the turbulence, the two-time tensor $\langle N(\mathbf{k}_0,t)\vartheta^*(\mathbf{k}_0,0)\rangle$ in equation (8) vanishes, and

$$\Phi_{i\vartheta 1} = \frac{(2\pi)^3}{V}\iint ik_{0i} \langle N(\mathbf{k}_0,t)(u_m\vartheta,_m)^*_{\mathbf{k}_0,t_1}\rangle \; G_0^*(\mathbf{k}_0,t-t_1)\,d\mathbf{k}_0 dt_1 \tag{12}$$

where $\mathbf{k}_0$ has been substituted for $\mathbf{k}$ to avoid confusion later, and the G_0 function may be taken out of the angle brackets because it is explicit with respect to time. $N(\mathbf{k}_0,t)$ is given by (3), and the Fourier transform of $u_m\vartheta,_m$ is:

$$(u_m\vartheta,_m)_{\mathbf{ko}} = \int ik_{0m}u_m(\mathbf{k}_2)\vartheta(\mathbf{k}_0 - \mathbf{k}_2)d\mathbf{k}_2 \tag{13}$$

Equations (3) and (13) are now substituted in (12), and the statistical average of $N(\mathbf{k}_0,t)(u_m\vartheta,_m)^*_{\mathbf{k}0,t2}$ formed. Let this quantity be $\langle N\rangle$. Then:

$$\langle N\rangle = \iint \frac{ik_{0m}k_{0n}k_{0j}}{k_0^2} \langle u_n(\mathbf{k}_1,t)u_j(\mathbf{k}_0 - \mathbf{k}_1,t)u_m^*(\mathbf{k}_2,t_1) \; \vartheta^*(\mathbf{k}_0 - \mathbf{k}_2,t_1)\rangle d\mathbf{k}_1 d\mathbf{k}_2 \tag{14}$$

The two-time third-order tensor is expanded in cumulants thus:

$$\begin{aligned}
&\langle u_n(\mathbf{k}_1,t)u_j(\mathbf{k}_0 - \mathbf{k}_1,t)u_m^*(\mathbf{k}_2,t_1)\vartheta^*(\mathbf{k}_0 - \mathbf{k}_2,t_1)\rangle \\
&\quad = \langle u_n(\mathbf{k}_1,t)u_m^*(\mathbf{k}_2,t_1)\rangle\langle u_j(\mathbf{k}_0 - \mathbf{k}_1,t)\vartheta^*(\mathbf{k}_0 - \mathbf{k}_2,t_1)\rangle \\
&\quad + \langle u_n(\mathbf{k}_1,t)\vartheta^*(\mathbf{k}_0 - \mathbf{k}_2,t_1)\rangle\langle u_j(\mathbf{k}_0 - \mathbf{k}_1,t)u_m^*(\mathbf{k}_2,t_1)\rangle \\
&\quad + \langle u_n(\mathbf{k}_1,t)u_j(\mathbf{k}_0 - \mathbf{k}_1,t)\rangle\langle u_m^*(\mathbf{k}_2,t_1)\vartheta^*(\mathbf{k}_0 - \mathbf{k}_2,t_1)\rangle \\
&\quad + Q(t,t_1)
\end{aligned} \tag{15}$$

where the cumulant of the correlation on the right-hand side, $Q(t, t_1)$, is small for large time separations $t - t_1$. Now it is assumed that $Q(t, t_1)$ is zero for all time separations as, for example, in direct-interaction theory applied to the convection of a scalar. The discard of $Q(t, t_1)$ is precisely analogous to the fourth-order cumulant discard approximation previously made in DG1 in the calculation of $\phi_{i\vartheta 2}$. There it was also shown that the third term on the right-hand side of the equivalent expression to (15) is zero unless $\mathbf{k_0}$ is zero, and that it could be also discarded in the subsequent analysis.

The two-time velocity correlations are expressed in terms of the two-time energy-spectrum tensor $S_{ij}(\mathbf{k_1},t_1,t_2)$ [equations 27 & 28 of DG1]. Now the two-time $u_i\vartheta$ correlations are expressed similarly as:

$$< u_i(\mathbf{k_1},t_1)\vartheta^*(\mathbf{k_2},t_2) > = S_{i\vartheta}(\mathbf{k_1},t_1,t_2)\delta(\mathbf{k_1}-\mathbf{k_2}) \quad (16)$$

$$< u_i(\mathbf{k_1},t_1)\vartheta^*(\mathbf{k_1},t_2) > = (2\pi)^{-3}\, VS_{i\vartheta}(\mathbf{k_1},t_1,t_2) \quad (17)$$

where $S_{i\vartheta}(\mathbf{k_1},t_1,t_2)$ is the two-time scalar-flux spectrum function

$$\int S_{i\vartheta}(\mathbf{k_1},t_1,t_2)dk_1 = < u_i(\mathbf{r},t_1)\vartheta(\mathbf{r},t_2) > \quad (18)$$

Substitution in (14) then produces

$$< N > = \frac{2V}{(2\pi)^3}\int \frac{ik_{0m}k_{0n}k_{0j}}{k_0^2} S_{nm}(\mathbf{k_1},t,t_1)S_{j\vartheta}(\mathbf{k_0}-\mathbf{k_1},t,t_1)d\mathbf{k_1} \quad (19)$$

$S_{nm}(\mathbf{k}, t, t_1)$ is expressed in terms of the single-time energy-spectrum tensor $S_{nm}(\mathbf{k}, t)$ by making use of a result of Weinstock (1976) (also Weinstock 1981 Appendix A):

$$S_{ij}(\mathbf{k}, t, t_1) = S_{ij}(\mathbf{k}, t) \exp[-0.5k_ik_j <u_iu_j> (t - t_1)^2 - i\mathbf{kU}(t - t_1)] \quad (20)$$

for $t - t_1 \leq \tau_E$, where τ_E is defined by

$$S_{ij}(\mathbf{k}, t, t-\tau_E) = e^{-1}S_{ij}(\mathbf{k}, t) \tag{21}$$

and **U** is the mean velocity. For isotropic turbulence:

$$S_{ij}(\mathbf{k}, t, t_1) = S_{ij}(\mathbf{k}, t)\exp[-0.5k^2\upsilon_0^2(t - t_1)^2 - i\mathbf{k}\mathbf{U}(t - t_1)] \tag{22}$$

where $\upsilon_0^2 \equiv q^2/3$. $S_{j\vartheta}$ is treated similarly so that, from (12)

$$\Phi_{i\vartheta 1} = -2\int\int d\mathbf{k}_0 d\mathbf{k}_1 \frac{k_{0m}k_{0n}k_{0j}k_{0i}}{k_0^2} S_{nm}(\mathbf{k}_1, t) S_{j\vartheta}(\mathbf{k}_0 - \mathbf{k}_1, t)$$
$$\int_0^t dt_1 \exp\{-0.5[k_1^2 + (k_0 - k_1)^2]\upsilon_0^2(t - t_1)^2\} \tag{23}$$

The time integration is performed with respect to t_1 for large t and, with $\mathbf{k}_0 = \mathbf{k}_1 + \mathbf{k}_2$, the incompressibility condition is used to obtain:

$$\Phi_{i\vartheta 1} = -\sqrt{2\pi}\int\int d\mathbf{k}_1 d\mathbf{k}_2 \frac{k_{2m}k_{2n}k_{1j}(k_{1i} + k_{2i})}{\upsilon_0(\mathbf{k}_1 + \mathbf{k}_2)^2(k_1^2 + k_2^2)^{1/2}} S_{nm}(\mathbf{k}_1) S_{j\vartheta}(\mathbf{k}_2) \tag{24}$$

In DG1 we were able to deal with the equivalent expression for $\Phi_{i\vartheta 2}$ by expanding it in powers of the anisotropy. The spectrum and stress tensors were split into an isotropic part and an anisotropic deviation. This step is feasible for S_{nm} in (24), but not for $S_{j\vartheta}$. Consequently, the only way to proceed is to restrict the analysis to the case of isotropic turbulence for which (Batchelor 1953)

$$S_{ij}(\mathbf{k}) = (\delta_{ij} - k_i k_j/k^2)\, E(k)/4\pi k^2 \tag{25}$$

The main contribution to the k_1 and k_2 integrations in equation (24) come from $k_1 \approx k_2$, i.e., from the energy-containing region in the vicinity of which $E(k_1)$ and $S_{j\vartheta}(\mathbf{k}_2)$ have maximum values. It is assumed that these maxima of $E(k_1)$ and $S_{j\vartheta}(\mathbf{k}_2)$ occur at approximately the same wavenumbers. According to Weinstock (1981) [paragraph 7(d) and Appendix C] this assumption, while greatly facilitating the analysis, introduces errors of about 3% only.

The resulting integral is zero unless $i = j$, and when dT/dx_2 is the only non-zero gradient, $j = 2$. Further manipulation produces the final expression for $\Phi_{2\vartheta 1}$:

$$\Phi_{2\vartheta 1} = -\sqrt{2\pi}\iint dk_1 dk_2 \frac{k_1^2 k_2^2}{(k_1^2 + k_2^2)^{3/2}} \frac{E(k_1)}{3\upsilon_0} E_{2\vartheta}(k_2) \tag{26}$$

in which the scalar spectrum $E_{2\vartheta}(k_2)$ is the integral of $S_{2\vartheta}(\mathbf{k_2})$ over a spherical shell and

$$\int E_{2\vartheta}(k_2)dk_2 = \int S_{2\vartheta}(\mathbf{k}_2)d\mathbf{k}_2 = \langle u_2\vartheta \rangle \tag{27}$$

Comparison of (26) with Monin's approximation (1) then produces

$$C_{\vartheta 1} = \sqrt{\frac{\pi}{2}}\frac{q^2}{\varepsilon}\iint dk_1 dk_2 \frac{k_1^2 k_2^2}{(k_1^2 + k_2^2)^{3/2}} \frac{E(k_1)}{3\upsilon_0} \frac{E_{2\vartheta}(k_2)}{\langle u_2\vartheta \rangle} \tag{28}$$

2.3 Calculation of $C_{\vartheta 1}$

To calculate $C_{\vartheta 1}$ from (28) expressions are needed for the spectra $E(k_1)$ and $E_{2\vartheta}(k_2)$. For $E(k_1)$ we use the simple two-range model of Comte-Bellot and Corrsin (1966). In the permanent, low-wave-number part of the spectrum $E(k_1)$ varies as $k_1{}^m$, and as $k_1{}^{-5/3}$ in the inertial subrange. The two parts of the spectrum intersect at wavenumber k_L, which is roughly the reciprocal of the integral scale, and where $E(k_1)$ is a maximum. The two-range spectrum is

$$\begin{aligned} E(k_1) &= A\, k_1{}^m && \text{for } 0 < k_1 \le k_L \\ E(k_1) &= \alpha\, \varepsilon^{2/3} k_1{}^{-5/3} && \text{for } k_1 \ge k_L \end{aligned} \tag{29}$$

where $A = \alpha\varepsilon^{2/3}(k_L{}^{-5/3-m})$ and the condition

$$\int_0^\infty E(k_1)dk_1 = \frac{1}{2}q^2 \tag{30}$$

is satisfied by

$$q^2 \equiv 3\upsilon_0^2 = \alpha\varepsilon^{2/3}k_L^{-2/3}\left[\frac{2}{(m+1)} + 3\right] \tag{31}$$

While enough is known about the energy spectrum for it to be modelled in this way with some confidence, the shape of $E_{2\vartheta}(k_2)$ is unknown, this quantity having received scant attention from theoreticians and the corresponding one-dimensional spectrum being difficult to measure. The assumptions of the last section allow a way round this difficulty, thus. If the peak value of $E_{2\vartheta}(k_2)$ also occurs at k_L, and the main contributions to the integrals are made from this region, it is permissible to set $k_1 \approx k_L$, $k_2 \approx k_L$ so that

$$\frac{k_1^2 k_2^2}{(k_1^2 + k_2^2)^{3/2}} \approx \frac{k_L}{2\sqrt{2}} \tag{32}$$

and to remove this factor from the integral. Then, from (28)

$$\begin{aligned} C_{\vartheta 1} &\approx \frac{\sqrt{\pi}}{12}\frac{q^2}{\varepsilon}\frac{k_L}{\upsilon_0 < u_2\vartheta >}\int_0^\infty E(k_1)dk_1 \int_0^\infty E_{u2\vartheta}(k_2)\,dk_2 \\ &\approx \frac{\sqrt{3\pi}}{24}\frac{q^3 k_L}{\varepsilon} \end{aligned} \tag{33}$$

q^3k_L/ε is found from equation (31) in which the value of m is fixed by the decay rate of isotropic turbulence. Thus if q^2 decays as t^{-n}, it is easily shown that $n = 2(m+1)/(m+3)$ and $n = 1.2$ when $m = 2$, in accordance with the grid-turbulence data analysed by Comte-Bellot and Corrsin (1966). With $\alpha \approx 1.5$, q^3k_L/ε is evaluated as 12.9 and consequently, from (33), $C_{\vartheta 1} \approx 1.65$.

A more precise calculation is theoretically possible but it would be dependent on the shapes of both the spectra. Equation (33) has been integrated numerically for a two-range model for $E_{2\vartheta}(k_2)$ similar to that for $E(k_1)$. It is assumed that $E_{2\vartheta}(k_2)$, like $E(k_1)$, varies as k_2^m in the low-wavenumber range, and as $k_2^{-\gamma}$ for

$k_2 > k_L$. For m = 2, the integration produces $C_{\vartheta 1} \approx 3.33$, 1.82 and 1.46 for $\gamma = 5/3$, 7/3 and 9/3 respectively. It may be supposed that $E_{2\vartheta}(k_2)$ decreases faster than $k^{-5/3}$ for $k_2 > k_L$, as has been found in measurements of the shear-stress spectrum, which is mildly analogous. If this is the case, then a value of $C_{\vartheta 1}$ in the lower part of the range above would be appropriate. In the light of the approximate calculation made above, it appears that $C_{\vartheta 1} \approx 1.7$.

3. Measurements

Discrepancies in the published data (by Sirivat and Warhaft 1983 and Budwig, Tavoularis and Corrsin 1985) suggested that there was scope for further experimentation. New measurements are reported by Dakos (1990). These have been made in the 450 x 450 mm^2 section of an open-circuit blower wind tunnel. Nearly-isotropic turbulence was generated by a conventional biplane grid made of 5 mm diameter rods set at 25 mm centres. Objectives were to obtain data at higher Reynolds numbers than had previously been reached, and at distances further from the grid. The measurements were taken at up to 220 mesh lengths (220M) from the grid (previously 120 and 150) and with U_1 = 15 m/s, $R_\lambda \approx 40$ (previously around 25). The turbulent energy decayed as x_1^{-n} with $n \approx 1.28$ and, at x_1 = 170M, $\langle u_1^2\rangle/\langle u_2^2\rangle \approx 1.17$, a typical figure for grid turbulence. The transverse temperature gradient, dT/dx_2 = 4 ^oC/m, was imposed by an array of electrically-heated 0.127 mm Nichrome wires spaced at 25 mm and located at x_1 = 40M (in Warhaft and Lumley's terminology, a "mandoline").

The development of the flow is described by the following equations written in terms of dimensionless flow parameters

$$\frac{d}{d\xi}\frac{q^2}{U_1^2} = -\frac{2\varepsilon M}{U_1^3} \tag{34}$$

$$\frac{d}{d\xi}\left(R_{2\vartheta}\frac{u_2'}{U_1}\frac{\vartheta'}{\beta M}\right) = -\frac{\langle u_2^2\rangle}{U_1^2} - C_{\vartheta 1}\frac{2\varepsilon M}{q^2 U_1}\left(R_{2\vartheta}\frac{u_2'}{U}\frac{\vartheta'}{\beta M}\right) \tag{35}$$

where $\xi \equiv x_1/M$, $\beta \equiv dT/dx_2$ and $R_{2\vartheta} \equiv \langle u_2\vartheta\rangle/u_2'\vartheta'$. The Monin approximation (1) has been substituted for the pressure term in (35) and the viscous dissipation term has been discarded under the local-isotropy hypothesis. Other components of the pressure-temperature correlation contribute nothing in isotropic shear-free turbulence. Many of the more elaborate formulae for $C_{\vartheta 1}$ also reduce to (1) because b_{ij} and its invariants are zero in this case.

If $R_{2\vartheta}$ and $<\vartheta^2>/(\beta M)^2$ are assumed to vary as ξ^p and ξ^q respectively, then (35) requires $2p + q = 2 - n$ for $C_{\vartheta 1}$ to be independent of ξ. Similarly, the $<\vartheta^2>$ equation:

$$\frac{d}{d\xi}\left(\frac{<\vartheta^2>}{\beta^2 M^2}\right) = -2R_{2\vartheta}\frac{u_2'}{U_1}\frac{\vartheta'}{\beta M} - C_\vartheta \frac{2\varepsilon M}{q^2 U_1}\frac{<\vartheta^2>}{\beta^2 M^2} \tag{36}$$

requires that $q - 2p = 2 - n$. The conditions are satisfied for $p = 0$, $q = 2 - n$, a result which has also been obtained by Budwig et al (1985) by a different route. In the present experiment $(-R_{2\vartheta})$ reached a constant value of 0.65 for $\xi > 160$. In the range $140 < \xi < 210$ the increase in the temperature variance is closely fitted by

$$\frac{<\vartheta^2>}{\beta^2 M^2} = 0.014\left(\frac{x_1}{M}\right)^{0.8} \tag{37}$$

Then, for $p \approx 0$, $q \approx 0.8$, $n = 2 - 2p - q \approx 1.2$. This figure compares well with a "best-fit" value of $n \approx 1.28$. The transport term on the left-hand side of (35) contributes less than 5% to the $<u_2\vartheta>$ budget, which gives $C_{\vartheta 1} \approx 1.7$ at $\xi = 200$, in good agreement with the theoretical estimate.[2]

Shih and Lumley (1986) in modelling Sirivat and Warhaft's (1983) flow draw attention to the importance of the time-scale ratio $r \equiv (q^2/\varepsilon)/(<\vartheta^2>/\varepsilon_\vartheta)$. Their expression for $C_{\vartheta 1}$ ($0.5\phi^\theta$ in their notation) is

$$\phi^\theta = 2C_{\vartheta 1} = 1 + r + 1.1\, r^2 F^{1/2} \tag{38}$$

where $$F = (1-\rho^2)/(1-\rho^2/3)^3 \tag{39}$$

[2] Not 1.44 as reported by Dakos and Gibson (1987)

and ρ is the correlation coefficient, $R_{2\vartheta}$ heretofore. The present experimental values subsituted in (38) give values of $C_{\vartheta 1} \approx 1$.

4. Concluding Remarks

A theoretical calculation has been made of the slow part of the pressure-scalar-gradient correlation in isotropic turbulence with a transverse scalar gradient. With assumptions about the shapes of the turbulence spectra it has been possible to estimate as about 1.7 the constant in Monin's approximation for this term. This value agrees closely with the results of new measurements, but not, with previously published data. Attention has been drawn, briefly, to considerable discrepancies in the data available now from three sources. These discrepancies remain unexplained, and it is concluded that there is still scope for further experimentation and analysis in this field.

References

[1] Batchelor, G.K. 1953 *The Theory of Homogeneous Turbulence,* Cambridge University Press.

[2] Budwig, R., Tavoularis, S. & Corrsin, S. 1985 *Temperature fluctuations and heat flux in grid-generated isotropic turbulence with streamwise and transverse mean-temperature gradients*, J. Fluid Mech. **153,** 441.

[3] Comte-Bellot, G. and Corrsin, S. 1966 *The use of a contraction to improve the isotropy of grid-generated turbulence,* J. Fluid Mech. **25,** 657.

[4] Dakos, T. 1990 *Fundamental heat transfer studies in grid generated homogeneous turbulence,* PhD thesis, Imperial College.

[5] Dakos, T. and Gibson, M.M. 1987 *On modelling the pressure terms of the scalar flux equations,* Turbulent Shear Flows **5**, Springer, p7.

[6] Launder, B.E. 1976 *Heat and Mass Transport,* in Turbulence (ed. P. Bradshaw) Topics in Applied Physics **12**, Springer, p231.

[7] Lumley, J.L. 1978 *Computational modelling of turbulent flows,* Advances in Applied Mechanics **18,** 123.

[8] Monin, A.S. 1965 *On the symmetry properties of turbulence in the surface layer of air,* Isv. Atmos. Ocean. Phys. **I**, 45.

[9] Rotta, J.C. 1951 *Statistische theorie nichthomogener turbulenz I & II,* Z. Phys. **129,** 547 and **131,** 51.

[10] Shih, T-H. and Lumley J.L., *Influence of timescale ratio on scalar flux relaxation: modelling Sirivat and Warhaft's homogeneous passive scalar fluctuations*, J.Fluid Mech. **162**, 211.

[11] Sirivat, A. and Warhaft, Z. 1983 *The effect of a passive cross-stream temperature gradient on the evolution of temperature variance and heat flux in grid turbulence,* J. Fluid Mech. **128,** 323.

[12] Weinstock, J. 1976 *Lagrangian-Eulerian relation and the independence approximation,* Phys, Fluids **19**, 1702.

[13] Weinstock, J. 1981 *Theory of pressure-strain-rate correlation for Reynolds-stress turbulence closures, Part 1. Off-diagonal element,* J. Fluid Mech. **105,** 369.

[14] Weinstock, J. 1982 *Theory of the pressure-strain-rate. Part 2. Diagonal elements,* J. Fluid Mech. **116,** 1.

LONG-TIME BEHAVIOR OF MATERIAL-SURFACE CURVATURE IN ISOTROPIC TURBULENCE

S. S. Girimaji

Analytical Services & Materials Inc.
NASA Langley, Hampton, Viginia 23666

ABSTRACT

The behavior at large times of the curvature of material elements in turbulence is investigated using Lagrangian velocity-gradient time series obtained from Direct Numerical Simulations of isotropic turbulence. The main objectives are: to study the asymptotic behavior of the pdf curvature as a function of initial curvature and shape; and, to establish whether the curvature of an initially plane material element goes to a stationary probability distribution. The evidence available in the literature about the asymptotic curvature-pdf of initially flat surfaces is ambiguous and the conjecture is that it is *quasi*-stationary. In this work several material-element ensembles of different initial curvatures and shapes are studied. It is found that, at long times the moments of the logarithm of curvature are independent of the initial pdf of curvature. This, it is argued, supports the view that the curvature attains a stationary distribution at long times. It is also shown that, irrespective of initial shape or curvature, the shape of any material element at long times is cylindrical with a high probability.

1. Introduction

Material surfaces have received a lot of attention recently (Pope, Yeung and Girimaji [1], Yeung, Girimaji and Pope [2], Drummond and Munch [3]) for the important role they play in the scalar-mixing process in turbulent flows. Following Pope [4], the above works address the more tractable problem of the evolution of infinitesimal –

rather than finite-sized – material elements in turbulence. One important issue that has not been settled with any degree of certainity is whether or not the curvature of initially plane surfaces evolving in turbulence go to a stationary probability distribution. Pope *et al* [1] suggest that the probability density function (pdf) of the curvature at long times may be *quasi*-stationary: the main body of the pdf assuming a stationary form, and the tail of the pdf at large values of the curvature growing, perhaps indefinitely, in time.

In this work we attempt to shed more light on the long-time behavior of the curvature-pdf using Lagrangian velocity-derivative time-series obtained from direct numerical simulations (DNS) of turbulence (*a la* Pope *et al* [1]). The main problem with inferring stationarity from the calculations in a straight-forward manner is that, the onset of the asymptotic state coincides with statistical errors, in estimating the moments, becoming very large. It is unclear whether the trends observed in the moments at these large times are due to the physics of evolution or due to statistical errors. One obvious method of obtaining unambiguous statistics from such a simulation would be to have a very large number of material elements in the calculation. Rather than that, in this work we study the long time behavior of several material-surface ensembles each with a different initial curvature and shape. This approach is motivated by the following reason. If the curvature pdf is really quasi-stationary, then the non-stationary part of the pdf is most likely to be a function of the initial curvature and shape (see Section 2, equations (9) and (10) and below). Conversely if the asymptotic pdf is found to be independent of the initial conditions, it is very likely that the long-time pdf is stationary.

We consider several ensembles of material surfaces. The initial curvatures and shapes of all elements in any given ensemble are identical, but, different for different ensembles. The initial radii covered in this study range from infinity to a small fraction of the Kolmogorov length scale. The initial shapes considered range from spherical to pseudo-spherical.

In Section 2, we present some relevant background, the governing equations, and some details of the numerical methods employed. In Section 3, we present the results along with some discussion. We conclude in Section 4 with a brief summary.

2. Equations and Numerical Methods

Consider an incompressible, statistically-stationary, and isotropic, field of turbulence characterized by the Kolmogorov time-scale τ_η, the Kolmogorov length-scale η, and the Kolmogorov velocity-scale v_η. Throughout the rest of the paper all the quantities (e.g. velocity gradients, time, strain-rate etc.) are suitably non-dimensionalized using the Kolmogorov scales. Consider the evolution of infinitesimal surface elements in the turbulent field. At any time t, let $\mathbf{N}(t)$ be the normal vector. To study the evolution of curvature Pope [4] introduces a time-evolving Cartesian coordinate system following each fluid particle. The 3-direction of the coordinate axes coincides with the surface normal $\mathbf{N}$ at all times. The 1- and 2-axes are on the tangent plane and rotate with the surface element about the normal $\mathbf{N}$. Let $\mathbf{h}$ be the 2-dimensional second-order tensor of the curvature associated with the surface with instantaneous eigenvalues k_1 and k_2. The velocity vector and strain-rate tensor following the surface element are $\mathbf{U}(t)$ and $\mathbf{s}(t)$. The second derivative of velocity following the fluid particle is $U_{i,jk}$, where subscripts after the comma denote differentiation with respect to that spatial coordinate. The strain-rate experienced by the element in the tangential plane is,

$$a \equiv s_{11} + s_{22}. \tag{1}$$

We define two scalars to completely characterize the curvature tensor: the characteristic-curvature C; and, the shape-factor S. The characteristic-curvature is a measure of the magnitude of the curvatures associated with the surface and is defined as,

$$C \equiv (k_1^2 + k_2^2)^{\frac{1}{2}}. \tag{2}$$

The shape-factor is a quantitative measure of the shape of the surface element. It is defined as,

$$S = \frac{k_s}{k_l}, \tag{3}$$

where, k_s and k_l are the smaller and larger in magnitude of the eigenvalues of the curvature tensor, $\mathbf{h}$. The shape-factor by definition is confined to the range $[-1, +1]$ and its value indicates the shape of the surface element: $S = +1$ means that the two principal curvatures are equal in magnitude and sign, implying a spherical surface element; $S = -1$ implies a pseudo-spherical (saddle-shaped) surface element; and, $S = 0$ means that the element is cylindrical in shape. Intermediate values of the shape-factor represent combinations of the above shapes. The shape-factor of a flat surface is indeterminate.

In the time-dependent coordinate system defined earlier, the curvature tensor ($\mathbf{h}$) evolves according to ([4]),

$$\frac{dh_{\alpha\beta}}{dt} = U_{3,\alpha\beta} - ah_{\alpha\beta} - (s_{\gamma\beta}h_{\alpha\gamma} + s_{\gamma\alpha}h_{\beta\gamma}). \tag{4}$$

Here and elsewhere in the paper, the Greek subscripts refer to components along the tangential plane and hence can only take values of 1 or 2. The equation for characteristic curvature can be derived from the above equation:

$$\frac{dC}{dt} = b - dC, \tag{5}$$

where,

$$b(t) = U_{3,\alpha\beta}h_{\alpha\beta}/C, \tag{6}$$

and,

$$d(t) = a + 2s_{\alpha\beta}h_{\beta\gamma}h_{\gamma\alpha}/C^2. \tag{7}$$

In eq.(5) the b-term represents the bending of the surface by the second derivative of velocity. The d-term represents the stretching of the surface by the velocity derivative. It is important to note that, only the bending term can initiate curvature in an initially flat surface. In statistically stationary turbulence b and d are likely to be stationary random variables.

The solution to the curvature equation (5) can be written in the following integral form:

$$\begin{aligned} C(t) &= e^{-\int_0^t d(t')dt'} \times \int_0^t b(t')e^{\int_0^t d(t'')dt''}dt' \\ &+C_0e^{-\int_0^t d(t')dt'}. \end{aligned} \tag{8}$$

In turbulence b and d are not known functions of time and hence only a numerical integration of the curvature equation is possible.

We analyze the above solution (equation 8) for the case when b and d are constants, to illustrate some important features of the behavior of material-element curvature in turbulence. Since the bending term initiates the curvature on an initially plane surface and causes the curvature to go from zero to positive values, only positive values of b are physical for this analysis. Positive d corresponds to the situation when the surface is being stretched by the turbulence and negative d to the situation of surface compression. In isotropic turbulence although the mean tangential strain-rate , $\langle a\rangle$ (similarly $\langle d\rangle$), is positive ([2], [5]) negative values of a are also possible. So negative and positive values of d are permitted in this analysis.

When both b and d are positive, equation (8) implies,

$$\lim_{t\to\infty} C(t) = \lim_{t\to\infty}[\frac{b}{d}(1-e^{-d\,t}) + C_0 e^{-d\,t}] = \frac{b}{d}, \tag{9}$$

where, C_0 is the initial curvature. When b is positive and d negative, the curvature is,

$$C(t) = -\frac{b}{d}(e^{-d\,t} - 1) + C_0 e^{-d\,t}. \tag{10}$$

From solutions to the constant-coefficient curvature equation we draw the following important inferences about the behavior of curvature in turbulence:

1. If b and d are both bounded (it is generally accepted they are), and, if the effect of positive-d dominates, then the curvature is bounded in time. This means that the curvature statistics are either periodic in time or go to stationary values at long times. Since the former is unphysical, the curvature will have to go to a stationary distribution. Further, this asymptotic stationary distribution will be independent of the initial curvature, since the effect of the initial curvature decays exponentially in time. Conversely, if at long times the moments of the curvature are independent of the initial curvature, then we can say that the curvature is statistically stationary - asymptotically.

2. If the effect of negative-d dominates, the curvature will grow indefinitely in time, possibly at an exponential rate. In this case, the effect of initial curvature grows rather than decays in time. So, if the pdf of curvature is non-stationary, then this pdf is most likely to be a function of the initial curvature. In other words, if the long-time pdf of curvature is *quasi*-stationary, then the non-stationary part of the pdf will depend on the initial curvature.

The above two inferences motivate this study of material surfaces with different initial characteristic curvatures and shapes.

In this study, we integrate the curvature equation (4) with the Lagrangian time-series of the velocity field obtained from DNS data. We perform such calculations for several ensembles of elements. The initial pdf of characteristic curvature of each ensemble is a delta function but at different values of curvatures for different ensembles. We discuss the numerical methods involved further below in this Section and present our results in the next Section.

2.1 Numerical Methods

The calculations are performed in two parts. First, direct numerical simulation of turbulence is performed from which the Lagrangian time series of velocity and its derivatives are extracted. Then using the time series of the velocity and its derivatives, the curvature equations are integrated.

Direct Numerical Simulations. The DNS data are obtained from a realization of turbulence whose parameters are similar to the one used in Pope *et al* [1]. For the sake of completeness some important details of the direct numerical simulations are provided below.

Direct numerical simulation of an incompressible isotropic turbulence field is performed using a modified version of the pseudo-spectral code of Rogallo [6]. The turbulence is maintained stationary by adding energy at the low-wave numbers in a way that leaves the high wave-numbers relatively unaffected (Eswaran and Pope [7]). A uniform mesh of 64^3 nodes is used. The kinematic viscosity is chosen to yield a spatial resolution adequate for resolving velocity second-derivatives. The spatial resolution is characterized by $k_{max}\eta$ where k_{max} is the largest resolved wave-number. It was demonstrated in Pope *et al* [1] that a $k_{max}\eta$ value of 1.5 resolves the second derivatives with less than 6% error. A constant time step of $\Delta t = 0.052\tau_\eta$ corresponding to a Courant number of $\frac{1}{2}$ is used. The resulting Taylor-scale Reynolds number is approximately 38.

Starting from a specified initial spectra the turbulence is allowed to evolve until the Eulerian field becomes statistically stationary. Then at time $t = 0$ a total of $I = 8196$ randomly placed material elements are released. The velocity and the velocity first- and second-derivative time-series following the fluid particles are computed using the algorithm of Yeung and Pope [8]. The total length of the time series is approximately $110\tau_\eta$.

Calculation of Curvature. To each fluid particle in the ensemble we attach a notional surface element. The initial orientation of the surface element with respect to the velocity-gradient tensor is randomly chosen. Then knowing the velocity first- and second-derivatives following each element, the surface curvature equations (4) are integrated using second-order Runge-Kutta schemes to yield the components of the curvature tensor from which characteristic curvature (C) and shape-factor (S) are calculated.

The calculations of various ensembles differ only in the initial values of characteristic curvature and shape factor. The initial locations, orientations and the velocity field time-series are exactly the same for all the ensembles. For details on the direct numerical simulation refer to Yeung and Pope [8] and for the accuracy of surface curvature integrations refer to Pope *et al* [1].

3. Long-time behavior of surface elements

The two quantities that are of interest are the characteristic curvature (C) and shape-factor (S). In Table 1 we provide the initial characteristic curvature and shape of the different ensembles studied. The initial normalized- (by Kolmogorov length scale) characteristic curvature ranges from zero to about 35. All of the principal shapes (flat, cylindrical, spherical, and pseudo-spherical) are represented.

Statistics of Characteristic Curvature. The values that characteristic curvature can take with finite probability covers a very large range and hence for expediency we deal with the logarithm of curvature. In Figure 1, the evolution of the mean of the natural-logarithm of curvature of various ensembles are plotted. The evolution at early times are understandably different. But after a time of about $90\tau_\eta$, the means of all the ensembles appear to converge and settle down to a stationary value. It is very important to note that, the initial curvature of many ensembles considered are much larger (about 70 times for some cases) than the asymptotic mean value. So, the asymptotic mean value is approached from both lower and higher values of initial curvature. In Figure 2, the evolution of the second raw-moment of the logarithm of curvature of different ensembles are plotted. Again, at long times ($\approx 100\tau_\eta$) the moments apparently converge to a stationary value. The evolution of the standardized third, fourth, and fifth moments are plotted in Figures 3, 4 and 5, respectively. Again, irrespective of the initial characteristic curvature or shape, the moments all appear to converge to identical values for all ensembles. The asymptotic values of the standardized third, fourth, and fifth moments appear to be 1.3, 5, and 18, respectively.

Statistics of Shape-Factor. The evolution of the mean of the shape-factor of various ensembles are compared in Figure 6. The initial evolution is again different for different shapes. However after as little time as $25\tau_\eta$, the mean values of various ensembles are the same to within statistical error. The asymptotic mean value is about -0.1, which is in agreement with the value that Pope *et al* [1] obtained for initially flat surfaces. The evolution of higher raw-moments (second through fifth) of the shape factor are shown in Figures 7 – 10. Again the moments of the various ensembles converge to identical values at times beyond $25\tau_\eta$. From the behavior of the even moments it is clear that, the probability of the the asymptotic shape factor being near-zero is high. So we conclude that, the material elements at long-times are mostly cylindrical in shape.

Inference. The results of this Section show that at long times, the (first five) moments of the shape-factor and the logarithm curvature i) appear to asymptote to constant values, and, ii) the asymptotic state is independent of the initial characteristic curvature or shape of the surface elements. From the direct and the implied evidence it is fairly safe to infer that, the pdf of the logarithm of curvature – and hence of curvature itself – is independent of initial curvature. A word of caution is in place here. We have only examined the moments of the logarithm of curvature, which are not very sensitive to the tail of the curvature-pdf.

4. Summary

The objective of this article was to examine the behavior at long times of the pdf of the curvature of material elements in turbulence. Following Pope [4], we simplify the problem of finite-sized surfaces to one of studying infinitesimal surface elements. We use the Lagrangian velocity-field time-series obtained from direct numerical simulations of turbulence for our analysis. Several ensembles of material elements are considered. Each ensemble is characterized by a different initial characteristic curvature and shape.

The two main observations of this paper are:

1. At long times, the (first five) moments of the logarithm of the characteristic curvature of various ensembles (of different initial conditions) converge to the identical stationary values. From this we conclude that in isotropic turbulence, the asymptotic pdf of characteristic curvature of material surface elements go to a stationary form that is independent of initial shape or curvature.

2. At long times, the (first five) moments of the shape factor of various ensembles converge to identical values. Irrespective of whether the initial shape is spherical, pseudo-spherical or flat, the asymptotic shape of a material element is cylindrical with a high probability.

The evidence shown and the conclusions drawn in this paper are certainly no proofs that the asymptotic evolution of the curvature pdf is either stationary or independent of initial conditions. Our calculations however clearly indicate that such might be the case and make as powerful a claim as can be made by a statistical study.

Acknowledgements. This work was supported by U. S. Air Force Office of Scientific Research (grant number AFOSR-88-0052). Computations conducted during the research were performed on the Cornell National Supercomputer Facility, which is supported in part by the National Science Foundation, New York State, the IBM Corporation and the members of the Corporate Research Institute.

References

[1] Pope, S.B., Yeung, P.K. and Girimaji, S. S., Phys. Fluids A, **1** (12), 2010 (1989).

[2] Yeung, P. K., Girimaji, S. S. and Pope, S. B., Combust. & Flame, **79**, 340 (1990).

[3] Drummond, I.T. and Munch, W., Distortion of Line and Surface Elements in Turbulent Flows. Submitted to J. Fluid Mech.

[4] Pope, S.B., Int. J. Engng. Sci. **26**, 5, 445 (1988).

[5] Girimaji, S. S., and Pope, S.B., J. Fluid Mech. **220**, 427 (1990).

[6] Rogallo, R. S. *Numerical Experiments in Homogeneous Turbelence.* NASA TM 81315 (1981).

[7] Eswaran, V. and Pope, S. B., Comput. Fluids. **16**, 257 (1988).

[8] Yeung, P. K. and Pope, S. B., J. Comp. Phys. **79**, 373 (1988).

Table 1: Initial Conditions of various ensembles.

Ensemble Code	C_0	Shape Factor, S	Shape
$E1$	0.0	–	Flat
$E2$	0.687	1.0	Spherical
$E3$	0.687	0.0	Cylindrical
$E4$	6.87	1.0	Spherical
$E5$	6.87	0.0	Cylindrical
$E6$	34.3	-1.0	Pseudo-spherical
$E7$	34.3	0.0	Cylindrical
$E8$	34.3	1.0	Spherical
$E9$	34.3	0.0	Cylindrical

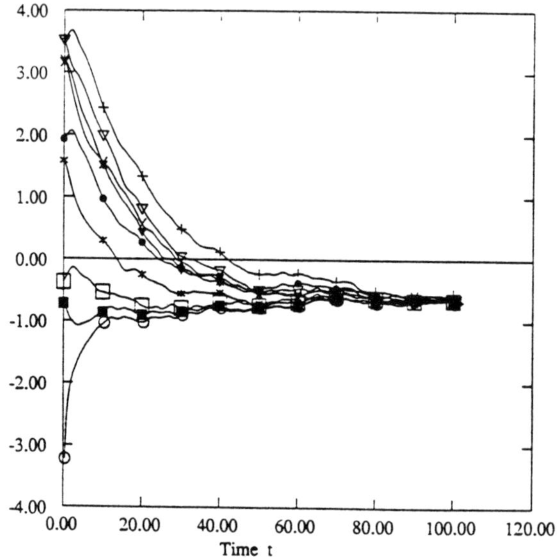

Figure 1: Evolution of mean ($\langle lnC \rangle$) of the logarithm of curvature. (○) – E1, (□) – E2, (□) – E3, (○) – E4, (⋆) – E5, (▽) – E6, (▽) – E7, (+) – E8, and (X) – E9.

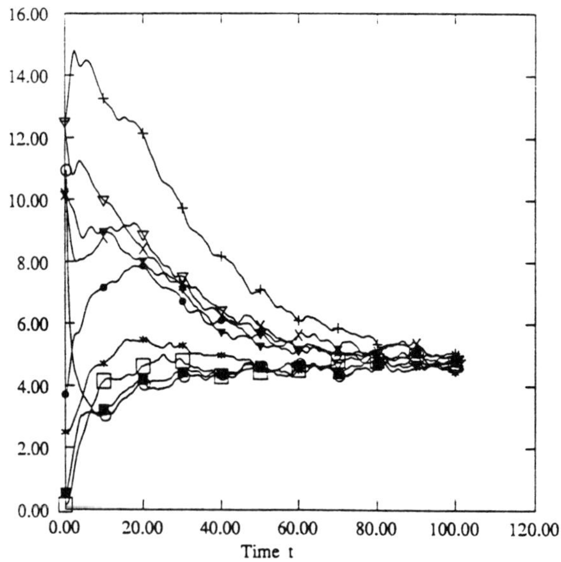

Figure 2: Evolution of $\langle (lnC)^2 \rangle$. Same legend as Figure 1.

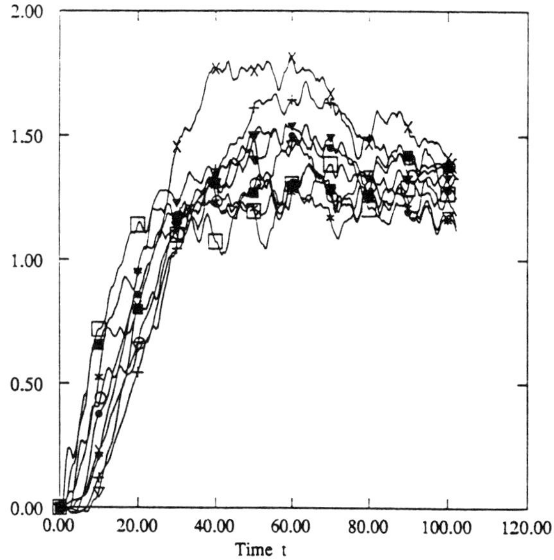

Figure 3: Evolution of the standardized Third Moment of logarithm of Curvature. Same legend as Figure 1.

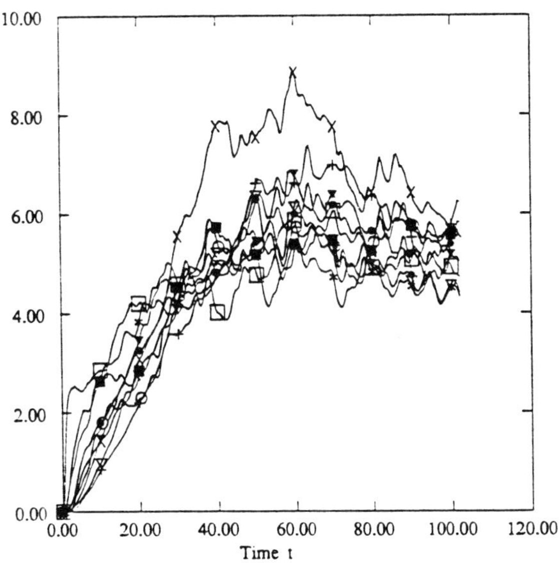

Figure 4: Evolution of the standardized Fourth Moment of logarithm of Curvature. Same legend as Figure 1.

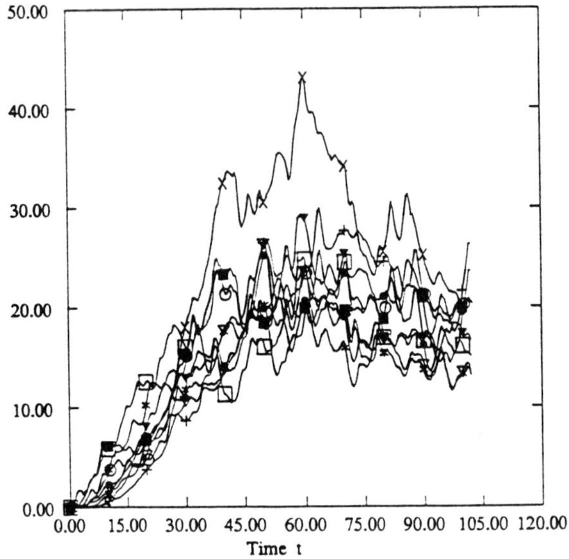

Figure 5: Evolution of the standardized Fifth Moment of logarithm of Curvature. Same legend as Figure 1.

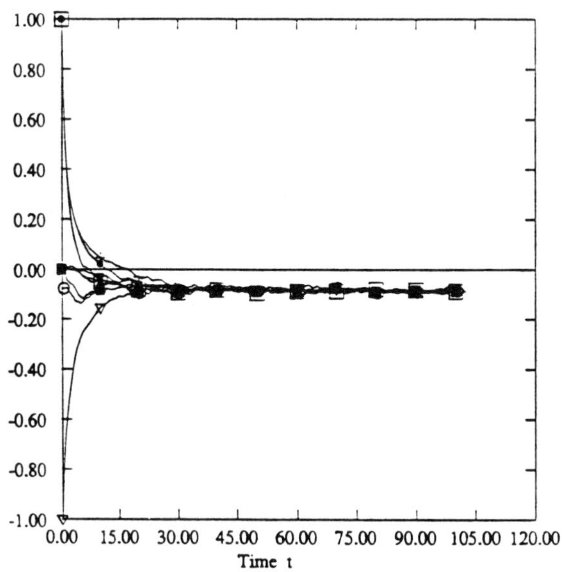

Figure 6: Evolution of the mean ($\langle S \rangle$) of the Shape factor. Same legend as Figure 1.

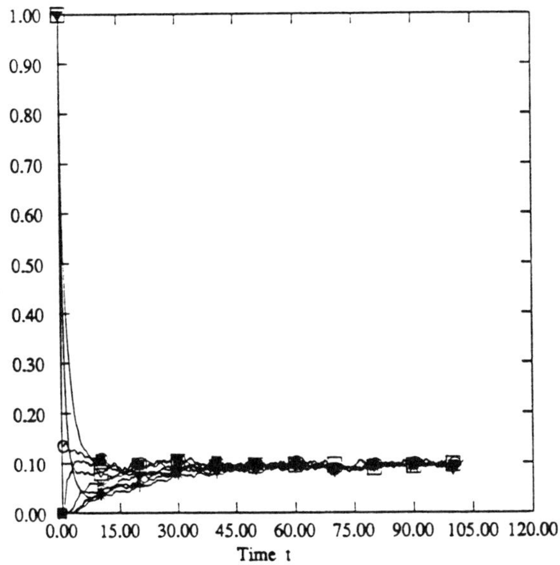

Figure 7: Evolution of $\langle S^2 \rangle$. Same legend as Figure 1.

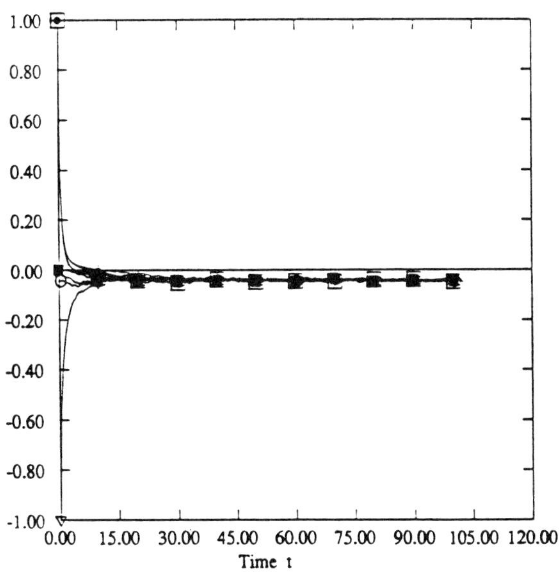

Figure 8: Evolution of $\langle S^3 \rangle$. Same legend as Figure 1.

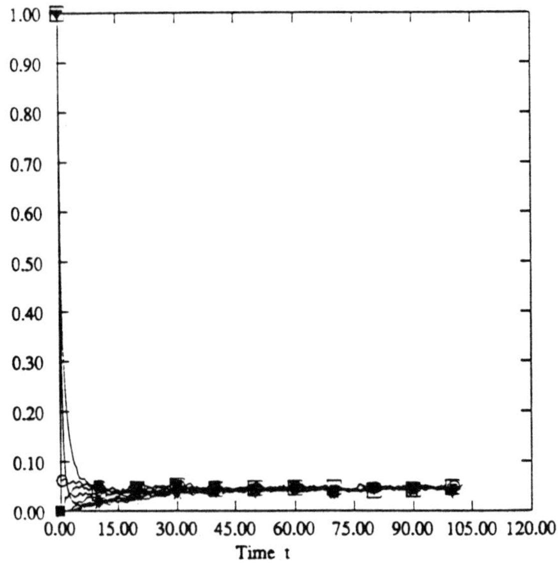

Figure 9: Evolution of $\langle S^4 \rangle$. Same legend as Figure 1.

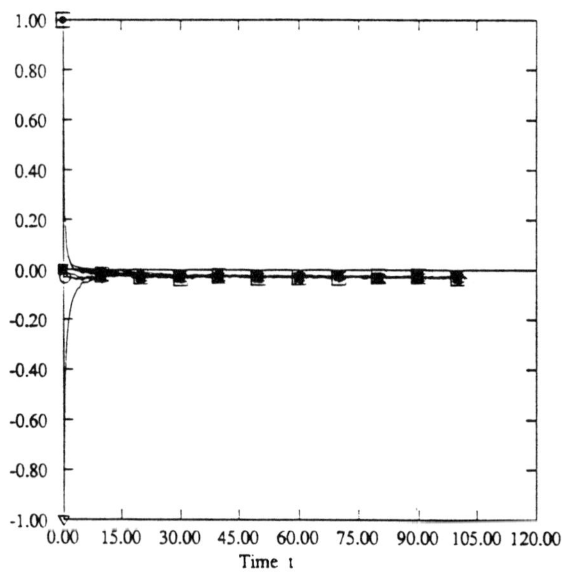

Figure 10: Evolution of $\langle S^5 \rangle$. Same legend as Figure 1.

MICROBUBBLE DRAG REDUCTION ON AN AXISYMMETRIC BODY UNDER AN APPLIED AXIAL PRESSURE GRADIENT

Steven Deutsch and Henry Clark III

The Applied Research Laboratory
The Pennsylvania State University
Post Office Box 30
State College, PA 16801

ABSTRACT

The influence of both a favorable and unfavorable axial pressure gradient on the reduction of skin friction drag on a 87 mm diameter microbubble body has been studied at the 12-inch diameter water tunnel at ARL Penn State. The unfavorable gradient separated the flow at approach speeds above 4.6 m/s. The gas flow rate at which separation ensued was roughly independent of tunnel speed. The favorable pressure gradient was found to inhibit the drag reduction mechanism substantially.

1. Introduction

It has been shown (Madavan, Deutsch and Merkle, 1984) that the injection of gas to form microbubbles in a liquid turbulent boundary layer can reduce the skin friction drag by as much as 80%. In a manner reminiscent of polymer drag reduction, bubbles are most effective when they are near the buffer region (Pal, Merkle and Deutsch, 1988). They appear to kill the turbulent production scale eddies there (Madavan, Deutsch and Merkle, 1985). The technique has been found to be relatively insensitive to the kinds of perturbations that may be simply constructed in the laboratory (e.g., the injection method or the addition of surfactant).

In the current experiment, we consider the microbubble phenomenon under the influence of both an adverse and a favorable axial pressure gradient in the hope of inducing a first order effect.

2. Approach

All experiments were run in the 12-inch diameter water tunnel at ARL Penn State. Experiments were performed on an 87 mm maximum diameter axisymmetric body. This axisymmetric body, which is shown in Figure 1, differs from the model described by Deutsch and Castano (1986) only in that it had a longer injection section (40 mm as opposed to 6.35 mm) and a shorter nose. The overall length of the body was preserved at 632 mm.

The body was equipped with a 273 mm long, isolated, cylindrical section which served as a skin friction force balance. By using an extremely stiff shear member as the force element, we were able to keep the gaps quite small (.127 mm and .254 mm at the upstream and downstream ends respectively). The balance section, which was 196 mm from the nose of the body, was knifed-edged and this together with the small gaps helped to reduce pressure effects on the force measurement. In addition to the force balance, a separate, interchangeable cylindrical section was instrumented with static pressure taps in order to test the axisymmetry and axial pressure gradient consistency as a function of gas injection. Fourteen taps were equally spaced at the top of the model, while eight taps were placed on the each of both sides and the bottom. Because of the way in which these pressure tap lines were brought out of the tunnel, the pressure gradient measurements could only be made for the zero pressure gradient case.

Two acrylic liners were used to create the axial pressure gradients. The liners were constructed of a clear acrylic, so that laser doppler measurements could be made through them. The liner which generated the favorable gradient tapered from a maximum radius of 152.3 mm at the nose of the body to a minimum radius of 135.9 mm at the downstream end. This liner resulted in a 28% increase in dynamic pressure in the streamwise direction. To produce the adverse gradient, the liner first narrowed the tunnel area to 86% of its initial open area and then gradually widened to a radius of 149.8 mm at the downstream end of the body. The minimum tunnel area occurred at a distance of 109.7 mm from the nose of the body. This second liner resulted in a 20% decrease in dynamic pressure over the region of adverse pressure gradient. Calculated dynamic pressures as a function of distance from the nose are shown in Figure 2. The location of the force balance is also shown in the figure.

3. Results

Some typical static pressure gradient results and their velocity equivalents are shown in Figure 3 as a function of gas volume flow rate. The results show that, although the scatter of the data is increased with gas injection, the flow remains axisymmetric and there is not an apparent change in the pressure gradient. These results support the static pressure measurements presented by Deutsch and Castano (1986), which were made on the tunnel wall. The fact that pressure gradient did not change with gas injection is important, in that it implies that the ratio of skin friction reduction should be accurate whether or not the absolute value of skin friction is measured accurately, by the balance, in a pressure gradient.

A Laser Doppler Velocimeter was used to measure velocity profiles at three locations on the balance section for each of the favorable, unfavorable and zero static pressure gradients. These measurements were made at 164 mm, 291 mm and 417 mm downstream of the trip wire at nominal tunnel speeds of 4.6 and 10.7 m/s. A typical inner variable plot is shown in Figure 4. Characteristic boundary layer parameters, for each of the pressure gradients, is given in Table 1 for both free stream speeds noted above. Although it is not possible to make these measurements while bubbles are being injected, the no injection measurements are quite useful. It is possible to calculate the local wall shear stress using a fit to the law of the wall and this data, when integrated, may be compared to the balance data. In general, we have found that the balance data and integrated velocity data agree to within better than 10%. This agreement gives us some additional confidence in the balance measurements.

Skin friction drag reduction ratio, which is defined as the skin friction drag with injection divided by the skin friction drag before injection, is plotted against gas volumetric flow rate in Figures 5-9. Nominal free stream velocities are 4.6, 7.6, 10.7, 13.7 and 16.8 m/s for Figures 5, 6, 7, 8, and 9, respectively. Each of the figures present the results for the zero, favorable and unfavorable gradients. The results presented in Figures 5-9, for the zero pressure gradient flow, are typical of those observed earlier by Deutsch and Castano (1986). At a freestream speed of 4.6 m/s, the flow is dominated by a double vortex structure that convects the bubbles out of the boundary layer, and allows for very little drag reduction. This phenomenon has been fully described by Deutsch and Pal (1990). Maximum drag reduction increases with increasing tunnel velocity through about 10.7 m/s, after which it begins to fall off.

Results for the case of the favorable pressure gradient are markedly different. Again there is no evidence of a double vortex structure at and above speeds of 7.6 m/s. The amount of drag reduction, however, is significantly less than that produced in the zero pressure gradient flow. In fact, over the entire speed range, the maximum drag reduction is somewhat less than 20%. For the case of the unfavorable gradient, we see approximately 60% maximum drag reduction compared to about 20% for the zero pressure gradient case at the lowest speed. At velocities of 7.6 m/s and above, the flow is clearly separated. The separation is indicated by both the balance measurement going to zero and by visual observation of the flow field. Gas flow rates just larger than $4x10^{-3}$ m^3/s induce the separation. This gas flow rate at separation seems to be fairly independent of tunnel speed.

4. Conclusions

The observation that injection of gas into a turbulent boundary layer induces separation is not an unexpected result. It does bolster the hypothesis that the presence of the bubbles in the boundary layer results in a decrease in the velocity gradient at the wall. The fact that the gas flow rate at which separation appears is not a strong function (if at all) of gas volume flow rate is significant but not well understood. The observation that a favorable streamwise gradient inhibits the drag reduction process is a surprising result. This is particularly true as the applied gradient is, by all measures, very weak. Further work in this area, which may well shed light on the physics of drag reduction and turbulent boundary layers under the influence of applied streamwise pressure gradients, would seem worthwhile.

Acknowledgements

This work was supported under Contract No. N00014-87-K-0196 under the direction of Mr. James A. Fein, Office of Naval Research.

References

[1]. Deutsch, S. and Castano, J.,"Microbubble Skin Friction Reduction on an Axisymmetric Body," Physics of Fluids, 29,11,3590,1986.

[2]. Deutsch, S. and Pal, S., "Local Shear Stress Measurements on an Axisymmetric Body in a Microbubble Boundary Layer," to appear in the December issue of the Physics of Fluids.

[3]. Madavan, N.K.,Deutsch, S. and Merkle, C.L.," Reduction of Turbulent Skin Friction by Microbubbles," Physics of Fluids, 27,2,356,1984.

[4]. Madavan,N.K., Deutsch, S. and Merkle, C.L.,"Measurements of Local Skin Friction in a Microbubble Modified Turbulent Boundary Layer," Journal of Fluid Mechanics, 156,237,1985.

[5]. Pal, S., Merkle, C.L., and Deutsch S.,"Bubble Characteristics and Trajectories in a Microbubble Boundary Layer," Physics of Fluids,31,4,744,1988.

Table 1. Characteristic boundary layer parameters for each pressure gradient

Location[+] (mm)	U_∞ (m/s)	δ (mm)	δ^* (mm)	θ (mm)	Re_θ	u* (m/s)
Zero Pressure Gradient						
164	4.94	4.67	.869	.560	2754	.190
164	11.03	5.44	.954	.642	7083	.389
291	4.80	6.59	1.09	.740	3545	.191
291	10.97	7.07	1.11	.793	8699	.402
417	4.84	7.93	1.26	.877	4242	.191
417	11.10	8.51	1.19	.879	9758	.412
Favorable Pressure Gradient						
164	5.17	4.37	.752	.504	3884	.243
164	12.35	4.63	.754	.520	6410	.463
291	5.59	5.86	.822	.587	3290	.233
291	13.00	5.74	.782	.580	7536	.500
417	5.89	6.65	.884	.659	3884	.243
417	13.70	6.40	.822	.607	8305	.522
Adverse Pressure Gradient						
164	5.84	4.71	.900	.610	3563	.217
164	13.25	4.63	.834	.572	7567	.464
291	5.50	6.68	1.12	.802	4406	.212
291	12.88	6.56	1.01	.736	9476	.465
417	5.22	8.49	1.35	.984	5142	.201
417	12.44	8.36	1.27	.932	11598	.443

*measured from trip wire

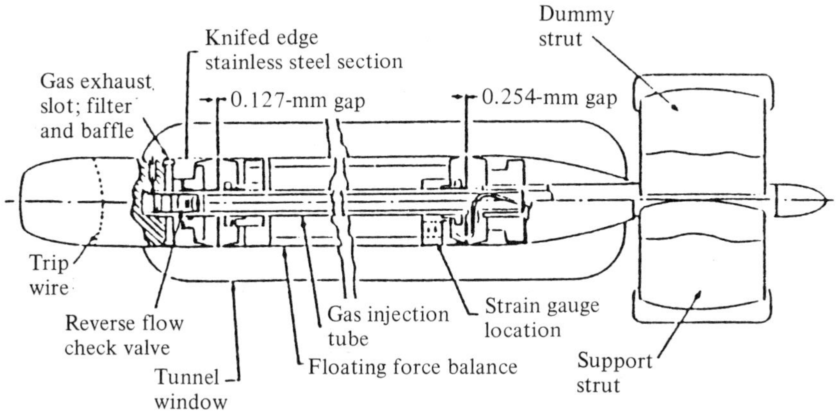

Figure 1. Schematic of the 87 mm diameter axisymmetric skin friction model.

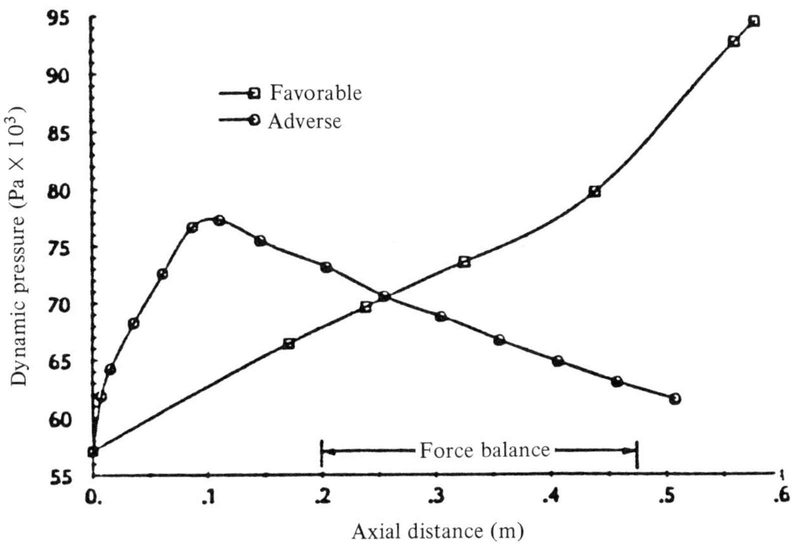

Figure 2. Dynamic pressure versus axial distance for favorable and adverse pressure gradients.

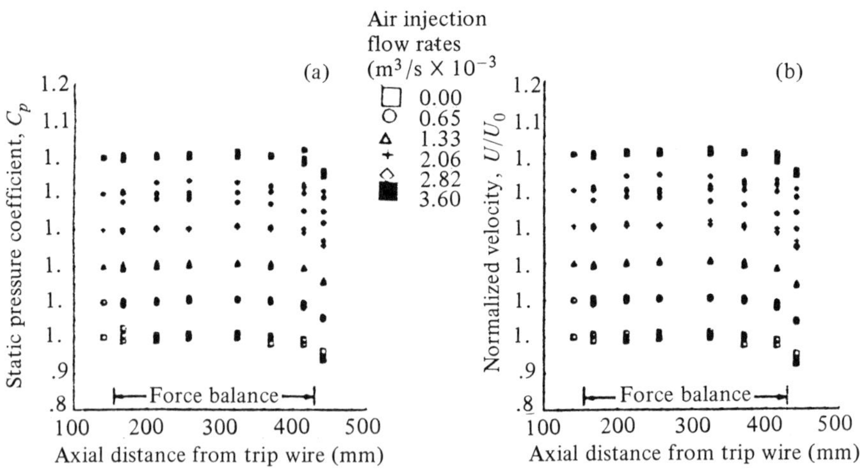

Figure 3. C_p (a) and U/U_0 (b) with air injection for $U_\infty = 16.8$ m/s. Note the shifted axes.

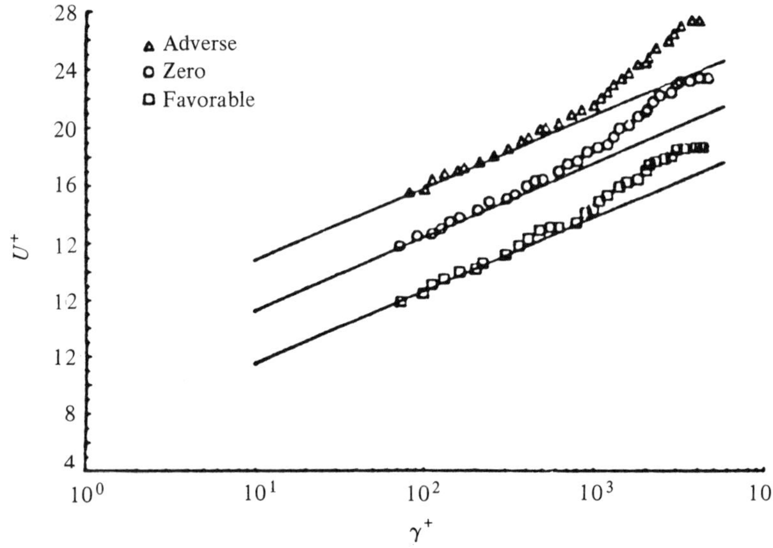

Figure 4. Normalized LDV profiles in inner variables 417 mm downstream from the trip wire at $U_\infty = 10.7$ m/s. Note the shifted axis for U_∞.

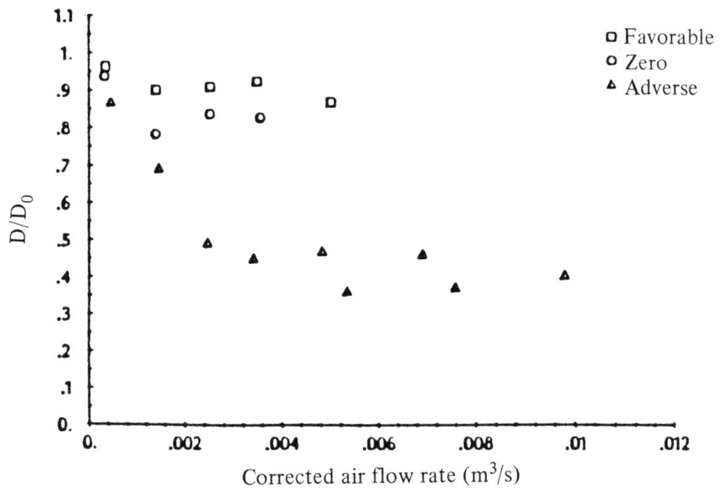

Figure 5. Drag ratio versus corrected air flow rate for each pressure gradient at $U_\infty = 4.6$ m/s.

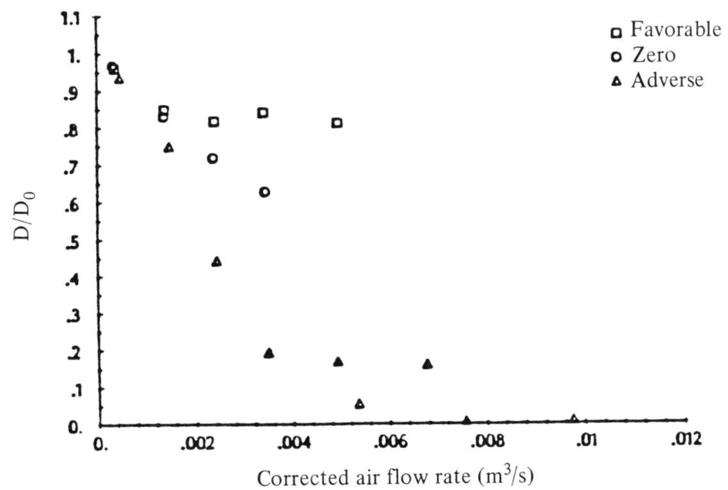

Figure 6. Drag ratio versus corrected air flow rate for each pressure gradient at $U_\infty = 7.6$ m/s.

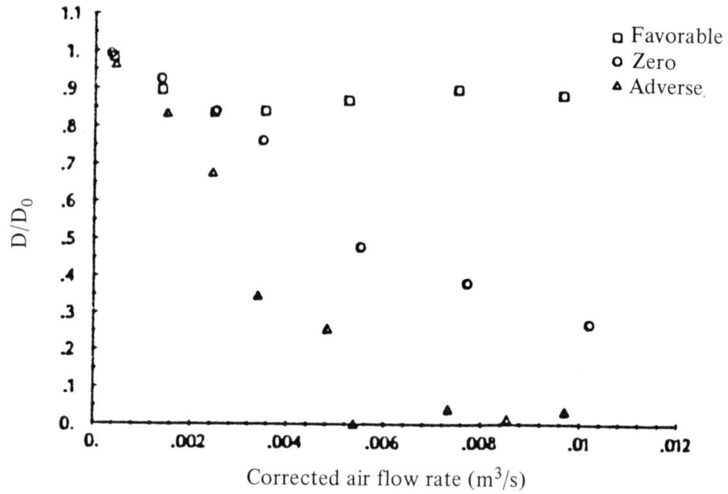

Figure 7. Drag ratio versus corrected air flow rate for each pressure gradient at $U_\infty = 10.7$ m/s.

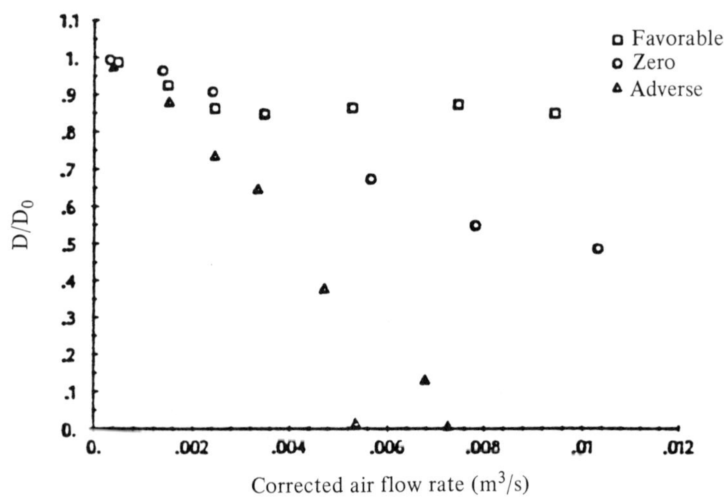

Figure 8. Drag ratio versus corrected air flow rate for each pressure gradient at $U_\infty = 13.7$ m/s.

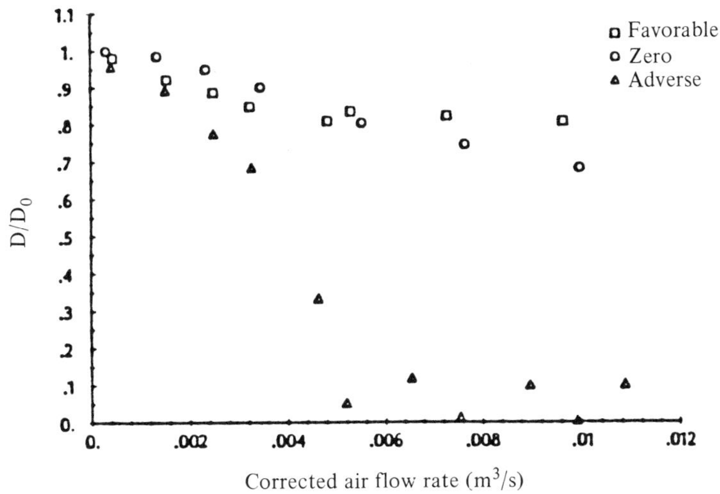

Figure 9. Drag ratio versus corrected air flow rate for each pressure gradient at $U_\infty = 16.8$ m/s.

TRILINEAR COORDINATES IN FLUID MECHANICS

Hassan Aref [1]
Institute of Geophysics and Planetary Physics
and
San Diego Supercomputer Center
University of California at San Diego
La Jolla, CA 92093

ABSTRACT

When in a physical problem three variable quantities add to a constant, it is possible to introduce a convenient geometrical representation of the variables in terms of trilinear coordinates. This usually permits an overview of the ranges of variation of the variables that is useful for qualitative analysis, and may point to quantitative insights that would be difficult to obtain in any other way. A brief historical discussion of trilinear coordinates is given, and examples of their use in solving flow problems are cited, including laminar flow in a pipe; the interaction of point vortices; and Lumley's representation of the eigenvalues of the anisotropy tensor for turbulent flow.

1. Introduction

There is a simple geometrical fact that is often useful in the graphical representation of data arising in physical problems. The geometry of the matter is this: Consider an equilateral triangle $A_1A_2A_3$ of height h (Fig.1). Let the distances of a variable point from the lines A_iA_j be denoted h_k, where ijk is any permutation of 123. Let the sign of h_k be positive or negative according to the

[1] Work supported by NSF/PYI Award MSM 84-51107.

location of the point and the direction of the arrows indicated in Fig.1. Thus, the point Q shown in Fig.1 has positive values of h_1 and h_2, but a negative value of h_3. All points inside the triangle $A_1A_2A_3$ have positive values of h_1, h_2 and h_3. The "simple fact" is the result that *for any point the sum of h_1, h_2 and h_3 equals h, the height of the equilateral triangle:*

$$h_1 + h_2 + h_3 = h. \tag{1}$$

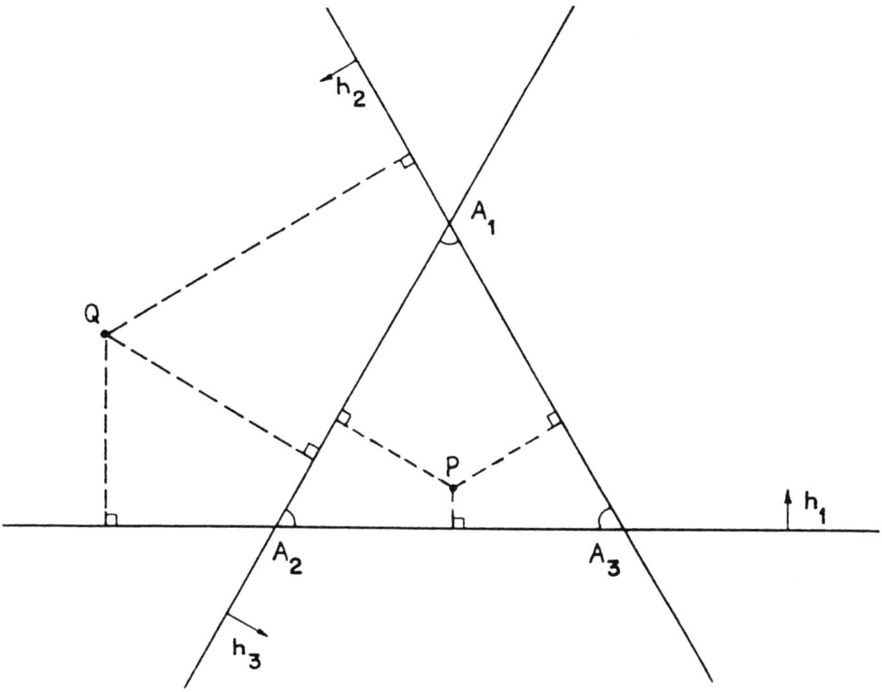

Figure 1: Definition of trilinear coordinates.

Several proofs may be given. Figure 1 provides an easy geometrical proof: The area of ΔA_iA_jP is $\frac{1}{2}$ s h_k ,where s is the common length of the three sides of the triangle (and, once again, ijk denotes a permutation of 123). The three areas of ΔA_1A_2P, ΔA_2A_3P and

ΔA_3A_1P sum to the area of $\Delta A_1A_2A_3$. Thus, For any location of P, $\frac{1}{2}$ s $(h_1+h_2+h_3)$ equals $\frac{1}{2}$ sh (the area of $\Delta A_1A_2A_3$). The argument is easily extended to the cases when one (or more) of h_1, h_2, h_3 is negative (such as point Q in Fig.1).

A simple algebraic argument arises by noting that $\nabla h_i = \mathbf{n}_i$, i=1,2,3, where the $\mathbf{n}_i$ are the unit normals indicated by the arrows in Fig.1. Then,

$$\nabla(h_1 + h_2 + h_3) = \mathbf{n}_1 + \mathbf{n}_2 + \mathbf{n}_3 = \mathbf{0} \tag{2}$$

by straightforward vector addition. Knowing that the sum of the h's is a constant, the value of the constant is easily found to equal h by letting P coincide with a vertex of the triangle.

If we describe the location of a variable point in the plane by the triple (h_1,h_2,h_3) rather than by a pair of cartesian coordinates or a set of polar coordinates, it is natural, in spite of the obvious redundancy, to think of the triple as a new set of coordinates, which may in fact sometimes be useful for writing more symmetrical expressions. When (h_1,h_2,h_3) are used in this way, we call them *trilinear coordinates*.

The reason for discussing this topic on this occasion is purely personal and has to do with one of my first meetings with John Lumley. John had just come to Cornell as Willis H. Carrier Professor of Engineering. I was a graduate student there in the Department of Physics. John had just completed the paper with G. R. Newman on the return to isotropy of homogeneous turbulence (see Lumley 1978) and had presented some of this material in his class on turbulence which I attended. The three eigenvalues of the anisotropy tensor sum to zero and thus constitute a limiting form of trilinear coordinates. In his paper Lumley (1978) called these eigenvalues b_1, b_2, b_3 and used them (see Fig.4 of Lumley, 1978) for a graphical representation of different turbulent flow states.

Meanwhile, I was working on the problem of three interacting point vortices and had discovered - as, unknown to me at the time, had Synge (1949) some 30 years earlier - that some of the integrals of motion in that problem could conveniently be taken into account by using trilinear coordinates (Aref 1979). By pure coincidence I had also called the variables in question b_1, b_2, b_3. On October 26, 1977 I gave a presentation of this work to a seminar on fluid mechanics and turbulence that was co-chaired by John and Mark Nelkin. As I recall both John and I were rather bemused at the coincidence of our chosen, somewhat unconventional, mathematical descriptions and their similar geometrical representations. For this reason I have always felt that I first met John Lumley in connection with trilinear coordinates! It therefore seemed most appropriate to trace their broader use in fluid mechanics in this presentation.

The subject, of course, is classical. A very clear and comprehensive exposition was given in the treatise of Ferrers, published in 1866. Modern books tend to mention trilinear coordinates only in passing. In fact, my restriction to an equilateral triangle is unnecessary. In any triangle we can consider the distances α, β, γ from the sides a, b, c (respectively). Then, it is an easy extension of the argument given above to prove that

$$a\alpha + b\beta + c\gamma = 2\Delta \tag{3}$$

where Δ is the area of the triangle. Sometimes $x_1 = a\alpha/2\Delta$, $x_2 = b\beta/2\Delta$, $x_3 = c\gamma/2\Delta$, are used in place of α, β, γ. These "areal coordinates" sum to unity.

Ferrers' (1866) treatise develops a number of results from which I quote without proof:

- The trilinear coordinates of the center of the inscribed circle all equal $2\Delta/(a + b + c)$.

- The trilinear coordinates of the centroid of the triangle are $(2\Delta/3a, 2\Delta/3b, 2\Delta/3c)$.
- The distance, r, between two points with trilinear coordinates $(\alpha_1,\beta_1,\gamma_1)$ and $(\alpha_2,\beta_2,\gamma_2)$ is given by:

$$r^2 = \frac{abc}{4\Delta^2} \times \{a \cos A\,(\alpha_1 - \alpha_2)^2 + b \cos B\,(\beta_1 - \beta_2)^2 + c \cos C\,(\gamma_1 - \gamma_2)^2\} \tag{4}$$

where A, B, C are the angles of the triangle. In particular, for an equilateral triangle:

$$r^2 = \frac{2}{3} \{(\alpha_1 - \alpha_2)^2 + (\beta_1 - \beta_2)^2 + (\gamma_1 - \gamma_2)^2\} \tag{5}$$

- Every straight line may be represented by an equation of the first degree in the trilinear coordinates. Conversely, every equation of the first degree represents a straight line.
- Every curve represented by an equation of the second degree in the trilinear coordinates is a conic section.
- The circle circumscribing a triangle is: $\frac{a}{\alpha} + \frac{b}{\beta} + \frac{c}{\gamma} = 0$.

2. Applications of Trilinear Coordinates

The mathematics of trilinear coordinates is both interesting and aesthetically appealing. The utility of trilinear coordinates arises from the essentially automatic representation of a "sum rule" that the geometrical representation allows. For example, in the theory of relativistic scattering of point particles, with two particles in both the initial and final state, S. Mandelstam in 1958 utilized a trilinear representation of certain combinations of momentum and energy to provide a very useful graphical representation, now usually referred to as the *Mandelstam variables* and the *Mandelstam plane*. A detailed discussion may be found in the text by Berestetskii, Lifshitz & Pitaevskii (1971).

An amusing application, almost in the vein of recreational mathematics but with a tenuous fluid mechanical connection, is discussed by Coxeter & Greitzer (1967) in their delightful book. They call it the "three-jug" problem and it concerns "the division of a liquid into stated portions with what appear to be inadequate measuring devices." For example: Two men have a vessel filled with 8 pints of beer, and two empty vessels with capacities of 5 pints and 3 pints, respectively. They wish to divide the eight pints of beer equally by pouring beer from vessel to vessel either filling or emptying (or both) a vessel each time. Is this possible? (The answer is "yes". One strategy is to fill the 5-pint vessel leading to the distribution 3:5:0 between 8-pint, 5-pint and 3-pint vessels. Next fill the 3-pint vessel from the 5-pint vessel leading to 3:2:3. The following steps are 6:2:0, 6:0:2, 1:5:2, 1:4:3 allowing the desired result 4:4:0 to be obtained. Figure 2, from Coxeter & Greitzer (1967), illustrates this solution and a different, eight-step solution.)

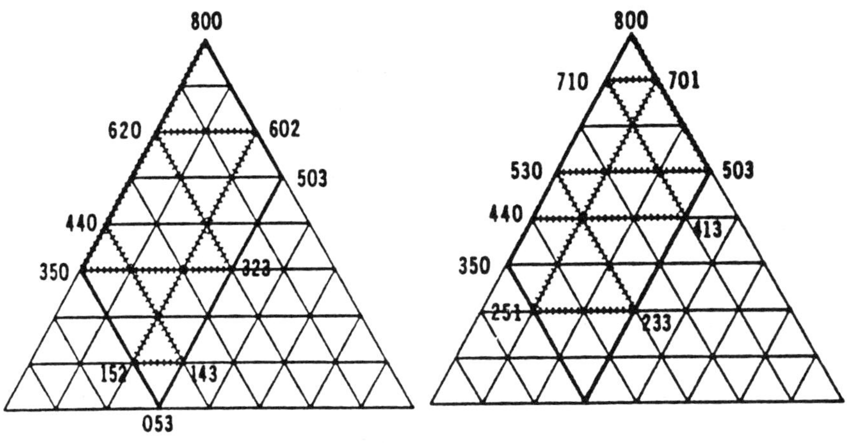

Figure 2: The "three-jug" problem. Trilinear coordinate representations of two solutions to the problem of dividing 8 pints equally using vessels with volumes 8, 5 and 3 pints.

I can do no better than cite a few lines from their description: "As a welcome relief from the ordinary squared paper, used for plotting points with given Cartesian coordinates, one can sometimes buy 'triangulated' paper, ruled with three systems of parallel lines dividing the plane into a tessellation of small equilateral triangles. Such paper is convenient for plotting points that have given *trilinear* coordinates with respect to a (large) equilateral triangle... These coordinates are ideal for representing any situation in which three variable quantities have a constant sum... One such situation arises when h pints (or ounces) of a liquid are distributed into three vessels so that there are x pints in the first, y in the second, and z in the third. The operation of pouring liquid gradually from the second vessel into the third is represented by a motion of the point (x,y,z) along a line x=constant in the direction in which y decreases while z (correspondingly) increases. If each vessel can hold h pints, each coordinate can take any value from 0 to h, and we have the (trivial) problem in which the domain of operations is the whole triangular region... Of far greater interest is the problem where $h \geq a > b > c$. Now the three given vessels have capacities a, b, c, and the problem is to measure out a stated quantity d of liquid by repeatedly pouring from one vessel into another, either emptying the former or filling the latter (or possibly doing both things at once). The domain of operation...may be a (regular or irregular) hexagon... but may in special circumstances reduce to a pentagon, trapezoid, parallelogram or (as we have seen) the whole equilateral triangle." They go on to consider examples of measuring problems that can be solved, and some that cannot, depending on the relationships of domain and available paths. There are interesting connections to the problem of billiards inside the domain in question, or, equivalently, the trajectory of a ray of light bouncing off perfectly reflecting mirrors.

Vortex dynamics

Let me now come to the vortex dynamics application that I alluded to above. If instead of considering the cartesian coordinates in the plane of three point vortices, one uses the lengths of the sides in the triangle spanned by the vortices, d_{12}, d_{23}, d_{31}, and writes the equations of motion for these quantities, then a system of three ODEs emerges, that, except for the orientation of the triangle, contain only the sides, and which have two integrals:

$$\Gamma_1\Gamma_2\, d_{12}^2 + \Gamma_2\Gamma_3\, d_{23}^2 + \Gamma_3\Gamma_1\, d_{31}^2, \tag{6}$$

$$\Gamma_1\Gamma_2 \log d_{12} + \Gamma_2\Gamma_3 \log d_{23} + \Gamma_3\Gamma_1 \log d_{31}. \tag{7}$$

Here the quantities Γ_i are the circulations of the vortices. The first quantity, (6), is related to the angular momentum of the fluid, the second to the kinetic energy. Three ODEs with two integrals leads to an integrable system, so that the motion of three interacting point vortices is, in fact, integrable, a point that was already clear to Kirchhoff and others in the 1870's, and on which a considerable amount of work was done by Gröbli (1877; for a discussion of the interesting history see Aref, Rott & Thomann, 1990).

Trilinear coordinates may be introduced by using (6) with some rescaling to make the base triangle equilateral. Details are in the reference already given (Aref 1979) and for an equivalent approach in the much earlier but nearly forgotten paper by Synge (1949). I do not want to reiterate this material here, but simply show the diagram that results for the symmetrical case of three identical vortices (Fig.3). Here $b_1 \propto d_{23}^2$, etc., the circle marks the subdomain of the triangle that corresponds to physically realizable three-vortex configurations, and the curves marked by θ=values correspond to motions of the three-vortex system at a fixed value of the energy (7), i.e., they are level curves of the function $b_1 b_2 b_3$.

Let me remark that recent work by Eckhardt (1988) and Rott (1990) implies that also certain cases of the four-vortex problem can be analyzed using such trilinear plots. (This work, in turn, has an interesting precedent in the 1898 thesis of Goryachev, a student of Zhukovsky.) A basic result of these works can be summarized in the following way: Let the vortex positions be designated z_1, z_2, z_3, z_4 in the flow plane thought of as a complex plane, and let their circulations be Γ_1, Γ_2, Γ_3, Γ_4. In general, this system will have an integral, analogous to (6),

$$L = \Gamma_1\Gamma_2\, d_{12}^2 + \Gamma_2\Gamma_3\, d_{23}^2 + \Gamma_3\Gamma_1\, d_{31}^2 + \Gamma_1\Gamma_4\, d_{14}^2 + \Gamma_2\Gamma_4\, d_{24}^2 + \Gamma_3\Gamma_4\, d_{34}^2. \tag{8}$$

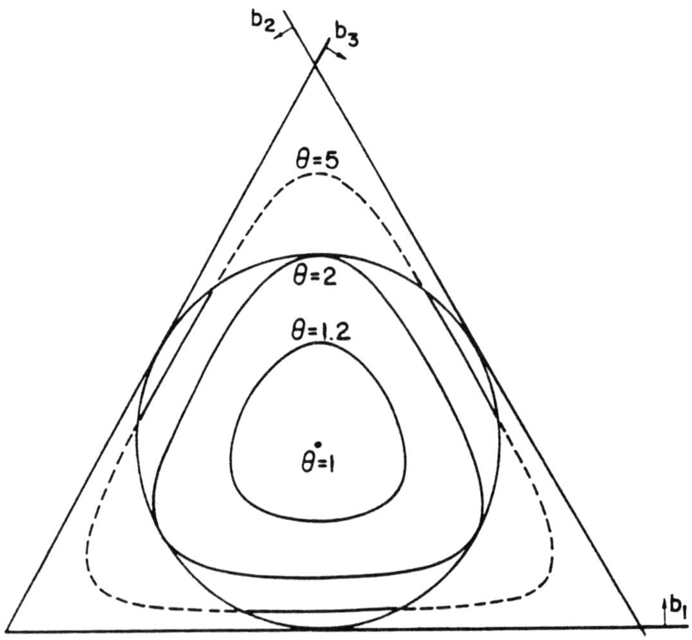

Figure 3: The "phase plane diagram" for the motion of three identical vortices (after Aref 1979). This diagram may also be interpreted as the equivelocity contours for laminar flow in a pipe with equilateral triangle cross-section.

However, if the sum of the circulations vanishes,

$$\Gamma_1 + \Gamma_2 + \Gamma_3 + \Gamma_4 = 0, \tag{9}$$

and the linear impulse vanishes,

$$\Gamma_1 z_1 + \Gamma_2 z_2 + \Gamma_3 z_3 + \Gamma_4 z_4 = 0, \tag{10}$$

then the four-vortex problem has the additional integrals:

$$L_1 = \Gamma_2\Gamma_3\, d_{23}^2 + \Gamma_2\Gamma_4\, d_{24}^2 + \Gamma_3\Gamma_4\, d_{34}^2, \tag{11a}$$
$$L_2 = \Gamma_3\Gamma_1\, d_{31}^2 + \Gamma_1\Gamma_4\, d_{14}^2 + \Gamma_3\Gamma_4\, d_{34}^2, \tag{11b}$$
$$L_3 = \Gamma_1\Gamma_2\, d_{12}^2 + \Gamma_1\Gamma_4\, d_{14}^2 + \Gamma_2\Gamma_4\, d_{24}^2, \tag{11c}$$
$$L_4 = \Gamma_1\Gamma_2\, d_{12}^2 + \Gamma_2\Gamma_3\, d_{23}^2 + \Gamma_3\Gamma_1\, d_{31}^2. \tag{11d}$$

Now, between four points in the plane there are *a priori* six distances. Between these there is one relation from Euclidean geometry. Another relation, the analogue of (7), arises from energy conservation. A third is given by (8). In general, we can expect no more, and the four-vortex problem is non-integrable. However, with the additional conditions (9) and (10), and the resulting integrals (11a-d), not all of which are independent from those just listed, integrability it re-instated.

We may now eliminate all distances that reference vortex four from the description using the first three of Eqs.(11). An integrable system of ODEs involving only the distances d_{12}, d_{23}, d_{31} results, and trilinear coordinates can be reintroduced using (11d).

Pipe flow

As my final example I return to the classical, viscous flow problem of Poiseuille but for a pipe with an equilateral triangle cross-section. In the preceding examples the choice of an equilateral triangle for representing the variables was a matter of convenience. In the present problem, however, it is the symmetry of the equilateral triangle that allows the simple solution that we shall obtain, which, so far as I am aware, does not generalize to other triangular cross-sections.

Steady, incompressible, viscous flow in a straight tube, extending along the z-axis, reduces to the solution of the PDE for the z-component of velocity, w(x,y), where x and y are the transverse coordinates. The balance between pressure gradient and viscous terms in the absence of fluid acceleration gives the problem

$$\nabla^2 w = -\Delta p/\eta L \tag{12}$$

where ∇^2 is the Laplacian in the xy-plane. The constant on the right hand side contains the pressure drop, Δp, over a length of pipe L, and the fluid viscosity, η. The boundary condition is that w must vanish at all points of the cross-section of the pipe (a curve in the xy-plane). Taylor (1937) discusses how this problem can be solved by suspending a soap film over the mouth of the pipe and applying a slight excess pressure to one side. The height of the film then solves (12) so long as the curvature is not too great.

The solution of (12) for a two-dimensional channel and a circular tube are, of course, very well known. In both cases w at a point P in the cross-section is proportional to the product of the distances of P from the two boundaries. Thus, for a pipe of circular cross-section we get the Poiseuille profile, which at radius r from the centerline is proportional to $(R + r)\,(R - r)$. This "distance-product rule" has very limited applicability (for example, it does not hold for a pipe of rectangular cross-section), but amusingly it does apply for a pipe with equilateral triangle cross-section. In such a pipe w is proportional to $h_1 h_2 h_3$ (using my original notation for the trilinear coordinates).

The calculation is straightforward:

$$\nabla(h_1h_2h_3) = \mathbf{n}_1h_2h_3 + h_1\mathbf{n}_2h_3 + h_1h_2\mathbf{n}_3$$

$$\nabla^2(h_1h_2h_3) = \nabla\bullet(\mathbf{n}_1h_2h_3+h_1\mathbf{n}_2h_3+h_1h_2\mathbf{n}_3) = 2\mathbf{n}_1\bullet\mathbf{n}_2h_3+ 2\mathbf{n}_2\bullet\mathbf{n}_3h_1+ 2\mathbf{n}_3\bullet\mathbf{n}_1h_2$$

In an equilateral triangle $\mathbf{n}_1\bullet\mathbf{n}_2=\mathbf{n}_2\bullet\mathbf{n}_3=\mathbf{n}_3\bullet\mathbf{n}_1=-\frac{1}{2}$ (see Fig.1). Hence, using (1),

$$\nabla^2(h_1h_2h_3) = -\,h. \tag{13}$$

Comparing this to (12) we see that

$$w = \frac{\Delta p}{\eta L h}\,h_1h_2h_3. \tag{14}$$

The boundary condition is clearly satisfied as well.

The velocity profile

$$w = \frac{\Delta p}{\eta L h}\,(h_1h_2h_3 - c) \tag{15}$$

gives the flow in a tube with boundary $h_1h_2h_3 = c$. As $c \to (h/3)^3$ we recover the Poiseuille flow solution in a tube of circular cross-section.

Figure 3 may now be viewed in an entirely different light. The pipe cross-section is the bounding triangle. The θ-curves are iso-axial-velocity contours (the circle is now irrelevant), or, alternately, a family of cross-sections within which laminar pipe-flow is given by (15). These curves (cubics in cartesian coordinates) would be height contours of a soap film made to span the triangular cross-section with a small excess pressure applied on one side (Taylor 1937)!

3. Conclusion

We are accustomed to the notion that adapting the coordinates to the geometry may greatly facilitate the solution of a problem. Thus, in the example of flow in a pipe with an equilateral triangle cross-section, the introduction of trilinear coordinates yields a particularly simple form of the solution.

In many problems where a "triangular geometry" does not enter, the constraint of three quantities adding to a fixed sum, nevertheless, suggests the introduction of a representation employing trilinear coordinates in some phase or parameter space. The advantage now is not in adapting the coordinates to the physical geometry, but rather the essentially automatic way in which the constraint is accommodated. This feature has already proven useful in several instances in fluid mechanics. Future applications are sure to arise.

Acknowledgements

Discussions with N. Rott, H. Thomann and V. V. Meleshko on the history and mathematics of few-vortex motion are gratefully acknowledged.

It is a pleasure to offer this paper to John Lumley for his 60'th birthday celebration.

References

[1] Aref, H. 1979 "Motion of three vortices." *Phys. Fluids* **22**, 393-400.
[2] Aref, H., Rott, N. & Thomann, H. 1990 "Gröbli's solution of the three-vortex problem." UCSD preprint.
[3] Berestetskii, V. B., Lifshitz, E. M. & Pitaevskii, L. P. 1971 *Relativistic quantum theory*. Part 1. Pergamon Press, Oxford, §68.
[4] Coxeter, H. S. M. & Greitzer, S. L. 1967 *Geometry revisited*. The Mathematical Association of America, Washington D. C., Sec. 4.6.
[5] Eckhardt, B. 1988 "Integrable four-vortex motion." *Phys. Fluids* **31**, 2796-2801.

[6] Ferrers, N. M. 1866 *An elementary treatise on tri-linear coordinates, the method of reciprocal polars and the theory of projections.* Second ed., Macmillan & Co., London, 182pp.

[7] Goryachev, D. N. 1898 "On some cases of the motion of rectilinear, parallel vortices." *Proc. Imperial Moscow University. Phys.-Math. series,* 106pp. (In Russian)

[8] Gröbli, W. 1877 *Specielle Probleme über die Bewegung geradliniger paralleler Wirbelfäden.* Zürcher und Furrer, Zürich. 86pp.

[9] Lumley, J. L. 1978 "Computational modeling of turbulent flows." *Adv. Appl. Mech.* **18**, 123-176.

[10] Rott, N. 1990 "Constrained three- and four-vortex problems." *Phys. Fluids A* **2**, 1477-1480.

[11] Synge, J. L. 1949 "On the motion of three vortices." *Can. J. Math.* **1**, 257-270.

[12] Taylor, G. I. 1937 "The determination of stresses by means of soap films." In *The Mechanical Properties of Fluids.* Blackie & Sons Ltd., London, 382pp.

AN ALTERNATIVE APPROACH TO DISTURBANCES IN BOUNDARY LAYERS

P.F. Easthope[1] *and W.O. Criminale*[2]

Department of Applied Mathematics
University of Washington
Seattle, Washington

ABSTRACT

By modelling the boundary layer on a flat plate as a piece-wise linear velocity profile it is possible to analyze disturbances in the flow in a systematic manner. The approach is that of an initial-value, boundary-value problem but, unlike classical normal modes employed in stability theory, the solutions here can be obtained in closed form and they are non separable. A specific example is treated where the vertical component of the perturbation velocity is prescribed as a localized pulse initially. The amplitude is then depicted as a function of time and the coordinates of the plane of the flat plate at a fixed vertical location. The role of three-dimensionality and the initial transient period of development - heretofore unknown - are both shown to be of significant importance in the dynamics. An argument is given to strongly suggest that the origin of large-scale oscillations known to be prevelant in turbulent shear flows may well be explained using these bases.

1. Introduction

Formally, the means of investigating perturbations in any shear flow is as an initial-value, boundary-value problem, but even the linear system does not lend itself to a tractable framework. The classical method of solution has been by use of a normal mode decomposition. This approach results in a non self-adjoint differential

---

[1] New address: Advanced System Architectures, Ltd, Alton, GU34 4PU, England.

[2] This work was supported in part by AFOSR Grant AFOSR-89-0404 and DOE Grant DE-FG06-88ER25061 A001.

equation that also requires singular perturbation techniques in boundary layers (cf. Lin 1955, Betchov and Criminale 1967, Drazin and Reid 1981). The result is an eigenvalue problem and the eigenvalues that are found are those of a discrete spectrum, thereby hardly sufficient to represent an arbitrary disturbance at zero time. There is also a continuous spectrum for this flow and this addition makes the problem complete. The work of Mack (1976) dealt directly with the calculations of both spectra for the case of the boundary layer. Gustavsson (1979) demonstrated the need to add the continuous spectrum.

Certainly, using the Laplace integral transform can be done with this problem and both eigen spectra are to be expected. As demonstrated by Gustavsson and Hultgren (1980) and Gustavsson (1979), the difficulty of the inversion adds to the ones already cited. Essentially, only implications, i.e. the substantiation of the continuous spectrum, can be ascertained by this route.

The boundary layer has been the center of most interest with respect to initial-value problems. This is because experiments have been abundant and, as a prototype flow, this case is of major interest. Three works serve to illustrate what can be achieved using the conventional approach. First, Criminale and Kovasznay (1962) and later, Gaster and Grant (1975), and Gaster (1975) all made attempts to track an initial pulse in the boundary layer (specified in terms of the normal components of the disturbance velocity). Although some allowances were made for change in the Reynolds number, the results obtained were essentially those of one normal mode. Just as normal modes are best suited for asymptotic analysis ($t \to \infty$) and determination of stability, the ensuing wave packet mirrors what can be determined from such limited data. It is well known that, during the early transient period, the motion can be governed by far different modes, including all the damped normal modes. The existence of a continuous spectrum makes this part of the dynamics even more lacking since it has never been incorporated. In short, the early picture is left unknown.

The more recent work of Criminale and Drazin (1990) has established an alternative means of analysis for initial-value problems. If mean flows are modelled by a piece-wise linear variation, then it is possible to obtain closed form solutions for the respective dependent variables. These solutions contain all of the information of both the discrete and continuous spectra (when such features are possible), making it possible to represent any arbitrary disturbance. It is important to note that the solutions are non separable and hence a single normal mode cannot be a solution of the initial-value problem.

Criminale and Drazin examined several prototypical parallel shear flows under inviscid conditions. In each case, the temporal behaviour was determined for initial conditions where either the velocity (vertical component) or one component of the vorticity was given. This is a sizeable advantage. Moreover, the technique avoids the difficulty of the singular perturbation calculation and even the critical layer so manifest in normal modes. It is clear from the results that structure can play a predominant role in the early period, even driving the perturbations to nonlinearity before any exponentially growing solution will reach critical values. This point was well demonstrated by these authors.

The time dependence of the disturbance variables is akin to what has been found from rapid distortion theory in turbulence (cf. Townsend 1976) but adjusted due to the fact that boundary conditions had to be met as well. It should be noted that, even in these circumstances, the turbulence response problem is linear for strong rates of shear (cf. Lee and Hunt 1990). As a result, for large Reynolds number, the laminar and turbulent wall regions for boundary layers have much in common. Large scale oscillations (small wave numbers) and their origins play central roles for both stability theory (these scales are unstable) and in turbulence dynamics. Understanding the origins is at least conceptually possible using this approach, as was suggested by Criminale (1987).

The work presented here is a fully viscous development and follows Easthope (1984) but is complimentary to Criminale and Drazin (1990). Lee and Hunt (1990) also considered the boundary layer by rapid distortion theory but were much more interested in effects on turbulent energy spectra than intimate details of the flow variables. It will be seen that large amplitude growth is possible in the early period, three-dimensionality is crucial to the development, and overall bodes well for a signature of streakiness.

2. Problem Statement

The problem to be treated is that of linearized disturbances in the boundary layer in a flat plate. The mean flow is taken in the x direction and depends on y only, the vertical coordinate; z is the transverse axis.

The non-dimensional linearized equations for small disturbances can be written as

$$\frac{\partial u}{\partial t} + U\frac{\partial u}{\partial x} + \frac{dU}{dy}v = -\frac{\partial p}{\partial x} + \varepsilon^2\nabla^2 u \tag{2.1}$$

$$\frac{\partial v}{\partial t} + U\frac{\partial v}{\partial x} = -\frac{\partial p}{\partial y} + \varepsilon^2\nabla^2 v \tag{2.2}$$

$$\frac{\partial w}{\partial t} + U\frac{\partial w}{\partial x} = -\frac{\partial p}{\partial z} + \varepsilon^2\nabla^2 w \tag{2.3}$$

$$\frac{\partial u}{\partial x} + \frac{\partial v}{\partial y} + \frac{\partial w}{\partial z} = 0 \tag{2.4}$$

where $\varepsilon^2 = 1/Re$, Re being the Reynolds number, ∇^2 is the Laplacian operator, and u,v,w and p are the velocity and pressure perturbations respectively.

In order to attempt an analytic solution to these equations, the mean flow $U(y)$ is approximated by a piece-wise linear profile:

$$U(y) = \begin{cases} \sigma y & y < H \\ 1 & y \geq H \text{ so that } \sigma H = 1, \end{cases}$$

and H can be regarded as the boundary layer thickness.

An initial disturbance is taken as

$$u_i(x,y,z,0) = Q_i\,\delta(x - x_0)\delta(y - y_0)\delta(z - z_0),$$

where u_i stands for u,v,w.

The boundary conditions are these of impermeability and non-slip on the plate itself:

$$u = v = w = 0 \quad \text{on} \quad y = 0.$$

All components must also vanish as $y \longrightarrow \infty$.

It is now assumed that Fourier transforms may be taken in the x and z directions, which implies that the disturbances are bounded both downstream and transversally.

Therefore define:

$$\hat{u}_i(\alpha;y;\gamma;t) = \iint_{-\infty}^{\infty} e^{i\alpha x} e^{i\gamma z} u_i(x,y,z,t)\,dxdz$$

as the transformed velocity components.

Taking the transforms of equations (2.1) to (2.4), and combining into a single equation in $\hat{v}$, we have:

$$\left[\frac{\partial^2}{\partial y^2} - \tilde{\alpha}^2\right]\frac{\partial \hat{v}}{\partial t} - i\alpha U(y)\left[\frac{\partial^2}{\partial y^2} - \tilde{\alpha}^2\right]\hat{v} + i\alpha\frac{d^2U}{dy^2}\hat{v} = \varepsilon^2\left[\frac{\partial^2}{\partial y^2} - \tilde{\alpha}^2\right]^2\hat{v} \tag{2.5}$$

$$\left.\begin{aligned} &\text{with } \hat{v}(\alpha;y;\gamma;0) = \underline{q}\,\delta(y - y_0) \\ &\hat{v}(\alpha;0;\gamma;t) = \frac{\partial \hat{v}}{\partial y}(\alpha;0;\gamma;t) = 0 \\ &\text{and } \hat{v},\ \frac{\partial \hat{v}}{\partial y} \longrightarrow 0 \text{ as } y \longrightarrow \infty. \end{aligned}\right\} \tag{2.6}$$

with $\underline{q} = \frac{Q_2}{2\pi} e^{i\alpha x_0} e^{i\gamma z_0}$, and $\tilde{\alpha}^2 = \alpha^2 + \gamma^2$.

The form of $U(y)$ suggests that solutions to (2.5) may conveniently be found by splitting the y domain into two zones; above and below $y = H$. Letting the superscripts (0) and (i) represent outer and innter solutions respectively, (2.5) desolves into:

$$\left[\frac{\partial}{\partial t} - i\alpha\sigma y\right]\left[\frac{\partial^2}{\partial y^2} - \tilde{\alpha}^2\right]\hat{v}^{(i)} + 2i\alpha\sigma\frac{\partial \hat{v}^{(i)}}{\partial y} = \varepsilon^2\left[\frac{\partial^2}{\partial y^2} - \tilde{\alpha}^2\right]^2\hat{v}^{(i)} \tag{2.7}$$

$$\left[\frac{\partial}{\partial t} - i\alpha\right]\left[\frac{\partial^2}{\partial y^2} - \tilde{\alpha}^2\right]\hat{v}^{(0)} = \varepsilon^2\left[\frac{\partial^2}{\partial y^2} - \tilde{\alpha}^2\right]^2\hat{v}^{(0)} \tag{2.8}$$

The d^2U/dy^2 term has now vanished from these

equations, being of the form:

$$\frac{d^2U}{dy^2} = -\sigma\delta(y - H),$$

but will influence the matching conditions at $y = H$.

In fact, $\hat{v}$, $\frac{\partial\hat{v}}{\partial y}$ and $\frac{\partial^2\hat{v}}{\partial y^2}$ can be shown to be continuous across $y = H$, but

$$\left.\frac{\partial^3\hat{v}}{\partial y^3}\right|_{H^-}^{H^+} = -\frac{i\alpha\sigma}{\varepsilon^2}\hat{v}(H,t) \qquad (2.9)$$

The main domain of interest for this paper is the inner layer; the solution for $\hat{v}^{(o)}$ is of lesser importance. It would therefore be useful to avoid solving for both $\hat{v}^{(i)}$ and $\hat{v}^{(o)}$ and carrying out the matching across $y = H$.

Equations (2.7) and (2.8) in fact contain information on the behaviour of $\hat{v}^{(i)}$ toward $y \to H$, and this can be seen as follows.

Write equation (2.7) in the form:

$$\left[\frac{\partial^2}{\partial y^2} - \tilde{\alpha}^2\right]\left[\frac{\partial\hat{v}^{(i)}}{\partial t} - i\alpha\sigma y\hat{v}^{(i)}\right] + 2i\alpha\sigma\frac{\partial\hat{v}^{(i)}}{\partial y}$$
$$= \varepsilon^2\left[\frac{\partial^2}{\partial y^2} - \tilde{\alpha}^2\right]^2\hat{v}^{(i)}.$$

This can be recast as

$$\left[\frac{\partial}{\partial y} - \tilde{\alpha}\right]\left\{\left[\frac{\partial}{\partial y} + \tilde{\alpha}\right]\left[\frac{\partial\hat{v}^{(i)}}{\partial t} - i\alpha\sigma y\hat{v}^{(i)} - \varepsilon^2\left[\frac{\partial^2}{\partial y^2} - \tilde{\alpha}^2\right]\hat{v}^{(i)}\right]\right\}$$
$$= -2i\alpha\sigma\frac{\partial\hat{v}^{(i)}}{\partial y}.$$

Regarding, for the moment, the right-hand-side in the nature of a forcing function, this can be integrated once:

$$\left[\frac{\partial}{\partial y} + \tilde{\alpha}\right]\left\{\frac{\partial\hat{v}^{(i)}}{\partial t} - i\alpha\sigma y\hat{v}^{(i)} - \varepsilon^2\left[\frac{\partial^2}{\partial y^2} - \tilde{\alpha}^2\right]\hat{v}^{(i)}\right\}$$

$$= E(t)e^{\tilde{\alpha}y} + \int_y^H 2i\alpha\sigma e^{\tilde{\alpha}(y-\theta)} \frac{\partial \hat{v}^{(i)}}{\partial\theta}(\theta,t)d\theta \quad (2.10)$$

and $E(t)$ is an arbitrary function of time, and the fixed limit on the integral can be taken without loss of generality.

If exactly the same operation is now carried out on (2.7), we have:

$$\left[\frac{\partial}{\partial y} + \tilde{\alpha}\right]\left\{\frac{\partial \hat{v}^{(o)}}{\partial t} - i\alpha\hat{v}^{(o)} - \varepsilon^2\left[\frac{\partial^2}{\partial y^2} - \tilde{\alpha}^2\right]\hat{v}^{(o)}\right\}$$

$$= F(t)e^{\tilde{\alpha}y}, \quad (2.11)$$

and $F(t) = 0$ if the condition on $\hat{v}^{(o)}$ as $y \longrightarrow \infty$ is to be satisfied.

Matching (2.10) and (2.11) across the layer $y = H$, and using the continuity and jump conditions (2.9), it can be seen that:

$$\varepsilon^2\left[\frac{\partial^3\hat{v}^{(i)}}{\partial y^3} - \frac{\partial^3\hat{v}^{(o)}}{\partial y^3}\right]\Bigg|_{y=H} = -E(t)e^{\tilde{\alpha}H}.$$

Or,
$$\hat{v}^{(i)}(H,t) = \frac{E(t)}{i\alpha\sigma}e^{\tilde{\alpha}H} \quad (2.12)$$

This equation implies implies that $\hat{v}^{(i)}(y,t)$ should behave like $e^{\tilde{\alpha}y}$ as $y \longrightarrow H$ from below.

Solutions to equation (2.7) are now sought, with the initial and boundary conditions (2.6), and the expectation that

$$\hat{v}^{(i)} \sim e^{\tilde{\alpha}y} \quad \text{as} \quad y \longrightarrow H.$$

3. The Inner Solution

Let us, for the moment concentrate on the inner solution. Equation (2.6) clearly supplies two solutions for $\hat{v}^{(i)}$ by inspection, namely,

$$A(t)e^{\tilde{\alpha}y} \quad \text{and} \quad B(t)e^{-\tilde{\alpha}y},$$

where A(t) and B(t) are arbitrary function of time.

The remaining two solutions are given by:

$$\frac{\partial K}{\partial t} - i\alpha\sigma y K = \varepsilon^2\left[\frac{\partial^2 K}{\partial y^2} - \tilde{\alpha}^2 K\right],$$

where
$$K = \left[\frac{\partial^2}{\partial y^2} - \tilde{\alpha}^2\right] \hat{v}^{(1)}.$$

If a new dependent variable $\Gamma(y,t)$ is defined as

$$\Gamma(y,t) = e^{\epsilon^2 \tilde{\alpha}^2 t} e^{-i\alpha\sigma y t} e^{\alpha^2 \epsilon^2 \sigma^2 t^3/3} K(y,t), \quad (3.1)$$

then Γ satisfies the equation:

$$\frac{\partial \Gamma}{\partial t} - 2i\alpha\varepsilon^2\sigma t \frac{\partial \Gamma}{\partial y} = \varepsilon^2 \frac{\partial^2 \Gamma}{\partial y^2}.$$

This can be cast into the familiar diffusion equation by the transformation of coordinates:

$$\tau = t$$
$$\eta = y + i\alpha\varepsilon^2\sigma^2 t^2 \quad (3.2)$$

so that

$$\frac{\partial \Gamma}{\partial \tau} = \varepsilon^2 \frac{\partial^2 \Gamma}{\partial \eta^2} \quad (3.3)$$

In order to solve equation (3.3), let us introduce an image source δ-function below the plate. That is,

$$\hat{v}^{(i)}(y,0) = \underline{q}\Big\{\delta(y - y_0) + \delta(y + y_0)\Big\}.$$

Then,

$$\Gamma(\eta,\tau) = \underline{q}\left[\frac{\partial^2}{\partial y_0^2} - \tilde{\alpha}^2\right]\left\{\frac{e^{-(\eta-y_0)^2/4\epsilon^2\tau}}{2\varepsilon\sqrt{\pi\tau}} + \frac{e^{-(\eta+y_0)^2/4\epsilon^2\tau}}{2\varepsilon\sqrt{\pi\tau}}\right\}.$$

Reverting to the original co-ordinates (3.2), and to K:

$$K(y,t) = \underline{q}e^{-\epsilon^2\tilde{\alpha}^2 t}e^{-i\alpha\sigma y t}e^{\alpha^2\epsilon^2\sigma^2 t^3/3}\left[\frac{\partial^2}{\partial y_0^2} - \tilde{\alpha}^2\right] \times$$

$$\left[\frac{e^{-(y-y_0+i\alpha\epsilon^2\sigma\tau^2)^2/4\epsilon^2 t}}{2\varepsilon\sqrt{\pi t}} + \frac{e^{-(y+y_0+i\alpha\epsilon^2\sigma\tau^2)^2/4\epsilon^2 t}}{2\varepsilon\sqrt{\pi t}}\right].$$

It is now only necessary to solve for $\hat{v}^{(i)}$ from $\left[\frac{\partial^2}{\partial y^2} - \tilde{\alpha}^2\right]\hat{v}^{(i)} = K$, and add the two inviscid $e^{\pm\tilde{\alpha}y}$ solutions.

The calculations here are tedious but straightforward; details can be found in Easthope (1984).

Eventually,

$$\hat{v}^{(i)}(y,t) = \underline{q}e^{-\epsilon^2\tilde{\alpha}^2 t}e^{-\alpha^2\epsilon^2\sigma^2 t^3/3}$$

$$\times\left[e^{i\alpha\sigma y_0 t}\left\{M(y - y_0,t) + \alpha\sigma t(\alpha\sigma t - 2i\tilde{\alpha})J_1(y - y_0,t) + \alpha\sigma t(\alpha\sigma t + 2i\tilde{\alpha})J_2(y - y_0,t)\right\}\right.$$

$$\left. + e^{-i\alpha\sigma y_0 t}\left\{M(y - y_0,t) + \alpha\sigma t(\alpha\sigma t - 2i\tilde{\alpha})J_1(y + y_0,t) + \alpha\sigma t(\alpha\sigma t + 2i\tilde{\alpha})J_2(y + y_0,t)\right\}\right]$$

$$+ A(t)e^{\tilde{\alpha}y} + B(t)e^{\tilde{\alpha}y} \tag{3.4}$$

Here,

$$M(y,t) = \frac{1}{2\varepsilon\sqrt{\pi t}}\,e^{-(y-y_0-i\alpha\epsilon^2\sigma t^2)^2/4\epsilon^2 t}$$

$$J_1(y,t) = \frac{e^{\epsilon^2\tilde{\alpha}^2 t}}{4\tilde{\alpha}} e^{-\tilde{\alpha}y} e^{i\alpha a\epsilon^2\sigma t^2}$$

$$\operatorname{erfc}\left\{\varepsilon\sqrt{t}\left[\tilde{\alpha} - y/_{2\varepsilon^2 t} + i\alpha\sigma t/_2\right]\right\},$$

and erfc(z) is the complementary error function;

$$J_2(y,t) = \overline{J_1(-y,t)},$$

an overbar indicating complex conjugate.

This solution satisfies the initial condition, and enforcing the two boundary conditions gives $A(t)$ and $B(t)$:

$$A(t) = -\tfrac{1}{2}\underline{q}\; e^{-\epsilon^2\tilde{\alpha}^2 t} e^{-\alpha^2\epsilon^2\sigma^2 t^3/3}\left[e^{i\alpha\sigma y_0 t}\left\{M(-y_0,t)\right.\right.$$

$$\times\left[1 + \frac{y_0}{2\varepsilon^2\tilde{\alpha}t} - \frac{3i\alpha\sigma t}{2\tilde{\alpha}}\right]$$

$$\left. + 2\alpha\sigma t(\alpha\sigma t + 2i\tilde{\alpha})J_2(-y_0,t)\right\}$$

$$+ e^{i\alpha\sigma y_0 t}\left\{M(y_0,t)\left[1 - \frac{y_0}{2\varepsilon^2\tilde{\alpha}t} - \frac{3i\alpha\sigma t}{2\tilde{\alpha}}\right]\right.$$

$$\left.\left. + 2\alpha\sigma t(\alpha\sigma t + 2i\tilde{\alpha})J_2(y_0,t)\right\}\right],$$

$$B(t) = \tfrac{1}{2}\underline{q}\; e^{-\epsilon^2\tilde{\alpha}^2 t} e^{-\alpha^2\epsilon^2\sigma^2 t^3/3}\left[e^{i\alpha\sigma y_0 t}\left\{M(-y_0,t)\right.\right.$$

$$\left.\times\left[1 + \frac{y_0}{2\varepsilon^2\tilde{\alpha}t} - \frac{3i\alpha\sigma t}{2\tilde{\alpha}}\right] + 2\alpha\sigma t(\alpha\sigma t - 2i\tilde{\alpha})J_1(-y_0,t)\right\}$$

$$+ e^{-i\alpha\sigma y_0 t}\left\{M(-y_0,t)\left[1 + \frac{y_0}{2\varepsilon^2\tilde{\alpha}t} - \frac{3i\alpha\sigma t}{2\tilde{\alpha}}\right]\right.$$

$$\left.\left. + 2\alpha\sigma t(\alpha\sigma t - 2i\tilde{\alpha})J_1(y_0,t)\right\}\right].$$

It remains to examine the form of (3.4) as $y \to H$. If $\varepsilon^2 t \ll 1$ and $\gamma^2 \gg \varepsilon^2 t$, $M(y \pm y_o, t)$ will be exponentially small.

Hence,

$$\hat{v}^{(i)} \sim q e^{-\alpha^2 \epsilon^2 \sigma^2 t^3/3} \frac{\alpha\sigma t}{2\tilde{\alpha}} (\alpha\sigma t - 2i\tilde{\alpha}) e^{i\tilde{\alpha}\alpha\epsilon^2\sigma t^2}$$

$$\times \left\{ e^{i\alpha\sigma y_o t} e^{-\tilde{\alpha}(y-y_o)} + e^{-i\alpha\sigma y_o t} e^{\tilde{\alpha}(y-y_o)} \right\}$$

$$+ A(t)\, e^{\tilde{\alpha}y} + B(t) e^{-\tilde{\alpha}y} \quad \text{as} \quad y \to H.$$

That is, toward the outer edge of the inner layer, the solution is dominated by inviscid solutions of the form $e^{\pm\tilde{\alpha}y}$, as might have been anticipated.

This is not quite of the correct form, since the $e^{-\tilde{\alpha}y}$ term will not match the required solution at $y = H$, but (3.4) is taken to be an adequate approximation for investigative purposes, particularly in the near wall region.

4. The Other Velocity Components, Pressure

Having obtained an expressed for $\hat{v}^{(i)}$, it is possible to determine $\hat{u}, \hat{w}, \hat{p}$ and the vorticity components.

In particular from (2.4)

$$i\alpha\hat{u} + i\gamma\hat{w} = -\frac{\partial\hat{v}}{\partial y} \tag{4.1}$$

The pressure can be obtained from (2.1), (2.3) and (4.1):

$$i\tilde{\alpha}^2\hat{p} = -\frac{\partial^2\hat{v}}{\partial y\partial t} + i\alpha\sigma y\frac{\partial\hat{v}}{\partial y} + \alpha\sigma\hat{v}$$

$$+ \varepsilon^2\left[\frac{\partial^2}{\partial y^2} - \tilde{\alpha}^2\right]\frac{\partial\hat{v}}{\partial y}. \tag{4.2}$$

Similarly, the vertical component of vorticity, $\hat{\Omega}_y$ is given by

$$\hat{\Omega}_y = -i\gamma\hat{u} + i\alpha\hat{w}, \tag{4.3}$$

and satisfies the equation

$$\frac{\partial\hat{\Omega}_y}{\partial t} - \frac{i\alpha\sigma y}{\tilde{\alpha}}\hat{\Omega}_y - \varepsilon^2\left[\frac{\partial^2}{\partial y^2} - \tilde{\alpha}^2\right]\hat{\Omega}_y = i\gamma\sigma\hat{v}. \tag{4.4}$$

Once the behaviour of $\hat{v}$ has been determined, (4.2) permits a solution for the pressure and (4.4) determines $\hat{\Omega}_y$. Besides the forced solution for $\hat{\Omega}_y$ there is the homogeneous solution where $\hat{v} = 0$; this solution has been termed the Squire mode by some authors. Finally, (4.1) and (4.3) can be inverted to give

$$\hat{u} = \frac{1}{\tilde{\alpha}^2}\left[i\alpha\frac{\partial\hat{v}}{\partial y} + i\gamma\hat{\Omega}_y\right]$$

and

$$\hat{w} = \frac{1}{\tilde{\alpha}^2}\left[i\gamma\frac{\partial\hat{v}}{\partial y} + i\alpha\hat{\Omega}_y\right]. \tag{4.5}$$

5. Computational Results

Equation (3.4) summarises the characteristics of the vertical velocity disturbance in y and t, in Fourier space for x and z. The more detailed behaviour of the flow can only be revealed by numerical inversion of the Fourier transforms in α, γ.

This requirement for computer solutions necessitated the replacement of the initial-value δ-function by more computationally tractable Gaussians. The form of equation (3.4) remained essentially unaltered.

A number of calculations were carried out with the aim of investigating the behaviour of the disturbance in both space and time; the results are illustrated in Figures 1 to 8. The Reynolds number is here taken to be 800; x and z mark the downstream and transverse directions respectively.

Figures 1 to 6 illustrate the disturbance at about

one-tenth of the boundary layer thickness ($y = 0.106$ in dimensionless units), for a sequence of dimensionless times, 0.01, 0.1, 0.25, 0.5, 1.0, and 10.0.

The non-dimensionalisation parameters are the free-stream velocity and the boundary-layer thickness. This implies that the time scale is in fact extremely short, and we are therefore displaying only the very early history of the disturbance. Fortunately, the more interesting behavioural aspects are manifest within this span; as computational constraints set limits on the feasible values of t.

The perspective plots show a gradual evolution from a nearly symmetric initial disturbance, to a longitudinal extension with laterally symmetric peaks, to a more uniform trough-like disturbance. The twin peaks are reminiscent of the phenomena observed by Gaster and Grant (1975), while the overall longitudinal stretching is as predicted by Criminale and Kovasznay (1962).

This spreading of the wave-packet can be more easily appreciated from Figure 7, which shows clearly the laterally symmetric expansion, approximately as $\sqrt{t}$. Initially, the wave-packet accelerates more rapidly in the downstream direction, but by about $t \sim 10$, is again symmetric.

Figure 8 shows the structure of the wave-packet at about one-third of the boundary-layer thickness ($y = 0.35$), at $t = 1$. This shows more structure than was apparent nearer the plate. Clearly, the physical mechanism is changing as well where vorticity is giving way to pressure by a dominant balance. It should also be recalled that the solution is only approximate in this outer region as well.

Some of the Gaster and Grant experimental findings have here been duplicated, most notably the double transverse maxima. Other features are, however, seemingly different. For example, the oxbow shape of the wave-packet observed by Gaster and Grant is here replaced by a longitudinal ellipse form. But there is little or none of the normal modes involved in these pictures as would be expected for longer time.

It is relevant to ask what happens if a more 2-D initial disturbance were used, rather than a point pulse. This can be viewed as an attempt to mimic the vibrating ribbon used by Klebanoff, Tidstrom and Sargent (1962). The numerical results obtained showed a wave-packet with considerable lateral spread, but the transverse maxima were absent. It is clear that the form of the initial disturbance is critical in determining the shape of the resulting disturbance.

The numerical results also indicated a very rapid growth in the amplitude of the wave-packet at early times. This can in fact be demonstrated analytically, by taking an approximation to (3.4) as $t \to 0^+$.

Using Gaussian initial sources instead of δ-functions, we have (in physical space):

$$v(x,y,z,0^+) \approx \frac{Q_2}{4\pi\beta} e^{-r^2/4\beta} \left\{ \frac{e^{-(y-y_0)^2/4\lambda}}{2\sqrt{\pi\lambda}} + \frac{e^{-(y+y_0)^2/4\lambda}}{2\sqrt{\pi\lambda}} \right\}$$

$$- \frac{Q_2}{4\pi^2} \frac{e^{-y_0^2/4\lambda}}{\sqrt{\pi\lambda}} \iint_{\alpha,\gamma=-\infty}^{\infty} \left\{ e^{-i\alpha(x-x_0)} e^{-i\gamma(z-z_0)} \right.$$

$$\left. \times\ e^{-\tilde{\alpha}^2\beta} \cosh(\tilde{\alpha}y)\, d\alpha d\gamma \right\},$$

where $$r^2 = (x - x_0)^2 + (z - z_0)^2, \qquad (5.1)$$

and β and λ control the sharpness of the initial Gaussian peaks in (x,z) and y respectively.

Equation (5.1) demonstrates firstly the symmetric nature of the disturbance in x,z - as also illustrated in Fig. 1 - but also that the main contributor to the initial growth is the $e^{\tilde{\alpha}y}$ term in $\cosh(\tilde{\alpha}y)$. In fact, if $\beta = \lambda$, the main contributor to v in (5.1) can be shown to be of the form

$$e^{(y^2 - y_0^2 - r^2)/4\beta},$$

and clearly has a powerful effect for locations within a hyperboloid of revolution extending vertically above the source.

6. Conclusions

The shear wall dynamics within a modelled boundary layer have been solved explicitly by an alternative means of linear analysis. In effect, this work is an extension of the inviscid theory presented by Criminale and Drazin (1990) for initial-value problems to encompass viscous boundary conditions. And, because the bases of the technique is like that of rapid distortion theory, the concepts are valid for either a laminar or a turbulent wall layer.

In terms of laminar boundary layers it has been shown that acute three-dimensionality (structure) can give rise to rapid growth of the perturbatioons which could lead to nonlinearity before the normal means of development of unstable Tollmien-Schlichting waves. In addition, the wave-packet characteristics shown double maxima and an elongated streaky solution in the downstream direction. This early transient period is unlike the predictions that have been made from the asymptotics of limited normal mode decomposition.

Interestingly enough, there is yet another important result. Even though the normal component of the velocity, v, will decay after rapid rise, the other components of the velocity can increase algebraically as a power of t. This can be seen at once if v is considered with no x-dependence. Clearly, $\hat{\Omega}_y$ will increase at least linearly with t and consequently u,w as well. Such a finding has already been reported by Landahl (1980) using a completely different approach.

The analysis shows that it is the large scales (small wave numbers) that are most energetic in this study. Larger values of the shear rate mean larger values of the Reynolds number and thus the problem is compatible to turbulent conditions. Can streaky structure or large scale oscillations have their origins in this manner? It is highly suggestive. Consider the ratio that has been used to define the streaky patterns in boundary layers. This is

$$\rho = \lambda_z v^+/\nu$$

where λ_z is the lateral scale, ν is the kinematic viscosity and $v^+ = [\nu(dU/dy)_{y=0}]^{\frac{1}{2}}$ with all quantities dimensional. In terms of the definitions presented here

$$\rho = \frac{\lambda_z}{H} [\sigma \, Re_H]^{\frac{1}{2}}$$

where $U'(0) = \sigma$. At the location $y/H = 0.106$, $\lambda_z/H = 0.84$ and, thus for $\sigma = 1$, $Re_H = 800$ then $\rho = 23.75$ or about one-quarter the reported value. Incorporating more turbulent values for both σ and the

Reynolds number, however, easily raises the value of ρ to 100. Further out in the shear layer, the ratio of λ_z/H decreases indicating that there is a strong effect of the wall and the need for vorticity dynamics.

Finally, an interesting use of this strategy is in progress. If each of the solutions for the perturbations velocity, namely u,v,w, are replaced by $\mathring{x},\mathring{y},\mathring{z}$ and taken as particle paths, a nonlinear non-autonomous system results. Solutions of the coupled three equations bear all of the aspects that are now common place in dynamical systems. We believe the solutions of this system will go a long way to understanding what is seen by various visualization methods such as dye, hydrogen bubbles or smoke and give greater insight into the problem of mixing. This work is in progress.

Acknowledgements

One of the authors (WOC) would like to gratefully acknowledge the hospitality extended while a guest at the Center for Turbulence Research, NASA Ames Research Center and Stanford University.

References

[1] Betchov, R. and Criminale, W.O., 1967, *Stability of Parallel Flows*, Academic Press.

[2] Criminale, W.O. and Kovasznay, L.S.G., 1962, The growth of localized disturbances in a laminar boundary layer, J. Fluid Mech. *14*, 59-80.

[3] Criminale, W.O., 1987, Understanding the origins of coherent structure, Proceedings, 2nd Inth. Symp. on Transport Phenomena in Turbulent Flows, Tokyo.

[4] Criminale, W.O. and Drazin, P.G., 1990. The evolution of linearized perturbations of parallel flows, Stud. Appl. Math. *83*:2, 123-157.

[5] Drazin, P.G. and Reid, W.H., 1981, *Hydrodynamic Stability*, Cambridge Univ. Press.

[6] Easthope, P.F., 1984, An alternative approach to layer disturbances, Ph.D. Diss., University of Washington.

[7] Gaster, M. and Grant, I. 1975, An experimental investigation of the formation and development of a wave-packet in a boundary layer, Proc. Roy. Soc. London A *347*, 253-269.

[8] Gaster, M., 1975, A theoretical model of a wave packet in the boundary layer on a flat plate, Proc. Roy. Soc. London A *347*, 271-289.

[9] Gustavsson, L.H., 1979, Initial value problem for boundary layer flows, Phys. Fluids, *22*(9): 1602-1605.

[10] Gustavsson, C.H. and Hultgren, L.S., 1980, A resonance mechanism in plane couette flow, J. Fluid Mech., *98*, 149-159.

[11] Klebanoff, P.S., Tidstrom, K.D., and Sargent, L.M., 1962, The three dimensional nature of boundary layer instability, J. Fluid Mech., *12*, 1-34.

[12] Landahl, M.T., 1980, A note on an algebraic instability of inviscid parallel shear flows, J. Fluid Mech., *98*, 243-251.

[13] Lee, M.J. and Hunt, J.C.R., 1990, The Structures of sheared turbulence near a plane boundary, Turbulent Shear Flows, *7*: 1-20, Springer Verlag.

[14] Lin, C.C., 1955, *The Theory of Hydrodynamic Stability*, Cambridge Univ. Press.

[15] Mack, L.M., 1976, A numerical study of the temporal eigenvalue spectrum of the Blasius boundary layer. J. Fluid Mech. *73*, 497-520.

[16] Townsend, A.A., 1976, *The Structure of Turbulent Shear Flow*, Cambridge Univ. Press.

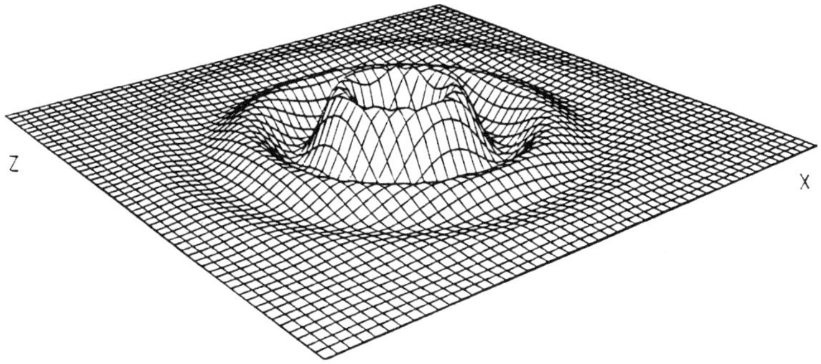

Figure 1 Pulse at y = 0.106, t = 0.01

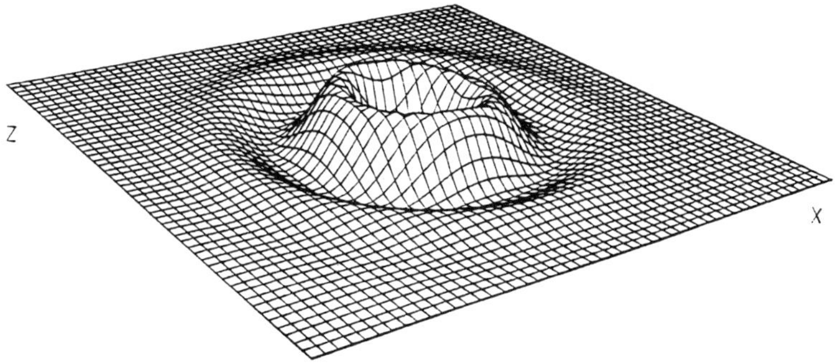

Figure 2 Pulse at y = 0.106, t = 0.1

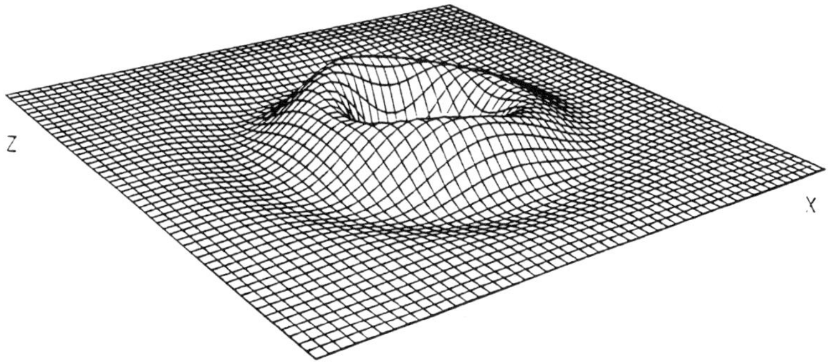

Figure 3 Pulse at y = 0.106, t = 0.25

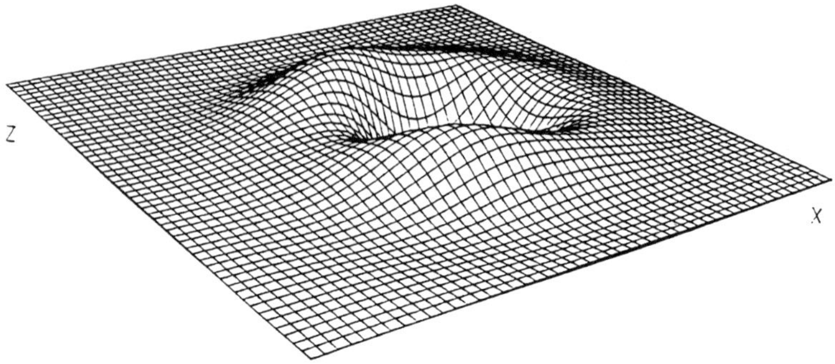

Figure 4 Pulse at y = 0.106, t = 0.5

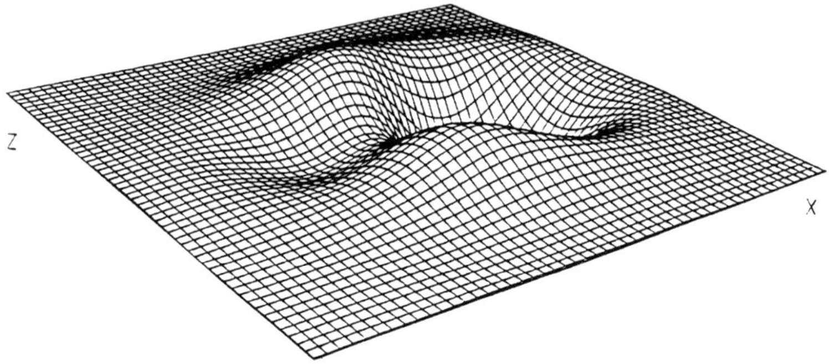

Figure 5 Pulse at y = 0.106, t = 1.0

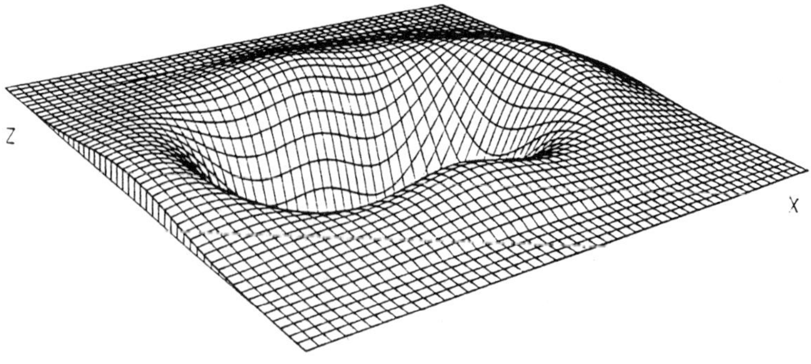

Figure 6 Pulse at y = 0.106, t = 10.0

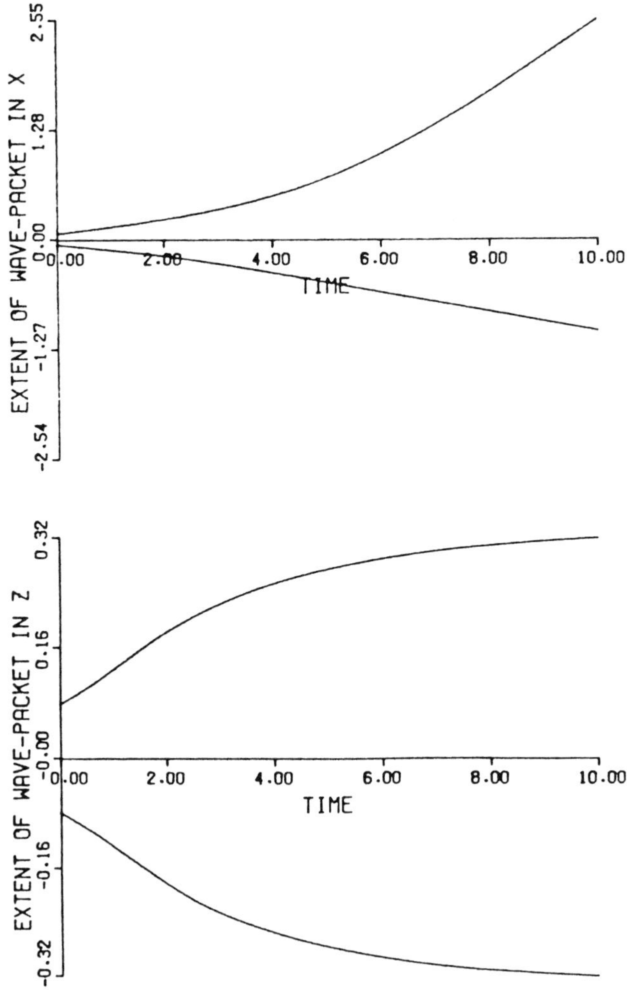

Figure 7 Spread of wave packet at y = 0.106

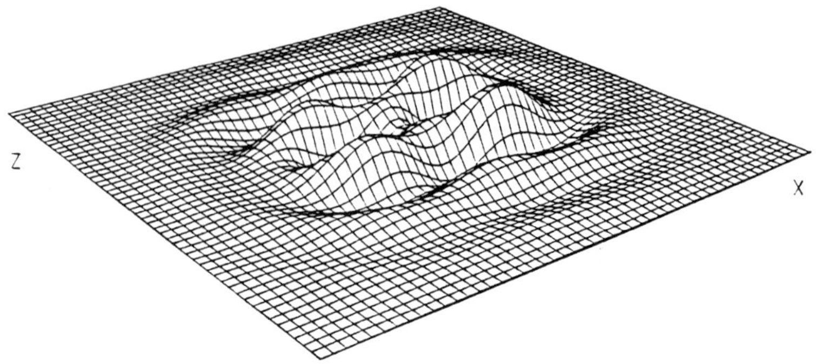

Figure 8 Pulse at y = 0.35, t = 1